Power Transmission Handbook

Content compiled by the

Power Transmission Distributors Association

CORE PURPOSE

"The Power Transmission Distributors Association, consisting of distributors and manufacturers who serve customers through distribution and encompassing the mechanical, electrical/electronic, fluid power and related technologies, is committed to advancing distribution and strengthening members to be successful, profitable and competitive in a changing market environment."

PTDA

POWER TRANSMISSION
DISTRIBUTORS
ASSOCIATION

230 W. MONROE ST., STE. 1410
CHICAGO, ILLINOIS 60606-4703
(312) 516-2100 FAX (312) 516-2101
E-Mail: ptda@ptda.org
www.ptda.org

This Handbook is intended to provide substantially accurate information regarding the subjects contained herein. Although every effort has been made to ensure its accuracy, PTDA and all other third party providers make no representations or warranties, express or implied, to any user or entity as to the accuracy, timeliness or completeness of the information contained herein. Users are cautioned that the information provided is current, but subject to change based on unforeseen and unforeseeable changes in technology beyond PTDA's control.

Printed in U.S.A.

FIFTH EDITION

ISBN 978-0-9817331-6-6 (Volume 1)
ISBN 978-0-9817331-9-7 (2 Volume Set)

Table of Contents

A Note on the Arrangement of Chapters

The Power Transmission Distributors Association has arranged most chapters of the *Power Transmission Handbook* alphabetically.

These are the exceptions:

- **"Fundamentals,"** Chapter 1, Provides the reader with basic knowledge essential to understanding later chapters.
- **"Adjustable Speed Drives,"** Chapter 12, and **"Controls and Sensors,"** Chapter 13, follow **"Motors,"** Chapter 11, because these chapters are closely interrelated, and information contained in each chapter should be mastered before moving on to the next.
- **"Sealants & Adhesives,"** Chapter 14, **"Accessories,"** Chapter 15, **"Lubrication,"** Chapter 16, **"Vibration Analysis,"** Chapter 17, are positioned last.

PREFACE 5

FUNDAMENTALS .... CHAPTER 1
- Introduction 9
- Force & Friction 9
- Torque 9
- The Circle 10
- Revolutions per Minute 11
- Power 12
- Efficiency 15
- Summary 15

BEARINGS .... CHAPTER 2
- Introduction to Bearings 19
- Plain Bearings 20
- Rolling Element Bearings 24
- Linear Bearings 34
- Mounted Bearings 36
- Bearing Standards 39
- Application and Selection 40
- Additional Information 42

BELT DRIVES .... CHAPTER 3
- Introduction 45
- Concept 45
- Belt Types 45
- Belt Standards 58
- Application and Selection 59
- Key Terms 59
- Summary 59
- Additional Information 59

CHAIN DRIVES .... CHAPTER 4
- Introduction 63
- Characteristics 63
- Types of Chain 64
- Engineering Class Chain 72
- Chain Rating 77
- Load 77
- Environment 77
- Sprockets 77
- FDA & USDA Acceptance 82
- Lubrication 82
- Caution 83
- Chain Guards 84
- Application and Selection 84
- Additional Information 84

CLUTCHES & BRAKES .... CHAPTER 5
- Introduction: Usage of Clutches and Brakes 87
- Friction Types 87
- Magnetic and Electromagnetic Designs 90
- Mechanical Lock-up Interfaces 91
- Actuation Methods 94
- Common Application Notes 96
- Selection Criteria 96

CONVEYORS AND COMPONENTS .... CHAPTER 6
- Introduction 99
- Unit Handling Equipment 99
- Special Unit Handling Applications 104
- Overhead Conveying Equipment 106
- Bulk Handling Equipment 109
- Special Concentrations of PT Components. 116
- Special Considerations 120
- Additional Information 122

COUPLINGS & U-JOINTS .... CHAPTER 7
- Introduction 125
- Rigid Couplings 125
- Flexible Couplings 126
- Coupling and U-Joint Standards 148
- Application and Selection 151
- Additional Information 151

GEARS .... CHAPTER 8
- Introduction 155
- Open Gears 155
- Enclosed Gears 165
- Lubrication 169
- Gear Rating Standards 169
- Application and Selection 170
- Additional Information 171

Table of Contents –

HYDRAULICS & PNEUMATICS . . . . . CHAPTER 9
- Introduction 175
- Hydraulic Pumps 175
- Pneumatic Compressors 178
- Valves 180
- Actuators and Motors 183
- Hydrostatic Drives 186
- Fluids and Conductors 188
- Additional Information 192

LINEAR MOTION . . . . . CHAPTER 10
- Introduction 195
- General Selection Criteria 195
- Linear Guidance 198
- Linear Propulsion 205
- Actuators 213
- Linear Motion Control Systems 221
- Key Terms 222
- Summary 223

MOTORS . . . . . CHAPTER 11
- Introduction 227
- Common Factors 227
- AC Motors 232
- DC Motors 235
- Servomotors 238
- Step Motors 239
- Application and Selection 240
- International Standards 242

ADJUSTABLE-SPEED DRIVES . . . . . CHAPTER 12
- Introduction 245
- Why Use Adjustable-Speed Drives? 245
- Mechanical Adjustable-Speed Drives 246
- Eddy-Current Drives 253
- Wound-Rotor Drives 254
- DC Adjustable-Speed Drives 255
- AC Adjustable-Speed Drives 256
- Servo Drives 264
- Key Terms 264
- Drive Performance Summary 265
- Engineered Drive Systems 265

CONTROLS & SENSORS . . . . . CHAPTER 13
- Introduction 269
- Motor Starters 269
- Control Devices 275
- Programmable Controllers and Computers 276
- Operator Displays 280
- Sensors 281
- Control Enclosures 289
- Codes and Standards 293

SEALANTS & ADHESIVES . . . . . CHAPTER 14
- Introduction 297
- Types of Sealants and Adhesives 297
- Key Terms 300
- Additional Information 300

PT ACCESSORIES . . . . . CHAPTER 15
- Introduction 303
- Shaft Seals 303
- Drive Tensioners 305
- Retaining Rings 307
- Shaft-Locking Devices 309
- Shaft Collars 311
- Keystock 312
- O-Rings 312
- Locknuts 313
- Single-Point Lubricators 313
- Additional Information 314

LUBRICATION . . . . . CHAPTER 16
- Introduction 317
- Lubrication 317
- Grease 324
- Solid Lubricants 325
- Selection Criteria 328
- Key Terms 329
- Additional Information 329

VIBRATION ANALYSIS . . . . . CHAPTER 17
- Introduction 333
- Theoretic Basis 333
- Rotating Machine-Fault Diagnostics 337
- Balancing 343
- Vibration Analysis of Structures 347
- Summary 350
- Key Terms 350

APPENDIX . . . . .
- Career Opportunities in the PT/MC Industry 357
- Common Conversion Factors 359
- Common Formulas 363
- Associations That Interface with the PT Industry 367
- Credits 371
- Index 373

Preface

The ***Power Transmission Handbook*** material has four broad objectives: (1) motivate the learner who is being introduced to new products, technology, and concepts; (2) provide experience in solving problems (using SI and common units) and presenting solutions in a logical manner; (3) introduce the learner to subject areas that are common to most PT/Motion Control practitioners and that require the application of fundamental engineering concepts; and (4) develop a basic knowledge base necessary to effectively solve open-ended problems through an applications process.

The material in the ***Power Transmission Handbook*** is written in a manner that allows for individual home study or for presentation in an organized classroom setting.

The ***Power Transmission Workbook*** offers a series of true-false, multiple-choice, and short answer questions for each of the 17 chapters. Requiring learners to complete the answers to these questions increases their potential to understand and retain the information by four times.

The supporting ***DVD series*** expands the learning experience by visually bringing to life the product(s) as used in the application. Visual training increases comprehension of the concepts covered through the easy-to-follow demonstrations and explanations. Featuring colorful computer graphics and concise field application footage, the videos (one for each chapter of the Handbook) make learning much more enjoyable and effective. Each DVD follows along with its corresponding chapter of the Handbook. Additional information on the DVD series can be obtained by contacting the Power Transmission Distributors Association.

The valuable content of the Power Transmission Handbook, Workbook and Videos—plus much more—is also available in a computer-based training tool. ***Power Transmission Interactive*** uses text, graphics, video, animation, voice prompts and interactive quizzes, to teach the basics of 14 power transmission/ motion control product categories and applications.

Chapter 1

Fundamentals

Chapter Objectives

When you have finished Chapter 1, "Fundamentals," you should be able to do the following:

1. Understand the meaning of the terms force, friction, torque, rpm, work, and power as they relate to the industry.
2. Understand how the principle of conservation of energy applies to power transmission.
3. Identify losses in a system given the efficiency of the components.
4. Quantify the important parameters using the basic equations of power transmission.

HALF PRICE BOOKS®

Half Price Books
1835 Forms Drive
Carrollton, TX 75006
OFS OrderID 39481330

Thank you for your order, ABE TP Orders!

Thank you for shopping with Half Price Books! Please contact Support@hpb.com. if you have any questions, comments or concerns about your order (738041538)

Visit our stores to sell your books, music, movies games for cash.

SKU	ISBN/UPC	Title & Author/Artist	Shelf ID	Qty	OrderSKU
S471012208	9780981733166	PTDA Power Transmission Handbook 5th Editi Power Transmission Distributors Association	DB 37.25.6	1	

SHIPPED STANDARD TO:
ABE TP Orders
800 Avondale Ave
Columbus OH 43212
help@textbookrush.com

ORDER# **738041538**
Abebooks

Introduction

Power **T**ransmission/**M**otion **C**ontrol (PT/MC) specialists concern themselves primarily with the manipulation and control of mechanical power (also called shaft power) using the principles of physics.

Force & Friction

> **FORCE**
> A force is a push or a pull acting on an object.

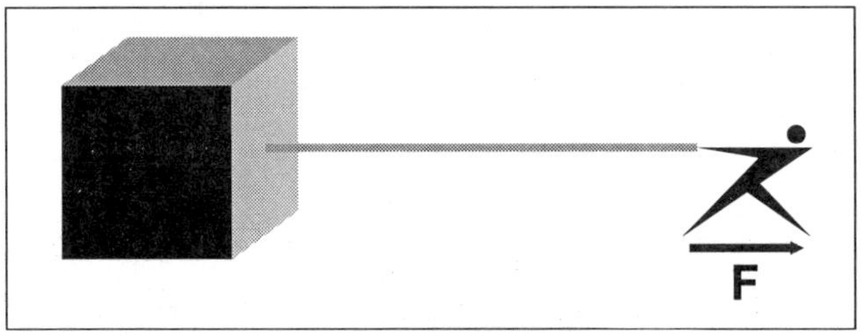

Figure 1-1: Force

In the real world, there is usually more than one force acting on any given object. It is useful to resolve these forces into a single force, which we call the "net force" or the "resultant force." In Fig. 1-2, the net force on the box is 25 lbs toward the right of the page.

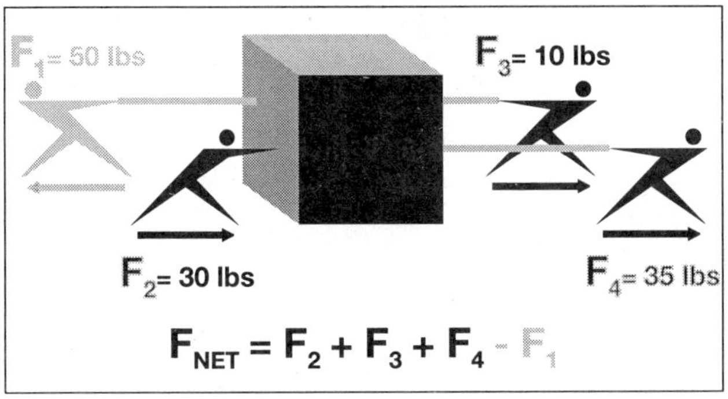

Figure 1-2: Net force

Provided the net force is sufficient to overcome friction, motion will occur in the direction of the net force. Friction is responsible for the resistance that you feel when you try to push something in order to set it in motion and that you must overcome in order for motion to occur. Once the object is moving, friction causes it to want to stop moving—such that you must keep overcoming friction if you want to keep the object moving.

Friction is very important in what we do. We treat friction as a force that opposes any attempt to initiate motion or, if the object is already in motion, tries to stop the object from moving by pushing in the direction opposite to that of the movement.

The force of friction always acts in the direction *opposite* to that of motion.

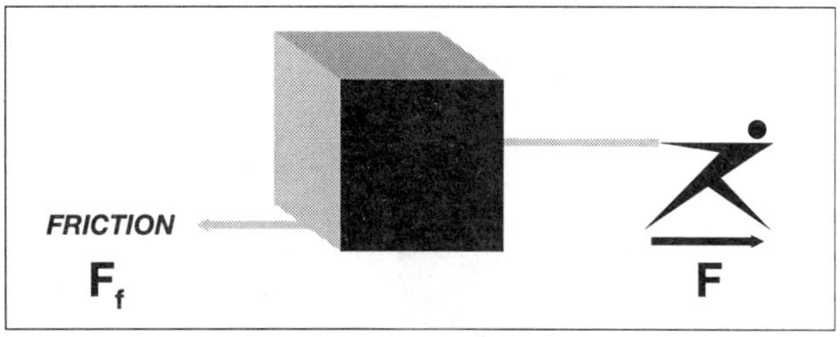

Figure 1-3: Friction

Torque

> **TORQUE**
> Torque is the tendency of a body to rotate as a result of a force being applied at some distance away from the center of rotation.

Torque has two components: a force and a distance. The distance component of torque is often called the lever arm.

In fact, T = F x D or, in words, torque is equal to the force applied multiplied by the distance to the center of rotation.

> In Imperial Units, torque is denoted in lb-in or lb-ft.
> In SI Units, torque is denoted in Nm (Newton Meters).

A good way to visualize torque is to look at a wrench acting on the head of a bolt.

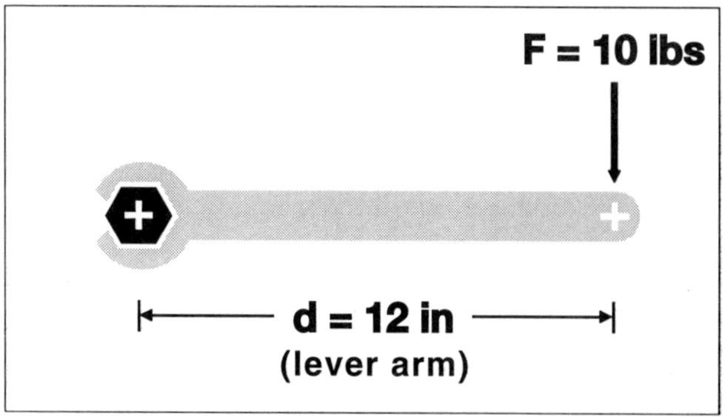

Figure 1-4: Torque

The torque applied to the bolt in this case is 10 lbs x 12 in = 120 lb-in.

Important note: *The distance of the lever arm is always measured at a right angle (90 deg) to the force..*

We can illustrate the resulting torque as follows:

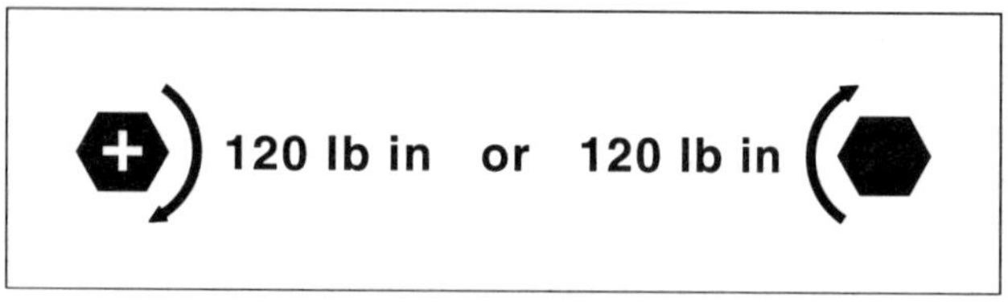

Figure 1-5: Resulting Torque

Observe that the torque is directional and can be *clockwise* (CW) or *counterclockwise* (CCW). The circular arrows serve to indicate the direction of the torque. In both cases, the torque is shown as acting in a *clockwise* direction on the bolt (see Fig. 1-5).

Let us consider the following example:

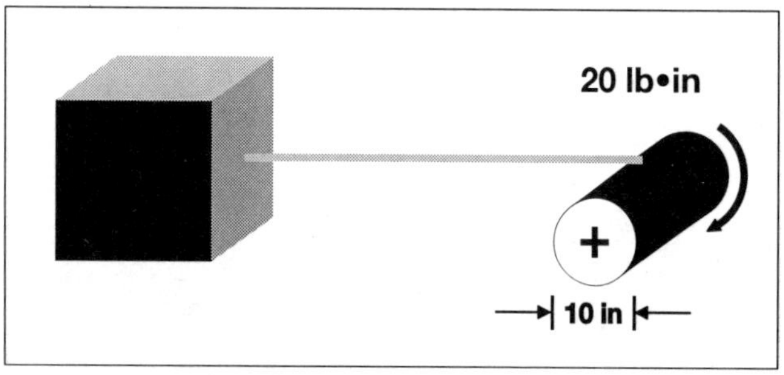

Figure 1-5: Example E1

Example E1

One end of a string is attached to a box, while the other end is attached to the outside of a drum, which measures 10 inches in diameter. A torque of 120 lb-in is applied to the center of the drum. Calculate the force exerted on the box by the string.

Solution: T = F x D

120 lb-in = F x 5 in

F = 120 lb in/5 in = 24 lbs

Example E2

Let us assume that the 120 lb-in of torque on the drum was generated by the wrench in Fig. 1-4. If we measure the wrench to have an effective length of 10 inches, what force would need to be exerted at the end of the wrench handle in order to produce the 120 lb-in of torque?

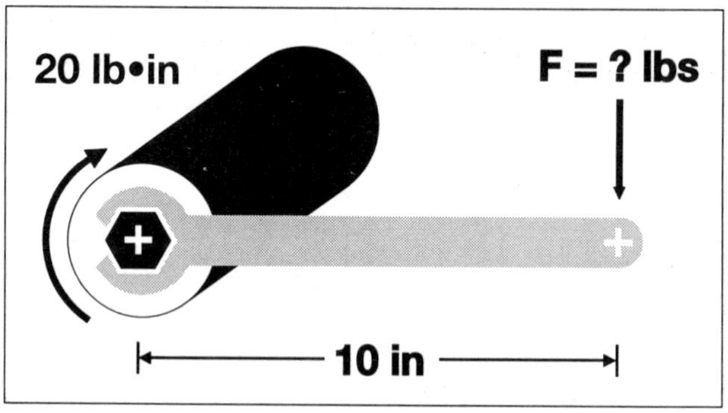

Figure 1-5: Example E2

Solution: T = F x D

120 lb-in = F? x 10 inches

F = 120 lb-in/10 in = 12 lbs

The Circle

Because many power transmission components are circular—such as pulleys, sprockets, drums, and gears—it is important that we are comfortable with the properties of the circle.

Let's take a minute and review these properties.

D (the diameter) = 2 x r (the radius)

C = the circumference = π x D = π D

Or C = π x 2 x r = 2 π r

π = 3.1416 (it is the ratio of the circumference of a circle to its diameter as discovered by ancient Greek mathematicians)

A = Area of the circle = π x r x r = π r^2

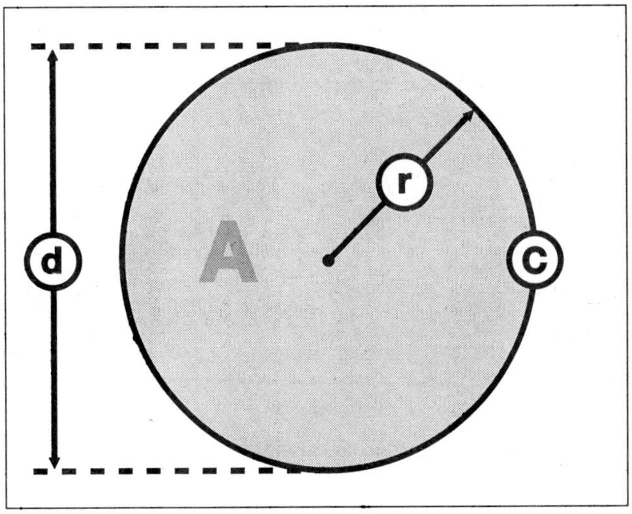

Figure 1-6: Circle

Example E3

To illustrate the importance of the properties of the circle, let us look again at the example of the drum with the string attached to a box. If, as in Fig. 1-7, we use a wrench to turn the drum, calculate how far the box will advance if we turn the drum exactly one revolution.

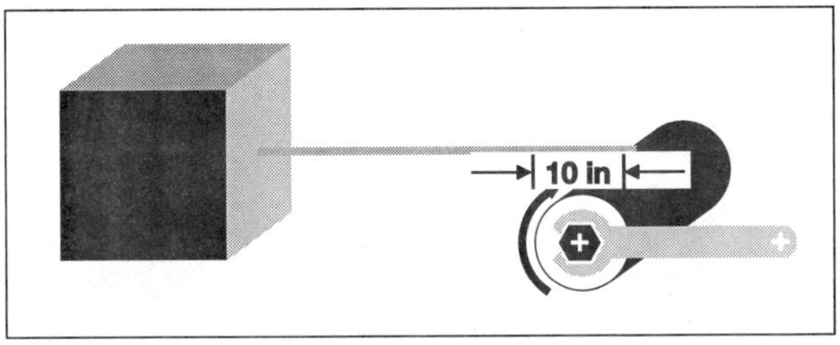

Figure 1-7: Example E3

Solution: One revolution of the drum would cause the string to do one wrap of the drum (visualize one circumference of the circle) and therefore pull the box forward by one wrap or circumference.

C = π D= (3.1416) x (10 in) = 31.416 inches

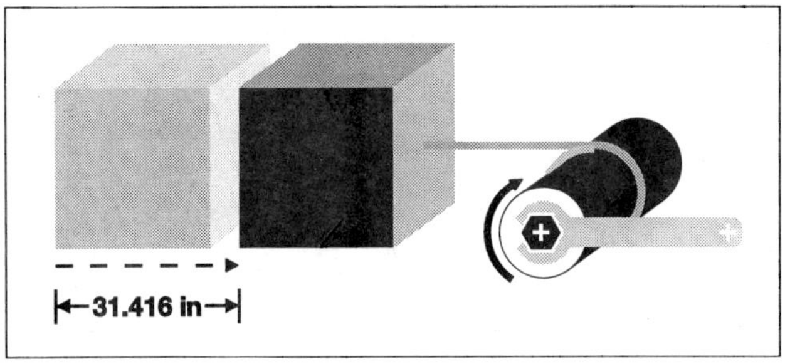

Figure 1-8: Example E3 answer

Revolutions per Minute

RPM

RPM stands for Revolutions Per Minute.

- In one revolution, linear distance travelled = circumference of circle = TTD π
- In "N" revolutions, linear distance travelled = TTD x N = TT D N

Revolutions per minute (rpm) is the most frequently used measure of the rate of rotation of anything that is turning (e.g., a shaft, a gear, a sprocket, a pulley, a drum, the hands of a clock).

Example E4

Looking again at the system in Fig. 1-8, we have figured out that one revolution of the drum advances the box approximately 31 in.

a) What if a customer asked you to design a similar system whereby a box was moved across his plant (which measures 500 ft) in a time of one minute. At what speed should the drum turn?

Solution: One revolution moves the box 31 in, or 31/12 in = 2.583 ft.

It follows that to move 500 ft, we need to turn 500/2.583 = 193.57 revolutions.

Because we wish this movement to occur in one minute, then the drum must turn 193.57 **r**evolutions **p**er **m**inute, or <u>193.57 rpm</u>.

b What if he now decided he wanted the box moved 500 ft in 20 seconds?

Solution: *Hint:* A good trick is to calculate the revolutions per second, and then multiply by 60 to convert back to revolutions per minute.

In one revolution, the box advances 2.583 ft (see part a). To move 500 ft, we need to turn 193.57 revolutions (from part a).

We need to do this in 20 seconds. Therefore, we need to turn 193.57 revolutions in 20 seconds.

In <u>one</u> second, we need to turn 193.57/20 = 9.679 revolutions.

If we turn 9.679 revolutions in one second, then we will turn 9.679 x 60 = 580 revolutions in one minute, or <u>580 rpm</u>.

Power

POWER

Power is the rate at which work is done, or, when referring to a source of power, a measure of how much work the device is able to do in a given amount of time.

"Work" in PT/MC terms refers to the transfer of energy to a body by the application of a force that moves the body in the direction of the force. It is calculated as the **product** of the **force** and the **distance through which the body moves**.

Example E5

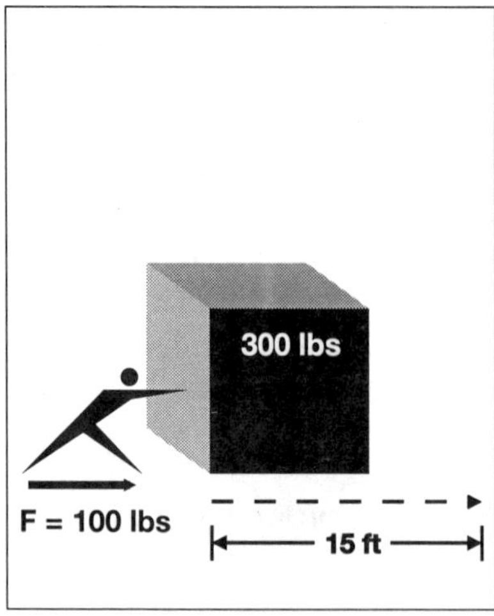

A 100 lb force is applied to move a 300 lb box a horizontal distance of 15 feet at a constant speed.

Work done = 100 lbs x 15 ft

W = 1,500 lbs-ft

Example E6

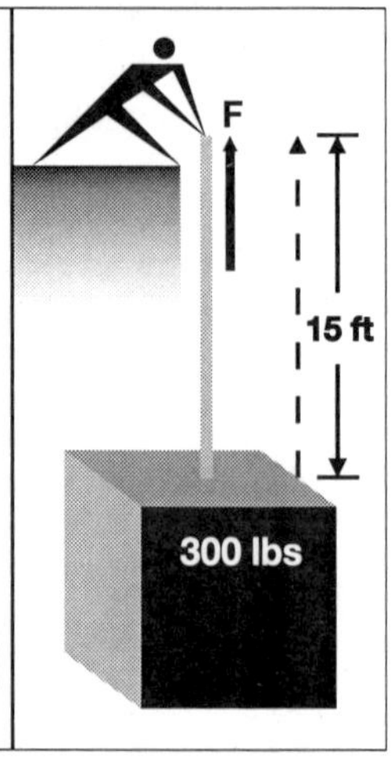

An upward force is applied to lift a 300 lb weight to a height of 15 ft.

Work done = 300 lbs x 15 ft

W = 4,500 lbs-ft

Work is therefore measured in units of force and distance, such as lbs-ft.

Calculate the work done in the following examples:

Note that the amount of work done is <u>independent</u> of <u>how long</u> it took to do it.

Power measures <u>the rate</u> at which work is done.

Example E7

Using the scenario from example E5, if the work is accomplished in 10 seconds, then the power exerted is:

1) $\text{Power} = \dfrac{100 \text{ lbs} \times 15 \text{ ft}}{10 \text{ secs}} = \dfrac{150 \text{ lbs-ft}}{\text{sec}}$

HORSEPOWER

Horsepower (HP) is the standard unit of power in the imperial system and is equal to 550 lbs-ft/sec, or 33,000 lbs-ft/min.

And therefore the power exerted in example E5 revisited is

$$\textbf{Power}_{10\,sec} = \frac{150}{550} = \underline{\textbf{0.27 HP}}$$

If however the work is accomplished in 2 minutes instead, then the power exerted is

$$\text{Power} = \frac{100 \text{ lbs} \times 15 \text{ ft}}{120 \text{ secs}} = 12.5 \frac{\text{lbs-ft}}{\text{sec}}$$

and

$$\textbf{Power}_{120\,sec} = \frac{12.5}{550} = \underline{\textbf{0.023 HP}}$$

The units commonly used in PT/MC are HP (in imperial units) or watts (in SI units, where 1,000 watts = 1 kilowatt or 1 kw).

Useful formulas for manipulating power are:

$1 \text{ HP} = 33{,}000 \frac{\text{lbs-ft}}{\text{min}}$	converts the rate of doing work into Horsepower.
$1 \text{ HP} = 0.746 \text{ kw}$	conversion from SI to Imperial units.

$$1 \text{ Watt} = \frac{1 \text{ Newton} \times 1 \text{ meter}}{1 \text{ second}} = \frac{1 \text{ N-m}}{\text{s}}$$

In the PT/MC industry, we are most often interested in shaft power. Shaft power is characterized by torque and speed.

So who decided that one horsepower should equal 33,000 lbs-ft/min?

The term horsepower was coined by James Watt, who invented the steam engine and wanted to sell it to replace horses, which at the time were used to do this kind of work. Watt determined that the amount of work an average work horse could do was 33,000 lbs-ft per minute (drawing coal from a coal pit). He therefore coined the term one horsepower to correspond to 33,000 lbs-ft per minute. Thus, if you purchased a six horsepower steam engine from Mr. Watt, it could replace the work of six horses. The SI officials were so impressed by Mr. Watt's invention that when they needed a label for measuring units of power, they chose the watt.

The most commonly used formula is that which relates power to torque and speed (rpm) as it would apply to a rotating shaft.

This formula is:

$$\text{HP} = \frac{\text{T} \times \text{rpm}}{63025}$$

where the torque is expressed in lb-in.

This is the most important formula in the power transmission/motion control industry.

This formula, along with the basic physical principle of conservation of energy, allows PT/MC experts to analyze all PT/MC systems.

The principle of conservation of energy (adapted to our purposes) essentially states that the total amount of energy in the universe is constant. Therefore, energy cannot be created or destroyed (though it can change forms).

Therefore, we must account for all of the power in a system so that power in equals power out.

Note that it takes power to create heat, noise, and light and to overcome friction such as rubbing. Some of the power injected into the systems we are analyzing inevitably gets used for these purposes, rather than serving to do what we require (such as lifting an elevator). These misdirected amounts of power are referred to as losses.

Because of friction, we must allow for these losses in our calculations and designs.

The Horsepower formula can be used to determine the power required to accomplish a given task.

Example E8

Let's go revisit the problem of moving a box across 500 ft

a) in one minute

or

b) in 20 seconds

and calculate the power required to do the job.

Solution: a)

$$HP = \frac{T \times rpm}{63025} = \frac{120 \text{ in-lbs} \times 193.57 \text{ rpm}}{63025} = 0.369 \text{ HP}$$

Torque from Fig. 1-5 and rpm from example E4 a)

b)

$$HP = \frac{T \times rpm}{63025} = \frac{120 \text{ in-lbs} \times 580 \text{ rpm}}{63025} = 1.10 \text{ HP}$$

Torque from Fig. 1-5 and rpm from example E4 b)

The Horsepower formula can also be used to calculate the amount of torque at any point in a drive system, when the power is known.

This is important because many components that we may need to specify, such as shaft couplings and torque limiters, are rated according to how much torque they can transmit.

Shown in the next column is a typical drive system consisting of a motor connected to a gear reducer using a v-belt drive.

Since the horsepower is known and since power is constant throughout the system, the equation can be used to determine the torque at each shaft in the system as follows (ignoring any losses for the time being).

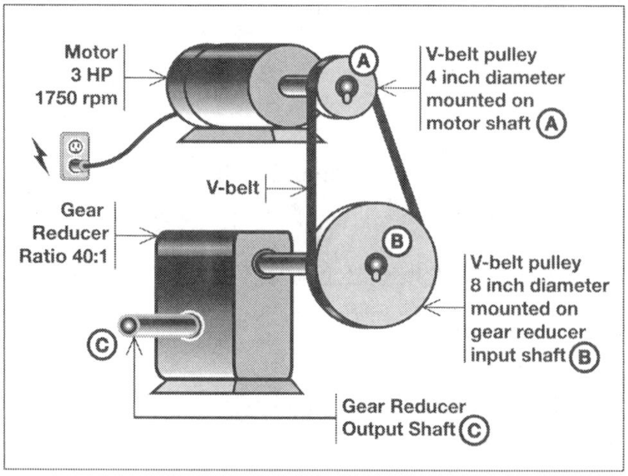

Figure 1-9: Typical drive system

Example E9

Motor shaft "A"

HP = 3 HP (given), rpm = 1750 (given), using the formula, we calculate the torque:

$$torque_A = \frac{3 \text{ HP} \times 63025}{1750 \text{ rpm}} = 108 \text{ lb-in}$$

Reducer input shaft "B"

HP = 3 HP (given), calculate the new rpm:

$$1750 \times \frac{4in}{8in} = 875 \text{ rpm}$$

Using the formula, we calculate the torque

$$torque_B = \frac{3 \text{ HP} \times 63025}{875 \text{ rpm}} = 216 \text{ lb-in}$$

Reducer output shaft "C"

HP = 3 HP (given), calculate the new rpm:

$$875 \times \frac{1}{40} = 21.9 \text{ rpm}$$

Using the formula, we calculate the torque

$$torque_C = \frac{3 \text{ HP} \times 63025}{21.9 \text{ rpm}} = 8634 \text{ lb-in}$$

Efficiency

EFFICIENCY

$$\text{Efficiency } (\eta) = \frac{\text{Power Out}}{\text{Power In}}$$

Losses in power transmission components are usually stated in terms of their "efficiency."

The efficiency of a power transmission component represents "that percentage of the power input which comes out in the form which is intended." For instance, a component that claims an efficiency of 90 percent will perform its function with no more than 10 percent loss.

Let us now do the same power calculations as example E9, taking into account the losses incurred along the way.

Example E10

Look at Fig. 1-9

In our sample system, we have two power transmission components that take the shaft power (speed and torque) from point A to point C:

1) The v-belt drive (between the motor and the gear reducer)—efficiency 0.93 (93 percent).
2) The gear reducer—efficiency 0.70 (70 percent).

Taking these efficiencies into account, we now get

Motor shaft A

HP= 3 HP (given), rpm = 1750 (given), using the formula, we can calculate the torque:

$$\text{torque}_A = \frac{3\text{ HP} \times 63025}{1750\text{ rpm}} = 108\text{ lb-in}$$

Reducer input shaft B

HP = 3 HP (given) x 0.93 = 2.79 HP

Using the formula, we can calculate the torque:

$$\text{torque}_B = \frac{2.79\text{ HP} \times 63025}{875\text{ rpm}} = 201\text{ lb-in}$$

Reducer output shaft C

HP = 2.79 HP (at input shaft) x 0.70 = 1.95 HP

Using the formula, we calculate the torque:

$$\text{torque}_C = \frac{1.95\text{ HP} \times 63025}{21.9\text{ rpm}} = 5612\text{ lb-in}$$

Taking efficiency into account, we are left with 5,612 lb-in torque at the reducer output shaft C rather than 8,634 lb-in, a difference of 35 percent from our answer in Example E9. (Example E9 is the same exercise done in Example E10 without taking efficiency into account.)

$$\frac{5612}{8634} = .65 = 65\text{ percent efficient}$$

As you can see, the losses are significant and cannot be ignored in the design of a PT/MC system.

Summary

We have now reviewed the basic concepts that apply to all PT/MC systems. The following chapters provide information about specific system components.

Chapter 2

Bearings

Chapter Objectives

When you have finished Chapter 2, "Bearings," you should be able to do the following:

1. Describe the history and background for bearings.
2. Explain the difference between plain bearings and rolling element bearings.
3. Explain the features of plain bearings.
4. Differentiate lubricated and self-lubricated plain bearings.
5. Describe applications where plain bearings would be considered.
6. Describe the various modes of lubrication for plain bearings.
7. Describe the features of rolling element bearings.
8. Describe the different types of rolling element bearings.
9. Differentiate ball and roller bearing applications.
10. Explain the purpose and effects related to the lubrication aspects of bearings.
11. Describe the types and features of mounted bearings.
12. Identify industry associations that recommend the various standards related to ball and roller bearing usage.

Introduction to Bearings

Bearings play a crucial role in the transmission of power to motion. Without bearings, there would be no efficient rolling or turning of shafts. There would be no trains, planes, or automobiles. When you think of anything that rotates, think of the bearings that are used to support the shaft, allowing rotation to occur. More importantly, by the end of this section, you should have a clearer understanding of what type of bearing is used in a given application and why.

Bearings have been traced back to use by the Phoenicians nearly 4,500 years ago. Remains of crude but effective bearings in the rudder shafts of sailing ships have been found. The bearing material used was lignum vitae, a natural wood with inherent lubricating oils. Surprisingly, wood bearings are still used today in selective applications. Over the years, significant advancements have been made in both design and materials to create the extensive product group broadly referred to as bearings. Phillip Vaughan, an inventor from Carmarthen, was the first to patent the rolling element ball bearing in 1794. In 1839, Isaac Babbitt was credited with inventing a system of pouring molten, low-melting-temperature metals into fixtures that supported shafts. These bearings are still used today and are referred to within the industry as Babbitt bearings. From primitive beginnings, bearings have evolved tremendously to the product forms we currently use.

Bearings are broadly grouped into one of two main categories: plain bearings and rolling element bearings. Regardless of grouping, bearings are implemented into various power transfer systems (equipment, machines, and devices) to accommodate and support both motion and force. Without bearings, there would be little control in mechanical motion. Force, or load combined with motion, creates friction on mating surfaces. Bearings are designed to help relieve this friction and allow for the smooth and effective transfer of motion.

There are generally three types of motion: rotary (turning on a fixed point), linear (sliding or rolling along an axis), and combined (linear and rotary). When motion combines with force in a cylindrical fashion, the resultant power or load is generally expressed as radial loading. Radial loads act at right angles to the shaft or axis of rotation, and axial or thrust loads are applied or act parallel to the axis of rotation.

To accommodate various application and design parameters, bearings have been specifically sized and rated. Bearing manufacturers use several important operational and physical criteria to support the sizing and rating process. These include: shaft size, metallurgical properties, space allotment, rotary or linear travel speeds, the type and amount of loading, specific environmental conditions, and special performance requirements.

Shafts are engineered to a specific size by material type to accommodate the combinations of loads, speeds, and application conditions. The size of a shaft may be expressed in both metric and standard measure (in.). Bearings are widely available in both measurements. Shafting of standard measurement sizes are most commonly found in British-based countries and the U.S. and Canada. Metric sizes are most commonly found in applications of bearings in the remainder of the world. Regardless of measurement type, the fundamental principles of use apply, and no real operational differences occur because of the type of measurement.
Speed is a major factor in the determination of which bearing best suits a particular application. The term velocity (V) is used as a reference for speed, and speed is generally expressed in revolutions per minute (rpm) in standard-measure applications and meters per second (m/s) in metric-calculated applications.

Load is another critical element in bearing selection and is expressed as pressure (P). The amount and type of load will determine the size and type of bearing. Speed times load equals the combined pressure and velocity, and the resultant factor is expressed as PV. This factor is very important in the design and determination of a bearing. Manufacturers' literature is extensive and should be consulted to determine the proper bearing.

Manufacturers have also calculated many of the factors required to select the proper bearing for an application. A simplified bearing selection process is provided by most manufacturers to help improve the accuracy and performance of bearings, based on the given conditions of the application. Consult the information contained within this guide and the manufacturer's literature to assist in the understanding and selection of bearings.

Environmental or ambient conditions play a very important role in the selection, implementation, and survival of

bearings. To help the performance of bearings, special attention is directed to the individual components and overall construction of the bearing.
These special features partially include seals, lubricants, construction materials, and the physical design parameters.

Now that we have reviewed the background and basics of bearings, understanding the makeup and types of bearings is important. There are two groups of bearings: plain and rolling element. Both have inherent capabilities to effectively manage the demands of today's power transmission system requirements, if properly applied. Each type of bearing has a specific and intended use. Both plain and rolling element types have limitations and advantages when applied to any particular application. Bearing manufacturers and distributors are more than willing to assist with the selection of the correct bearing for an application. These selection services are generally free of charge and welcomed.

Plain Bearings

The term plain bearing generally refers to a non-rolling element-designed bearing. The term plain (or spelled as "plane") has been applied to reference two aspects of this bearing type. One is that plain bearings are designed to support shafts that are fixed on a particular plane, and the other expression gives reference to the simple design. Simplicity should not be interpreted as inferiority in reference to rolling element type bearings. Users also refer to plain bearings as bushings or sleeve bearings; however, a bushing is used to merely take up intermediary space, and a sleeve is a fitting or fixture.

Plain bearing types are generally cylindrical designs that are commonly called straight bore plain bearings. If the design incorporates a perpendicular surface, the bearings are called flanged type plain bearings. The inside race, or shaft bore, and the outer race, or housing bore, may be used as either the bearing surface or the retaining surface. The flange of a plain bearing may be used as a seating or positioning component as well as a thrust wearing surface (see Fig. 2-1).

One distinct advantage of a plain bearing is that any direction of rotation is acceptable. Radial, linear, axial, and oscillating motions can be accommodated by plain bearings.

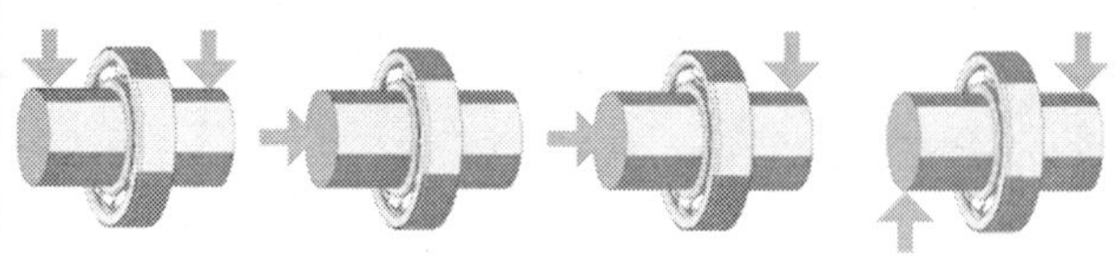

Figure 2-1: Types of Loads Acting On A Bearing

Plain bearings are made from a wide variety of materials. A broad separation of material groups is between metallic and nonmetallic types. A further separation of plain bearings is between unlubricated and self-lubricating types. Each type has a specific functionality and, if properly applied, will provide a long and trouble-free service. Manufacturers' literature and services are extensive and should be consulted for the proper application of bearings.

Metallic Plain Bearing

Metallic plain bearing materials include a wide range of metallic compounds, such as bronze, copper and aluminum alloys, lead and silver Babbitt, sintered iron, and copper powders, to name a few. By combining a variety of metals, hardness levels can be controlled to accommodate for higher loading and speed capabilities. (See Table 2-1.)

The following are advantages of metallic plain bearings.

1. A higher load carrying capacity, size for size, over nonmetallic types.
2. A wide temperature range, including operating continuously in a very high or low temperature range.
3. Structural integrity in large bore diameter applications under severe loading, torsion, and stress.
4. A cost-to-performance ratio benefits in large-sized bearings and those operating in unique or severe service applications.

Nonmetallic Plain Bearing

Nonmetallic plain bearing technology has advanced immensely in the last 20 years. This bearing type uses a wide variety of nonmetallic materials to provide excellent service over an equally wide range of applications. Materials include: polyacetyl, polyethylene- and resin-based plastics, carbon graphite, Teflon compounds, woven fibers, and even natural wood. Nonmetallic bearings are made by injection

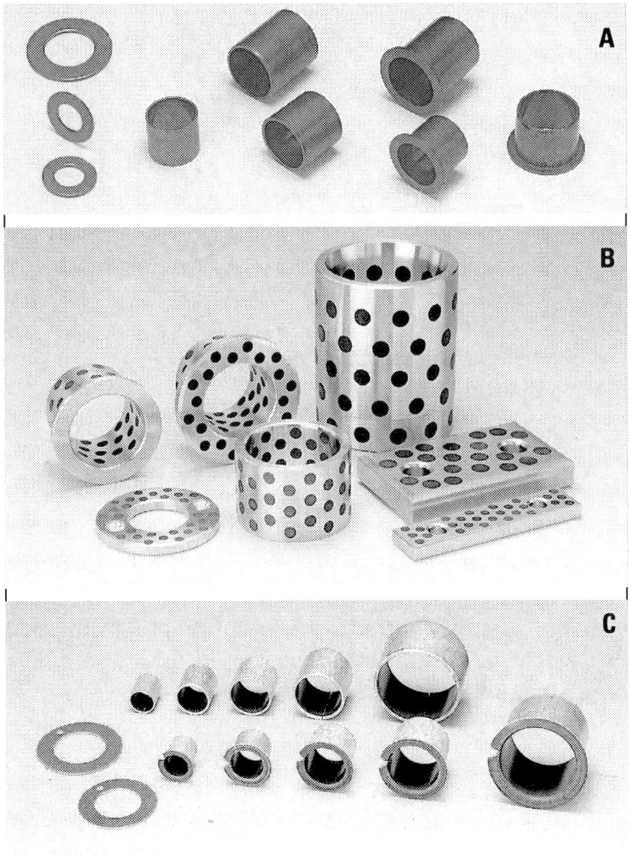

Figure 2-2: Three Types of Plain Bearings,
A: Plastic, B: Metallic, C: Composite

molding, compression molding, extrusion, machining, and winding, to name a few. Nonmetallic bearings offer similar performance benefits as seen in the metallic bearing groups. Handling shock and vibration is a performance feature that is found with nonmetallic bearings. Improvements made in the designs and compositions of materials have elevated the awareness and use of nonmetallic bearings in applications reserved specifically to other forms of bearings (see Fig. 2-2).

Lubrication

Plain bearings are generally separated by those requiring lubrication and by those that are self-lubricating. Friction is minimized by maintaining a very thin film of lubricant between the stationary and rotating members (see Fig. 2-3).

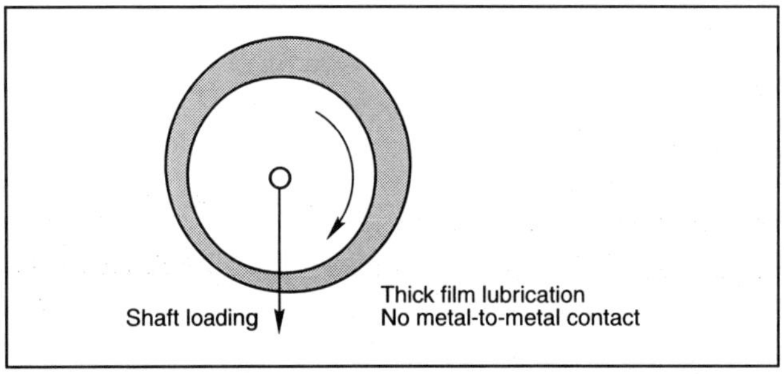

Figure 2-3: Scematic of Lube Film Between Shaft And Bearing

TABLE 2-1 — Comparison of plain bearing alloys

Bearing material	Load carrying capacity, lb/in.²	Maximum operating temperature, F	1 = Best, 5 = Worst			
			Compatibility*	Conformability & embeddability**	Corrosion resistance	Fatigue strength
Tin-base babbitt	800-1,500	300	1	1	1	5
Lead-base babbitt	800-1,200	300	1	1	3	5
Three-component bearings, babbitt surfaced	2,000-5,000+	300	1	2	2	2
Cadmium base	1,500-2,000	500	1	2	5	4
Copper-lead	1,500-2,500	350	2	2	5	3
Lead bronze	3,000-4,000	450	3	4	4	2
Tin bronze	4,000+	500+	5	5	2	1
Aluminum alloy	5,000+	250	4	3	1	2
Silver (overplated)	5,000+	500	2	3	1	1

* Compatibility indicates anti-welding and anti-scoring characteristics.

** Conformability and embeddability represent ability to compensate for misalignment and ability to absorb foreign particles without scoring and wearing.

Plain bearings can be lubricated initially when installed in equipment or where a source of lubricant is supplied by design. Properly selected greases or dry film lubricants are less likely to flow away from the friction surfaces under operational conditions. Using oil as a lubrication source generally requires a continual supply to maintain a proper lubrication film. In the case of self-lubricated bearings, inherent lubricants are supplied from the bearing itself, and therefore, supplemental lubrication demands become less likely or unnecessary.

The following are modes or degrees of plain bearing lubrication:

- Boundary — bearing and shaft rub together with only a thin film of lubricant on the surfaces. Grease-lubricated bearings generally operate this way.
- Mixed-film — part of the load (where shaft and bearing are closest) is supported on a boundary film and the remaining part is supported by hydro-dynamic pressure.
- Full-film or hydrodynamic — a thick, continuous film of self-pressurized lubricant separates the shaft from the bearing with no metal-to-metal contact.
- Hydrostatic — external pumps supply pressure to the lubricant, providing a full film to prevent metal-to-metal contact.

The first two levels of lubrication, boundary and mixed-film, can be upgraded to full-film lubrication by increasing lubricant viscosity, improving shaft finish, decreasing the load, or increasing speed.

The load-carrying ability of hydrodynamic or full-film bearings increases with speed. Because there is no metal-to-metal contact, full-film bearings offer the least friction and require less energy to rotate.

Hydrostatic bearings exhibit high load capacity at all speeds. Load and speed are limited only by the ability of the external pump to supply enough pressure and supply a flow rate sufficient to carry away generated heat. But these bearings are impractical in many applications because an external pump, motor, and other support devices are required.

Lubrication grooves are machined in the inner race of metallic and nonmetallic bearings to assist the even disbursement of lubricant over the load-bearing surface. In general, when a bearing has a length-to-bore diameter ratio greater than 2:1, grooves should be considered. (See Fig. 2-4.)

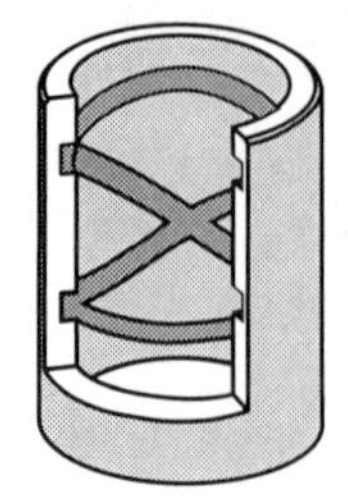

Figure 2-4: Plain Bearings with Four Groove Patterns Shown

A thru hole is drilled through the bearing wall at the center location of the bearing opposing the lubrication entry point. The purpose of this hole is to allow the entry of grease or high-viscosity oils into the groove pattern. Designs may also incorporate a lubrication hole drilled into the center of the shaft that supports the bearing. In this design, grease is pumped through the end of the shaft, which exits at the center of the bearing on the inner race of the bearing. A series of grooves are machined at a depth, one quarter of the wall thickness, to allow the lubricant to flow both longitudinally and axially along the bearing's surface. Another form of retaining lubrication is to have dimples created in an even pattern over the entire bearing surface to allow the lubricant to be trapped in the dimples. This prevents the lubricant from being forced out of the bearing.

Sintered bearings are made from powdered metals. These include copper, cast iron, and white metal. Powdered metal is compressed into the size and shape of the bearing. Within the mixture of powdered compounds are materials that are intended to disintegrate under extreme temperatures. The parts are passed through a sintering furnace and subjected to very high temperatures, turning the brittle solid powdered structure into a much harder and very porous structure. The bearings are saturated with oil, and, when in use, the heat draws out the oil to create an oil film to reduce friction and dissipate heat. When the rotation stops and the bearing cools, the oil is reabsorbed into the pores of the bearing. Auxiliary lubrication is still recommended with this type of bearing. This bearing type has limitations when the bearings are mounted in a vertical position or in environments of extreme dust and moisture.

Self-Lubricated Plain Bearings

Advancements in the field of self-lubricated bearings have made this group the most popular and effective plain bearing type in recent years. These bearings can perform in applications where no other bearing type can. Such applications include extreme temperatures, heavy moisture concentration (including full submersion), and corrosive environments. The term self-lubricating refers to this bearing type having a lubricant of some nature as an inherent or component part of the actual bearing structure. These inherent lubricants are usually solid-based lubricants such as carbon graphite, polytetrafluoroethylene (PTFE), oils, and molybdenum disulfide (MoS2). In extreme temperatures or in the presence of fluids, conventional petroleum or synthetic oil/grease lubricants have difficulty withstanding these conditions. Scientifically perfected inherent lubricants remain in the bearing loading zones and can withstand extreme conditions.

Some of the more common designs of self-lubricated plain bearings include:

Solid Lubricant Liners

These bearings usually have a cylindrical metallic or woven fiber body to provide rigidity and structural strength. A thin layer of durable, low-friction material is bonded to the intended bearing surface of the bearing body. The bonding methods are varied and include pressure rolling with an adhesive, pressure rolling a low-friction material into a sintered bronze substructure (see Fig. 2-5), spray-coating a thin layer, fusing a liner under extreme heat, and pressing a thin sintered layer or sleeve into the rigid body structure.

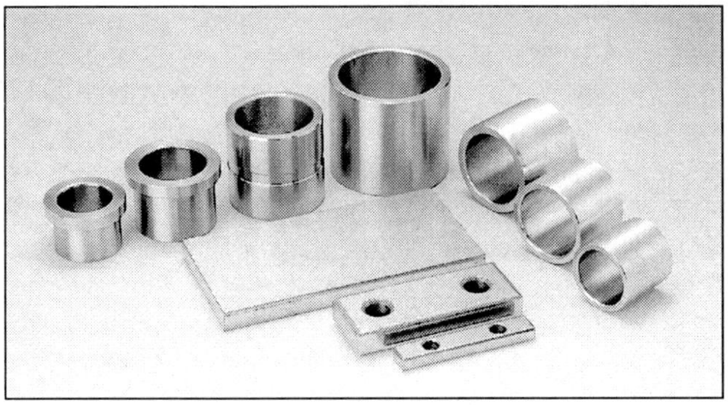

Figure 2-5: Steelback Type

Lubrication Plugs and Filled Grooves

Bronze materials made from a variety of alloys have holes drilled through the walls of the bearing. These holes are filled with a lubricating material plug usually made of carbon graphite or Teflon (PTFE). These plugs perform two valuable functions. As the rotating shaft turns, it wears minute particles of the bearing body material and the lubrication plug, mixing these particles together and forming the film of lubrication. The heat that is generated in the bearing is also readily dissipated through these lubrication plugs in the bearing wall, allowing the bearing to run cooler and with greater efficiency (see Fig. 2-6).

Figure 2-6: Self lubricating Metallic Plug (Polkadot Type) Bearing

Self-lubricated Bearing Plastics

Special compositions of plastic materials are formed into bearing shapes with lubricating agents as part of the structure of the plastic. The lubricants are synthetic and natural oils, Teflon, graphite, molybdenum, and other manufacturers' proprietary materials. Plastic bearings are lightweight and cost-effective compared to most other forms of plain or rolling element bearings. They offer maintenance-free, corrosion-resistant service and perform in environments where other bearing forms that require grease or oil lubrication cannot.

Applications Suited to Plain Bearings

Oscillating shaft applications

These are applications where the shaft generally does not turn in a full rotational pattern. An example is a gate hinge or a flapper on a piece of equipment that moves only 10–120 deg. from the home position and then returns to the starting position. Oscillations speeds can be low (1–10 per minute) or very high (several hundred per minute). Rolling element bearings are not suited for oscillation applications. This is because the design of the rollers in a rolling element bearing is intended to turn in a full rotational pattern, which accommodates the rolling frictional aspect as well as ensures proper lubrication through this rolling action. The one exception is when a full complement of needle rollers is used in conjunction with a hardened sleeve or hardened shaft.

Extreme Service Environments

Applications of an extreme nature are those outside of normal conditions. These make it very difficult to find any bearing type that will provide a reasonable service life.

Rotating shafts that are fully submersed in liquids of some nature, such as a shaft support bearing in a sewage treatment facility: These bearings operate directly under the sewage waste water that is corrosive and abrasive. Applications where the bearings are exposed to concentrated amounts of dust and grit are usually associated with aggregate operations, mining, and/or the cement industry. Extreme service conditions related to temperature are prevalent. Self-lubricated plain bearings can operate in low temperature applications such as freezers, chemical plants, and the food industry. Low-temperature bearings can operate effectively at temperatures reaching -200° F to -300° C. There are also a great many high-temperature applications, such as gypsum board plants, wood and brick kilns, bakeries, and the steel industry. These applications can reach extreme temperatures, and modern self-lubricated bearings can accommodate temperatures reaching +900° F or 480° C. It is important to remember that bearings for extreme service applications should be carefully selected and that manufacturers' literature is extensive and should be consulted.

Rolling Element Bearings

Background and History

Rolling element bearings support load and motion through a variety of roller designs contained within an assembly of races. The earliest recovered example of a bearing is a wooden ball bearing supporting a rotary table from the remains of the Roman Noemi ships in Lake Noemi, Italy in 40 A.D. Leonardo da Vinci is often credited with drawing the first roller bearing in 1500 A.D. However, Agostino Ramalli is the first to publish sketches of radial and thrust bearings. The first patent was issued to Philip Vaughan in 1794. Henry Timken, a visionary and manufacturer of carriages, patented the tapered roller bearing in 1898.

Rolling element bearings generally consist of one or two cylindrical rows of steel balls or rollers positioned between inner and outer rings. Grooves machined in the inner and outer races match the profile of the rolling elements and act as guides. There may be a separator or cage to keep the rolling elements equally spaced and to allow adequate lubrication to help the bearings reach the desired service life. Special steels with very hard surfaces allow bearings to rotate millions of revolutions under adverse conditions and with great forces acting upon them. Some bearing types have seals that fit between the races to keep the lubricant within the bearing and prevent foreign matter from entering the bearings. The two most common issues related to premature failure in bearings relate to lubrication and particulate contamination.

Ball Bearings

In general, ball bearings cost less than roller types and are normally used for light to moderate loads. These bearings have a small area of contact (referred to as "point contact") between balls and raceways. This arrangement enables them to operate at high speeds with minimal fatigue damage and less heat generation than roller bearings. There are two general classes of bearings, which are separated by the finish of the rolling elements and races. The unground type of ball bearing is a lesser precision bearing intended for moderate service. Lower rotational speed applications and thrust applications not requiring smooth, noise-free operation may be candidates for this bearing type. The unground description relates to the lack of finish and concentricity of the balls and the reduced precision of the raceways. Cost is a factor, and in low-demand applications, these bearings may be sufficient to manage the application. In all other bearing types described in this section, the types of bearings will be referred to as precision ground bearings.

Radial Ball Bearings (Precision Ground)

There are two basic types of radial ball bearings: the non-filling-slot, or Conrad type (see Fig. 2-7), and the filling-slot, or maximum-capacity type (see Fig. 2-8). The non-filling-slot or Conrad bearing has a deep, uninterrupted

Figure 2-7: Cutaway Deep Groove Ball (62/6300 Series)

Figure 2-8: Cutaway Max Capacity Showing Filling Slot

raceway on the inner and outer rings. It carries high radial loads plus moderate thrust loads in either direction.

The filling-slot bearing contains more balls than an equivalent non-filling type and, therefore, has a higher radial load capacity. Because of the filling slots, however, thrust loads must be lighter than standard and applied only in combination with a heavier radial load. When using filling slot or "maximum capacity" ball bearings, the side of the bearing with the filling slot must be facing the side that the thrust originates from. This will ensure that the thrust force will not damage the bearing. From these two basic types, other ball bearings have evolved to handle specific applications. For example, the double-row deep-groove ball bearing (see Fig. 2-9) handles higher radial loads than a single-row bearing. It should be noted, however, that a double-row deep-groove ball bearing will not accommodate greater speeds. The speed ratings are the same for single- and double-row ball bearings.

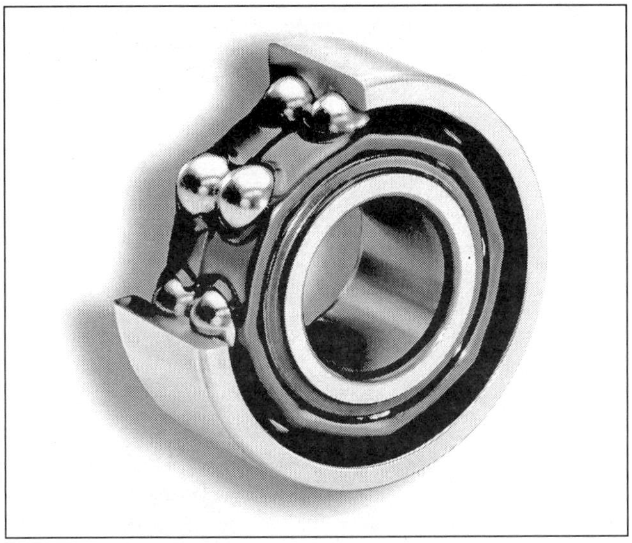

Figure 2-9: Double Row Cutaway (32/5200 Series)

The other type, the self-aligning radial ball bearing, compensates for misalignment between shaft and housing bores. These are available in both internal and external self-aligning versions. The internal type has two rows of balls between a grooved, double-row inner ring and an outer ring with a spherical outer race (contacting the balls). They are capable of accommodating shaft misalignments generally fewer than 3 degrees and supporting light shaft loads that vibrate and deflect under loading conditions. External self-aligning bearings are identical to single-row deep-groove bearings except that the outer ring has a spherical outer surface that mates with an identical opposing spherical raceway (see Fig. 2-10).

Figure 2-10: Internal Self Aligning Ball

Radial ball bearings have internal clearances between the raceways and balls to compensate for the effects of press fitting the bearings on shafts or into housings. The clearance (gap or space) also compensates for thermal expansion and contraction due to wide temperature changes. The temperatures surrounding a bearing or apparatus are referred to as ambient temperatures. Before mounting, check the manufacturer's specifications for required shaft and housing tolerances. It is common to press fit the bearing on, or into, the rotating member. For example, a bearing in an idler pulley would typically have a press fit on the outer ring.

Manufacturers offer a guide for the maximum safe operating speed of ball bearings. This speed value (DN) for inner ring rotation is the product of bearing bore diameter (mm) and shaft speed (rpm). Typical values range from about 200,000 DN to 750,000 DN, depending on the type of ball bearing and lubricant (grease or oil).

Angular-Contact Bearings

The single-row angular-contact bearing accepts a higher thrust load in combination with a moderate radial load as compared to deep-groove ball bearings. The comparison made between angular-contact bearings and deep-groove ball bearings is that the deep-groove types are designed to manage a greater radial load than thrust load. It should be noted, however, that some thrust loading can be handled by deep-groove bearings. In angular-contact bearings, the ratio of thrust to radial load depends on the angle of contact between the races and the bearing axis. Available designs can handle thrust loads from 150 percent of the radial load to more than 300 percent. To accommodate thrust in one direction, angular-contact bearings are mounted singularly or in sets of two or more bearings. Paired bearings are generally mounted either back-to-back or face-to-face to accommodate thrust loads in both directions. In certain applications, particularly machine tool spindles, bearings can be mounted in tandem to maximize axial stiffness and optimize speed capabilities. Double-row angular-contact bearings generally handle greater radial loads than single-row bearings of a similar diameter. (see Fig. 2-11)

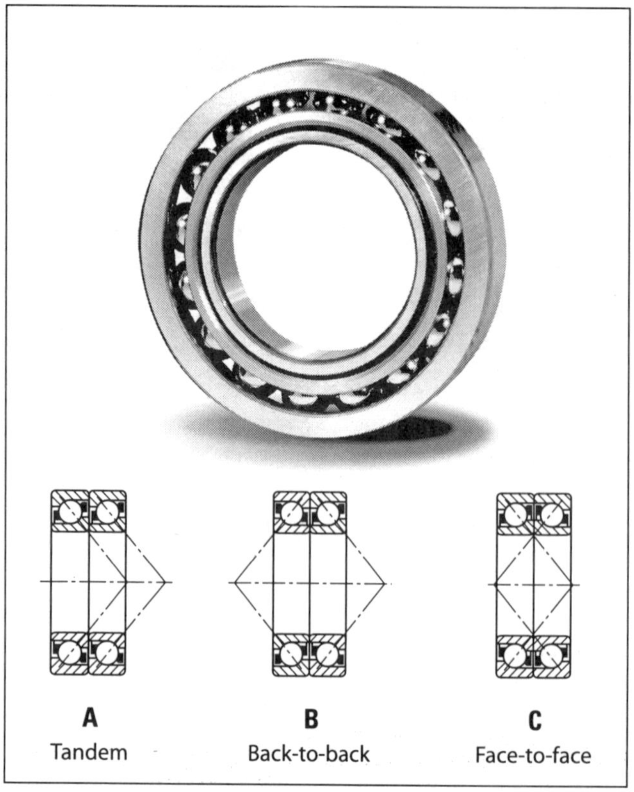

Figure 2-11: Angular Contact Single (7200 Series), A: Angular Pair Face to Back, B: Angular Pair Back to Back, C: Angular Pair Face to Face

Thrust Ball Bearings

Thrust ball bearings are intended to carry primarily thrust loads (loads running parallel to the axis of rotation) and provide axial shaft location. These bearings differ from other types in that the space between rings is oriented perpendicular to the axis of rotation (see Fig. 2-12 & 2-13). A thrust ball bearing generally contains two races, either flat or grooved, separated by steel balls with or without ball retainers in a separable or retained assembly. The races are generally hardened to resist wear. The degree of hardness of the races varies by manufacturer, and the races may be surface or thru hardened.

Thrust bearings are classified by the race contour (ball contact surface), which is either flat or grooved.

Flat-race flat-seat bearings (see Fig. 2-12) consist of two flat washer type races and a ball-retainer assembly. This assembly carries thrust loads without restraining shaft oscillation or flexing. Best suited for light loads, these bearings have a load capacity about one-third of that of comparable grooved-race bearings.

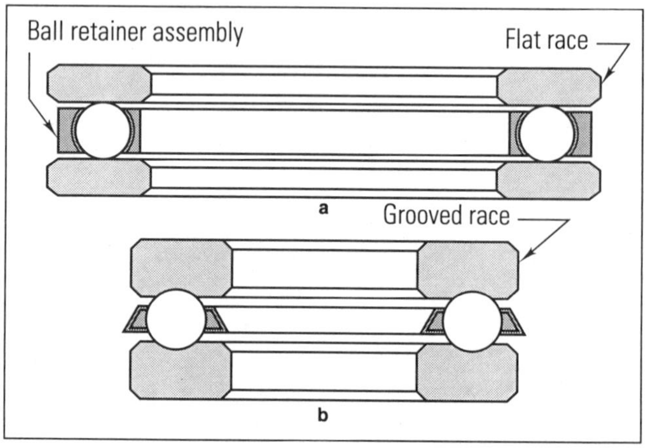

Figure 2-12: Flat Seat Thrust Flat Raceway

Grooved-race flat-seat bearings (see Fig. 2-13) are the most common type of thrust ball bearings. They consist of two grooved washer-type races, one shaft-mounted and one bore-mounted, plus a ball-retainer assembly. The shaft-

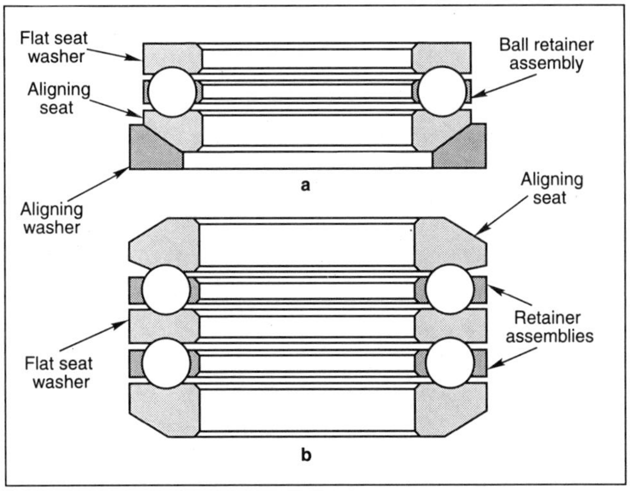

Figure 2-13: Flat Seat Thrust Grooved Raceway

mounted race is generally the race that rotates, and the housing-mounted race is generally the stationary race.

Other thrust bearings for special applications include banded and aligning types. Banded types have both stationary and rotating grooved races and are encased in a band. They are commonly used where the bearing's outer circumference must be protected from contamination, for blind installation, or where separating forces cause substantial axial motion of bearing components.

Aligning-type grooved-race bearings are available in single- and double-acting types. A single-acting type consists of a flat-seat washer and a ball-retainer assembly, plus a spherical aligning seat and matching aligning washer. The aligning members compensate for misalignment due to shaft deflection or mismatch and provide uniform load distribution through the bearing. The aligning washer is soft to enable proper seating through wear.

The double-acting type has two retainer assemblies, and it carries thrust loads in either direction.

Precision Bearings

Precision bearings are bearings that look like non-precision bearings, except precision bearings operate with a great deal more accuracy. This group or class of bearing has separate designations that relate specifically to the amount of precision, usually expressed as having an ABEC classification. ABEC stands for ANNULAR BEARING ENGINEERING COMMITTEE of the ABMA. This group establishes and maintains the standards for this class of bearing.

What is different about precision bearings over standard class bearings? Precision bearings are made to much more exact machining and fabricating processes. They generally are not made in high volume automated conditions as non-precision bearings are. The components that make up a precision bearing are made to much more exact sizes which reduce the amount of play or clearance between the parts. The balls or rollers are made more concentric or they are rounder. The races are ground smoother and rounder to match the rounder elements. The cages or retainers are made more exact and of higher grade materials that maintain their shape and reduce friction better.

We need precision bearings and Super Precision bearings for applications that demand exact rotation, high accuracy, limited end play (rigidity) , smooth running and a capability of achieving higher speeds. Precision bearings are made under a graduated class rating system. This system or grading called ABEC Class defines the maximum limits for eccentricity. Machine tools, machine spindles and precision instrumentation require bearings with a high level of accuracy and precision. The more exact we can make bearings the more precise we can make parts and equipment. Precision bearings are able to achieve faster speeds for a longer period of time due to the ease in which all of the components of a precision bearing function together. This type of bearing is very expensive size for size as compared to a non-precision class bearing. As much as 50 times the price. So it is obvious that precision bearings are only used where they are truly needed for performance. Precision bearings are made in most types of bearings, such as : ball, tapered, angular contact, cylindrical and needle as examples.

In keeping with precision bearings, a very popular group of bearings are miniature bearings. Precision miniature bearings are used extensively in the instrumentation, aerospace, medical tool and medical components, prosthetic limbs and high speed precision tools industries. Miniature bearings are available in both standard class and the precision to super precision classes. Miniature ball or roller bearings are made in straight bore, flanged style and special sizing. The bores sizes range from as small as 1mm or 0.039" to 12mm or ½" in diameter. The next time you go to the dentist, imagine how small and how fast the bearings supporting the drill/polish tools are.

Table 2-2 shows the machining limits that ensure under the specific classes of bearings listed in the ABEC scale the eccentricity (how round the bearing turns) will not exceed these limits.

Table 2-2 — Bearing machining limits

Class	Max. Eccentricity Inch	Max. Eccentricity Metric
ABEC 1	0.000295"	0.0075mm
ABEC 3	0.000197"	0.0050mm
ABEC5	0.000138"	0.0035mm
ABEC 7	0.000098"	0.0025mm
ABEC 9	0.000070"	0.0012mm

Unground ball bearings

The bearings described previously are precision-ground, which provides high-speed and high-load-carrying capabilities. As previously mentioned, unground types of thrust and roller bearings play a role in various applications. To review, unground bearings have fewer parts, cost less, and can reduce installation costs. Because tolerances are much looser than in precision bearings, extra parts for locking the bearing on the shaft are often unnecessary. Unground bearings are used in non-precision bearing applications.

Like precision types, these unground ball bearings are available in radial, thrust, and combination types, but all of these are best suited to light loads and low speeds. A typical unground bearing is made of a machined or stamped inner race, hardened steel balls, and an outer race of one or more machined or stamped parts. Because surfaces are unground and some parts are stamped, the bearing has more radial and axial freedom than a precision bearing.

Thin-section Bearings

Where space, weight, and maintenance of a lower gravitational center is important, thin-section bearings are often used. Available in both ball and roller types, these bearings have a lower inertia than conventional types of equal bore diameter. They require smaller envelopes, which reduce both the drive weight and size. Cross-sectional area of thin-section bearings remains constant within a series, regardless of bore diameter.

Thin-section bearings have a lower load capacity than comparable conventional bearings. But, they enable lighter, more compact designs than conventional types, even extra-light bearings.

Thin-section bearings are designed for light to medium-duty drives that operate at medium and slow speeds. They are not well-suited for heavy-duty or high-speed drives operating continuously.

Table 2-3 — Weight comparison between extra-light and thin-section bearings

Bore, in.	Thin-section weight, lb	Extra-light weight, lb	Weight savings, lb
2.0	0.038	0.430	0.392
4.0	0.19	2.76	2.57
6.0	0.28	10.60	10.32
8.0	3.5	27.0	23.5
12.0	5.2	85.0	79.8
16.0	12.3	170.0	157.7
24.0	19.0	443.0	424.0
32.0	25.0	1,000.0	975.0
40.0	31.0	1,779.0	1,748.0

Because rolling elements and races are so small, thin-section bearings must be properly supported in the drive assembly. Be sure that axial, radial, or moment deflection of the thin-section bearing does not prohibit its use. Also, imperfections in the bore or shaft diameter may reduce life or increase torsional drag of the bearing. The use of a clean and stable lubricating medium is very important. Particulate contamination plays a critical role in the survival of thin-section bearings.

Roller Bearings

Roller bearings have more of rolling surface area in contact with the inner and outer races and generally support heavier loads than comparably sized ball bearings. We discussed the contact of ball bearings, and it was said to be a point contact. In roller bearings, due to the configuration of the rolling elements, the contact type for this bearing group is called line contact. Roller bearings (see Fig. 2-11) handle moderate to heavy loads and are capable of handling high-impact or shock loads. They exhibit less elastic deformation than ball bearings because contact stresses for rollers are lower for a given load due to the increased rolling contact surface area.

Roller types are divided into four main categories: cylindrical, needle, spherical, and tapered.

Cylindrical roller bearings

Of the four types, cylindrical bearings have the highest combined radial load and speed capacity capability. A non-locating type of cylindrical bearing (see Fig. 2-14) enables axial movement of the inner and outer ring to accommodate both thermal and axial expansion of the shaft and tolerances in the assembly. Cylindrical bearings with shoulders on the inner and outer rings (see Fig. 2-15) accommodate some thrust loading, but the limiting factors are the amount of heat generated and the amount of heat that can be dissipated.

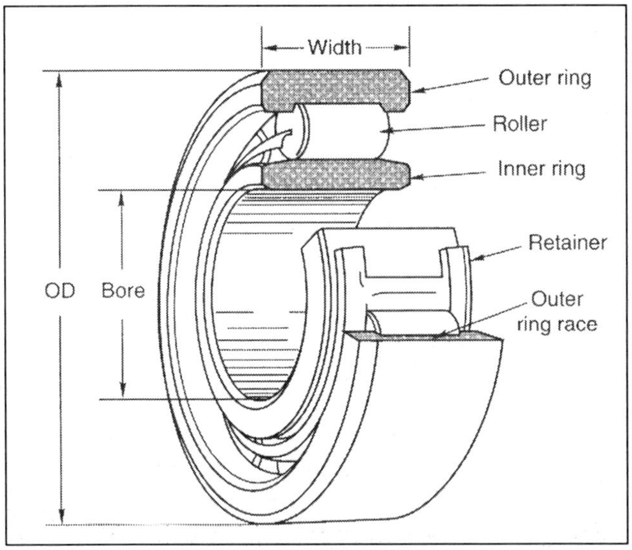

Figure 2-14: Cylindrical Roller Bearing Cut Away Separable Inner

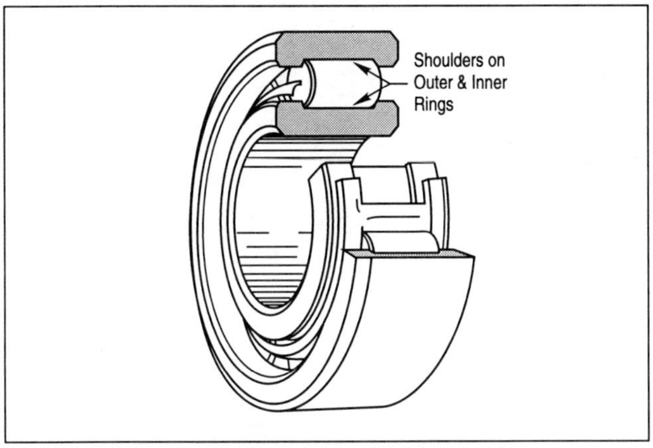

Figure 2-15: Cylindrical Roller Bearing Cut Away with Thrust Shoulders

The speed of a cylindrical roller bearing may be limited by several factors. These include the roller length-to-diameter ratio, grade of precision, roller guidance, type of cage, type of lubrication, shaft installation, housing accuracy, and heat dissipation capability of the mounting assembly. For general use, a roller length equal to the roller diameter provides the best balance of load and speed capabilities. The limiting speed of this roller bearing is usually considered equal to that of a comparable series ball bearing. Due to all the variables that can affect the speed limits of this type of roller bearing, manufacturers should be consulted to ensure true bearing life.

In many applications, bearings must position a shaft in housing so that the shaft rotates freely with minimum radial and axial movement. To do this, the bearings generally support the shaft at only two points—usually at each end of the shaft. One bearing is mounted in a fixed or secured pivot point at one end of the shaft. This bearing is called the "fixed bearing." The method of fixture may be by pressing the bearing inner race to the shaft or incorporating a locking collar or sleeve device to secure the bearing to the shaft. The opposing end of the shaft may have a bearing mounted in a "floating bearing" position. Floating generally refers to the bearing assembly's ability to allow the axial forces to pass through the bearing. These axial forces are generally caused by expansion and contraction caused by heating and cooling. If both ends were in a fixed position, the expansion/contraction forces could physically destroy the bearing. Generally, the ability to manage the float in a bearing assembly is the simple omission of a fixing ring during assembly.

Though roller bearings support more load than ball bearings, roller types are more sensitive to misalignment. The line contact of the rolling elements provides more surface area to manage the load; however, maintaining even contact along the line can be difficult if the mounting area is not true. For example, angular misalignment between the shaft and housing causes non-uniform loading of rollers, which reduces bearing life. Poor bearing alignment on the shaft causes misaligned inner and outer rings, even with unloaded bearings. An external force or load acting upon the bearing during and after assembly may cause rolling element misalignment. Manufacturers have created assemblies that make installation easier and more efficient. They are also a valuable source for assistance in properly mounting bearings. They should be consulted in difficult cases. Cylindrical bearings require a hardened shaft surface to run on. These bearings are available with both fixed assembly and separable assembly inner rings. It should be noted that some applications have the cylindrical bearings with the rolling elements running on the application shaft. In these cases, the mating shaft must be hardened, generally 63–68 RC on the Rockwell C scale.

Needle Roller Bearings

Though similar to cylindrical roller bearings, needle bearings have a much smaller diameter-to-length ratio. (The word "needle" was originally used when the roller length was at least six times the diameter.) There are generally two types of needle bearings. The first, cage separated types, have a bearing retainer or element separator as part of the assembly. This type of bearing is designed to accommodate higher speeds and more precision rolling. The second type of needle bearing assembly is a full complement style. The needle roller elements are positioned in the outer ring so that each needle is up against the other. They are not separated like the cage style element. This style of needle bearing is also referred to as a drawn cup needle bearing.

Needle bearings can carry heavy radial loads in a minimal amount of radial space. Their load capacity is higher than most single-row ball or roller bearings of comparable outside diameter (OD). This bearing type permits the use of larger, stiffer shafts for a given OD and provides a low-friction rolling bearing in about the same space as a plain bearing.

In the basic drawn-cup needle bearing (see Fig. 2-16), the outer race is a thin, drawn cup with a hardened surface. The term "drawn" refers to a manufacturing method that produces the outer race. Roller ends are shaped so that lips on the outer race keep them from falling out. Because the outer race is thin, it must be installed in a correctly sized and properly backed-up housing to transmit load effectively. Generally, a hardened shaft acts as the bearing's inner race, although an inner race can be supplied if the shaft cannot be hardened.

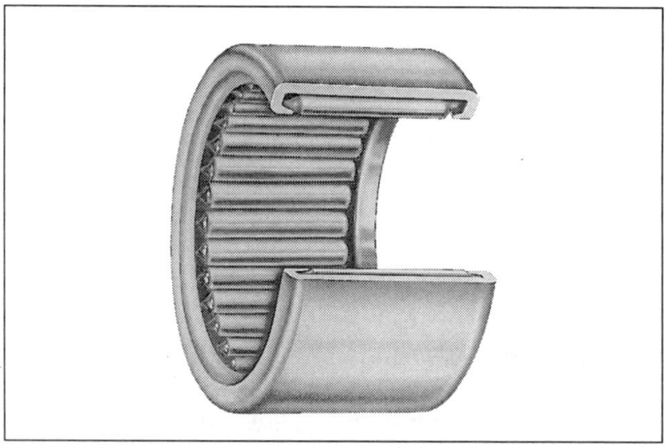

Figure 2-16: Drawn Cup Needle Bearing

The grease-retained, drawn-cup needle bearing (see Fig. 2-17) is applied without an inner ring by fitting the rollers directly onto a hardened shaft. This design is not used as extensively as the basic version because rollers may fall out if the shaft is removed. Also, the grease that retains rollers in the cup is incompatible with some applications. This bearing has slightly higher load-carrying capacity than the basic type because the grease-retained rollers have spherical ends. A caged needle roller bearing (see Fig. 2-18) is designed for heavy-duty, high-speed applications. The heaviest-duty version of this type has a machined cage. Both machined outer-race and drawn-cup caged bearings have sufficient internal voids to allow pre-greasing the bearing for lifetime lubrication. Although the assembly and greasing are designed for a full-life service, bearing performance and life can be extended with periodic lubrication. The more difficult the conditions and the more extreme the service, the more often it should be lubricated.

Figure 2-17: Grease Retained Drawn Cup Needle Bearing

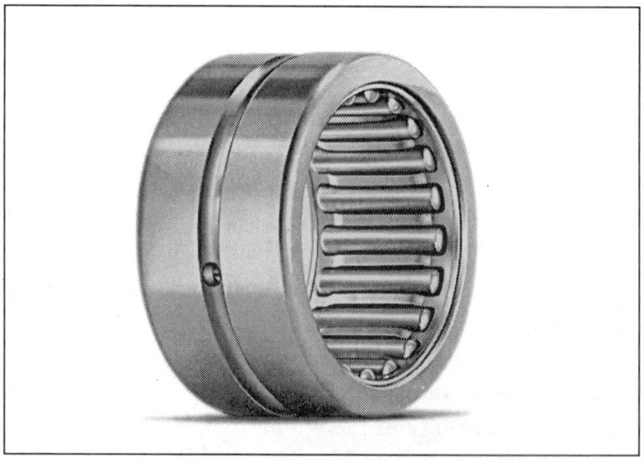

Figure 2-18: Caged Machined Needle Bearing

When selecting needle roller bearings, consider the following guidelines:

The most compact and economical arrangement uses the equipment shaft as the bearing's inner race. When the shaft is used as the inner race, it is important to check manufacturer's shaft requirements for the following:

1. Case hardened or through hardened steel.
2. Sufficient shaft size to prevent shaft deflection.
3. Shaft diameter tolerance.
4. Shaft taper tolerance.
5. Shaft out-of-roundness tolerance.
6. Shaft surface finish.
7. Shaft end chamfer.
8. Shaft sealing surface if seals are used.

Cam-Type and Load-Following Bearings

Another common adaptation of bearings is incorporated in a shaft and head configuration commonly referred to as cam-follower or load-follower bearings. Cam-type bearings can accommodate both symmetrical and elliptical motion rotation applications. The construction of this bearing type consists of an inner and outer race with a plain or rolling element bearing between the races. A stem or shaft extends from the inner race. This stem can be hollow to press onto a stub shaft or incorporate threads to be screwed into a corresponding threaded hole. If the roller head does not have a stub shaft extension, it is generally referred to as a yolk-type follower. If the inner race has been machined in a concentric or parallel configuration, this type of follower will manage symmetrical rotation. If the inner race has been machined in a specific off-center pattern, the follower will accommodate eccentric or elliptical rotation.

This bearing type is used for applications that require the attached shaft to follow a specific running alignment or support an extended load. They are particularly useful in applications where the load can be supported by the head as a guide roller or when a specific pattern of travel is required. In most cases, care is taken by the manufacturers to ensure sufficient hardness in the following components to reduce wear due to rolling and sliding friction (see Fig 2-19).

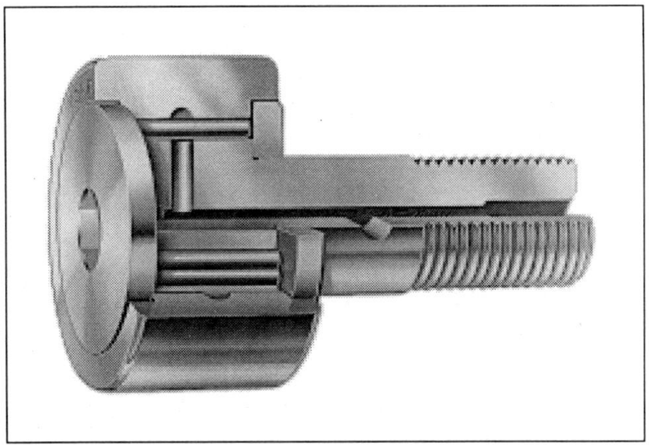

Figure 2-19: Stud Cam Follower

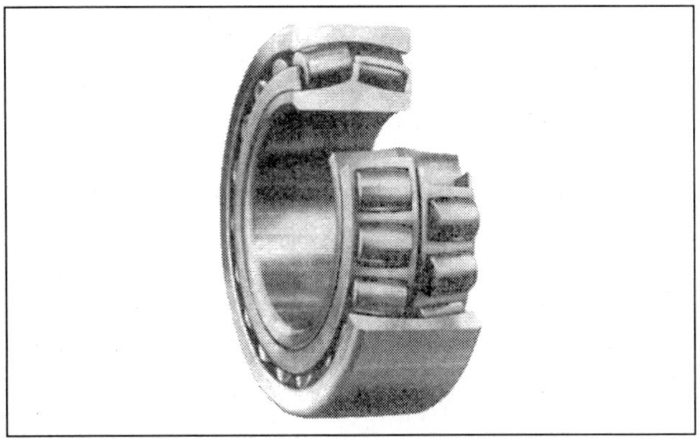

Figure 2-20: Cut Away Spherical Bearing

Spherical Roller Bearings

Usually containing one or two rows of rollers within a spherical outer raceway, spherical roller bearings (see Fig. 2-20) support high radial or combined radial and thrust loads. These bearings typically have rolling elements that are barrel shaped with flat ends and curved profile diameters. Because the outer raceway (roller contact) is spherical, these bearings have internal self-aligning capabilities. This allows spherical bearings to be used in heavy-duty applications where the need to maintain alignment can be accommodated by the bearing. They accommodate misalignment, usually from 0.5 deg. to 2 deg., with no decrease in rating or life. There is a single-row spherical bearing that is designed to manage heavy radial loads with a lesser thrust or axial load compliment. They are also suited for lighter applications where contamination is an issue. A seal can be used between the races, similar to those as shown in the radial ball bearing section.

Radial capacity of a spherical roller bearing depends on the ratio of radial-to-thrust load. For satisfactory operation, a bearing of this type should have 4 lb. of radial load for every 1 lb. of thrust load.

Spherical roller bearings are supplied with straight or tapered bores. Straight bore bearings may be installed on the shaft by a press fitting, shrink fitting, or mechanical gripping.

Press fitting: When any bearings, not just spherical bearings, are to be installed on a shaft, the need for proper retention or seating is critical. Press fitting is as the name suggests—using force and pressure to seat a bearing in the desired location so there will not be any independent rotation of the raceway being fit to the shaft. Generally, non-specialty shafts are made to the specific dimensional size, not larger or smaller. Bearings are generally made to a specific dimensional size with a very slight tolerance to allow them to fit on the shaft.

There are two major considerations to pressing bearings on shafts. The first is the type of service the bearing will be expected to perform in. Different press fitting requirements exist for different service requirements. A heavy or severe service requirement will call for a greater interference press fit than a lighter, low-shock application. The other major consideration is the effect press fitting a bearing has on the bearing clearance. The clearance in a bearing is the gap or play between the rolling elements and raceways. Care must be taken to ensure the proper resultant clearance remains in the bearing after press fitting.

Shrink or temperature fitting: This is a practice of heating the inner ring for shaft fits and cooling the outer race for housing fits. When bearings are heated, they grow or expand in size, and when they are chilled, they shrink and or contract in size. Fitting a bearing on a shaft is easily accommodated without pressure by heating the inner race with a source of heat. Using an electric induction heater or using heated water and oil solutions are common practices. The heat, when directed to the inner race, causes the bore diameter to grow, so slipping the bearing on the shaft is effortless. When the bearing cools, it will shrink, and the required fit interference will be established. When bearings need to be seated firmly in the housing bore, using liquid nitrogen or other freezing agents will reduce the outer diameter sufficiently to slip the bearing into position. After the bearing warms, the required retention force will be applied.

Press fitted bore bearings are either installed on a tapered shaft or a straight shaft with an adapter called a sleeve, nut, and washer (SNW).

Vibrating equipment, such as rock crushers and rock sorters, may require spherical bearings with very close outer race tolerances and inner races made with special steel types.

In most cases, spherical roller bearings lubricated with grease are limited to a maximum safe operating speed (DN value) of no greater than 100,000. Oil-lubricated bearings can operate up to 200,000 DN. Some spherical bearings under special conditions can operate successfully at 1 million DN. For speeds greater than these general guidelines, consult the bearing manufacturer.

Tapered Roller Bearings

A wide variety of equipment, including appliances, machine tools, mobile equipment, turbines, and industrial machinery employs tapered roller bearings in applications ranging from low to high speed. Tapered bearings carry heavy radial loads, thrust loads, or both.

Bearing components include an inner ring or cone, an outer ring or cup, and tapered rollers that are spaced and contained by a cage (see Fig. 2-21). The bearing race and roller angles can be matched to the load—shallow angles for predominantly radial loads and steeper angles for larger thrust loads.

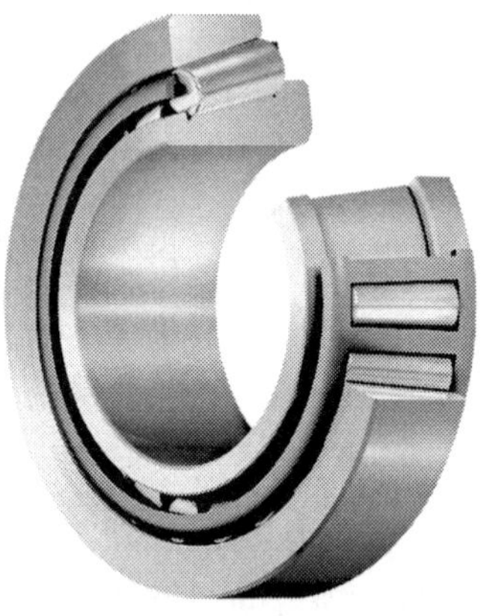

Figure 2-21: Cut Away Tapered Roller Bearing

Because tapered roller bearings support thrust loads in one direction, they are generally used in pairs facing opposite directions. End play is adjusted during mounting to allow for free rotation while preventing excessive radial play.

Tapered roller bearings can tolerate some minor misalignment because their internal clearance is adjustable during installation. This clearance can be optimized for a given application without remachining shafts or housings.

Precision-class tapered roller bearings are limited to a maximum radial run out (out-of-round condition) of 75 millionths of an inch. Super-precision bearings, for high-accuracy applications such as machine-tool spindles, limit radial run out to 40 millionths of an inch.

Cups, cones, and rollers for tapered roller bearings typically are case carburized, a process that produces hard, long-lasting contact surfaces capable of carrying heavy loads. The tough, ductile core endures heavy shock loads.

Tapered roller bearings are available in various types. The basic single-row bearing comes in many angles and roller lengths, providing a range of radial and thrust ratings. For more capacity, two-row bearings are used. Severe-service applications, such as rolling mills, require four-row bearings.

Bearing Lubrication

Adequate lubrication is essential for bearing life and efficiency and is considered the number one maintenance priority. Lubricants reduce the sliding friction, help dissipate the heat generated during use, and minimize rolling resistance due to deformation of rolling elements and raceways under load. By providing a protective film between the rolling elements and raceways, lubricants prevent wear, scoring, and seizure. Lubricants also help exclude contaminants and help protect the bearing components from corrosive elements associated with harsh and extreme environments. Bearings may be lubricated with either oil or grease. Although oil is preferred, grease is often used for convenience and ease of application. Synthetic and dry lubricants are suitable for severe applications.

Grease is a combination of petroleum or synthetic oil, a thickener, and assorted additives. It is normally applied in low- and moderate-speed applications. Because grease is more easily contained within a housing, it is well-suited to the food and chemical industries, where contamination of the end products must be avoided.

Many ball bearings are supplied factory-lubricated with grease for application in inaccessible locations or where field lubrication is impractical.

Oil is often used in high-speed or high-temperature applications where the lubricant is circulated through a bearing to remove heat. Though oil can be pumped and filtered, it is more difficult to seal and retain in the bearing housing. Oil viscosity, its resistance to flow, must be selected for the expected operating temperatures in both oil- and grease-lubricated bearings. In rolling-element bearings, excessive oil viscosity may cause skidding of rolling elements and high lubricant friction, leading to overheating and raceway damage. Insufficient viscosity may allow metal contact and cause bearing seizure.

Lubricating oils are classified as either petroleum or synthetic. Petroleum oils, often the least expensive, are widely used. Synthetic oils generally operate over a wider temperature range than petroleum types, but they can be more costly. Common types of synthetics include esters, ethers, halogenated compounds, silanes, and silicone polymers.
The rollers and races of tapered roller bearings are built on a cone principle where the apexes of the cones (rollers and races) meet at a common point on the bearing axis. This geometry permits tapered roller bearings to carry heavy radial or thrust loads or both.

Load on a tapered roller bearing is divided into a radial component (perpendicular to the axis) and a small roller-seating force. The seating force keeps the large end of each roller in contact with a rib on the cone, providing roller guidance and alignment. The shape of the cone helps to boost the tapered bearings performance.

Because the tapers of the rollers and races meet at a common apex on the bearing centerline, the rollers rotate with true rolling motion (no skidding) over the raceway, minimizing friction and wear.

Dry lubricants are used in low-speed, high-temperature applications such as furnaces and kilns. Usually, the solid is applied in a carrier fluid, which later burns off, leaving the solid lubricant. In other cases, lubricant is applied as a dry film to bearing raceways, and then it transfers to rolling elements in operation.

For rolling-element bearings operating in a housing, one-third to one-half of the diameter of the lowest rolling element should be submerged in oil. Otherwise, these bearings may overheat, causing lubricant deterioration, noisy operation, power loss, and premature bearing failure.

Bearing lubricants generally contain rust and oxidation inhibitors. However, the degree of corrosion protection varies widely with different lubricants. In cases of extreme contamination, a constant supply of pressurized lubricant to the bearing will purge the contaminants.

Linear Bearings

Linear bearings are a group of bearing types that provide low friction motion in one dimension. Unlike rotary motion bearings, linear bearings are intended to slide in a straight line, forward or backward. Designs may incorporate both rolling elements and plain sliding surface bearings. (see Fig. 2-22)

Figure 2-22: Round Shaft Linear Ball Bushing

Linear bearings usually consist of a fixed support component called a rail, way, or guide. Sliding on the support is a block, shoe, carriage, or slide, which moves in the direction of travel designed into the bearing application. Seals or wipers are usually part of the blocks to help keep contaminants out and wipe the rail as the slider block moves. The slide will have a profile that accommodates the particular support. The shapes of the slide, carriage, or block are usually one of three basic design types. These designs will be classified as a box way, dove tail, and single- or double-round rail. (see Fig. 2-23)

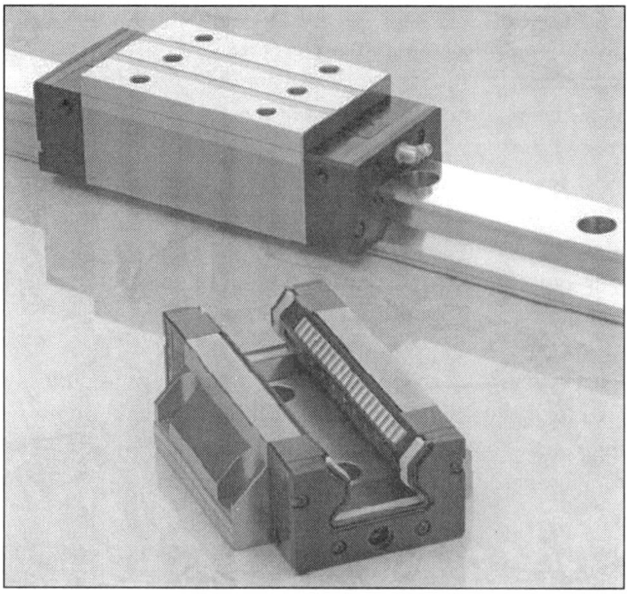

Figure 2-23: End View Dovetail Block

Linear bearings have increased in popularity in the last 20 years due to technological advancements related to the automation of equipment, the demands of process equipment, and the need for fast, precise actuation. The bearing portion of an assembly can incorporate both rolling-element and plain-bearing designs. Rolling-element designs incorporate precision steel balls in a non-recirculating (limited range of motion with very low friction) and recirculating (unlimited range of motion) configuration. A non-recirculating ball assembly uses a carrier of some type with an assembly of steel balls that are cage separated (like the radial ball bearings, only in a straight line as opposed to a spherical arrangement) or in a full complement arrangement. A full complement allows the balls to rub up against each other with no separation, while a cage-separated arrangement separates the balls and retains them in a precise alignment.

A recirculating design incorporates a path or channel within the slider that allows the balls to roll within the path or channel as the block travels over the support. This design was invented in the early 1950s by John Thompson, and although refined over the years, the principle of allowing the rolling elements free travel within the channel is still very popular today. Generally, there are two main formats used in the design of the ball contact within an assembly. These are the circular arch and the Gothic arch designs. The circular arch is a two-point contact design with a circular line contact, similar to that used by roller-bearing elements. The Gothic arch uses four contact points, more closely associated with the point contact as seen in radial ball bearings. The theory of the Gothic arch is to retain the balls effectively and reduce the contact area to lessen friction and allow a slightly freer travel.

Incorporating plain bearings is also a very popular linear bearing design in both round-rail and profiled-block assemblies. Essentially, this design of linear bearing assembly will incorporate a stationary support rail, and a precision-fit plain-bearing slider will envelope the support. These plain-bearing sliders use very low friction self-lubricating over the rails. The coefficient of friction is higher with plain bearings; however, advantages of increased load capacity, reduced weight, lower cost, corrosion resistance, and low maintenance, often make plain linear bearings the choice of designers.

A sub group of linear bearings is called linear motion wheels on rails. This type of bearing captures profiled channels (the application dictates which channel or rail will be stationary) with wheels acting as the bearing elements. In essence, these wheels are a roller with a ball or roller bearing assembly within the wheel. The materials used can be steel, plastic, or ceramic, to name a few. Cam-follower special-profiled bearings are commonly used as the bearing medium. This design is very simple and effective. The use of wheels or rollers in a linear rail system is common when telescopic motion is required, when misalignment capabilities are desirable, and when vibration and shock are present in an application. (see Fig. 2-24). For more information, see chapter 10.

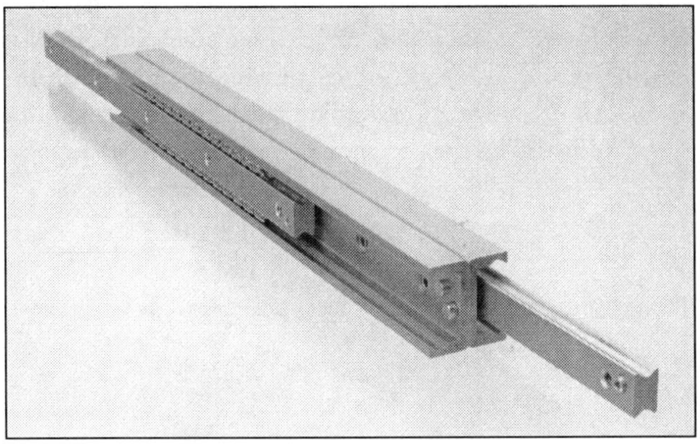

Figure 2-24: Ball Type Telescopic Rail

Lubrication of Linear Bearings

In rolling-element designed linear bearings, the type and amount of lubrication is dependent upon the application and the service required. Lubrication enhances the life and performance of the bearings. Usually, a thin film of light grease is packed into the channels surrounding the steel balls. Additional lubrication is applied in relationship to the use of the product. Manufacturers' literature defines the guidelines for lubrication types and intervals. The slider blocks may be fitted with grease fittings or oil ports.

The lubrication of plain-bearing type linear slides is generally not required. The self-lubricating properties of the plain bearings allow for a low friction, smooth operating action without the need of lubrication. Wipers and seals prevent foreign matter from entering the sliding surface and allow for long-life performance. In some instances, however, periodic lubrication applications may be necessary.

Seals

Seals are intended to keep the elements out of the bearing and hold the lubricant in the bearing. Contamination of bearings and the issues related to the actual lubricant used cause more bearing failures than simple metal fatigue. Contaminants such as dust and dirt, liquids, and corrosive agents are present in many applications. Seals are designed to prevent most of these contaminants from entering a bearing, thereby extending the normal expected life of the bearing.

Both contact and non-contact seals and shields are used in bearings to retain lubricants and exclude contaminants (see Fig. 2-25). Contact seals maintain contact with both rotating and stationary components to prevent lubricant loss. By contrast, non-contact types such as labyrinth seals pro-

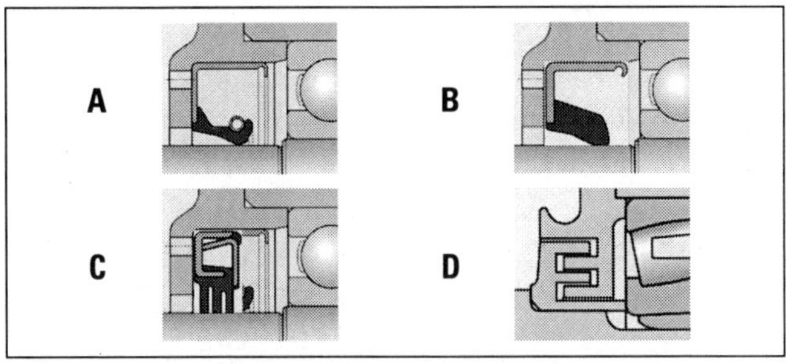

Figure 2-25: a) Oil Seal, b) Grease Seal, c) Triple Lip Seal, d) Labyrinth Seal

vide a clearance between rotating and stationary parts. This arrangement eliminates rubbing friction and heat buildup, making labyrinth seals suitable for high-speed operation. See Chapter 15. Seals are manufactured in a variety of styles and materials to help resolve the issues previously discussed. Seals in bearings are made to be both permanent and replaceable. In larger bearing assemblies, and when housing is incorporated, seals are generally both replaceable and available in a variety of materials. In the smaller, higher-volume ball and roller bearings, seals are fixed components that are not replaceable.

A shield is a metal or nonmetallic ring or plate that sets up a barrier and closes the gap between races. Shields do not usually contact the rotating race. There is a minute gap between the shield edge and the raceway. They are effective in keeping larger particles of contaminate out, and they help to retain most of the grease. In situations of very minute particulate or elevated temperatures, shields are less effective. A seal should be considered instead. Because shields do not contact the rotating race, there is less resistance to turning and less frictional heat generated. In low- or high-speed applications operating in a relatively clean environment, shields may prove to be an effective solution to help extend bearing life.

Mounted Bearings

Most types of plain- and rolling-element bearings are available as solid mounted units, which eliminates the need to select and assemble components. Installed in a sturdy housing, each bearing provides shaft support for radial, thrust, or a combination of loads. They also reduce friction in applications where machined bearing seats in the equipment frame are undesirable or impractical. Mounted bearing types include pillow blocks (both solid and split-housing), flanged cartridges, cylindrical cartridges, and take-up units. Pillow blocks, the most common type of mounted unit, have the plane of the mounting base perpendicular to the shaft and the bolting holes parallel to the shaft (see Fig. 2-26).

They are used on vertical or horizontal surfaces or on an inclined plane. Regardless of the mounting position, forces should be in the direction perpendicular to the base. Split housing styles are suitable where the need to dismantle the equipment or replace the internal bearing components on a

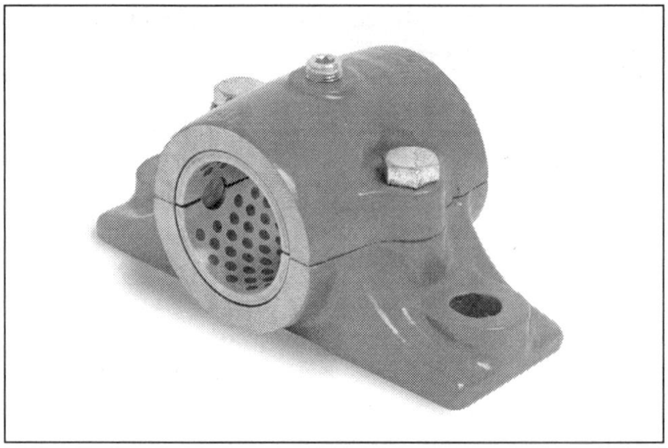

Figure 2-26: Split Plain Pillow Block Bearing

regular basis is required. Split housing and or split housing and split bearing combination proves effective on line shaft applications. This application usually consists of a long shaft that several components are driven from. It would be very time consuming to remove all of the apparatus from the line shaft. By implementing a split-bearing housing, replacement of failed bearings or line shaft components is the answer to an efficient change. Depending on the application, mounted bearings come with two and four mounting hole bases. Standard-duty are typically two-hole, whereas heavy-duty and high-shock loaded applications use four bolt hole bases to secure the housing to the pedestal.

Flanged cartridge blocks (see Fig. 2-27) are usually mounted on vertical surfaces where a shaft passes through a machine frame at a right angle. A piloted, flanged cartridge-type bearing is also available, which provides better mounting accuracy and more support than a basic unit. Cylindrical cartridges, like flanged cartridges, provide shaft support

Figure 2-27: Flange Cartridge Block

where the shaft axis is perpendicular to, and passes through, machined housings. Cartridge bearings use a machined counter bore that seats the bearing into the fixture with the flanged face being used to secure the bearing.

Take-up units (see Fig. 2-28) are used to adjust the center distance between shafts or where belt-tightening is required, such as in conveyor applications. Complete take-up units with frames are available for both horizontal and vertical adjustments and top or side mounting.

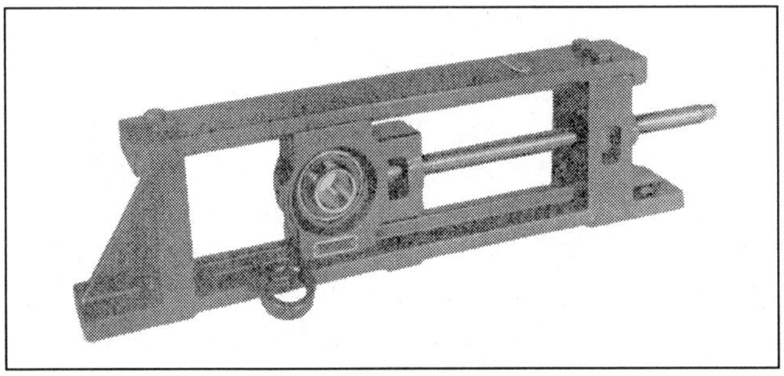

Figure 2-28: Take Up Assembly

Mounted Plain Bearings

Several types of mounted plain bearings are available: plain-bored, cast-iron, Babbitt-lined in both solid and split styles, flanged bearings, take-ups, solid-film, and self-lubricating bushed pillow blocks. A plain (journal) bearing mounting may consist of a bored housing with bearing liner or a split housing with a solid or split internal bearing. Any type of plain bearing may be incorporated into a housing style.

Hydrodynamic sleeve-bearing pillow blocks operate at low speeds in a similar way to grease-lubricated plain bearings, that is, with a boundary film. As speed increases, however, these units establish a full lubricant film, eliminating metal-to-metal contact. Typical applications include heavy-duty fans, turbines, and diesel generators.

Mounted Rolling-Element Bearings

Most ball bearings used with housing units incorporate a wide inner ring with an integral locking collar device. There are two common locking types: the set screw collar and the eccentric, or cam-locking, collar (see Fig. 2-29). Both types enable a slip fit over commercially ground shafting.

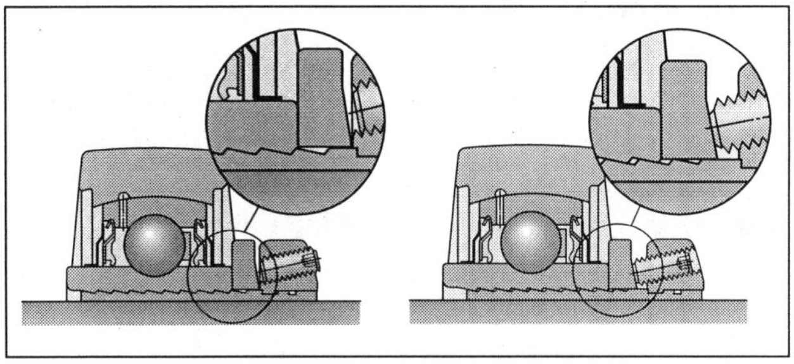

Figure 2-29: Ball Bearing Insert Bearings, Set Screw, Eccentric, Squeeze Lock

The set screw locking device has two set screws threaded into holes in an extension of the bearing's inner ring. The set screws are tightened onto the shaft after the bearing's base has been secured into position. Usually two set screws at 120 deg. of separation are used to create a concentric location of the shaft to the bearing. The set screws bite into the shaft to not only prevent slippage between the shaft and bearing, but also to help with axial movement. Some versions of this type of locking system use the inner ring of the insert bearing and/or incorporate a collar for added support. The set screws are threaded into the collar, and they contact the shaft through untapped holes in the inner ring extension.

The eccentric locking collar uses an extended inner ring of the bearing that contains a channel, eccentric to the shaft, and a machined channel in a collar that fits over the inner ring extension. The lock collar acts as a cam, or wedge, and secures the bearing to the shaft by rotating the collar in a direction opposite to the intended direction of rotation of the shaft. A set screw in the collar prevents the loosening of the collar during reverse rotation or under shock loading. This is a precaution and does not compensate for frequent reverse rotation. Eccentric locking collars should not be used for bidirectional applications. They do provide a very concentric fit of shaft to bearing and also do not damage the shaft. They also make removal or repositioning much easier than set screw locking insert bearing types.

Mounted ball-bearing inserts that fit into housings typically have two styles of outer race: a spherical outer surface that compensates for angular shaft misalignment and a straight or cylindrical outer ring that fixes the insert into a rigid position in housing. The cylindrical OD type with a slip-fit clearance allows axial movement due to shaft expansion. Cylindrical, spherical, and tapered roller bearings are also available in the mounted bearing styles. There are two styles of mounting roller bearings. One is the unitized cartridge, or insert type, and the second is the housing assembly type. In the unitized version, the inner ring, rolling elements, cages, seals, and outer ring are in a contained unit (see Fig 2-30). The required clearances for the rolling elements are preset and pre-lubricated.

Figure 2-30: Cutaway of Double Tapered Bearing Pillow Block

Some manufacturers provide a tightening ring to allow for the adjustment of the bearing's internal clearance. These bearings lock to the shaft with set screws through a collar. Due to heavier loads, these bearings require a more secure locking arrangement.

The second style of mounted roller bearings is the mounted bearing assembly. This style is used in larger shaft applications where critical shaft-to-bearing running clearances must be maintained. The housing is split into separate halves parted at the center line parallel to the base. A roller bearing, such as a spherical roller bearing, is fitted with a SNW in the inner bore. The sleeve is slightly larger than the shaft size and has slits machined in it and is threaded at the one end. The washer fits on the sleeve and the nut threads onto the sleeve. The bearing assembly is mounted onto the shaft into position and lowered into the housing base. As the nut is tightened, the sleeve locks to the shaft and tightens the assembly. Very thin, flat gauges called feeler gauges are used to determine the amount of diametral clearance to be removed from the bearing. (Diametral clearance is the gap between the rolling elements and the raceway. The clearance set up is necessary to ensure the proper rolling clearance will be maintained during operation.) Once the assembly has been properly fit to the shaft, the washer locks the assembly into place and prevents loosening. Grease is packed into the roller and cage area of the bearing, and the top housing half is bolted into position. The seals are added to complete the assembly.

Many large bearings are mounted and dismounted with hydraulic pressure between the inner ring and the shaft. Others are applied by thermal shrink-fitting onto the shaft. Complete split roller-bearing assemblies (see Fig. 2-31) are available for applications where replacement is difficult, such as with long shafts or shafts with tight-fitting components outside the bearing. All the bearing components are split into two halves so they can be removed and assembled without disturbing other components on the shaft. This style of mounted roller bearing is available in single- and double-row roller-bearing types. There is a slight deration of load-carrying capability in this style of bearing.

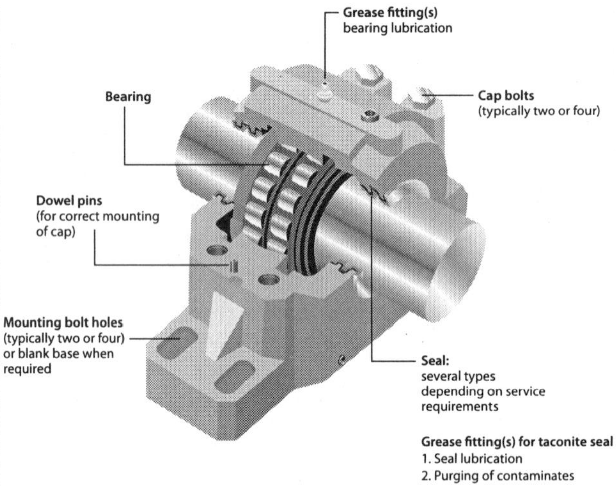

Figure 2-31: Spherical Pillow Block, Adapter Type Cutaway

Rod-end Bearings

Spherical and rod-end bearings (see Fig. 2-32) have a spherical-shaped ball that rotates within a race to accommodate a wide range of angular misalignment. They are commonly used in linkages and low-speed applications. They may consist of a spherical ball made from a variety of metals or nonmetallic compounds with a corresponding cylindrical

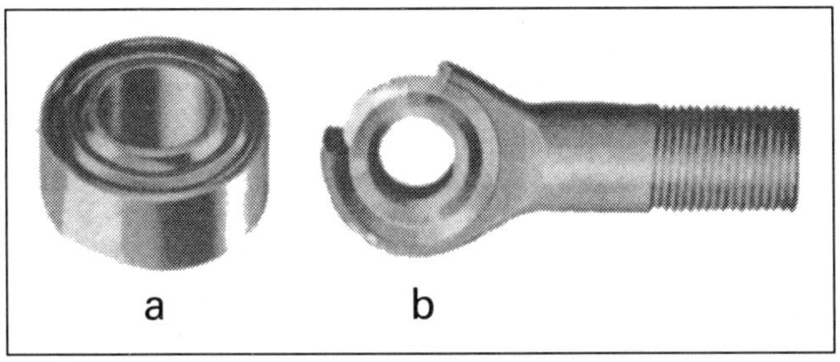

Figure 2-32: A: Spherical Plain Ball Bush B: Male Rod End

outer ring. This type is intended to be fitted into a housing bore within a machine or apparatus. It is common to refer to this series as a spherical plain bearing. Rod-end type bearings are generally a metallic or nonmetallic stem with a spherical ball mounted in one or both ends. If the shank, or stem of the rod, end is threaded on the outside surface, it is generally referred to as a "male" rod end. Inversely, if the shank is threaded on the inside of the extension, this rod end variety is generally referred to as a "female" rod end. (See diagram that shows com, male, and female types.)

Cam-Type and Load-Following Bearings

Another common adaptation of bearings is incorporated in a shaft and head configuration commonly referred to as cam-follower or load-follower bearings. Cam-type bearings can accommodate both symmetrical and elliptical motion rotation applications. The construction of this bearing type consists of an inner and outer race with a plain- or rolling-element bearing between the races. A stem or shaft extends from the inner race. This stem can be hollow to press onto a stub shaft or incorporate threads to be screwed into a corresponding threaded hole. If the roller head does not have a stub shaft extension, it is generally referred to as a yolk-type follower. If the inner race has been machined in a concentric or parallel configuration, this type of follower will manage symmetrical rotation. If the inner race has been machined in a specific off-center pattern, the follower will accommodate eccentric or elliptical rotation.

This bearing type is used for applications that require the attached shaft to follow a specific running alignment or support an extended load. They are particularly useful in applications where the load can be supported by the head as a guide roller or when a specific pattern of travel is required. In most cases, care is taken by the manufacturers to ensure sufficient hardness in the following components to reduce wear due to rolling and sliding friction (see Fig. 2-33).

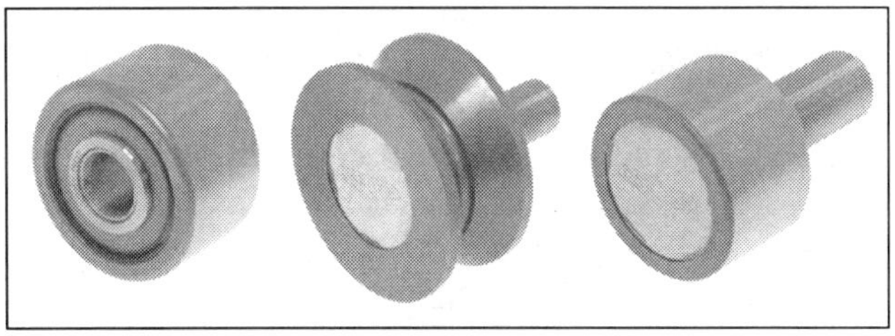

Figure 2-33: Cam Yoke, V Shape Outer Ring Load Roller with Stem, Large Cam Roller

Bearing Standards

A large number of standards and recommended practices for ball- and roller-bearing usage have been developed in the U.S. by the American Bearing Manufacturers Association (ABMA), formerly known as the Anti-Friction Bearing Manufacturers Association. These standards were subsequently adopted by the American National Standards Institute. In general, corresponding standards are published by the International Organization for Standardization.

ABMA standards cover various types of bearings (both standard and metric), plus terminology, tolerances and gauging, shaft and housing fits, mounting accessories, load ratings, metal balls, and roller bearing vibration and noise. ABMA also works with ANSI and the ISO in developing national and international bearing standards. Excerpts of many standards are included in bearing catalogs.

Standard		Normal				High
	ISO	P0	P6	P5	P4	P2
	JIS	Class 0	Class 6	Class 5	Class 4	Class 2
ABMA:	Ball Bearings	ABEC 1	ABEC 3	ABEC 5	ABEC 7	ABEC 9
	Roller Bearings	RBEC 1	RBEC 3	RBEC 5	–	–
Inch:	Tapered Roller Bearings	Class 4	Class 2	Class 3	Class 0	Class 00
Metric:	Tapered Roller Bearings	Class K	Class N	Class C	Class B	Class A

Table 2-4: Shows the differences between standard designations for precision class

The following is a list of common international standard designations.

- ISO = International Organization for Standardization
- ABMA = American Bearing Manufacturers Association
- ABEC = Annular Bearing Engineers' Committee
- RBEC = Roller Bearing Engineers' Committee
- JIS = Japanese Industrial Standards
- DIN = Deutsch Industry Norm
- BSI = British Standards Institute

Load and Life Ratings

The ABMA has two standardized load ratings for rolling-element bearings: the basic load rating and the static load rating. (Bearing catalogs often include details on how bearings are rated.)

The basic load rating, used for bearings that rotate, defines the load that a group of bearings can endure for one million revolutions. The static load rating is the load that corresponds to a total permanent deformation of rolling element and raceway of 0.0001 of the rolling-element diameter. Experience has shown that deformations up to this value have little effect on bearing operation. Load ratings for ball and roller bearings are given in ABMA standards No. 9 and No. 11, respectively.

The rating life, L_{10}, of a bearing is normally expressed as the life, in millions of revolutions, that 90 percent of a group of bearings will complete at a given load and speed before fatigue failure occurs. This life can also be given in hours of operation. The basic load rating can be used to predict the L_{10} life of a bearing under specific load and speed conditions.

Bearing life is usually inversely proportional to bearing load and speed. Thus, reducing the load or speed, or increasing the bearing size can substantially increase the bearing life.

Because it is virtually impossible to predict the exact life of a bearing, a safety factor should be used to minimize the possibility of premature or unexpected failure. Bearing replacement and machine downtime may cost far more than that for installing a larger bearing.

New philosophies of bearing fatigue are being evaluated within the bearing industry based on improvements in bearing steel metallurgy, the effects of lubricating film thickness, presence of contamination, and load zone. Therefore, the industry may adopt new life-prediction methods in the future. Please consult with each manufacturer to determine how to compute their published ratings.

Shaft and Housing Fits

To fully utilize the capacity of a bearing, the inner ring (ID) must properly fit the shaft and the outer ring (OD) must properly fit the housing bore. Fit selection depends on the load characteristics, bearing dimensions, temperature, heat expansion of the shaft, housing material and other factors.

The ABMA and ANSI have adopted a system that provides a choice of shaft and housing tolerances, which, when used with standard ball and roller bearings, provides a range of shaft and housing fits for light, normal, and heavy loads. The standard (ABMA No. 7) covers metric radial ball and roller bearings, except tapered roller bearings. In general, the ISO has published corresponding standards.

Application and Selection

Bearing selection is based on three major factors: load, speed, and life. But consideration must also be given to other operating and environmental conditions, as well as installation and operating costs. Here are some factors to consider:

- Load: Is it radial, thrust, or a combination of both? What is its magnitude? Is it, by nature, uniform, light shock, or heavy shock?

- Speed: What is its magnitude? Is it constant or variable? Unidirectional or Oscillatory? Bearing speeds are limited by tolerance grade, lubricant used, retainer design, and type of bearing seal.

- Required life: How long must the bearing operate? This is usually expressed as the number of hours of operation required at a given load and speed.

- Noise and vibration: Will the bearing operate where noise is undesirable or in high-speed equipment where balance is essential to prevent vibration?

- Misalignment: Is it present? To what degree?

- Temperature: What are the ambient and operating temperatures? Speed, load and external heat affect operating temperature.

- Environmental conditions: Are contaminants or corrosives present? To what degree and what types?

Once you have identified and evaluated these application constraints, compare the capabilities of various bearing types with the application requirements. Consider both operating and maintenance requirements, as well as cost.

Plain bearings are most often used where speed and load requirements are low and precision is not a concern. Benefits include low first cost and low assembly cost.

Rolling-element bearings can handle heavier loads and higher speeds. They have the lowest friction and require the least energy to turn. Load requirements usually determine the type of rolling-element bearing to use for a specific application. For example, predominant thrust loading requires either thrust, angular contact, or tapered roller bearings. Heavy radial loads call for roller bearings.

Ball bearings are typically applied in high-speed applications, whereas roller bearings are best suited for high loads. Tapered or spherical types offer similar features. However, spherical bearings have a dynamic alignment capability not present in tapered bearings. But tapered bearings have better thrust capacity.

To select mounted bearings, consider shaft size, radial and thrust load, load characteristics, speed, mounting limitations, and environment.

The first step in selecting a mounted bearing is to determine shaft size based on the bending and torsional loads. However, overhung loads or loads between widely spaced centers may call for large shaft diameters, even when bearing loads are light. If so, consider light-duty bearings or machining the shaft ends for smaller bearings. However, shafting is not always chosen on an optimum engineering basis. Design standardization, available stock sizes, and similar factors may dictate size.

The method of securing the bearing to the shaft is determined by considering cost vs. performance. Ease of installation is important to the builder and the maintainer. Sometimes, machine maintenance requires frequent bearing disassembly. In such cases, bearing assemblies that can be easily installed and disassembled should be used.

Choose the mounting unit with regard to its support structure and surroundings as well as its strength requirements. Also, consider installation factors such as clearance and structural members required for mounting. In most cases, housings transfer load to the structure by surface support and contact, while mounting bolts simply locate and secure. When housings are applied so mounting bolts supply support, carefully consider bolt size, bolt-hole fit, mounting procedure, ability of the housing to support a "pull-off" load, or any other factor that may affect capacity.

Also consider housing material strength and configuration. A gray-iron pillow block with a thick cross section and reinforcing web may be stronger than a thinner, steel housing, though the tensile strength of steel is greater.

Additional Information

Addresses and phone numbers for associations are provided in the Association Appendix.

Tedric A. Harris, *Rolling Bearing Analysis*, 3rd Edition, John Wiley & Sons, New York, 1991.

ANSI/ABMA Standard 7-1988, "Shaft and Housing Fits for Metric Radial Ball and Roller Bearings (Except Tapered Roller Bearings) Conforming to Basic Boundary Plans," American Bearing Manufacturers Association, Washington, D.C.

ANSI/ABMA Standard 9-1990, "Load Ratings and Fatigue Life for Ball Bearings," American Bearing Manufacturers Association, Washington, D.C.

ANSI/ABMA Standard 11-1990, "Load Ratings and Fatigue Life for Roller Bearings," American Bearing Manufacturers Association, Washington, D.C.

American National Standards Institute, New York, NY.

Bearings Specialists Association, Glen Ellyn, Ill.

Chapter 3

Belt Drives

Chapter Objectives

When you have finished Chapter 3, "Belt Drives," you should be able to do the following:

1. Have a basic understanding of belted power transmission drives.
2. Understand the history of belts.
3. Identify all of the basic types of v-belts, multi-rib belts, double-v, flat belts, and synchronous belts.
4. Know the characteristics and unique qualities of each of the various types of belts.
5. Understand where each type of belt is generally used and why.
6. List general application requirements for belt selections.

Introduction

Belt drives are typically used to transfer torque and/or facilitate speed change from one shaft, the driver, to a second parallel shaft, the driven. Some applications may use multiple driven shafts to connect drives over longer distances and to power multiple points in a drive system. Other belt drives may use belts capable of connecting shafts at right angles. In all applications, the drive consists of one or more belts and a sheave/pulley relationship. Tension between shafts keeps the belt in contact with the sheave/pulley surface faces, and the resultant frictional force provides the transfer of torque from the driver to the driven shaft. Differences in diameter dimensions of the driven sheave/pulley to the driver sheave/pulley, (the ratio) create a speed change that may be either an increase or decrease of the driven rotational speed in relationship to the driver rotational speed.

Drive belts are grouped in three basic classifications determined by their cross-sectional shape and design: v or wedge, flat, and synchronous or toothed. Each classification favors an application based on speed, horsepower, torque, distance between shafts, environment, space restrictions, shaft alignment, and cost, as well as other considerations that will be discussed later in this chapter.

Driver and driven "pulleys" are grouped in three basic classifications as well. The component being turned by the belt is commonly called a pulley; however, each type of belt requires a distinct "pulley" system. The v-style belt matches to a grooved pulley called a sheave. These are available in single to multiple grooves. The toothed-style belt matches to a toothed "pulley" called a sprocket, and these are available with various numbers of teeth, and the flat belt uses the classic drum-style pulley.

Belt drives have many advantages over other drive types like close coupled, roller chain, and gear drives. The following is a partial list of advantages:

- They are relatively economical to install and maintain.
- They are reliable and trouble-free.
- They are simple to design.
- They are easy to install.
- No lubrication is required.
- They lend themselves to a wide variety of applications.
- They can be adjusted for virtually any speed requirement.
- V-belt drives are quiet and operate smoothly.
- Pulleys are less expensive than comparable chain sprockets.
- They have long sheave/pulley life.
- Belts can act like a fuse or clutch in the system to prevent catastrophic machine failure.

Concept

The basic concept of belt drives has its origins in the lever, where a force is applied to the end of a lever arm, causing rotation about a pivot point at the other end of the lever. The force applied at the end of the lever, measured in pounds, is multiplied times the length of the lever, in feet, to yield a turning "moment" at the pivot point, measured in foot-pounds. This turning moment is commonly referred to as torque. If you take this lever idea a step further, and think of an infinite number of levers crowded together and acting around the same pivot point, like spokes of a wheel, you have a basic belt-and-pulley concept model. The turning force is being provided by the frictional forces of the belt, acting around the rim of the pulley. Tension in the belt keeps it in contact with the pulley surface so the frictional forces can act without slippage.

Belt Types

Flat Belts

Some of the earliest belt and pulley systems took the form of large wooden wheels with flat, leather belts. During the Industrial Age of the late 1800s and early 1900s, very large pulley and belt systems connected steam engines or water wheels to long shafts that would stretch the entire length of a factory. Along these "line shafts," any number of pulleys could be fastened in place, with belts that dropped down

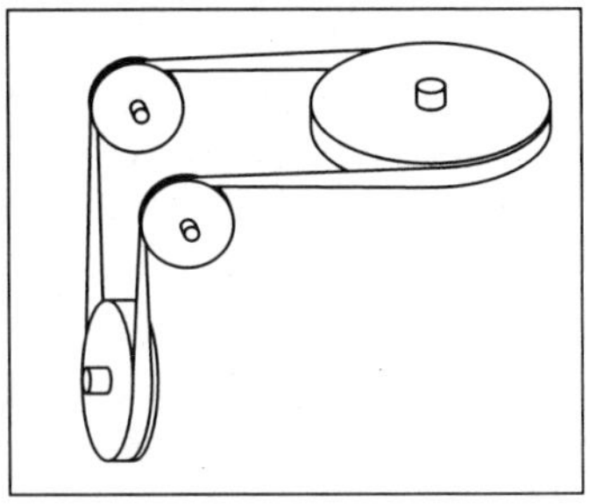

Figure 3-1: Flat belts have the flexibility to go around corners for special drive configurations.

to drive various types of machinery. To vary and adjust for different machine speeds, pulleys of various sizes speed up or speed down the turning rate of the line shaft.

The disadvantage of the flat belt is that it relies on high tension in order to keep the belt in firm contact with the pulley and a high friction force to keep it from slipping. In the late nineteenth and early twentieth centuries, it was customary to use belt dressings: liquids that increased the stickiness of the belt to help prevent slip. A slipping belt could easily walk off the edge of the pulley, resulting in significant maintenance requirements and down time. Pulleys must be crowned and often have flanges to keep the belt centered and prevent it from walking off the pulleys.

Flat belt systems are still in use today; they have the advantage of being flexible, can bend easily around small pulley diameters, and are adaptable to serpentine and twisted drives. They are made in a wide variety of lengths and widths, and can be truly endless or open-end. The open-end variety is typically available in rolls from which sections are cut to length and either vulcanized together or spliced together with fasteners.

Rope Belts

In the late 1800s, cotton and hemp ropes were also being used as belts, and the pulleys were fitted with v-shaped grooves for the ropes to ride in. This kept them from walking off the pulley, and the wedging action of the rope in the grooves allowed for higher normal forces at lower belt tension. In most cases, ropes allowed for more economical drives, longer center distances, and lower maintenance costs.

With the advent of electric motors and internal combustion engines, power transmission systems were able to be placed more remotely and not rely on huge, centrally located line shafts. Machines became more self-contained and could be placed virtually anywhere, provided electricity or other fuels were available. Rope and leather belts were not as well adapted to these demanding drives, as tension had to be greatly increased to overcome slip. Belt drives were becoming more compact and able to carry higher horsepower loads at faster speeds and eventually replaced the rope drives in most applications.

V-Belts

At the turn of the twentieth century, the first v-shaped leather belt was introduced, and in 1917, John Gates introduced the first vulcanized rubber v-belt. This design allowed the belt to wedge tightly into the sheave groove as tension was applied to the belt, and higher torque values and higher drive speeds could be maintained with less slip. The "v" shape of the belt matches the "v" of the sheave grooves, with more contact area than round ropes, resulting in higher efficiency.

Theory

In order to get a better understanding of how v-belt drives work, let us take a look at the physics behind them. Let us first look at the frictional properties. As seen in Fig. 3-2, the force of gravity (G) is acting on a body (M) at rest on a flat surface. In order to move the body across the surface, a force (F) must be applied perpendicular to the gravitational force. As F increases, it will reach a point where it overcomes G, and the body moves. The ratio of F to G at the point of slippage is called the coefficient of friction (μ). It is relative to and specific to the material composition of the body with mass (M) and the composition of the surface it rests upon.

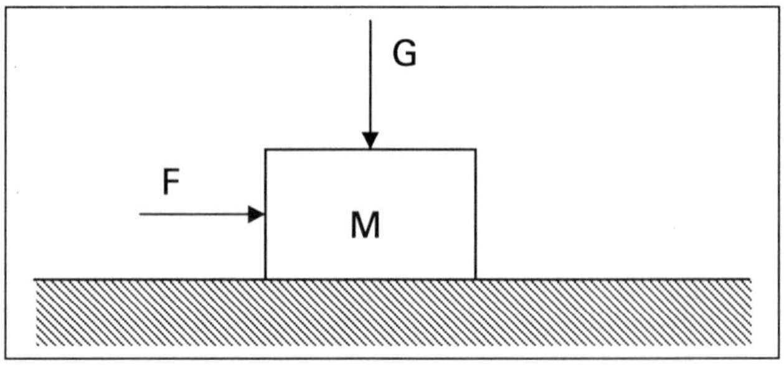

Figure 3-2: Frictional Properties

$$\mu = F \div G$$

From this, we can see that if the coefficient of friction is high, then F must be higher in order to overcome the resistance of G acting on M.

The wedging action of the belt in the sheave groove is illustrated in Fig. 3-3. The force (F) acting on the belt is provided by the applied tension in the belt and results in wedging forces directed perpendicular to the sheave groove surfaces.

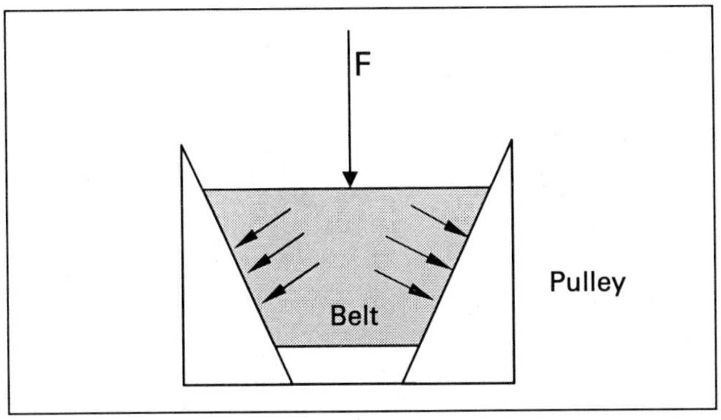

Figure 3-3: Wedging Action

For a rotating sheave system (see Fig. 3-4), the applied tension in the belt acts as the force to wedge the belt into the pulley groove. The tensions in the belt entering and exiting the sheave are known as T_1 and T_2, respectively. If T_1 and T_2 are equal, no rotation can occur, so a tension differential must exist if the system is to rotate.

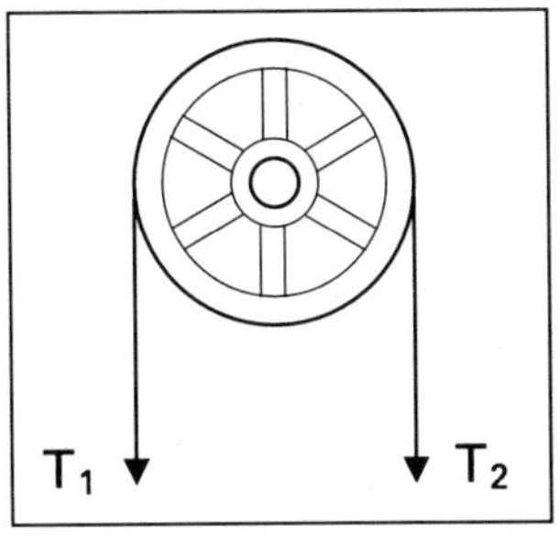

Figure 3-4: Belt Tension

If we add a second sheave (see Fig. 3-5), the belt becomes an endless member, and the entire sheave system could rotate indefinitely if a turning force, like a motor, is attached to the driver sheave.

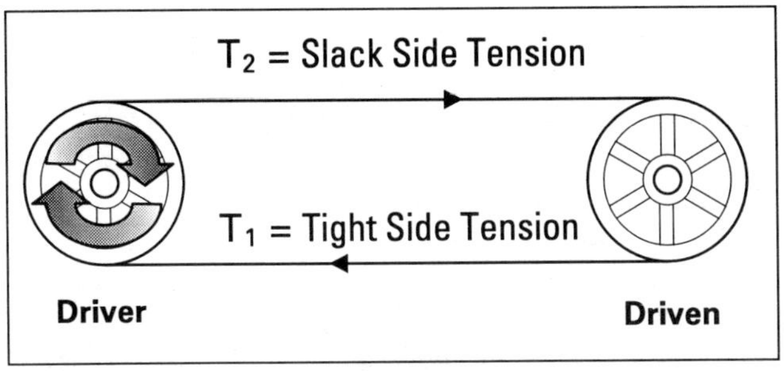

Figure 3-5: 2-pulley System

As the driver sheave rotates, the belt induces rotation in the driven sheave. Relative to the driver sheave, T_1 is referred to as the tight side tension, and T_2 is the slack side tension. The difference between these two values is called the effective tension, and the ratio of T_1 to T_2 is called the tension ratio.

$$T_e = T_1 - T_2$$

$$R_1 = T_1 \div T_2$$

In this 2-sheave system, the coefficient of friction is calculated from the following formula:

$$\mu = \frac{\ln\left(\frac{T_1}{T_2}\right)}{\alpha}$$

where T_1 and T_2 are the tight side and slack side tensions, respectively,
α is the angle of wrap on the pulley, measured in radians.

In a belt and sheave system, the tension ratio and the coefficient of friction must both be high enough to prevent slippage as torque is applied to the system via the sheave shafts. Flat belts require a 2:1 tight side to slack side tension ratio, as opposed to 5:1 for v-belts.

As a general rule, belt manufacturers pick materials with appropriate coefficient of friction properties to ensure friction properties that are not too high or too low. The tension is a calculated value, and end users should consult the engineering manuals published by belt manufactures to determine the proper tension, or consult directly with the manufacturer's belt design engineers.

V-Belt Types

V-belts are generally found in two broad construction categories: envelope and raw edge. Fig. 3-6 illustrates the main appearance differences. The envelope variety is fully enclosed with a rubber-impregnated fabric and resists exposure to oils, greases, dirt, water, and most chemicals.

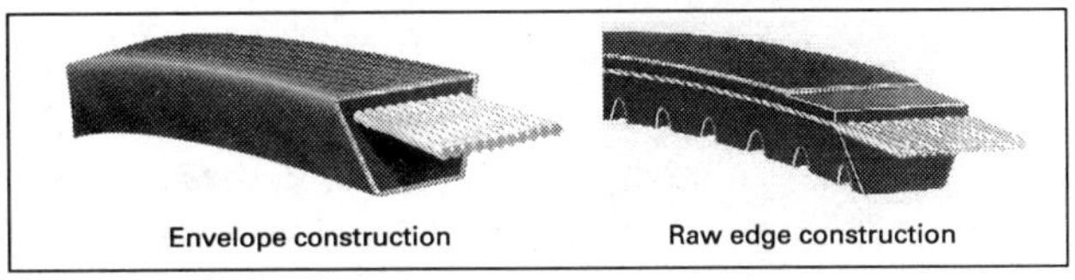

Figure 3-6: General Types of V-belts

The raw edge variety is essentially the same as the envelope construction without the envelope, although there is a fabric surface on the back of the belt to resist abrasion when running on a back-side idler sheave. The cushion rubber, below the cord line, provides the wedging action in the sheave groove and is generally fiber-loaded to provide support to the cord line and to resist abrasion from the sheave.

The basic belt construction is the same, whether envelope or raw-edge. Fig. 3-7 illustrates the various components of the belt and the function of each. In the envelope construction, the backing fabric is eliminated, and the entire belt carcass is wrapped in a fabric envelope.

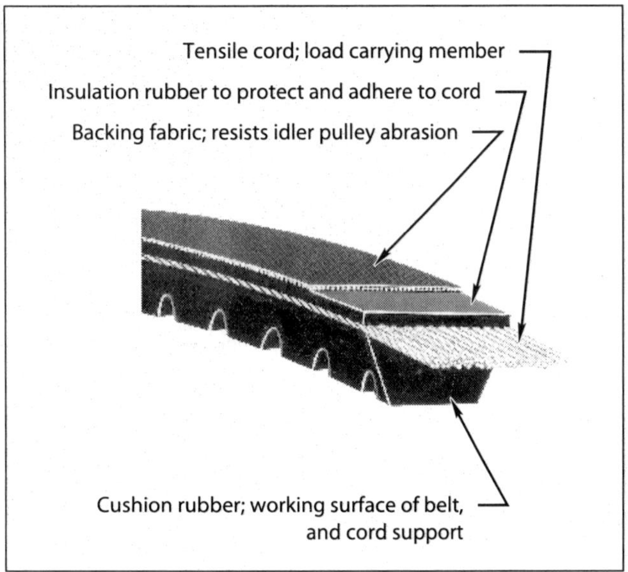

Figure 3-7: Parts of a V-belt

V-belts may be cogged or non-cogged (see Fig. 3-6). Cogs impart more flexibility to the belt when they are to be used with small diameter pulleys or back bending idler pulleys. Many different cog configurations are seen throughout the industry, from evenly spaced to randomly spaced, perpendicular across the belt, or angled.

Classical V-Belts

Classical v-belts have been around the longest and are used worldwide. They carry the ISO designations of W, Y, Z, A, B, C, D, and E, although the W, Y, and Z sizes are rarely used in North America. The ARPM standard only references the A through D sizes, which are quite common and comprise a majority of the total industrial v-belt sizes sold in North America.

The general part number format is the cross section size followed by the inside length in inches (e.g., B90 is a B-section of 90 inches inside length). General cross--sectional dimensions are shown in Fig. 3-8 below and are specifically addressed in ARPM standard IP-20 and ISO standards 4183 and 4184.

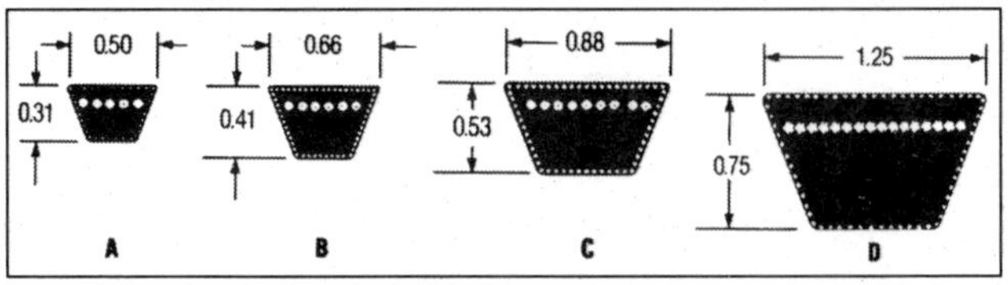

Figure 3-8: General V-Belt Sizes

Classical v-belts are capable of covering a load range from fractional (less than 1 HP) to 500 horsepower. They are less efficient than narrow-v belt systems and generally contribute to higher overhung bearing loads. However, classical v-belts have a high tolerance for poor operating conditions and unusual drive conditions such as reverse bending and twisted drives.

Because they are so entrenched in the industry, classical belts and sheaves tend to be more economical than narrow-v systems. They cover a wide range of sizes and are readily available through industrial distributors. This versatility is likely to keep them at the forefront of the belt industry for years to come.

Classical v-belts are available as single belts or banded sets, cogged or non-cogged, envelope or raw edge design, and as double-v belts. This last group is a special configuration with mirror image dimensions about a central tensile cord line. They are used in serpentine drives, where a single belt is used to drive several components, and can drive off of either the front or the backside of the belt. Fig. 3-9 shows the three most common classical double-v belts, as referenced in ARPM standard IP-21.

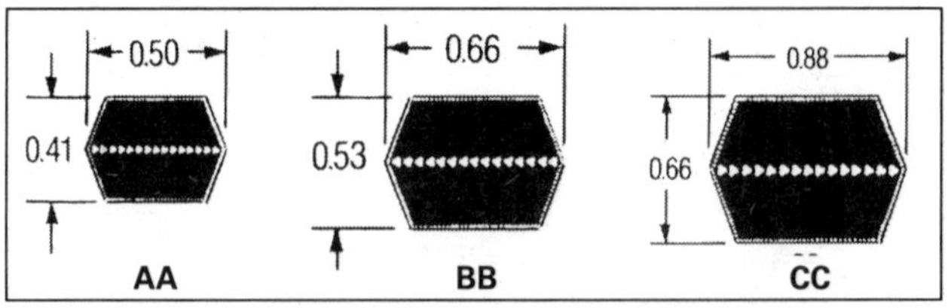

Figure 3-9: Double-V General V-Belt Sizes

Wedge V-Belts

One of the drawbacks of the classical belt line is that relatively few of the tensile cords are fully supported by the cushion in contact with the sheave. In 1959, the first "narrow" v-belt design was introduced. Fig. 3-10 illustrates the concept. The narrow, or "wedge," design provides more support to the center of the belt, has greater contact surface with the sheave, and allows for higher horsepower capability with a narrower drive footprint and reduced overhung bearing load.

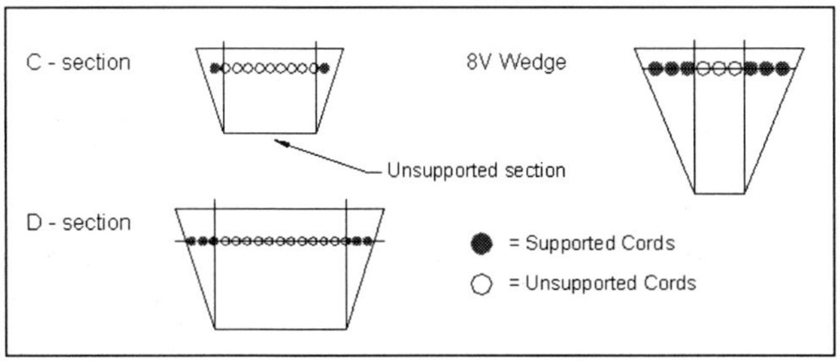

Figure 3-10: Tensile Cord Support

A typical 400 HP drive that uses twelve D-section classical belts could be replaced by eight 8V wedge belts (see Fig. 3-11). The result is far less overhung load on bearings and shafts, more economical sheaves, and fewer belts.

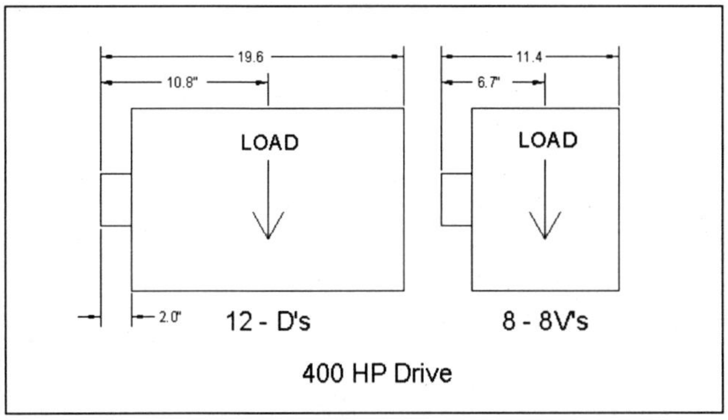

Figure 3-11: Comparing Classical Belt to Wedge Belt Overhung Load

ARPM standard IP-22 addresses the wedge line of v-belts, with dimensions and tolerances for standard sheaves. Reference dimensions are given for the belts (see Fig. 3-12). There is no ISO standard for wedge belts, as they are not a standard, global belt cross section; they are unique to North America.

Part number designations for the wedge belts are shown in terms of belt top width followed by length. The numerical prefix indicates the belt top width in one-eighth-inch increments, and the length is shown in nominal outside length in inches (e.g., a 5V500 part number indicates a

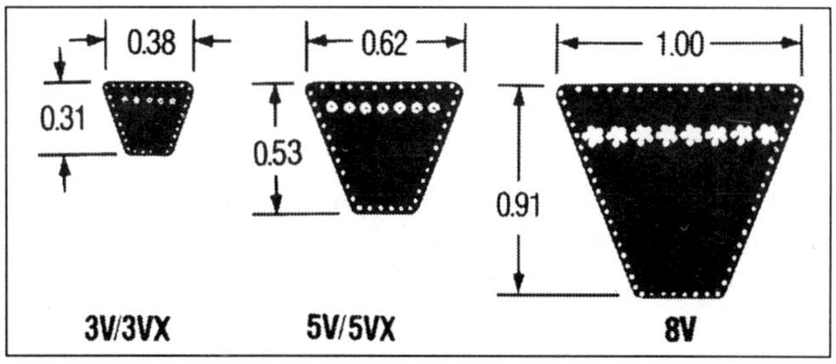

Figure 3-12: Wedge V-Belt Sizes

5/8" top width, wedge v-belt, of 50.0 inches nominal outside length).

Wedge belts can be either envelope construction or raw edge, cogged or non-cogged. The cogged version is indicated with an "X" in the part number, following the top width designation (e.g., a 3VX950 would be a 3/8 in. top width wedge v-belt that is cogged, of 95.0 in. nominal outside length).

Fractional Horsepower V-Belts

The light-duty, single v-belt line exists only in North America and countries that have adopted U.S. belt standards. These belts are designed to operate drives of less than one horsepower. They are identified with a 2L, 3L, 4L, or 5L prefix, which indicates top width in eighth-inch increments, followed by the belt length in inches of nominal outside length (e.g., a 3L400 part number identifies a 3/8 in. top width, light duty v-belt of 40.0 in. nominal outside length). ARPM standard IP-23 addresses the FHP belt line and offers the reference dimensions (see Fig. 3-13).

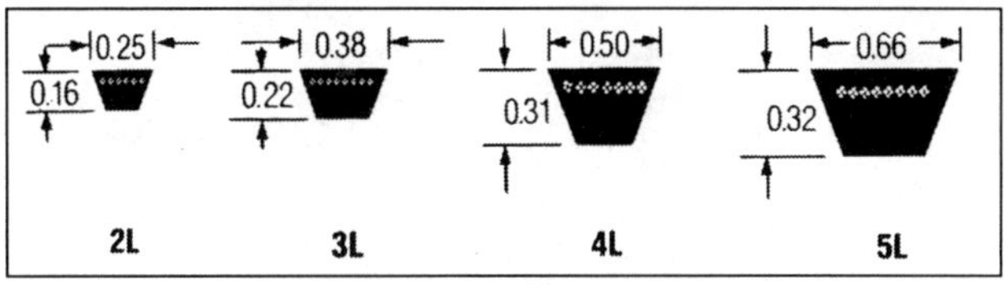

Figure 3-13: Fractional Horsepower V-Belt Sizes

The FHP 2L, 3L, 4L, and 5L top width dimensions are the same as the classical W, Y, A, and B top widths, respectively; the FHP belts are slightly thinner than their classical counterparts. Classical belts are often used in place of FHP belts if the latter is not available, but FHP belts should never be used to replace classical v-belts, as the FHP is not constructed of the same high grade materials that allow the classical v-belts to handle higher-horsepower applications.

As the name implies, these belts are used singly, on drives of 1 horsepower or less.

Because the same sheaves will accommodate either classical belts or their FHP counterparts, it is common in the industry to use the belt interchangeably, although this practice is not recommended and can have unexpected results.

Banded Belts

Many applications can experience instantaneous speed fluctuations and impulse loads that frequently cause single belts to whip and turn over in the sheave. To handle these severe applications, belt designers incorporate a common backing material across several single belts. These "banded" belts work in unison and ensure equal tension in all the belts. Rollover tendency is virtually eliminated. Banded sets can be used on very wide belt drives and are available in a variety of widths. Part number designations are the same as previously stated for both classical and wedge sizes, but there is a prefix that identifies the number of belts or ribs in the band (e.g., a 4/5VX1200 would be a four-belt banded set of 5VX1200 wedge belts).

Classical and wedge banded belts are designed to operate in the same multi-groove sheaves that would be used for multiple single belts. The same ARPM standards, IP20 and IP22, for classical and wedge also apply to classical and wedge banded belts. Fig. 3-14 illustrates common configurations for classical and wedge banded belts.

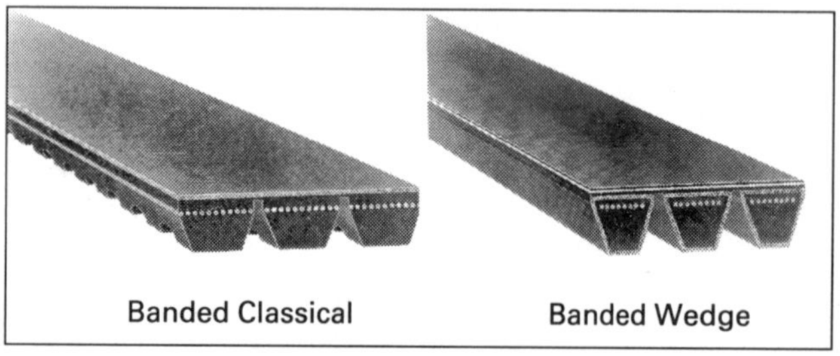

Figure 3-14: Banded Belts

V-ribbed Belts

This belt category is basically a flat belt with a corrugated driving surface—ridges running longitudinally around the inside surface of the belt. ARPM standard IP-26 and international standard ISO-9982 specify the groove-to-groove pitch distance, groove angles, and groove depth of v-ribbed drive sheaves, and the nominal belt dimensions shown in Fig. 3-15 for the common H, J, K, L, and M profiles.

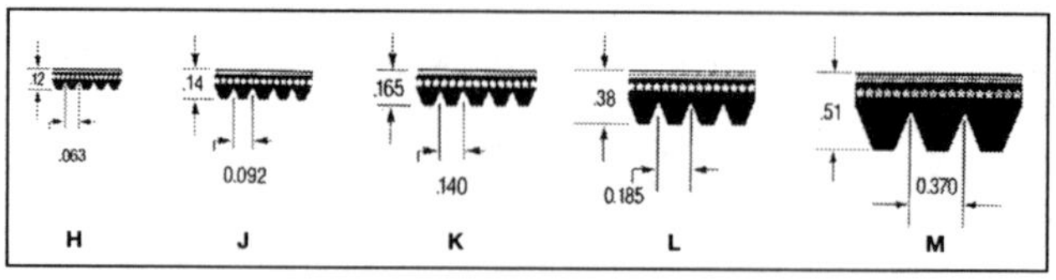

Figure 3-15: Banded Belt Sizes

Relatively thin, with a well-supported tensile member, these belts perform better than v-belts on drives with small diameter sheaves. They lend themselves well to serpentine sheave configurations and can drive flat pulleys from the backside of the belt as well; this is of particular importance in serpentine drives.

The low mass and high flexibility of v-ribbed belts allow them to tolerate small sheaves, reverse bending, high linear speed, and high speed ratios. The single unit construction eliminates any length-matching issues, similar to banded classical and wedge belt designs.

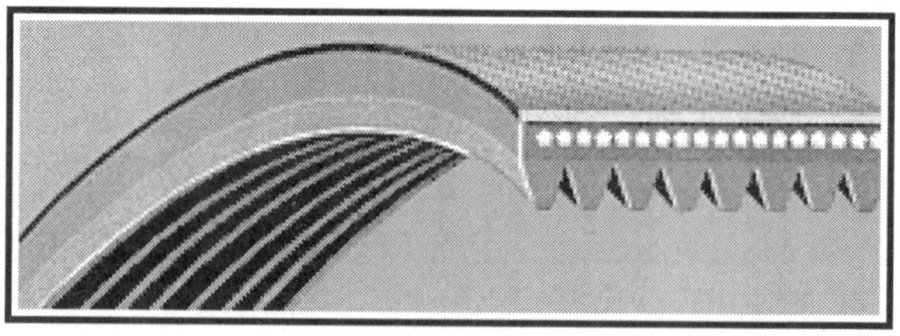

Figure 3-16: Banded Belt Profile

The H profile is used primarily in the appliance sector as the drive belt for horizontal tub dryers. They are also found in many small electric hand tools and consumer appliances like belt sanders and kitchen blenders.

The K profile has been used almost exclusively in the automotive sector as the prime mover for front-end accessory drives (FEAD). One belt connected to the crankshaft drives the alternator, water pump, power steering pump, A/C compressor, and any other auxiliary equipment, such as air pumps or alternator/starter hybrids.

The J, L, and M profiles can be used on any industrial drive that currently uses classical or wedge belts. Although available in most industrial distribution houses, the v-ribbed belts have only slowly grown in use. This is primarily due to the relative shortage of sheaves in the market place, as compared to the more common v-belt sheaves. The K profile is beginning to appear in more and more industrial applications, being driven by OE manufacturers who are looking for ways to reduce cost and drive size.

V-ribbed belts do not rely as much on wedging action to transmit torque as v-belts. Consequently, tensions tend to run higher; bearing and shaft selection is more critical than with classical or wedge v-belts. There is also a greater tendency for the belt to jump ribs in misaligned conditions.

Variable Speed Belts

This is a special class of two-sheave drive that can vary the driven sheave speed. Many industrial drives have some speed variation requirements, such as fans where air-flow adjustments need to be made. In their simplest form, the sheave width is adjusted manually while the drive is stopped. More complex designs adjust the speed while the drive is operating, either by manual control or through automatic sensing of speed and torque requirements of the drive. Fig. 3-17 shows a typical variable speed drive, where the sheave on the right is the adjustable sheave. One or both of the two drive sheaves is designed so that one sidewall of the sheave can be moved in or out to change the width of the sheave groove, and therefore, the pitch diameter of the belt (see Fig. 3-18).

Figure 3-17: Variable Speed Drive

As the sheave width changes, the drive belt rides either deeper or shallower in the sheave, allowing for an infinite range of speed ratios within the limits of the sheave's adjustment range.

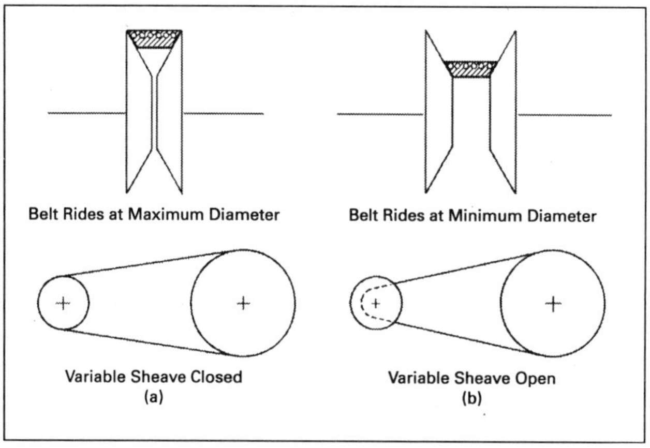

Figure 3-18: How a Variable Speed Drive Works

Either the driver sheave or the driven sheave can be the variable-pitch sheave, and in some cases both. With a single adjustable sheave, speed ratio changes of 20 percent to 40 percent are typical. If both driver and driven sheave are adjustable, speed ratio changes can range from 40 percent to 100 percent.

The amount of the speed ratio change can be calculated from the following formula:

$$P_d = \frac{B_b}{2\text{Tan}\frac{\alpha}{2}} - H_b$$

Where P_d = Pitch diameter variation
B_b = Belt top width
H_b = Belt thickness
α = Sheave groove angle

One can see that as top width increases and thickness decreases, a greater range of pitch diameter variation is possible.

Within certain limits, standard v-belts can be used in single and double variable-sheave drives; classical and light-duty v-belts are commonly used. Wedge belts are used much less frequently because they are thicker and do not allow as much pitch diameter change as the classical and light-duty belts. These drives can be single or multi-belt drives, depending on design power requirements. Where significant speed variation is required, up to 10:1, special variable speed drive belts are used. These are wider and thinner than standard v-belts, to allow for a wide range of pitch diameter changes. These are true "variable speed" belts.

Variable speed belts can be found in either envelope or raw edge constructions. Because they are typically wider and thinner than classical or wedge v-belts, they have more unsupported cords in relation to the sides of the pulley. Consequently, transverse stiffness must be designed in to keep them from collapsing in the sheave. Transverse stiffness can be designed in by either arching the back of the belt or by incorporating stiffening materials above or below the cord line. Typical configurations are shown in Fig. 3-19.

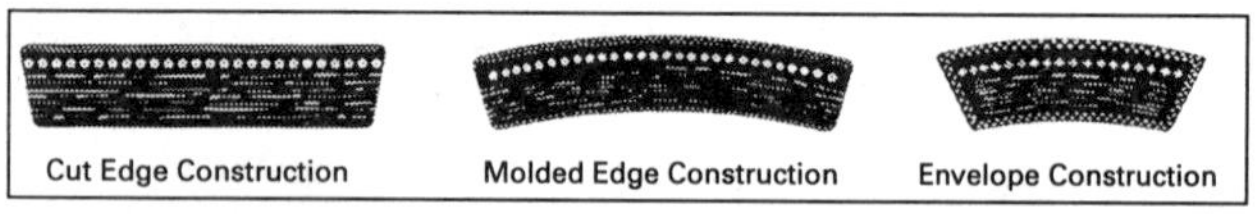

Figure 3-19: Different Types of Variable Speed Belts

The standard format for variable speed part numbers is found in ARPM standard IP-25. The following is an example:

3226V585

Where
32 = 32/16" top width
26 = 26° included angle
V = Variable Speed
585 = 58.5" Pitch length

Although variable-speed drives have a long history in the industry, the list of standard sizes is relatively short. The majority of variable-speed belts are custom sizes, designed for specific applications produced on a mass scale. The most popular sizes can be found in ARPM standard IP-25.

V-Belt Length

Belt length terminology is one of the most commonly confused and abused areas of belt technology. Misconceptions abound, and measure techniques are often misapplied. There are several systems for measuring belt length; some are related directly to the physical dimensions of the belt, and others are referenced off of the sheave dimensions. Measurements cannot always be converted easily from one system to the other and can be further complicated when also applying European standards.

Belt length can be designated as outside length (L_o), pitch length (L_p), datum length (L_d), effective length (L_e), or inside length (L_i). There is no common ground here, as different systems have been used over the years for each kind of belt.

Inside and outside length are the oldest and easiest systems of measurement; a tape measure was wrapped around the inside or outside circumference of the belt, and the measurement was read directly off the tape. This method is carried over from flat belt technology but is not very practical for v-belts, as neither the inside nor outside surfaces of the belt actually touch the pulley. Also, these lengths are not related to an installed condition, so the effects of applied tension are not taken into consideration. Because belts can elongate when tension is applied, they need to be shorter when measured with no tension, so that as tension is applied and the belt stretches, it will still fit within the limits of sheave movement.

Pitch length is the circumference of the belt at the pitch line or neutral axis of the belt. This is the horizontal plane through the belt where the length remains constant whether the belt is flexed forward or backward. It is a difficult dimension to quantify, as it relates to a zone that is inside of the belt and does not correlate to any exterior surface, either on the belt or on the sheave.

The datum length system is an international system and is described in ISO standard 1081; it is the same system previously known as the pitch length system. It defines the datum diameter as the diameter of the sheave at the datum width of the sheave groove. The datum length of the belt corresponds to the length of the belt at a level that is coincident with the datum diameter of the sheave. As with the pitch system, it can be difficult to assess and does not easily cross-reference to any other length-measurement systems.

The effective length system was introduced in an effort to compensate for the shortcomings of inside, outside, and pitch/datum length systems. The effective diameter of a sheave is the diameter at a specified groove width. In most cases, this is defined as the groove width at the outside circumference of the sheave, as it is easily accessible. The effective length of the belt is then defined as the circumferential length of the belt at the effective or outside diameter of the sheave, as measured at a specified tension. This last point is important, as it takes into consideration the true operating length of the belt with tension applied.

Most belt manufacturers design their belts so the plane of the tensile member rides at or very close to the top of the sheave groove while in operation and with tension applied.

This makes effective length measurement fundamentally easier than previous systems.

The fundamental principle of length measurement is illustrated in Fig. 3-20. It consists of two sheaves of equal diameter. One sheave is fixed in position, and the other is movable. Both sheaves are free to rotate.

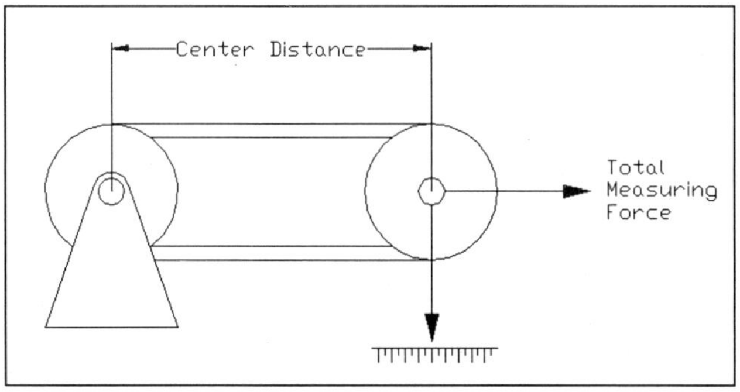

Figure 3-20: Fundamental Principle of Length Measurement

A measuring force is applied to the movable sheave as the belt is rotated. This rotation allows the belt to seat in the sheave groove and affords a dynamic measurement. The center distance between sheave shafts is read off of a graduated scale. As the belt is rotating, any variations in thickness will be revealed by swings in the scale reading. The length is calculated from the average of the minimum and maximum readings.

The belt length is calculated as:

$$L = 2C + \pi D$$

where L = Belt length
C = Average shaft center distance
D = Sheave diameter

ARPM and ISO standards specify the total measuring force to be applied as a function of belt cross section. The standards also specify the cross-sectional dimensions of the measuring sheave grooves.

The chart in Table 3-1 summarizes the length-measuring systems used for various v-belt types, as specified in various ARPM standards. Cross-reference charts are provided in the standards, where appropriate, to convert between systems.

Examples of the part number designation or each cross section type are as follows:

- A60 — A section v-belt with 60 in. nominal inside length.
- 5L500 — 5L light duty (FHP) v-belt with 50 in. nominal outside length.
- 5V950 — 5V wedge v-belt with 95 in. nominal outside length.
- 2332V585 — Variable Speed belt; 23/32 in. top width, 32 deg. included angle and 58.5 in. pitch length.
- BB60 — Classical B-section dual side v-belt with 60 in. approximate inside length.
- 180J5 — 5-rib J-section v-ribbed belt with 18 in. nominal outside length.

Conversions between length systems appear in published standards, where appropriate. If in doubt, or if a standard does not offer conversion factors, users should consult the belt manufacturer.

Table 3-1: ARPM Length Measuring Systems

BELT SECTION	PART NUMBER DESIGNATION	ARPM STANDARD	STANDARD LENGTH MEASURE SYSTEM
Classical	Inside length	IP-20	Datum system (L_d)
Light Duty	Outside length	IP-23	Effective system (L_e)
Wedge	Outside length	IP-22	Effective system (L_e)
Variable Speed	Pitch length	IP-25	Pitch system (L_p)
Hex	Inside length	IP-21	Datum system (L_d)
V-ribbed	Outside length	IP-26	Effective system (L_e)

Synchronous Belts

Synchronous belts, also known as timing belts, are a newer concept in power transmission belt evolution. They resemble a flat belt with evenly spaced, raised teeth that are perpendicular to the long axis of the belt. The belt teeth are designed to mesh with complimentary teeth on a "pulley" or sprocket, in a way similar to how a chain works. These belts combine the advantages of chain and gear with the advantages of v-belts, but without the limitations usually associated with these conventional types of drives. There is no stretch, no slip, and no metal-to-metal contact, so no lubrication is required. Synchronous belts are very versatile, with application opportunities on drives up to 600 HP and speeds from under 100 fpm to over 6,000 fpm.

In addition to power transmission systems, timing belts can also be used in linear indexing systems where very precise positioning is required. Timing belts offer important savings in weight, space, and construction without the sacrifice in efficiency. They are adaptable to almost any type of power transmission drive, from electric typewriters to heavy industrial milling machines and grinders.

Designed and manufactured with extreme care, with pitch, tooth depth, width, and other measurements accurate to a precise degree, synchronous belts are highly engineered products. Materials typically consist of high-strength tension members, specially compounded rubber and urathanes, for tooth strength and flexibility, and abrasion-resistant fabrics. They are designed to eliminate excessive heat build-up and operate efficiently.

Timing belts do not rely on friction to transmit torque; instead, they rely on positive engagement of belt teeth with complimentary teeth on a sprocket. This results in very little relative movement between the belt and the sprocket, unlike v-belts, which can slip in the sheave. Synchronous belts should be used where exact synchronization between the driver shaft and driven shaft is required. Synchronous drives deliver 100 percent of the intended rotational speed with 98-percent efficiency.

Nomenclature

Synchronous belts are made with a much higher degree of precision than v-belts, due to the exacting nature of synchronous drives. Dimensions have very tight tolerances, especially tooth-to-tooth pitch and pitch line location. Fig. 3-21 illustrates the most common dimensions and references used in synchronous belt terminology.

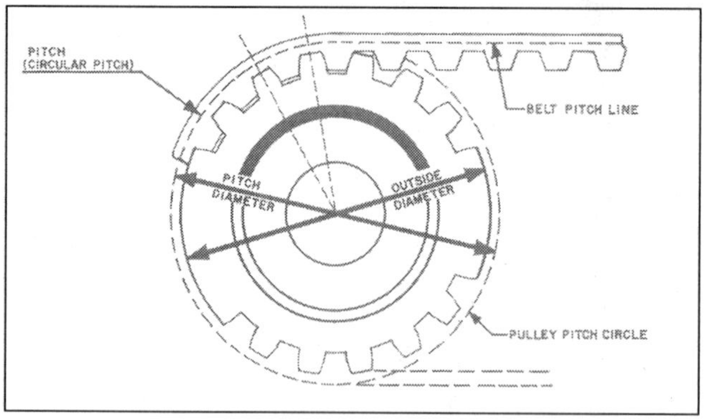

Figure 3-21: Synchronous Belt Dimensions and Terminology

The pitch line of the belt is the same concept as with v-belts; it is the horizontal plane of the belt where the length remains constant, whether the belt is flexed forward or backward. In synchronous belts, the pitch line is the defining reference point for belt length and must be coexistent with the pitch circle or pitch diameter of the sprocket in order for the belt and sprocket teeth to mesh properly and move together at the same precise speed. The generating tool racks used to cut teeth in synchronous sprockets are designed with this concept in mind.

History

Synchronous belt drives were first seen in the early 1940s. The tooth profile was a simple trapezoidal design, similar to that shown in Fig. 3-22.

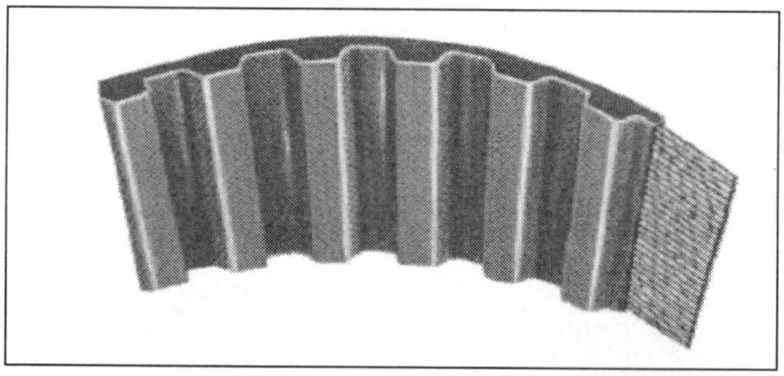

Figure 3-22: Synchronous Belt Profile

The trapezoidal design is still in use today and is sometimes referred to as an involute tooth profile. It is available in six standard pitches in the inch system. The letter designations for these sizes are as follows (see Fig. 3-23).

MXL	Mini Extra Light	0.080″ pitch
XL	Extra Light	0.200″ pitch
L	Light	0.375″ pitch
H	Heavy	0.500″ pitch
XH	Extra Heavy	0.875″ pitch
XXH	Double Extra Heavy	1.250″ pitch

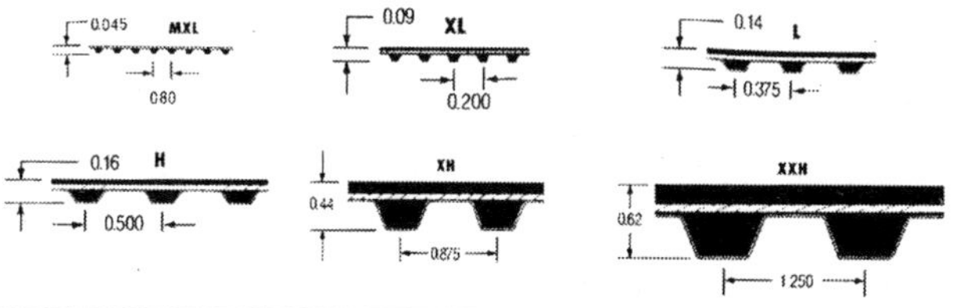

Figure 3-23: Synchronous Belt Letter Designations

The trapezoidal synchronous belt specifications are covered in ARPM standard IP-24 and international standards ISO 5294 and 5296, where it is referred to as the involute tooth profile. Major topics addressed include generating tool rack forms, sprocket tooth dimensions, belt dimensions, length measurement, power and torque formulas, and standard pitch lengths and tolerances—details that are beyond the scope of this handbook.

Part number designations for trapezoidal belts follow this format:

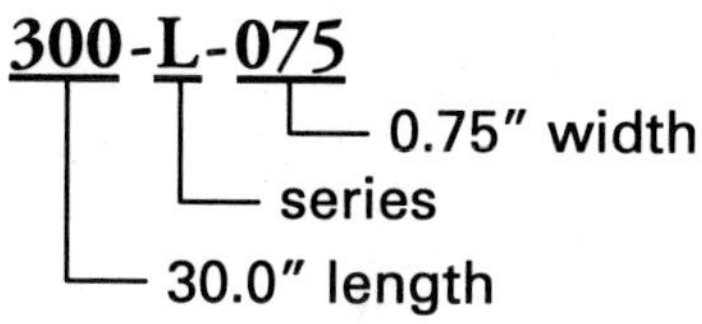

Synchronous belts are also available in dual-sided configuration, with teeth on both sides of the belt. These are often used on twisted or serpentine drives where driven shaft rotation must be reversed from the driver shaft.

In recent years, several new tooth profiles have appeared in the market. These new profiles are based on gear tooth design and utilize many of the features of gear tooth technology for reducing noise and increasing efficiency.

These designs are generally referred to as first-generation curvilinear tooth profiles, and the specifications are presented in ARPM standard IP-27. The ARPM breaks the curvilinear profiles into three primary types: H-Type, S-Type, and R-Type. The basic outlines of these three tooth forms are shown in Fig. 3-24.

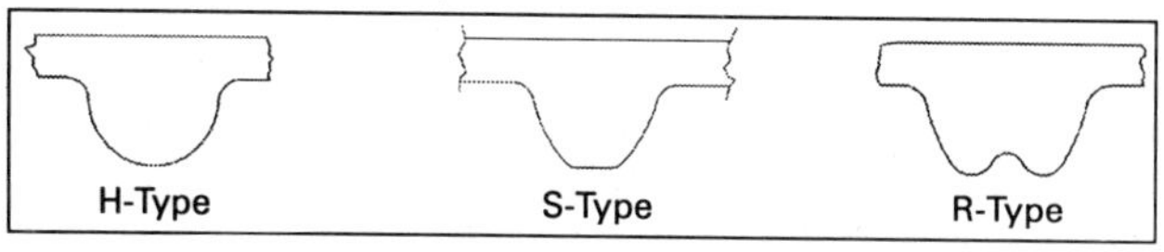

Figure 3-24: Three Primary Types Of Synchronous Belt Curvilinear Profiles

H-Type

The H-type tooth profile is commonly referred to as HTD. It has a round tooth profile, and its shape is defined by a group of radii centered a specific distance from the pitch line of the belt. Exact tooth dimensions are shown in Fig. 3-25 and Table 3-2.

The H profile is commonly available in 3mm, 5mm, 8mm, 14mm, and 20mm pitches, in a variety of widths and

Table 3-2: Nominal Belt Dimensions - "H" Type Profile - M M's

Belt Section	Pitch	h_s	h_d	h_t	"X" Ref.	"Y" Ref.	r_{bb}	r_{bt}	a
H3M	3	2.4		1.22	0.030	0.348	0.86	0.30	0.381
DH3M	3		3.2	1.22	0.030	0.348	0.86	0.30	0.381
H5M	5	3.8		2.08	0.051	0.582	1.50	0.41	0.572
DH5M	5		5.3	2.08	0.051	0.582	1.50	0.41	0.572
H8M	8	6.0		3.38	0.089	0.787	2.59	0.76	0.686
DH8M	8		8.1	3.38	0.089	0.787	2.59	0.76	0.686
H14M	14	10.0		6.02	0.152	1.470	4.55	1.35	1.397
DH14M	14		14.8	6.02	0.152	1.470	4.55	1.35	1.397
H20M	20	13.2		8.38	0.226	1.887	6.50	1.91	2.159

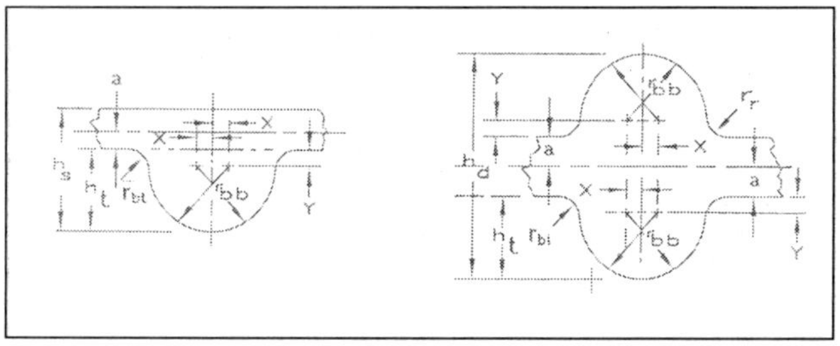

Figure 3-25: H-Type Tooth Profile

lengths. Part number nomenclature typically follows the following format:

560-H8M-30
- 30mm belt width
- 8mm pitch H-type tooth profile
- 560mm pitch length

A dual-sided version of this belt would have the following part number:

560-DH8M-30

Sprockets for H-type drives have the following general part number format:

P40-8M-20-SH

Where P = "Pulley" Sprocket
40 = 40 teeth
8M = 8mm pitch
20 = Maximum belt width in mm
SH = Bushing type (in this example, a SH bushing)

S-Type

The S-type tooth profile is commonly referred to as STPD. The tooth form is derived from circular-arc gear tooth geometry introduced in the 1920s and perfected by M.L. Novikov, a Russian Engineer. The Novikov design is the dominant circular-tooth gear design in use today. STPD belt drives are used extensively throughout Europe and Asia. North American usage is limited, but growing as more and more industrial equipment is imported from Europe and Asia.

S-type synchronous drives are typically available in 3mm, 4.5mm, 5mm, 8mm, and 14mm pitches, as shown in Fig. 3-26, in a variety of widths and lengths.

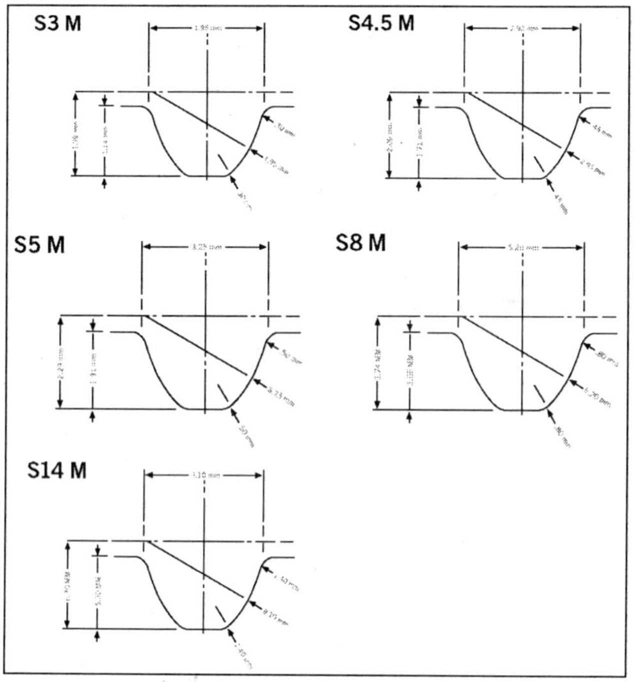

Figure 3-26: S-Type Tooth Profile

Part number nomenclature typically follows the following format:

100-S8M-375
- 375mm pitch length
- 8mm pitch S-type tooth profile
- 10mm belt width

A dual-sided version of the above belt would have the following part number designation:

100DS8M375

Where the "D" embedded in the part number indicates "dual-sided."

Sprockets are designated as follows:

S8M18X30

Where S8M = 8mm pitch STPD profile
18 = Number of sprocket teeth
30 = 30mm sprocket face width

R-Type

The R-type synchronous belt is a deep-profile parabolic tooth design that can be used in R-Type and/or H-type sprockets. A variety of belt length and width combinations are available in 5mm, 8mm, 14mm, and 20mm pitches, in single-side or dual-side configuration. Single-side variants are indicated in Fig. 3-27.

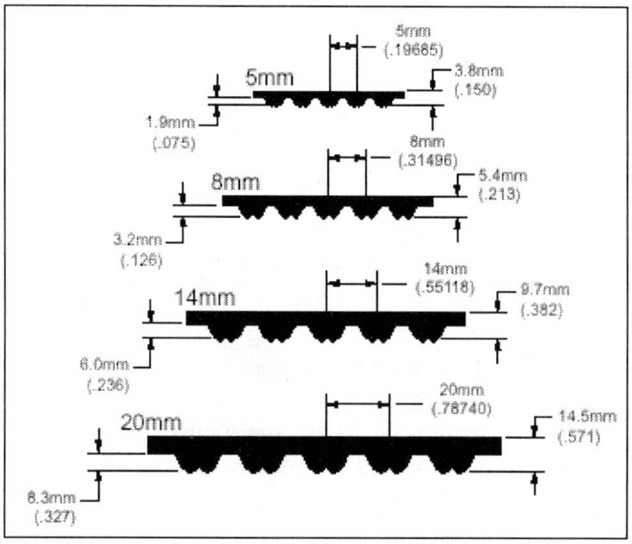

Figure 3-27: R-Type Tooth Profile

The R-type profile is readily identifiable by the indentation at the bottom of the tooth, which is designed to relieve stresses imposed on the tooth when it wedges into the mating groove of the sprocket.

Part number nomenclature for R-type belts typically follows the following format:

177 - 3M - 15
- 15 = 15mm belt width
- 3M = 3mm tooth pitch
- 177 = 177mm pitch length

The same belt in dual-sided configuration would have the following part number designation:

D177-3M-15

Sprocket part numbers for the R-type profile have the following form:

P44-20M-230-MPB

Where P = Sprocket
44 = Number of teeth
20M = Tooth pitch in mm
230 = Sprocket face width in mm
MPB = Bore type for mounting and bushing (in this case, MPB indicates Minimum Plain Bore)

The R-type tooth profile has found great acceptance in the automotive industry as cam timing drives, replacing chain. Timing belts are much lighter than chain, require less maintenance, and do not require lubrication. There are many variants of the R-type profile in the automotive sector, and great care must be taken when replacing these belts to make sure the correct profile is used.

Helical Offset Tooth

Recently, a new form of synchronous belt has appeared and is gaining acceptance throughout the industry. It is called a Helical Offset Tooth (HOT) and is derived from helical gear technology. It looks very different from any other timing belt, as shown in Fig. 3-28, and resembles a tractor tire tread.

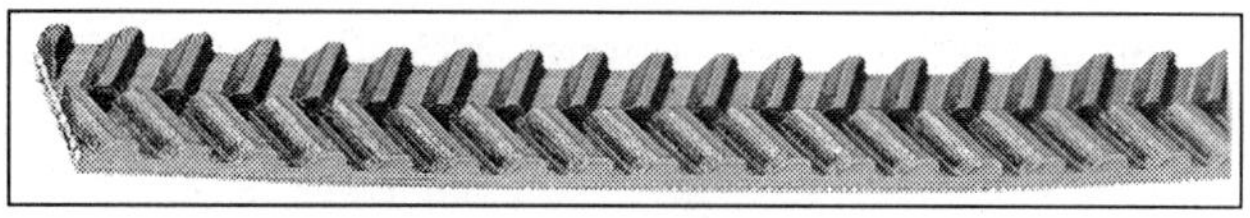

Figure 3-28: Helical Offset Tooth Profile

The HOT belt teeth are not perpendicular to the long axis of the belt. Rather, they are placed at an angle, similar to helical gears, except that the teeth on the right and left halves of the belt are angled opposite each other and are offset by half a pitch. The matching sprocket design is illustrated in Fig. 3-29, shown with a straight bore and key-way:

This arrangement allows the belt to self-center on the sprocket, without the aid of flanges, which are found on virtually all other conventional timing belt sprockets and pulleys. Because the teeth work like helical gears, with con-

Figure 3-29: Helical Offset Sprocket

tinual rolling engagement, vibration and noise are greatly reduced over straight-tooth synchronous drives.

The HOT profile relies on circular-arc Novikov gear geometry to describe the shape of the teeth, very similar to the S-type profile. The half-pitch side-to-side offset of the teeth utilizes noise-cancellation technology to further reduce the noise typically produced by the engagement of the belt teeth with the sprocket teeth, especially at high speeds.

The HOT profile is patented technology, available in the following pitches and widths:

8mm x 0.63"	14mm x 1.38"
8mm x 1.25"	14mm x 2.07"
8mm x 2.50"	14mm x 2.75"
	14mm x 4.13"

Belt Standards

The belt standards that are generally referenced today are those published by the following organizations:

- Association for Rubber Products Manufacturers (ARPM)
- International Standards Organizations (ISO)
- Society of Automotive Engineers (SAE)

Several other global organizations have published belt standards at one time or another, such as the British Standards Institute (BSI), Deutche Norman (DIN), Japanese Industrial Standards (JIS), and others. Most of these, though still in circulation, are either consistent with or have been superseded by either the ISO, ARPM, or SAE standards. Even the ARPM is working toward normalization with the ISO standards and is a voting member for approval of all ISO standards that pertain to belting.

The American Society of Agricultural Engineers publishes a belt standard for agricultural belts, but the ARPM is the principle author of that document.

The NIBA-The Belting Association publishes standards that pertain to conveyor belting and still lists standards for leather belting.

The Society of Automotive Engineers (SAE) publishes standards for automotive v-belts, v-ribbed belts, and timing belts.

The American Petroleum Institute (API) has published belt standards pertaining to refineries and oil fields. These primarily address anti-static and flame-resistance characteristics of belts used in the petroleum and chemical industries.

The British National Coal Board (NCB) has published standards pertaining to belts used in mining. This standard, like the API standard, is concerned with anti-static and flame-resistance characteristics.

There are many original equipment manufacturers that maintain their own published standards for belt dimensions and performance. Most reference industry standards published by the aforementioned organizations and also incorporate characteristics specific to their products and needs.

Note: Belt standards typically specify fully tolerable dimensions only for the pulleys. Specific values for groove dimensions, angle, minimum recommended diameters, and surface finish are included. General reference dimensions are given for the belt, but each belt manufacturer has their own philosophy regarding optimum belt dimensions and materials, and these final dimensions can differ from one belt manufacturer to another. Because of this, belts from different manufacturers may not be compatible and should never be installed together on multi-groove pulleys.

Application and Selection

Belt selection is based on application requirements such as horsepower or load, speed, life, space limitations, and need for precision, plus environmental conditions and cost (both initial and operating). These and other selection factors are described in manufacturers' design manuals.

After identifying the application requirements, match them with the appropriate belt type. Consider both operating and maintenance requirements.

Flat belts can bend easily around small-diameter pulleys or twist to go around corners. They are often used in serpentine drives.

V-belts have less stretch than flat belts and more lateral stability. They are often used because of their wide availability.

Synchronous belts precisely match the motion of driving and driven shafts for indexing or positioning applications. Compared to gear or chain drives, synchronous belt drives operate with less noise and without lubrication.

Key Terms

BSI: British Standards Institute

DIN: Deutsche Norman

JIS: Japanese Industrial Standards

ISO: International Standards Organization

ARPM: Association for Rubber Products Manufacturers

SAE: Society of Automotive Engineers

FHP: Fractional horse power

Coefficient of Friction: The ratio of the force that maintains contact between an object and a surface and the frictional force that resists the motion of the object.

Longitudinally: Placed or running the length wise.

Tensile Member: Capable of being stretched or extended.

Tooth Pitch: The distance between two corresponding points on adjacent screw threads or gear teeth.

Vulcanize: To improve the strength, resiliency, and freedom from stickiness and odor of (e.g., of rubber) by combining with sulfur or other additives in the presence of heat and pressure.

Summary

The advent of power transmission belt drives has had a profound impact on our lives since the time of the Industrial Revolution and will likely continue far into the future. Low in cost, easy to maintain, and adaptable to a wide variety of applications, belt drives are an efficient and economical means for transmitting power from one rotational component to another.

From low-power, low-precision applications to drives producing several hundred horsepower and high-precision positioning systems, there is a v-belt, multi-v belt, variable speed belt, or synchronous belt for virtually every requirement.

Additional Information

Addresses and phone numbers for associations are provided in the Association Appendix.

Belt Selection and Application for Engineers, edited by Wallace D. Erickson, Marcel Dekker Inc., New York, 1987.

Association for Rubber Products Manufacturers (ARPM), Indianapolis, IN (Belts).

Mechanical Power Transmission Association (MPTA), Naples, FL. (Sheaves and sprockets).

NIBA - The Belting Association, Waukesha, WI. (Conveyor & elevator belting).

Society of Automotive Engineers (SAE), Warrendale, Pa.

Chapter 4

Chain Drives

Chapter Objectives

When you have finished Chapter 4, "Chain Drives," you should be able to do the following:

1. Describe the advantageous characteristics of chain drives.
2. Identify the features of roller chain.
3. Identify the features and benefits of self-lube roller chain.
4. Be familiar with "heavy series" and other high-strength roller chains.
5. Understand the various corrosion-resistant roller chains.
6. Be familiar with several common specialty roller chains.
7. Identify the features of double-pitch roller chain.
8. Identify the features of silent chain.
9. Differentiate various standard chain attachments.
10. Explain the applications for engineered steel class chain.
11. Describe the categories and applications of three types of chain lubrication.
12. Describe the application information needed to design a drive.

Introduction

Roller chains have been used in many power transmission and conveying applications for well over 100 years. Most people will be familiar with roller chain use in bicycle and motorcycle applications, for example, but they are also used much more extensively, such as in many industrial applications not obvious to most people.

Roller chains perform many critical functions everyday in industrial "behind the scenes" applications, from food processing, packaging, and beverage can manufacturing, to printing publications and oil field exploration. All of these applications are vital to our everyday lives, yet many of these applications are relatively unknown to the consumer who purchases and enjoys the end-products produced using this essential technology.

Chain drives for industrial applications employ the same principle as on a bicycle — they transmit power from one sprocket to another via a chain. But chain drives for these applications typically operate at much higher power levels than those for bicycles. And these drives are available with characteristics suitable for a wide variety of speeds, torques, and operating conditions.

Chain drives also generally reduce speed and increase torque from an input shaft to an output shaft. Some drives serve as a conveyor. The long strand of chain that pulls a rollercoaster train up the first hill is a familiar example of the latter.

Roller chains have evolved from basic carbon steel construction and now incorporate many new features and materials. Today's technology offers state of the art self-lubing, corrosion resistant, and lightweight material options that make roller chains even more versatile than ever before. These technological advances have expanded the potential for this already diverse product.

The objective of this chapter is to introduce the basic roller chain fundamentals as well as the various new technologies.

Characteristics

Chain drives operate in tension. Teeth on the driving sprocket engage the chain, applying a tensile load that pulls the chain links and powers the driven sprocket.

Only a few sprocket teeth are required for effective engagement, which allows higher speed ratios than are usually attainable with belt drives. Non-synchronous belts (those without teeth, such as V-belts and flat belts) require a substantial minimum arc of contact between belt and sheave to develop enough friction to avoid slippage. Loads greater than that for a single-strand chain can be easily accommodated by multiple-strand chains.

The advantageous characteristics of chain drives include the following:

- Versatility: roller chains can be easily cut to length from bulk quantities such as 100-foot reels for use as required.
- Variety: roller chains are available in limitless variations of attachment styles, spacing, and materials.
- Positive engagement: there is no slippage between driving and driven shafts because power transmission is positive over sprocket teeth.
- Flexibility: roller chains are capable of bending back over sprockets while maintaining positive engagement and avoiding slippage.
- Durability: roller chains provide long operating life because flexure and frictional contact occur between long-wearing, hardened bearing surfaces separated by an oil film.
- An ability to operate in hostile environments, such as high ambient temperatures and moist, oily, and corrosive atmospheres—especially where chain is made of stainless steel, bronze, or other stable materials.
- Unlimited shelf life because metal chain ordinarily does not deteriorate with age and is unaffected by the sun or reasonable ranges of heat and moisture as long as a protective factory oil coating remains intact.
- Chain drives generally can be repaired or replaced without disturbing associated equipment.
- Roller chains can be sized to handle large HP loads and speeds, well beyond the capability of other options such as belts.

Disadvantages of chain drives include:

- Standard chains may have higher operating noise levels than belts or gear drives, but low noise roller and silent chain options are available.
- Like belts, roller chains will eventually wear out and have a potential for elongation from wearing of contact surfaces on links and sprocket teeth.

- They have slightly lower speed limitations than those for belt and gear drives.
- There is a need to replace sprockets because of wear when worn chain is replaced. (V-belt sheaves undergo very low wear and usually do not need to be replaced when belts are replaced.)

Types of Chain

A wide variety of standard and non-standard chain and sprocket designs are available. The American Society of Mechanical Engineers (ASME) publishes standards, prefix ASME B29, for chains and sprockets. The standards cover transmission and conveyor chains as well as standard sprocket tooth form dimensions, pitch diameter tables, and sprocket measuring procedures.

Various types of sprockets are available, including cast iron, powder metal, flame-cut steel plate, machined metal, and composite materials. To protect against shock or overload, special-purpose sprockets are available with shear pins, overload clutches, and other devices.

Roller Chain

Standard single-strand roller chain, the best-known and first chain to be dimensionally standardized by ASME (ASME B29.1), contains pins that pivot inside bushings. The pins are usually press fit into pin link plates, and bushings are press fit into roller link plates (see Fig. 4-1). A free-turning roller surrounds each bushing to provide rolling engagement and contact with sprocket teeth. Thus, a strand of roller chain is an assembly of alternating pin links and roller links.

The distance between flexing joints (pin centers) in roller chain is the pitch, a characteristic that distinguishes one chain size from another. The larger the pitch, the larger the links and the higher the load rating. Although a small-pitch chain carries less load, it provides smoother, quieter operation and is capable of operating at a higher speed than a chain of larger pitch. For high-speed, high-load applications, multi-strand small-pitch roller chains may be used.

A length of chain is made into a continuous strand by using a special type of pin link, called a connecting link (see

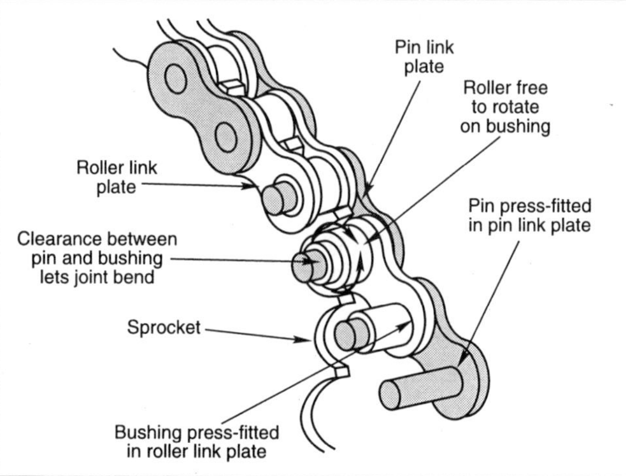

Figure 4-1: Single-strand standard roller chain conforms to ASME B29.1. Pin links are shaded to differentiate them from roller links.

Fig. 4-2). The connecting link's pins are inserted through each exposed bushing hole of the roller link at each end of the strand. Cotter plates or a spring clip (shown) keep the connecting link's pins from working loose. These cotter pins or clips are necessary for manual assembly and disassembly.

Because there is no way of connecting one pin link directly to another, or one roller link directly to another, standard strands of chain have an even number of pitches — an equal number of pin links and roller links. However, an offset link (see Fig. 4-2) can be installed to produce a strand with an odd number of links. An offset link is a combination of a roller link and a pin link. When a strand of chain is taken apart, one half of the offset link attaches to the strand's pin link, and the other half attaches to its roller link. So, adding an offset link to a strand that has an even number of pitches produces a strand with an odd number of pitches.

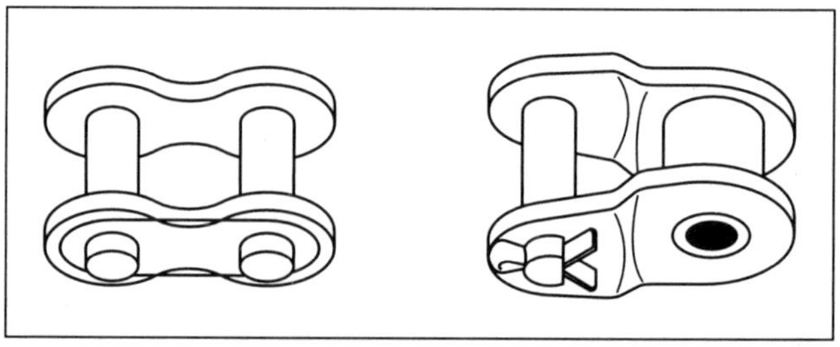

Figure 4-2: Connecting link, left, makes a length of chain into a continuous strand. Offset link, right, allows constructing strand of chain with odd number of pitches. One half of offset link mates with roller link, other half mates with pin link.

Standard roller chain is available with pitches from 1/2 (12.7 mm) to 3 in. (76.2 mm). Designation numbers for chains are obtained by multiplying the pitch value by 80. Thus, a 1/2 in. (12.7 mm) pitch chain is a No. 40 chain, a 5/8 in. (15.87 mm) is No. 50, etc. A zero in the designation indicates a normal-weight chain. A 1 in the designation indicates a lightweight chain.

Chain drives are also available with 1/4 in. (6.35 mm) and 3/8 in. (9.52 mm) pitches, but with a rollerless construction. With these so-called rollerless, bushed chains, the external surface of each link-bushing contacts the sprocket teeth directly. These chains carry the designations No. 25 and No. 35, respectively. The 5 in. each designation indicates the rollerless design.

A hyphenated 2 suffixed to a chain number denotes a double strand; 3, a triple strand; etc. For example, No. 60-2 designates two strands of No. 60 chain (normal weight 3/4 in. (19.05 mm) pitch), 60-3 indicates a triple strand, and 60-4 a quadruple strand.

Self-Lube Roller Chain

Self-lube roller chains are excellent options for applications where lubrication is not possible due to potential contamination or accessibility issues, or where it is preferable to eliminate the costly and time-consuming practice of lubricating. Self-lube roller chains contain their own lube supply and require no maintenance during their service life. The most common style of self-lube roller chains contain a sintered (powdered metal porous) bushing containing special high performance lubricants, which maintain a coating of lube on the pin and bushing components as the chain operates and heats up during use (see Fig. 4-3). When the operation of the chain stops, the lube is absorbed back into the bushing where it is stored for future use.

There are also self-lube roller chains with special seals between the link plates made of felt material, which prevent contamination from entering the bushing, allowing the chains to be used in harsh environments containing very fine contaminant materials (see Fig. 4-4).

Self-lube roller chains are suitable for use up to 302 °F (150 °C), and there are high temperature versions containing food grade lube suitable for use up to 446 °F (230 °C), excellent for baking oven applications. These types of roller chains are available in ASME, British Standard, and attachment styles, making them extremely versatile.

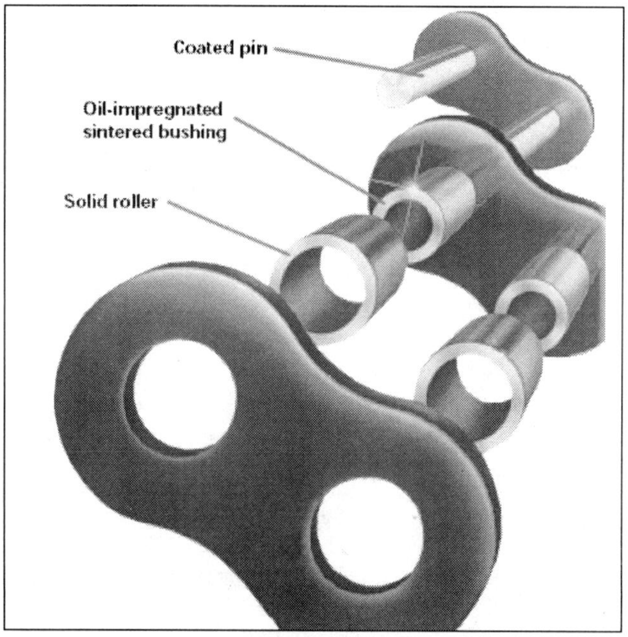

Figure 4-3: Self-lube roller chain schematic.

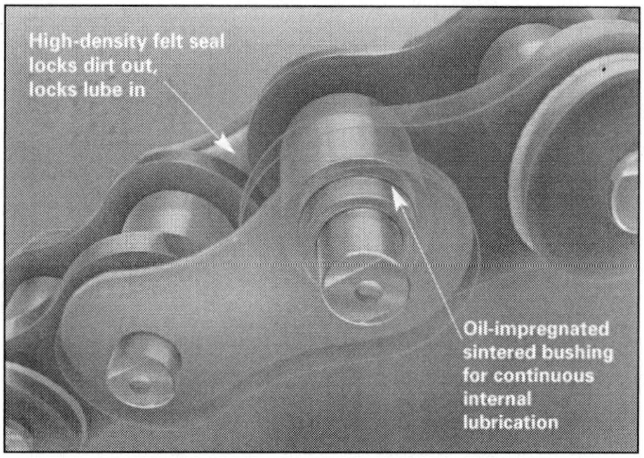

Figure 4-4: Self-lube roller chain with felt seals between link plates designed to prevent contaminant from entering the bushing and blocking the flow of lube.

Heavy and Super Series Chains

Heavy Series chains have a link plate thickness equal to the next largest ASME chain. They are used where space and weight limitations prohibit use of larger sizes. These chains have a higher working load rating than ASME standard roller chains and will withstand greater shock loads. They are available in single or multiple strands.

Nomenclature
Add suffix H to ASME Standard chain
Example: 60H

There are also Super Series roller chains in both "H" and "non H" series, designed to carry even greater loads and withstand greater shock loads. These chains are excellent options for applications where the output demands and loading have increased and retrofitting to a larger size or a multi-strand chain of the same size is cost-prohibitive or physically impossible due to space constraints.

Super Series chains are capable of providing working loads up to 60 percent greater and breaking strengths up to 40 percent greater than standard chains (see Fig. 4-5).

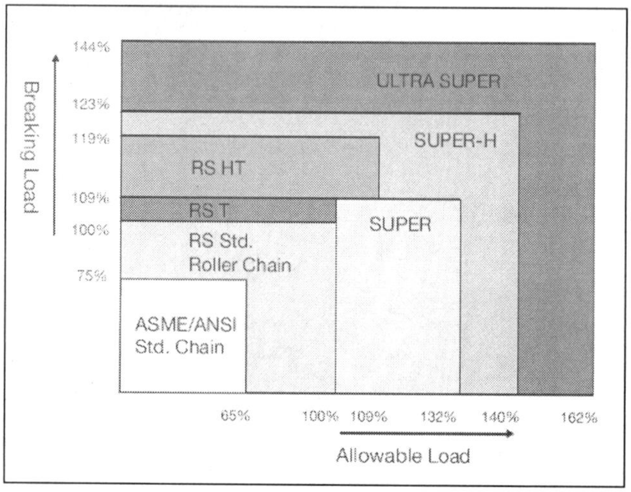

Figure 4-5: Heavy Duty Super Series Performance Characteristics.

Corrosion Resistant Chain

Corrosion resistant chain includes stainless steel, nickel plated, plastic, and new technology using a zinc and nickel alloy that is mechanically applied to the chain components before assembly. See the "High Tech Coated Chains" section. Food processing plants, fertilizer plants, and waste management plants are large users of these types of roller chains.

Stainless steel is available in three standard alloys: 300 series (all 316 or all 304 material) for high resistance to acids and alkalis and and 600 series, which has heat treated pins, bushings, and rolls for increased hardness and maximum strength. The 300 series components are austenitic (un-heat treatable) and are lower in hardness. Although they provide the best corrosion resistance, they will wear faster than the 600 series chains.

Nickel plated chain is less expensive than stainless steel and is used in wash down and wet environments.

Tight and Slack Strands

The links of chain transmitting load often are said to be in the tension or tight strand. After the links rotate off the driving sprocket, they make up the slack strand because this length of chain carries no load and sags between the sprockets. In drives where a chain travels primarily in a horizontal path, the slack strand usually should be positioned under the tight strand. If the slack strand was positioned over the tight, excessive slack or whipping motion could cause the slack strand to contact the tight strand underneath it. This is an important consideration in applications with long center distances. Tensioners may be required in these applications.

There are three situations that do not lend themselves to locating the slack strand on the bottom:

- Application parameters may dictate, for example, that the driving sprocket rotate clockwise and be positioned to the left of the driven sprocket. This arrangement requires the slack strand to lie above the tight strand.
- In bidirectional applications, the driving sprocket rotates clockwise much of the time and counterclockwise the rest of the time. Here, the tight strand alternates from top to bottom.
- In drives where sprockets are oriented vertically, strands no longer lie in a top-and-bottom orientation. For applications such as these, tensioners should be used to keep chain drives running smoothly. Tensioners and their use are described in Chapter 15 – Power Transmission Accessories.

Double Pitch Roller Chain

Also known as extended-pitch chain, double pitch power transmission roller chain has dimensions listed in ASME Standard B29.3. These dimensions are the same as standard roller chain of comparable load capacity, except the pitch is doubled (see Fig. 4-6). Because a given length of double pitch chain contains only half as many pitches, it is lighter and less expensive than standard roller chain. It is especially suitable for long center distance applications.

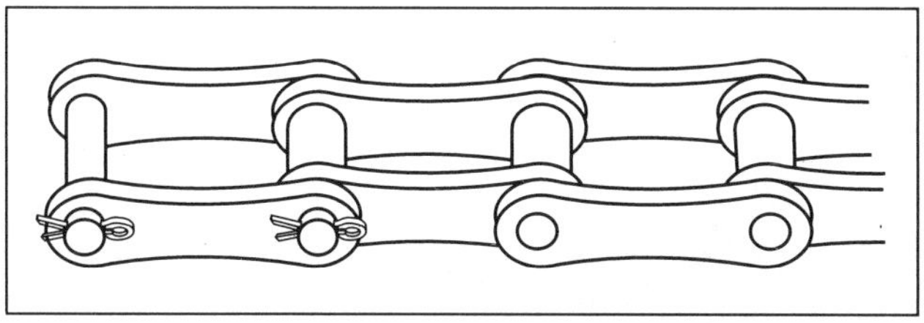

Figure 4-6: Double-pitch drive chain conforms to ASME B29.3. Dimensions are identical to standard roller chain, except pitch is doubled, which reduces weight and cost.

Because all dimensions are the same except pitch, double pitch roller chain offers essentially the same strength characteristics as standard roller chain. Double pitch chain is designated with a 20 preceding what would be a standard roller chain number. For example, 2050 designates 1 1/4 in. (31.75 mm) double pitch roller chain. The "2" designates double pitch. The "5" signifies 5/8in.pitch = .625 in. x 2 for double pitch = 1.25in.pitch total.

Another extended pitch drive chain is the double pitch conveyor chain. Double pitch conveyor chain (see Fig. 4-7) has straight side bars to prevent wear as it drags along a conveyor bed. These chains can be made with oversize rollers that act as wheels to elevate the chain off of its conveyor bed. This reduces wear and friction. This numbering system also uses 20 to indicate double pitch length. A letter "C" is put in front of the number to indicate conveyor series with small rollers, for example, C2050. If the chain has large rollers, the last 0 is replaced by a 2, for example, C2052. All conveyor double pitch chains larger than number 50 are heavy series chains; for example, number 60 chain is C2060H. As with standard roller chain, a great many attachments are available.

Sprockets for double pitch roller chain are either single-cut (one tooth per link) or double-cut (two teeth per link).

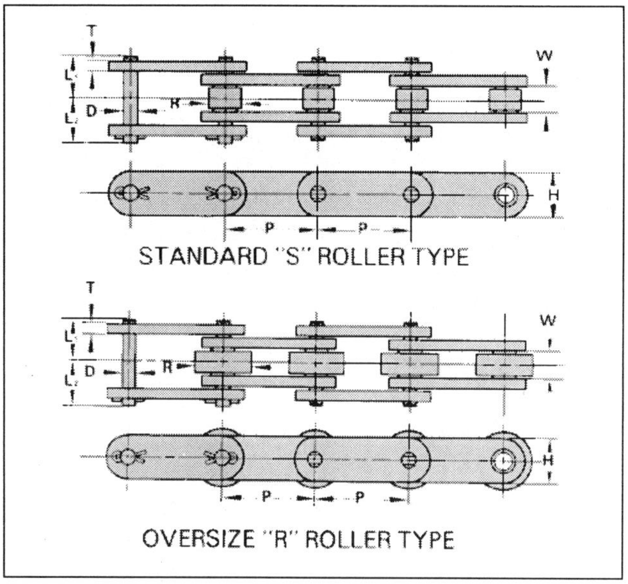

Figure 4-7: Double pitch conveyor roller chain conforms to ASME B29.4. This style chain was designed for conveyor applications where the chain will slide or roll over a surface.

The double-cut sprocket is actually a standard roller chain sprocket for 30 teeth or greater, and special for less than 30 teeth. Using sprockets for standard roller chain with an odd number of teeth reduces sprocket wear, because teeth that engage rollers on one revolution lie between rollers on the next. So, sprocket teeth engage chain rollers once for every two revolutions of the sprocket if the sprocket has an odd number of teeth.

When a sprocket has an even number of teeth, every other tooth will engage rollers on every revolution. Half the sprocket teeth will undergo full-time wear and half will undergo no wear. Once half the teeth have worn, loosening the chain allows advancing the sprocket one tooth space so the unworn teeth will engage chain rollers and the worn teeth will not.

Hollow Pin Chain

This series of chain has dimensions listed in ASME Standard B29.27. It has a hole through the chain pin, allowing for the installation of various types of attachments. The bore diameters are the same as the equivalent pitch chain pins in ASME B29.1 and B29.4. The single-pitch chains operate on ASME B29.1 sprockets and the double pitch on ASME B29.4 sprockets (see Fig. 4-8).

Standard Chain Attachments

Attachments allow chain to serve as more than just a means of transmitting power from one shaft to another. Attachments allow chain to act as a conveyor to move parts or even bulk materials from one location to another. The versatility of attachments opens opportunities for chain (especially roller chain) to be used more as a conveyor than as a means of power transmission. Attachments also can play an important role in control systems.

The accompanying illustration shows just a few of the standard attachments available from chain manufacturers. Use of these attachments is limited only by the imagination. For example, if a pair of parallel chains each used straight extensions on every other link, bar stock could be carried along the path of the chain between the attachments. With a single strand, mounting the attachments upside-down would allow inserting a hook into each hole to carry small parts. A bent extension presents an ideal medium for mounting a flat surface, or even containers, for transporting bulk materials.

In control functions, an extended pin can trip an electromechanical switch, actuate a lever or cam, or even generate an electronic pulse as it passes by a magnetic pickup or proximity switch. For example, a signal may be needed to lower tooling after a feed table has rotated 1/4 turn. If the sprocket attached to the table has 48 teeth, then 12 chain links would represent ¼ turn. Therefore, extended pin or straight extension attachments installed every 12 pitches would cause a proximity or magnetic switch to produce a signal for every ¼ turn of the table (sprocket).

Attachments can be installed on the chain either by the manufacturer or by a distributor. Obviously, strands of chain requiring a large number of attachments are best assembled at the factory.

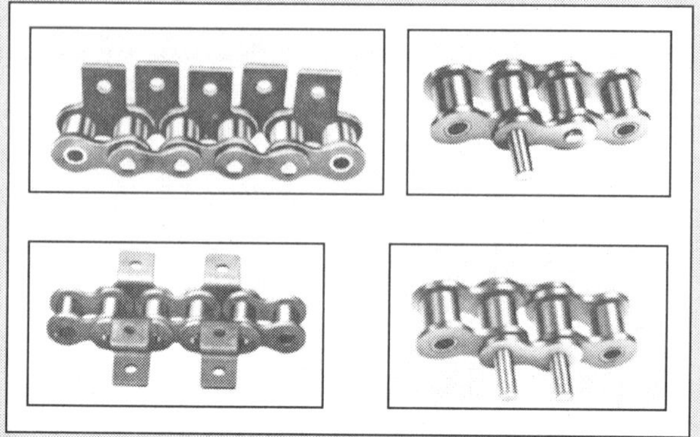

A few of the many different attachments used in standard roller chain. Various configurations are available — such as single, double, one side, or both sides — that can be used at every pitch, every other pitch, or at any multiple specified by the user.

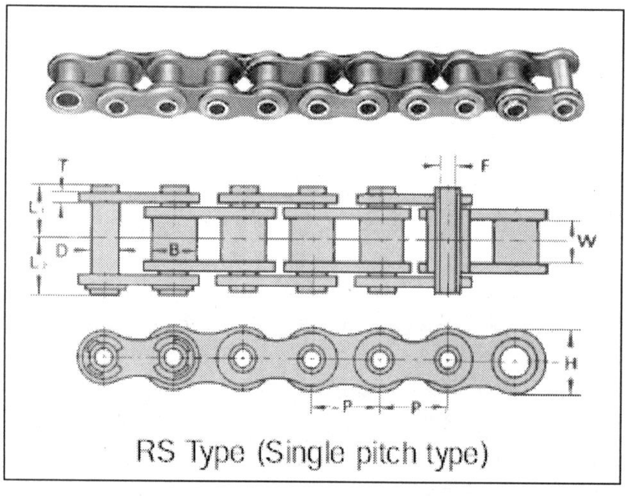

Figure 4-8: RS type (single pitch) hollow pin chain.

Many types of crossrods, pins, and custom attachments may be inserted at any point without removing the chain from the drive system. Also, they are sometimes assembled with modified attachment link plates directly on the pin link or roller-link pitch.

Side Bow Chain

Side bow chains are also known as curve chain and radius chain, and they are available in sizes 40 through 80. This series of chain incorporates extra clearance between pins, bushings, and linkplates to allow side-flexing around curves or 90 deg. twisting. Many manufacturers use standard roller links with shortened pitch pin plates and utilize undersize pin diameters. This construction will allow interaction with standard 40 through 80 sprockets. However, this construction produces a great deal of backlash when the chain stops. To overcome this result, some manufacturers use a barrel shape pin or an hourglass shape internal diameter bushing to achieve the same radius with normal backlash (see Fig. 4-9).

The most common attachments used for this series are bent tab or straight tab roller links. Extended pins are also available, but care must be observed as to which diameter is correct.

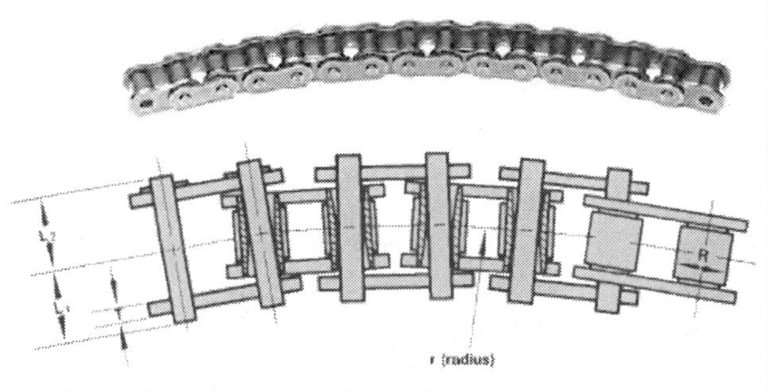

Figure 4-9: RS type (single pitch) side bow chain.

Leaf Chain

Also known as cable chain and counter-balance chain, leaf chain dimensions are listed in ASME Standard B29.8. These chains consist of roller chain type link plates, assembled on pins, without the use of bushings and rollers (see Fig. 4-10).

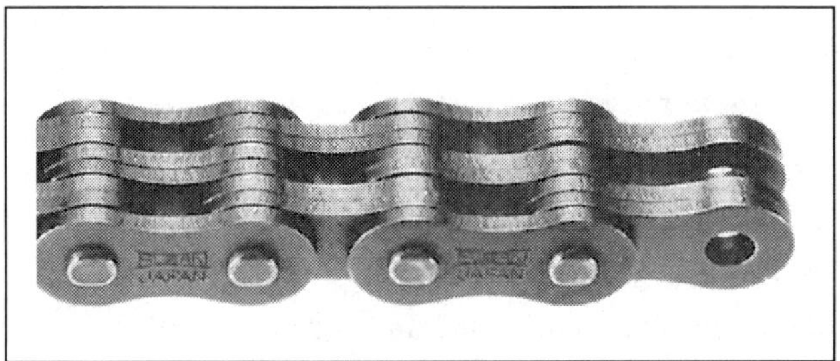

Figure 4-10: BL series leaf chain with 4 x 4 lacing.

This construction utilizes a sheave because there is no provision for sprocket teeth engagement. These are widely used as forklift truck mast chains as well as counterweight chains for machine tools and elevators, and they offer relatively high strength versus weight per foot than roller chains. They run over doors or similar lifting and balancing applications. They can be supplied with male or female terminations to accept female or male clevises, as desired. The pin diameters are press fit into the two outermost plates. All the remaining plates are loose fit. The distance between pin centers is the pitch, and that characteristic distinguishes one chain size from another. The larger the pitch, the larger the links and the greater the strength.

The original AL series uses the standard plate thickness, pin diameter, and pin plate contour (plate height) of the equivalent roller chain pitch. For example, a No. 80 has a 0.125 in. thickness plate, 0.3 12 in. pin diameter, and a maximum plate height of 0.820 in. The AL series has those same dimensions. The BL series was established some years later.

The BL8 plates use the heavy 0.156 in. thickness and have a maximum plate height of 0.950 in., equivalent to the 80H roller chain inside plates. The pin diameter is increased to the next roller chain size of No. 100 (0.375 in.).

The AL series was removed from the B29.8 Standard in 1977. However, it is still available by some manufactures as replacement chain.

The B29.8 chain number designation consists of the series prefix followed by 3 or 4 digits. For instance, inBL1034, the first letter indicates the A or B series and the letter "L" indicates leaf chain. The 10 represents 10/8ths of an inch pitch (1 1/4 in.). The 3 and 4 represents the lacing; 3 plates on the pin link and 4 plates slip on the inner link.

A very important consideration in the field is that leaf chain requires the proper lubrication. In order to achieve satisfactory service life, periodic lubrication must be provided. Like all bearing surfaces, the precision made, hardened steel, joint-wearing surfaces of leaf chain require a film of oil between articulating parts to prevent rapid wear. Generally, the heaviest (highest viscosity) oil that will penetrate the joints (30w-40w) is best. Applying oil to the external surfaces will help prevent rust, but oil must flow into the live bearing surfaces for optimum wear life. To prepare chain for oiling, the leaf chain plates should be brushed with a stiff brush or wire brush to clear the spaces between the plates so that oil may penetrate the live bearing areas.

Silent Chain

Quieter running and more flexible than conventional roller chain, silent (inverted tooth) chain (see Fig. 4-11) is available in 3/16 in. (4.76 mm) to 2 in. (50.8 mm) pitch sizes. The standard for inverted-tooth chain is ASME B29.2.

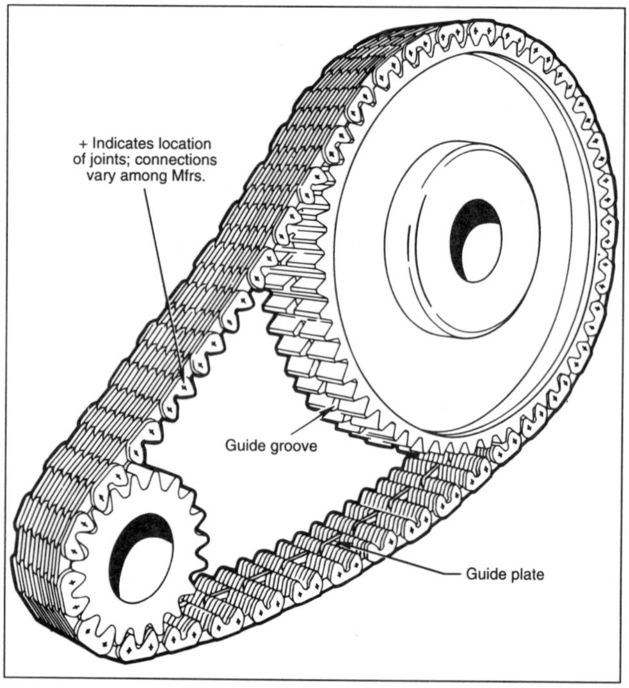

Figure 4-11: Silent (inverted-tooth) chain conforms to ASME B29.2. Guide plates in this chain ride in a circumferential groove in the center of the sprocket faces.

Silent or inverted tooth chain is made up of many links positioned very close together. Because of this, silent chain is capable of carrying far more horsepower per inch of width than standard roller chain. Silent chain is designed with teeth built into the chain that mesh with teeth in the silent chain sprocket. They run very smoothly and with little vibration.

Silent chain is used on applications ranging from fractional horsepower for3/16 in. pitch to 1,000 HP for 2 in. pitch.

The several different types of silent chain construction prohibit mixing chains in a strand, but a chain of the correct size usually will operate on sprockets from a different manufacturer. Load capacity is increased by widening silent chain rather than using multiple strands.

For silent chain with a 3/8 in. (9.52 mm) pitch or greater, an SC prefix in the chain size designation indicates conformance to ASME B29.2. The first one or two digits following the SC indicate pitch in 1/8 in. increments, followed by width in 1/4 in. increments designated by the next two digits. For example, SC 1012 designates ASME standard silent chain with a 1 1/4 in. (31.75 mm) pitch and a 3 in. (76.2 mm) width.

For silent chain with a 3/16 in. (4.76 mm) pitch, the numbers following the 03 (which identifies the chain as 3/16 in. (4.76 mm) pitch) indicate the total number of chain links wide. Width (thickness) of each link is approximately 1/32 in. (0.79 mm). Therefore, SC0314 designates a 3/16 in. (4.76 mm) pitch silent chain that is approximately 7/16 in. (11.11 mm) wide.

Two common techniques are used to keep silent chain from sliding axially off of sprockets. In one technique, center link plates ride in circumferential grooves in the sprocket. In another, side link plates hold the chain in place on the sprocket.

Because silent chain has a flat back, it lends itself to conveying applications in addition to power transmission.

High Tech Coated Chains

High tech coated chains (see Fig. 4-12) have the following attributes:

- Triple protected corrosion resistant carbon steel chain (plates, bushings, and rivet pins).
- Durable double-coated rollers and extended D-1 pins.
- Coated and sealed with three layers to minimize corrosion resistance and maximize service life in difficult applications.
- Same high strength as standard ASME chain.

These types of chains are excellent alternatives to using more expensive stainless steel and will provide much longer service life than stainless in wet environments. They are not intended to replace stainless steel in very corrosive chemical or wash down environments, but in wet environments they are ideal.

Figure 4-12: High Tech Coated Chains.

Plastic Chains

Plastic chains are available in a variety of materials suitable for various temperature ranges, chemical exposure, low friction, and anti-static. Plastic chains are often constructed using stainless steel pins or pins and plates (see Fig. 4-13). They are quiet, lightweight, corrosion resistant, and excellent choices for operation in clean or food grade environments.

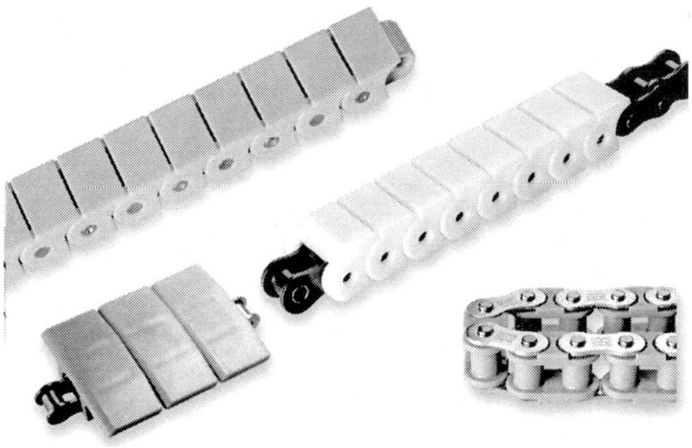

Figure 4-13: Plastic Chains

Tabletop Chain

A very common type of chain used in many plants today is tabletop chain. Tabletop chain is a chain that could run over a sprocket of different materials and has a flat surface on one side. This flat surface enables material to move from one area to another. The flat surface could be made of plastic, steel, stainless steel, and a variety of other material. The flat surface is called a plate and is part of the chain or connected to the chain (see Fig. 4-14).

Other applications could require all parts of the tabletop chain to have any of the following characteristics: chemical resistant, heat stabilized, melt resistant, or ultra-violet resistant. While these options may be available, they are not com-

Basic Plastic Flexing Chain

Some tabletop chains are available up to 12 inches of top width

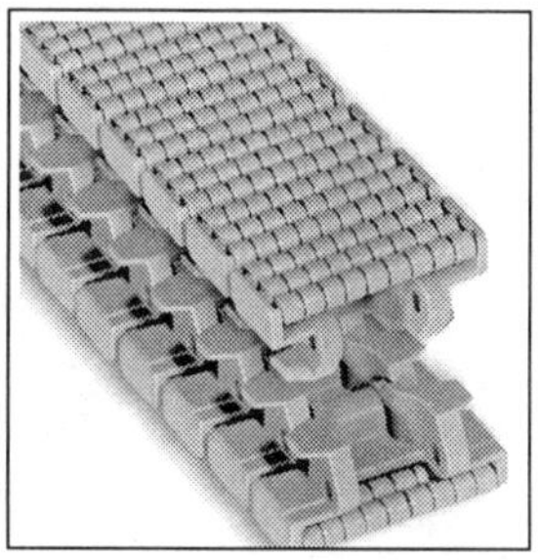

Tabletop Plastic Chain with larger carrying surface

This tabletop chain has many added features for industry

Tabletop chain can be made with stainless steel roller chain

Figure 4-14: Tabletop chains

mon in every chain. Consult with the chain manufacturer for applications requiring any of the above specifications.

Specialty Roller Chains

Figure 4-15: Thermoforming "sticker" chain.

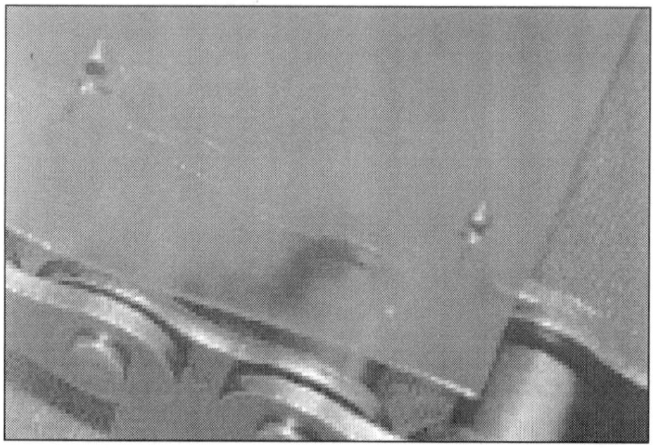

Figure 4-16: Thermoforming "sticker" chain shown with plastic film.

Thermoforming chains, also known as "sticker" chains, are used to convey plastic film through the forming process. They typically use two strands in parallel and contain sharp spikes to pierce the film, allowing it to be conveyed.

Free flow chains allow products to be conveyed and also to accumulate when necessary. While accumulating, the chain continues to travel freely underneath the product being accumulated. A very unique free flow chain design utilizes rolls of different diameters, which allows the product being conveyed to actually travel 2.5 times faster than the chain itself. This style chain requires the use of specialized guides to obtain the 2.5 times speed effect. This design allows for accumulation as well.

Figure 4-17: Free flow chain using multi-diameter rolls to achieve a 2.5 times speed effect.

Engineering Class Chain

Engineering class chains are typically large in pitch (normally greater than 2 in.), and compared to smaller roller chain, they are designed with higher safety factors and more clearances. Although there are many standard sized engineering chains, they are often custom designed to work in a special application or machine. Engineering chains have been mass produced for over a century in the United States. Most innovations are from the addition of specialized attachments, materials, seals, or plating for specific applications and environments.

Engineering chains are rated by working load and ultimate strength. The working load of the chain should not be exceeded during its operation. The load on the chain, chain tension, environment, and service factors should be evaluated when selecting or designing a chain for use in an application. Although engineering chains are designed in all shapes and sizes, some are specified by ANSI B29 standards or ISO/DIN standards. Most engineering chains can be categorized as one of the following:

Drive Chain

Drive chains are used for power transmission and are constructed with a small roller to improve interaction with the sprocket. There are a large variety of sizes of Engineering Class Drive Chains. Many of these sizes are specified by ASME B29.

They are constructed with pins, bushings, roller and sidebars.

Some drive chains have straight sidebars. Others have offset sidebars.

The bushings and pins are press fit into the sidebars. The bushings articulate on the pins and the rollers spin about the bushings. Standard pitch ranges from 1.654" to 7.50" and the Working Load ranges from 2485lbs to 37,000lbs.

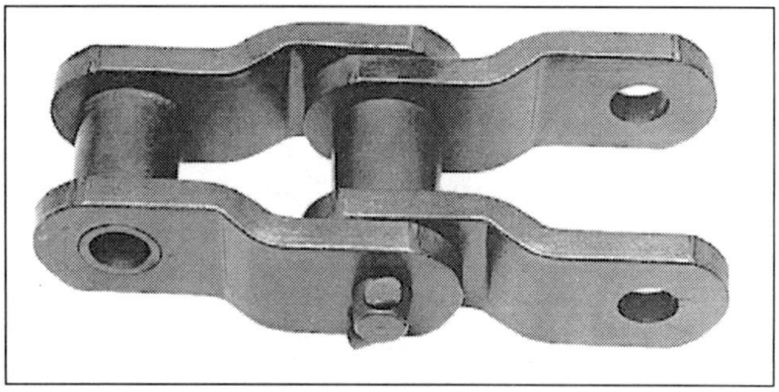

Figure 4-18: Drive Chain

High quality Drive chains are designed with abrasion resistant pins, bushings, and rollers because drive applications are relatively high speed have short distances between sprockets compared to conveyor applications. The components are normally made of hardened carbon or alloy steel.

Figure 4-19: High quality drive chain

Roller Conveyor Chain

Roller Conveyor chains with rollers that are used to convey bulk material, parts, or assemblies of parts. They are normally designed with attachments to connect the carrying device (top plate, pan, cradle, paddle, etc.) to the chain.

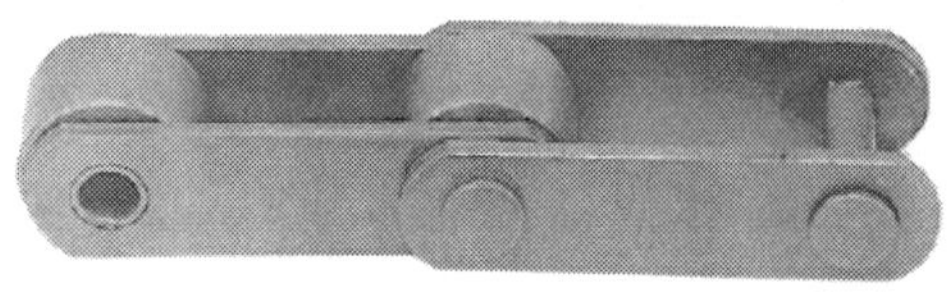

Figure 4-19: Roller Conveyor Chain

Figure 4-20: Roller Conveyor Chain Clear Iso View

Roller conveyor chains are constructed of pins, bushings, rollers, sidebars, and typically attachments. The pins and bushings are press fit into the sidebars. The bushings articulate around the pins and the rollers spin around the bushings. The type of steel and hardness of the components vary depending on the chain.

The pitch ranges from 1.654" to 24" and the Working Load ranges from 2100lbs to 22,300lbs. Most of these chains are sold with attachments. The type of attachment and spacing depends on the application. Some attachments are welded on in the field. Below is a list and description of the most common attachments for conveyor chain (see Fig. 4-21):

- A1: This is a tab protruding from the top of the chain to the right or left with one hole in it.
- A2: This is a tab protruding from the top of the chain to the right or left with two holes.

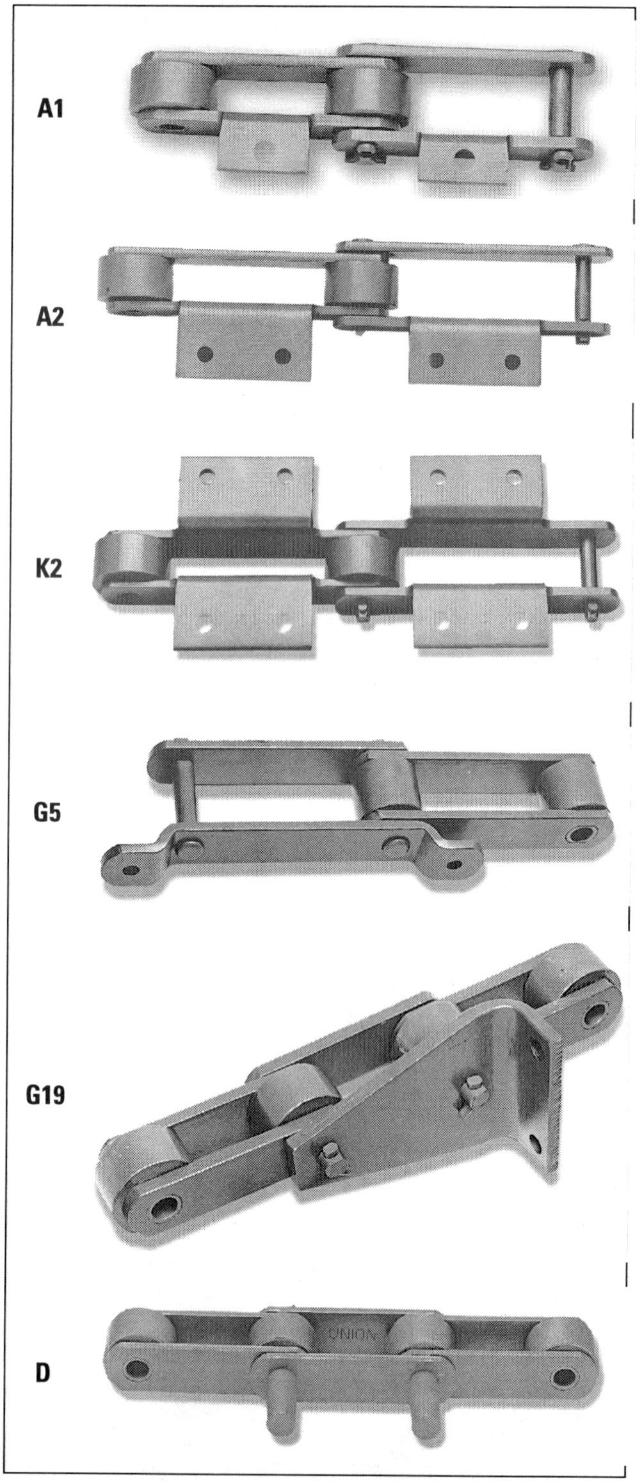

Figure 4-21: Some of the Attachments for Conveyor Chain

- K1: This is a tab protruding from the top of the chain to both sides with one hole in it.
- K2: This is a tab protruding from the top of the chain to both sides with two holes in it.
- G5: This attachment consists of offset extended outside sidebars with bolt holes in them. They can be on one or both side of the chain. These attachments are normally used to secure buckets in a bucket elevator with two strands of chains.
- G6: This attachment consists of bolt holes in the sidebars on one or both sides of the chain. These attachments are normally used to secure buckets in a bucket elevator with two strands of chain.
- G19: This attachment consists of a horizontally bent sidebar with bolt holes for mounting pushers between two strands of chain.
- G29: This attachment consists of angle brackets welded vertically to the sidebars. The brackets have holes in them to connect the carrying device (pusher, pan, paddle, etc.)
- A11: These are tabs welded horizontally to the sidebars. The tabs have bolt holes in them to connect the carrying device (pusher, pan, paddle, etc.)
- C: These are pivoting attachments that are bolted to A11 attachments. They provide a degree of freedom for the carrying device that is bolted between two strands of chain
- D: D attachments are extended pins that protrude out one or both sides of the chain.
- S: S attachments are vertical extensions of the sidebars used as pushers
- G: G attachments are extended sidebars on one side of the chain that serve as pan ends on apron pan conveyors
- High Sidebars: High sidebars are extended sidebars on both sides of the chain used to convey items such as pallets. Often top plates are welded to the high sidebars.

Bushed chain

Bushed chains are similar to roller conveyor chains except they do not have rollers. They are normally used to convey bulk material, parts, or assemblies of parts but sometimes used for power transmission. They are normally designed with attachments to connect the carrying device (bucket, top plate, pan, cradle, paddle, etc.) to the chain.

Bushed chains are constructed with pins, bushings and sidebars (no rollers). The pins and bushings are press fit into the sidebars. The bushings articulate on the pins. Common Bushed chains range in pitch from 2.609" to 7" and working loads from 2,500lbs to 25,000lbs

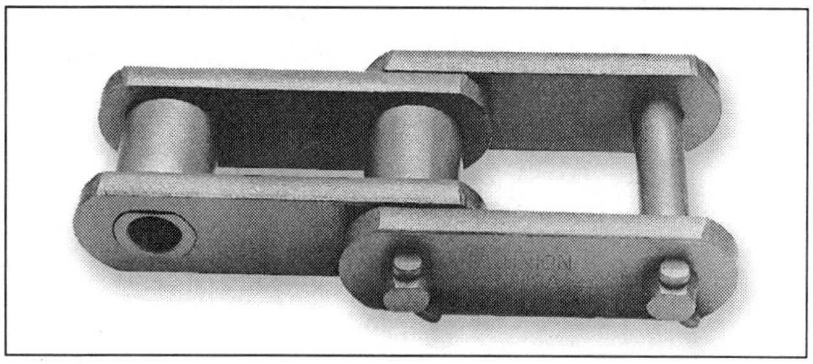

Figure 4-22: Bushed Chain

Cast Combination

Cast combination chains are similar to bushed chains except the inside links are cast. Cast combination chains have no bushings.

Welded Chain

The bushings are welded onto the sidebars of welded steel chains. Welded steel chains do not have rollers. They are often used to as conveyor chains and sometimes as drive chains. Most cast pintle chains have been replaced with welded steel chains.

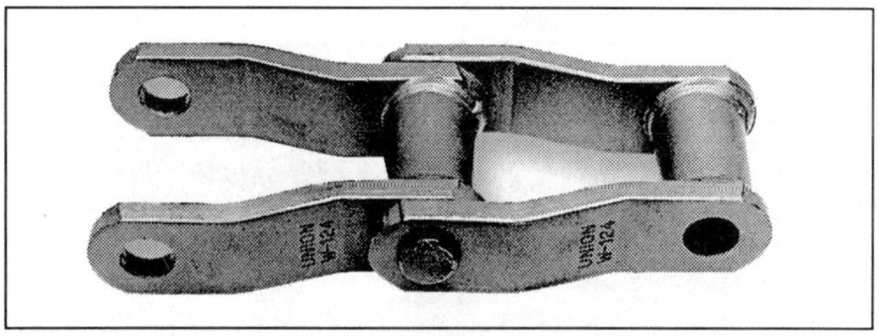

Figure 4-23: Welded Chain

Pins are used to couple the links together and are press fit into the sidebars. Attachments are normally formed or welded onto the sidebars. The links articulate around the pins. Common welded chains range in pitch from 2.609" to 9" pitch and working loads from 3,500lbs to 27,600lbs.

Cast Chains

Cast chains are similar to welded chains except the links are cast and the attachments are cast into the links instead of welded. The links are connected with the pins pressed into the sidebars.

Bar and Pin Chains

Bar and pin chains are designed for lifting, pulling, or dragging items. They lack rollers and bushings. They are commonly used to for high tension applications (draw benches, lifting devices, dam gates) or in-floor conveyors.

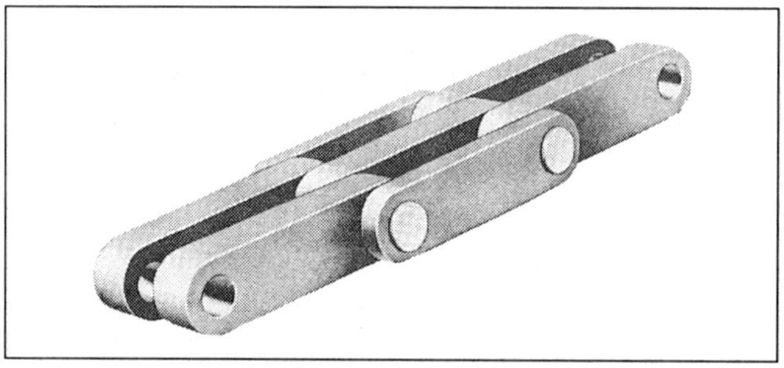

Figure 4-24: Bar and Pin Chain

Bar and pin chains lack bushings and rollers. They consist of a sidebars and pins. Often bearings are inserted into the sidebars to improve abrasion resistance. The pins are press fit into the outside bars. The inside bar articulates around the pins. Common Bar and Pin chains range in pitch from 2.125" to 18" and working loads from 4,000lbs to 30,000lbs. Some bar and pin chains are designed with extra lateral flexibility in order to flex around corners laterally (i.e. dairy floor chains).

Figure 4-25: Bar and pin chain used to carry milk crates

Figure 4-26: Bar and pin chain used to draw steel into tubing

Barloop Chains

Barloop chains are designed for economy, high strength, and flexibility. They lack rollers and bushings and are most commonly used in overhead conveyor trolley systems.

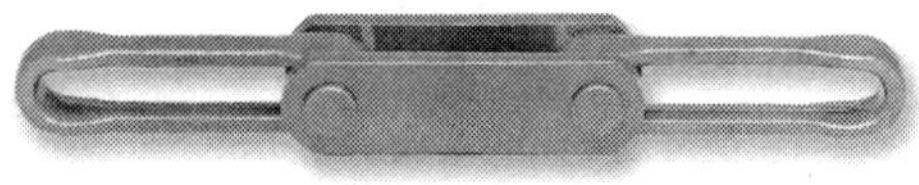

Figure 4-27: Barloop Chain

Barloop chains, sometimes called Drop Forged Rivetless chains, lack bushings and rollers. They consist of a sidebars and pins. The inside bars are forged loops. The pins and outside bars are normally forged. The chain can be hand assembled as the pins are not press fit into the outside bars. The pins are shaped to lock into the outside bars so they do not rotate. Common Barloop chains range in pitch from 3.014" to 9.031" and working loads from 2,000lbs to 10,800lbs. Barloop chains are designed with extra lateral flexibility in order to flex around corners laterally (i.e. overhead trolley conveyors). Several attachments are available that bolt into the loop link. The forged outside bars and pins can be replaced with stamped sidebars and press fit pins, which allows for design flexibility and material flexibility.

Detachable Link Chain

Detachable link chains lack pins, bushings, and rollers. The links are cast, forged, or molded. The links are designed to connect to each other to form a chain. The links can be detached easily for replacement.

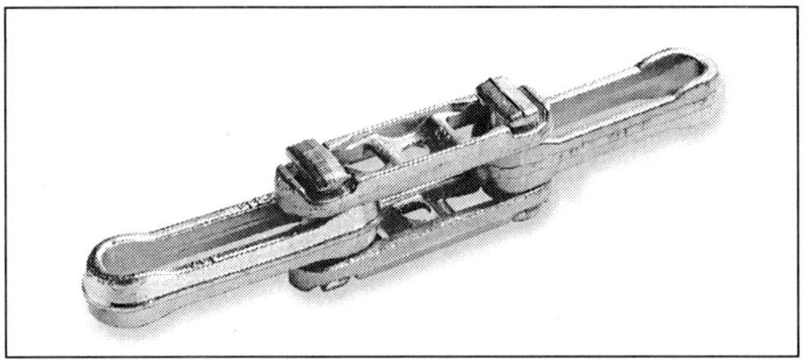

Figure 4-28: Detachable Link Chain

The links are designed to hand assemble and disassemble. There are no machined or press fit components so they lack precision. Detachable link chains are often found on old agricultural equipment but are still used on some new machines and conveyors. They range in a wide variety of shapes and sizes.

Plastic Engineering chains

Plastic Engineering class chains are used where tension is low and corrosion is high. These chains are typically used for conveyors in water or chemical applications. Each link is molded and held together with pins. They are available in a limited variety. They are used where steel or stainless steel chains will not work. They typically consist of molded links that are connected with press fit pins and cotters or retaining rings to keep the pins in place. The pins are typically stainless steel.

Chain Rating

Engineering chains are rated by their working load and ultimate strength. The ultimate strength is the force required to break the chain. This is a theoretical values based on the geometry of the chain and the properties of the material. Some manufacturers publish the average calculated ultimate strength (AUS). Others publish the minimum calculated ultimate strength. Working Load is calculated based on pin-bushing area. Working load, WL, is the maximum pull that can be exerted on a chain without exceeding 3,500 PSI bearing pressure. 3,500 PSI bearing pressure is a value recommended by the American Chain Association. In some applications, the working load is calculated as a fraction of the Ultimate Strength (1/7th or 1/10th are common).

Load

The load on the chain should not exceed the carrying capacity of the rollers and the tension (sometimes called chain pull) on the chain should not exceed the designed Working Load of the chain. The tension can be calculated following methods outlined in chain manufacturer's catalogs.

Environment

(corrosion, abrasion, temperature, hours of operation, contamination)

Chains should be selected with materials that can adequately handle the environment. Stainless steels are commonly used in hot or corrosive environments. Service factors should be factored into the chain selection criteria. Most manufacturer's formulas take into account service factors.

Sprockets

The following defines sprocket terminology and provides guidance to help select the best sprocket for a given application.

Sprocket Caliper Diameter

The term "caliper diameter" refers to the dimension measured from sprocket tooth valley to sprocket tooth valley on the opposite side (180deg.) of the sprocket. This dimension is used to measure the diameter of the sprocket plate (not including sprocket teeth). On sprockets that have an odd number of teeth, the measurement would be taken from the valley of one tooth to the valley as close to 180 deg. on the opposite side of the sprocket (see Fig. 4-29) for an illustration of how this dimension is measured).

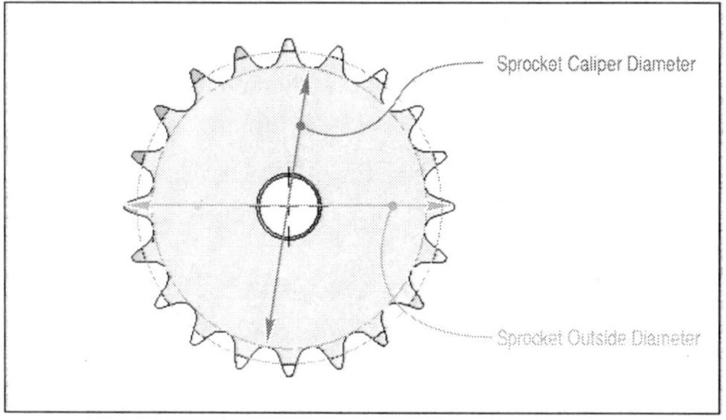

Figure 4-29: Sprocket Diameters

Sprocket Outside Diameter

Sprocket outside diameter is the measurement from sprocket tooth peak to sprocket tooth peak on the opposite (180deg.) side of the sprocket. For sprockets with an odd number of teeth, the measurement would be taken from the peak of one sprocket tooth to the peak of a tooth as close to 180 deg. as possible (see Fig. 29 for an illustration as to how this dimension is measured).

Maximum Bore Diameter

Maximum bore diameter refers to the maximum bore size diameter a sprocket can be machined to without compromising structural integrity. This term is associated with both "B" and "C" style sprockets (see Fig. 4-30).

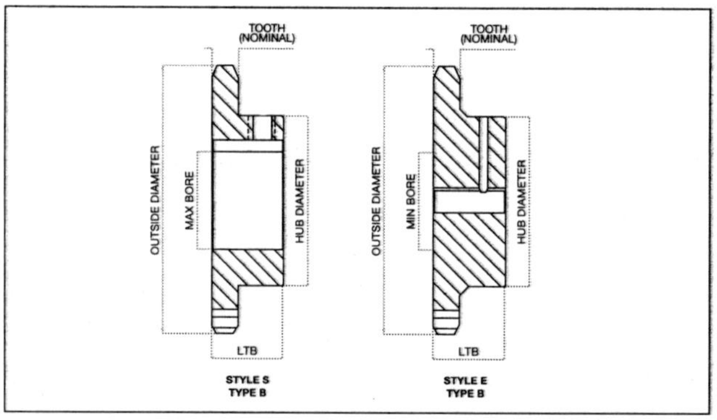

Figure 4-30: Sprocket Bore Diameter

Length Thru Bore

"LTB" refers to the inside hub diameter and the length to which it was machined. This machined length must be long enough to accommodate the proper size key way to withstand shear and torque stress induced by the rotating shaft (see Fig. 4-31).

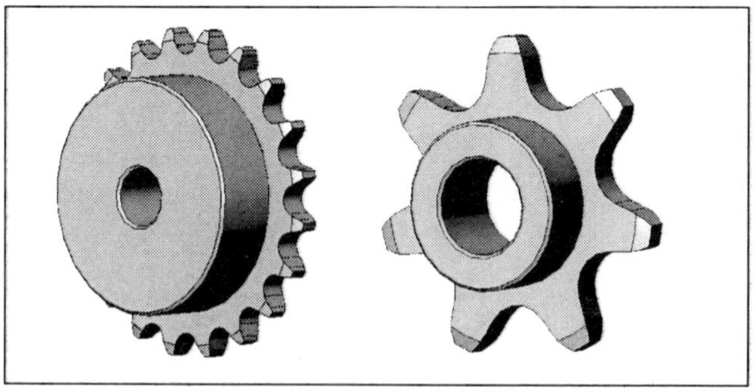

Figure 4-31: Sprocket LTB: Length Thru Bore

Plain Bore

Associated with "A," "B," and "C" style sprockets where there is no special machining performed to accommodate key ways or set screws, only a hole to accommodate shaft diameter (see Fig. 4-32).

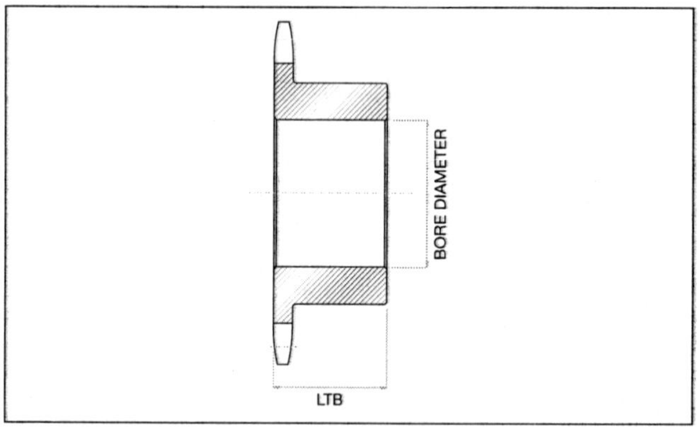

Figure 4-32: Sprocket Plain Bore

Finished Bore

Associated with "B" and "C" style sprockets where the inside diameter of the hub is machined with a standard key way and two set screws. Finished bore hubs can also be machined to specific requirements depending on the needs of the application (see Fig. 4-33).

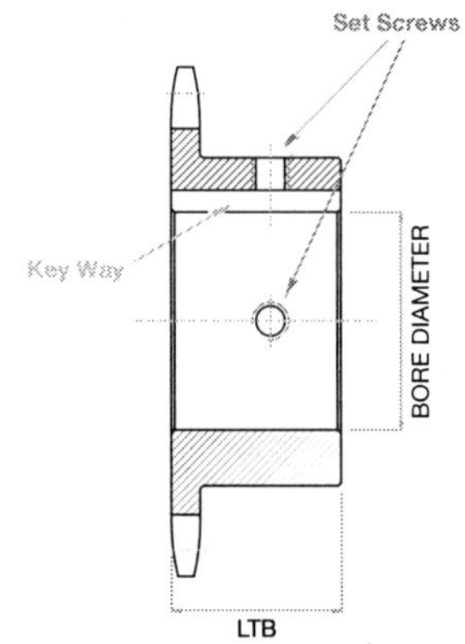

Figure 4-33: Sprocket Finished Bore

Sprocket Hub Style

Manufacturers provide various sprocket hub styles to meet the application requirements of the drive system (see Fig. 4-34). for the most common stock style hub configurations offered by many of the U.S. based sprocket manufacturers.)

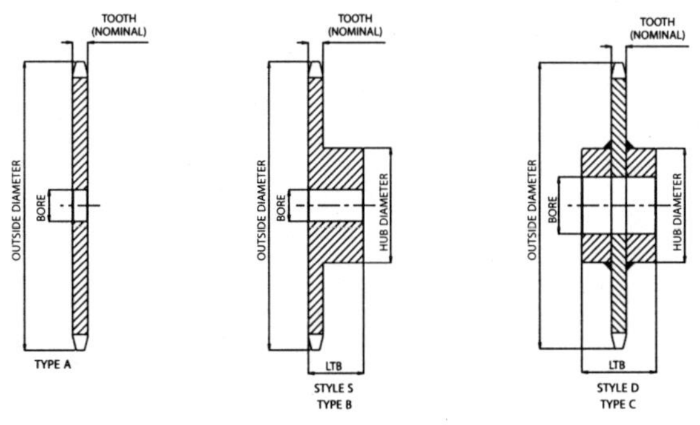

Figure 4-34: Stock Style Hub Configurations

Small and Large Roller Double Pitch Sprockets

These sprockets are similar to standard sprockets, with the exception that the number of sprocket teeth is reduced by 50 percent. Double pitch sprockets are used with small roller double pitch chain to accommodate the longer distance between rollers. This type of sprocket is primarily used in conveying applications where torque requirements are lower and long service life is critical. The primary advantage associated with double pitch sprockets is the fact that half the teeth engage per revolution, thus resulting in longer service life. Common chain sizes used with these sprockets include the following: 2040, 2050, 2060, 2080, and 2100, or 2042, 2052, 2062, 2082, and 2102 for larger diameter chain roller style sprockets. ((see Fig. 4-35 for an illustration of a standard and double pitch sprocket for comparison purposes.)

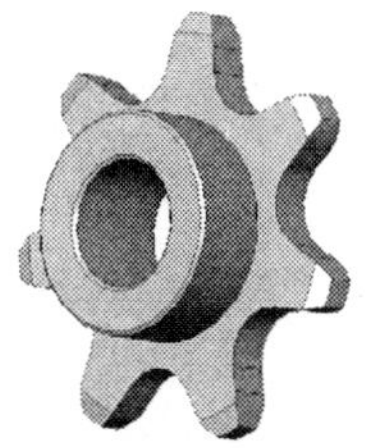

Figure 4-35: Double Pitch Sprocket

Multiple Strand Sprockets

This type of sprocket is commonly used in applications where higher torque and power requirements are needed, or where two or more items are powered by a common drive shaft. The spacing between the rows of teeth corresponds with the center-line of chain strands. Due to the added width of chain, the "LTB" of multiple strand sprockets is correspondingly longer. This style of sprocket is available in 40 - 160 in. chain pitch with plain, finished, or Taper Lock/QD style hubs (see Fig. 4-36).

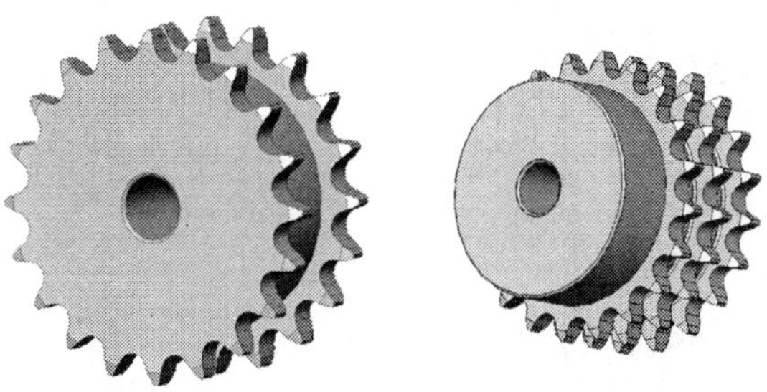

Figure 4-36: Multiple Strand Sprockets

DOUBLE PLUS® Sprockets

This type of sprocket is specifically designed to work with DOUBLE PLUS® chain. This style of sprocket is used in conveyor applications where product on the conveyor is propelled at twice the speed of the drive system powering the conveyor. The main benefit in incorporating DOUBLE PLUS® chain and sprockets is less noise and longer chain life (see Fig. 4-37).

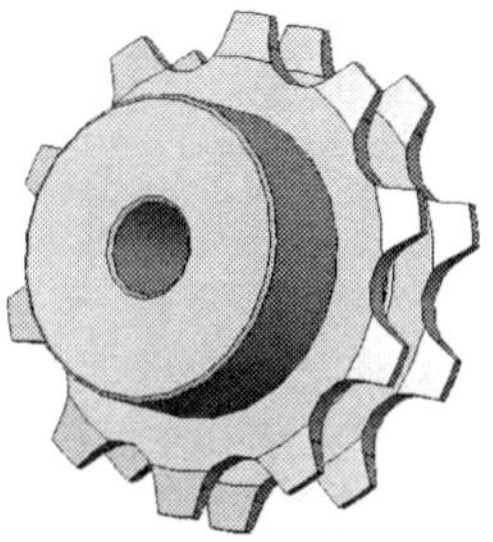

Figure 4-37: DOUBLE PLUS® Sprocket

QD® Sprockets

This type of sprocket is used in applications where higher working loads are prevalent and high clamp loading on the drive shaft is desirable. Sprockets with tapered bushings will fall into the QD®, Split–Taper, or Taper Lock® family. QD bushings are flanged and most commonly utilize large anchor bolts around the circumference of the flange to retain itself to the sprocket. However, this style of bushing can also be welded. The primary advantage of the QD® bushing is that it offers ease of installation and removal, provides superior clamp force, and aligns the sprocket 90deg. to the drive shaft to assure alignment (see Fig. 4-38).

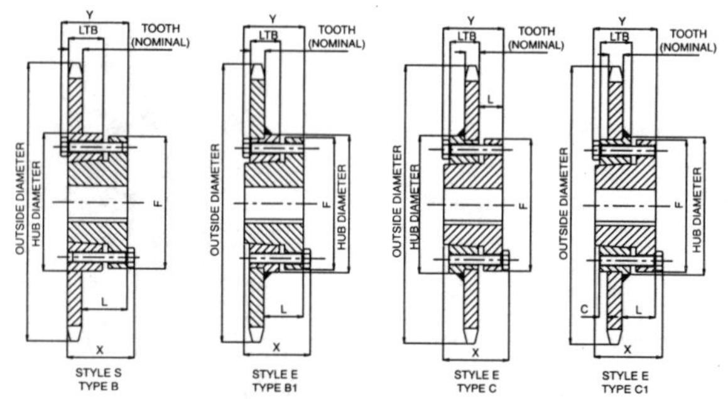

Figure 4-38: QD® Sprockets

TAPER-LOCK® Sprockets

Taper Lock® bushings are similar to QD® style bushings in that they both utilize a split through the taper and flange to provide a true clamp on the shaft that is equivalent to a shrink fit. This type of bushing is retained to the sprocket with a series of set screws on the OD of the bushing (parallel to the shaft), or it can be welded to the sprocket itself. With Taper Lock® bushings, there is no need for a set screw over the drive shaft key. Taper Lock® bushings offer flexibility in that it allows multiple sized bores for a single bushing size (see Fig. 4-39).

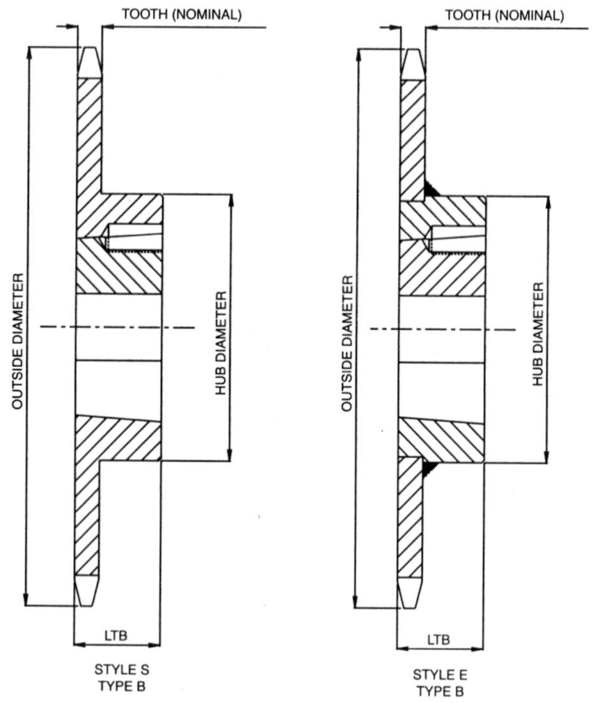

Figure 4-39: TAPER-LOCK® Sprockets

Steel Split Sprockets

Split sprockets are split through the entire radius of the sprocket to allow ease of installation and removal. The sprocket halves are held together by bolts located on either side of the hub. This particular style is available in chain pitch sizes of 40 through 240, and bore diameters of 3/4 in. through 6 in., depending on the chain pitch selected (see Fig. 4-40).

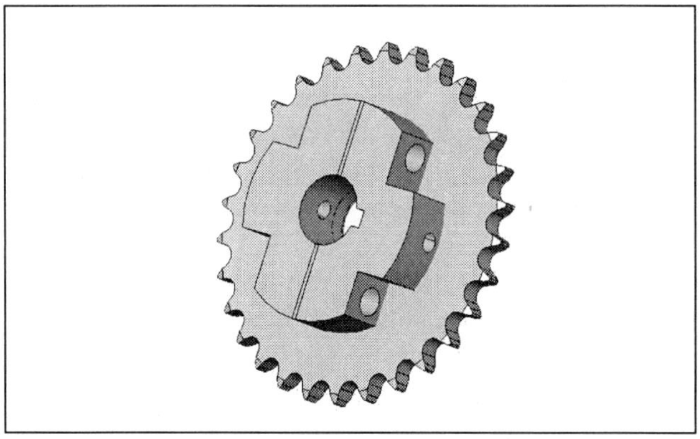

Figure 4-40: Split Sprocket

British Standard Sprockets

This style of sprocket is similar to an ANSI style sprocket with the exception of the fact that it is designed to propel British standard chain. British standard chain has slightly different dimensions with regard to chain pitch. ANSI standard chain pitch is measured in 1/8 in. increments (pin to pin), whereas British standard chain follows a 1/16 in. pin to pin spacing (see Fig. 4-41).

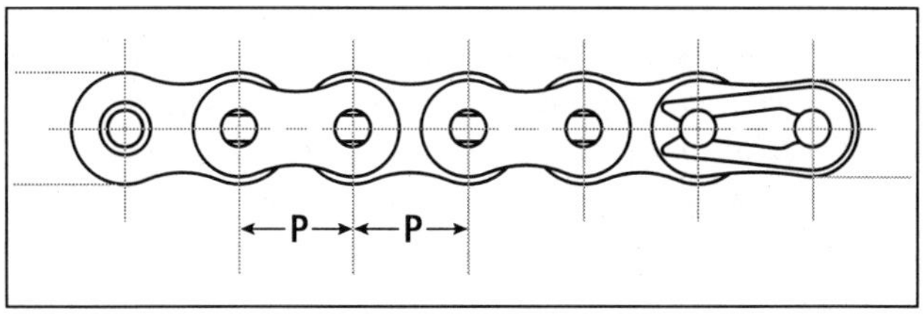

Figure 4-41: British Standard Sprocket

Double Single Sprockets

This type of sprocket is commonly used in applications where two or more items are powered by a common drive shaft. The space between the sprocket plates is wider than a multi-strand sprocket and allows two separate strands of chain to engage without contacting the other. With this type of sprocket, one strand of chain may exit in a different direction than the other. For example, one strand exits toward the ceiling, and the other runs parallel to the floor (see Fig. 4-42).

Figure 4-42: Double Single Sprocket

Idler Sprockets

This style of sprocket is used in applications where the chain may experience slack due to long lengths, non-adjustability of the drive shaft, or where the chain has to be guided around an obstruction. Use of Idler sprockets prevents chain whipping and uneven distribution of load. This type of sprocket may also be used in applications that experience directional reversal or whipping. In applications that experience chain whipping, having the Idler sprockets positioned on the outside of the chain is beneficial (see Fig. 4-43).

Figure 4-43: Idler Sprocket

FDA & USDA Acceptance

The Food and Drug Administration (FDA) accepts certain chain materials for direct contact with food. Stainless steel is the most accepted material for direct contact. Many thermoplastic materials are approved, and the qualifications are listed by the chain manufacturers. Before using a chain in an application requiring FDA approval, consult with the chain manufacturer.

The United States Department of Agriculture (USDA) Inspection Service certifies the design of the product for use in handling meat, poultry, and dairy products.

The chains accepted for handling food in meat and poultry plants are usually indicated by "USDA" in the chain specification pages under food handling qualifications. Chains accepted for use in dairy plants are indicated by "USDA Dairy." The acceptance will be listed as either approved for direct contact or for packaged products only. It is important that any chain specified for use in these areas be approved for that use and the manufacturer consulted.

Tabletop chains provide many added features, including side guards, pushers, and combs.

Lubrication

Proper lubrication is essential for a chain drive to achieve its potential design life. In fact, the type of lubrication employed may be the sole limiting factor in determining the horsepower capacity of a chain drive. Lubrication systems for chain generally fall into one of three categories: Type A–Manual or Drip, Type B – Bath or Disc, and Type C – Oil Stream.

In cases where conventional lubrication methods are impractical or uneconomical, permanently lubricated chain can be specified. Here, oil is applied to the chain during manufacture, and joints are subsequently sealed to keep the oil from leaking out. This technique increases chain cost but often is justified for drives that run intermittently at low speed with light loads, or where manual lubrication is not possible.

Type A

Type A lubrication involves shutting a drive down at regular intervals and manually applying oil to the chain. This method limits initial cost, but from a performance standpoint, it is the least desirable method of lubrication for several reasons:

- It reduces productivity because the machine must be shut down at regular intervals for lubrication.
- Workers are taken away from performing more productive tasks.
- Workers may waste lubricant by applying too much, which also creates a dirty and dangerous work environment. Others may apply too little, which minimizes the benefits of the lubricant. Or lubricant may not be applied frequently enough.

Perhaps the greatest disadvantage of manual lubrication is the "feast-or-famine" condition it creates. When lubricant is applied, the chain receives more than it needs (feast). Conversely, chain operates with an inadequate supply of lubricant for a period before the next application (famine). Thus, chain drives lubricated manually operate with either too much or too little lubricant most of the time.

Drives that operate at variable speeds aggravate the problem because the chain requires more frequent lubrication at high speed than at low. Unless application schedules account for different operating speeds, the feast-or-famine condition is even more pronounced.

Type B

Type B lubrication involves lubricating a chain (and sprocket) constantly, when operating, by routing it through a shallow pool of oil or by applying oil directly to the chain with a disc or brush.

The oil bath method (see Fig. 4-44) uses an enclosure around the chain with a supply of oil that collects at the bottom. The enclosure is positioned so that the chain wrapped around the lower sprocket submerges in the pool of oil as it travels around the sprocket. The oil coats the chain and carries away heat as oil is flung off the chain, and then the oil gradually works its way back down to the sump.

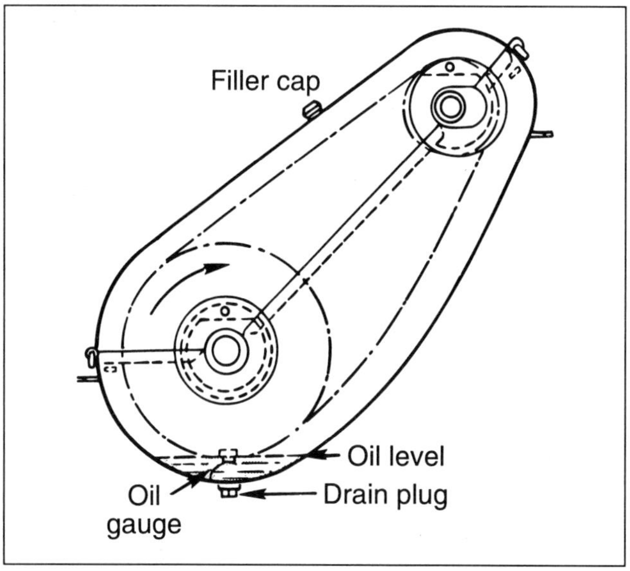

Figure 4-44: Oil-bath lubrication involves directing the chain into a pool of oil, which penetrates into joints of the chain to reduce friction and carry away heat.

The chain should not submerge completely in the pool; this would needlessly hurl the oil throughout the interior of the enclosure. Ideally, the chain should submerge so that the surface of the pool intersects the center of the chain pins at their lowest point.

The disc method of lubrication (see Fig. 4-45) applies oil to the chain as it passes by or rubs against a disc, wheel, brush, or other implement, usually near a sprocket. While this method does not apply oil as effectively as an oil bath, it is sometimes much more practical and economical than using an enclosure. Enclosures can become increasingly expensive as drive size increases. Furthermore, lubricant can be applied with a brush at more than one critical location, a useful technique for drives with long distances between sprockets.

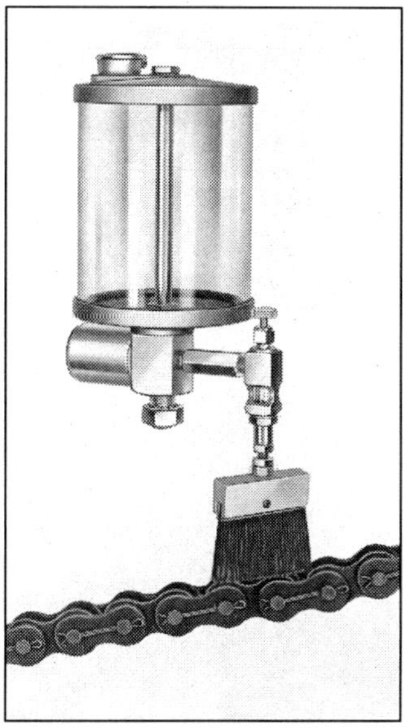

Figure 4-45: Disc or brush-type lubrication applies precise amounts of oil directly to the chain. This method is especially useful for lubricating chain and sprockets at several locations on large drives.

A variety of accessories are available to stop the flow of oil when the drive stops, signal when an oil reserve is low, and apply oil at different rates along the drive. A system of valves, switches, and tubing can be incorporated to apply just the right amount of oil to critical areas of several drives.

Type C

Type C lubrication uses a pump to force oil not only onto the chain, but into the working joints. This forced lubrication is preferred for high-speed drives because the oil works into the many crevices and joints in the chain to reduce friction, thereby reducing heat generation and wear, increasing efficiency, and extending life.

In addition to lubricating chain and sprockets, oil also carries away heat. Much of the heat then transfers to a drive enclosure much like that used with oil-bath lubrication but with tighter sealing. Heat transferred to the enclosure dissipates to the surrounding atmosphere. An oil pump mounted in the base of the chain drive enclosure forces oil through tubing that directs the jet toward critical areas of the drive. An auxiliary electric motor usually drives this pump. However, a small auxiliary chain drive connected to the main drive can drive the pump if electricity is unavailable, impractical, or undesirable.

Although this is the most costly method of lubrication, it is also the most effective for extracting the greatest performance and life out of the chain drive.

Caution

When connecting or disconnecting chain, make sure to do the following:

- Always lockout equipment power switch before removing or installing chains.
- Always use safety glasses to protect your eyes.
- Wear protective clothing, gloves, and safety shoes.
- Support the chain to prevent uncontrolled movement of chain and parts.
- Use of pressing equipment is recommended on larger chain. Tools should be in good condition and properly used.

- Do not attempt to connect or disconnect chain unless you know the chain construction, including the correct direction for pin/rivet removal or insertion.

Chain Guards

All roller chain drives should be enclosed by a chain guard to prevent injury. Applications where chain guards cannot be used are extremely dangerous and every precaution possible should be taken to prevent bodily injury.

Application and Selection

Regardless of the type or class of power transmission chain, the following information is needed to design a drive:

1. Type of input power source (electric motor, internal combustion engine, steam turbine, etc.).
2. Type of driven load (uniform load, moderate shock, heavy shock).
3. Power (HP) to be transmitted.
4. Full-load speed of fastest (usually the driving) shaft.
5. Desired speed of slowest (usually the driven) shaft.
6. Shaft diameters.
7. Center distance between shafts. (If distance is adjustable, what is the range of adjustment?)
8. Limits on space and position of drive.
9. Lubrication available.
10. Other special conditions, such as more than two sprockets, use of idlers (Chapter 15), abrasive or corrosive environments, extreme temperatures (hot or cold), and wide variations in load and speed.

These data, when used with published capacity ratings, allow a designer to select the proper drive for an application. ASME B29 standards contain capacity ratings, as do chain manufacturers' catalogs. Standards covering roller chain, silent chain, and detachable-link chain provide horse power capacity tabulations for a wide range of sprocket sizes and rotational speeds. Standards for steel detachable chain cover only allowable working loads. The standards and some manufacturers' literature contain both allowable working loads and horsepower tabulations for offset sidebar chain.

For most power applications, ratings for chain are based on a life of 15,000 hr, assuming alignment, lubrication, and maintenance requirements are met. It is best to apply generous service factors, as high as 1.7, especially if heavy shock loading is anticipated.

Speed ratios should not exceed 10:1 in a single reduction for roller or silent chain or 7:1 for other types. If needed, double-reduction drives (see Fig. 4-30) can be used to stay within these limits. For roller chain operating at low speed, the smaller (driving) sprocket usually can operate effectively with 12 to 17 teeth. At high speed, the smaller sprocket should have at least 25 teeth to avoid problems that can occur when a chain moving rapidly in a straight line has to suddenly wrap around a small-diameter sprocket.

Additional Information

Addresses and phone numbers for associations are provided in the Association Appendix.

Mechanical Power Transmission Association (MPTA), Naples, FL

American Society of Mechanical Engineers (ASME), New York, NY

Chapter 5

Clutches & Brakes

Chapter Objectives

When you have finished Chapter 5, "Clutches & Brakes," you should be able to do the following:

1. Describe the purpose and function of clutches and brakes.
2. Identify the features of friction types.
3. Identify the features of electromagnetic types.
4. Identify the features of pneumatic hydraulic types.
5. Contrast and compare various mechanical-lockup interface methods.
6. Describe the available methods of actuation, their operational characteristics, advantages, and limitations.
7. Describe the most important criteria for the right selection of a clutch or brake.

Introduction: Usage of Clutches and Brakes

Clutches and brakes are used in a diverse and wide-ranging number of commercial, industrial, and mobile equipment applications. Clutches are used to connect the prime mover (motor or engine) to the load in a manner that allows the torque transmission to be interrupted. Brakes are used to stop, hold, or slow a load.

Applications are very diverse. Clutches are used to engage a wide range of pumps, fans, and conveyors in industrial, mobile, and business machine applications. Torque values can range from a few inch pounds to more than more than 50 million inch pounds and unit diameters from as low as one inch or to more than 50 inches in size.

Brakes are used to stop and hold loads safely in applications such as stamping presses, escalators, or inclined conveyors. Combination clutch brakes are used in baggage handling conveyors and packaging systems where accurate and consistent cycling is valued. Mobile applications can include clutches and brakes that control overhead cranes, farm combines, and mower decks on garden tractors. Business machine applications are characterized by smaller size units with lower torque values for such applications as Automated Teller Machines (ATMs), voting machines, ticket dispensers, and similar light-duty applications.

When starting or stopping, clutches and brakes pass energy between an input shaft and an output shaft through their point of contact. This energy will show up as heat and will be equivalent to the kinetic (moving) energy imparted to the load (in the case of a clutch) or absorbed from the load (in the case of a brake). The transmission of torque can take place via frictional forces (the most common method), electromagnetic flux, or mechanical lockup. In each type of transmission, there will be a wide range of configurations including single friction disk, multiple disk, caliper disk, or drum.

Along with the different methods of transmission, clutches and brakes can be categorized by the actuation method, such as mechanical, electro-magnetic, pneumatic, or hydraulic. Given that there are many functional variables, it is not surprising that a wide array of product exists in today's marketplace. This can result in some selection confusion but can also provide a large number of choices so that the best solution for any given application can be made.

Selection criteria will be discussed in greater detail later in this chapter, but it is important to keep in mind that torque, response time, envelope size, envelope configuration, repeatability, cyclic requirement, and thermal capacity will all be factors to consider. Different methods of actuation will have different impacts on unit performance, and all are not equal for a given application.

In addition to the interface type, input and output geometry and actuation method, the type of mounting must also be considered. Shaft mounting, flange mounting, base mounting, or engine flywheel (PTO) mounting are all commonly found in the marketplace.

Friction Types

Friction type clutches and brakes are the most common for general purpose starting and stopping. They are manufactured with pneumatic, hydraulic, electric, or mechanical actuation methods. Friction designs may have single, dual, or multiple friction face engagement. Most types will be available for shaft, flange, engine flywheel, or foot mounting. The following are some common examples:

Fig. 5-1 is a side view of a shaft mounted, through-bore, single-interface clutch only. This unit is a pneumatic-actuated design with a pilot mounting flange so that a user can easily adapt a sheave or sprocket to the unit.

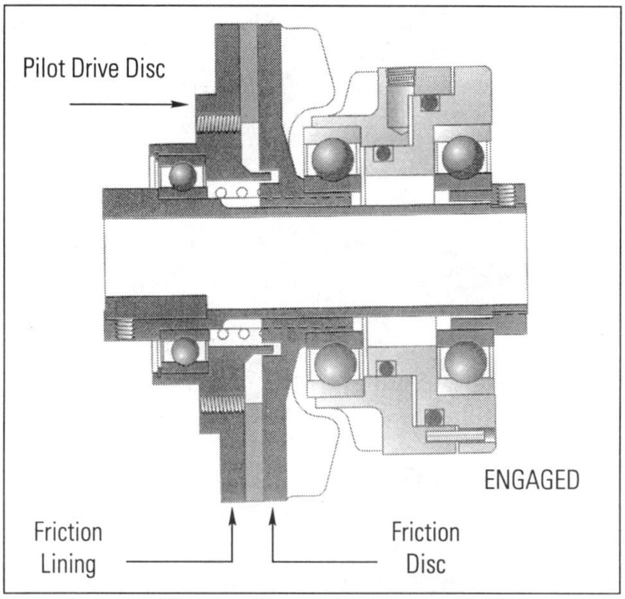

Figure 5-1: Single interface clutch

Fig. 5.2 shows a cutaway of a clutch/brake combination for motor shaft mounting. The unit has two sets of friction faces and would be electro-magnetically actuated.

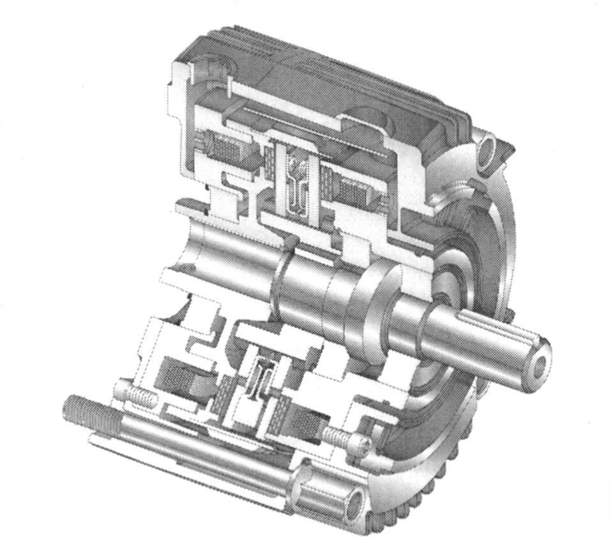

Figure 5-2: C-face mount clutch/brake

One of the most common industrial configurations of the friction-type units are those that are designed for use with C-face motors and gearboxes. The commonality of mounting dimensions from one manufacturer to another ensures that a unit designed for a given frame size will mount correctly regardless of which manufacturer built the unit. Figure 5.3 shows a cross section view of a pneumatically-actuated version of the unit shown in figure 5.2. Both provide a hollow bore input to mate with the motor shaft and both would have an output that meets the C-face configuration.

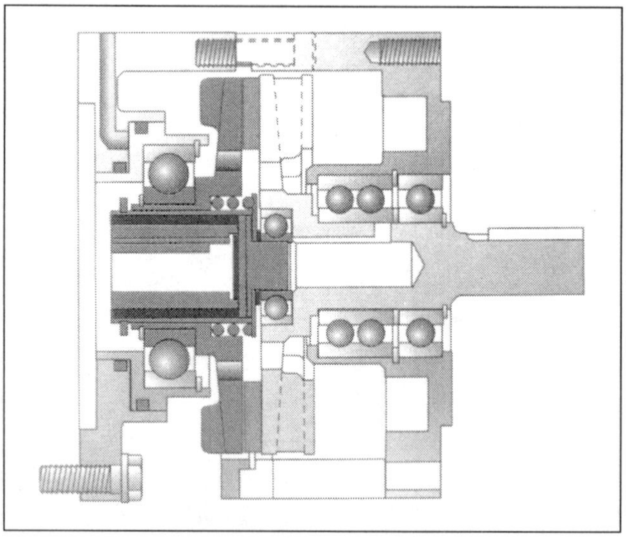

Figure 5-3: Flange mount clutch/brake

Figure 5-4 shows the common ways that C-face units can be attached to various types of equipment.

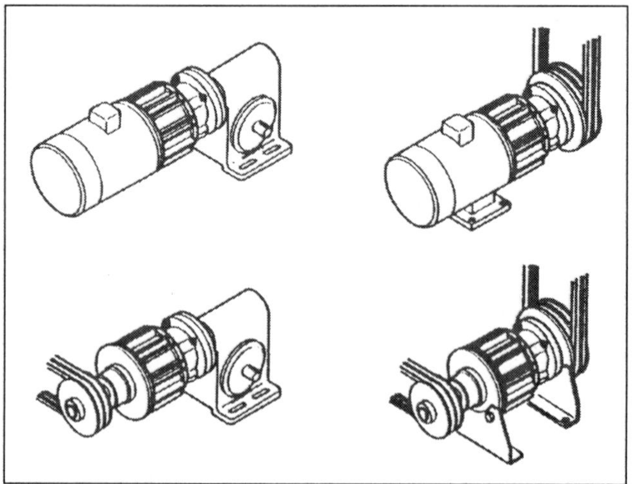

Figure 5-4: Clutch/brake, flange mounted

The units shown in Figures 5.2 and 5.3 are available in both open and totally enclosed housings. Food and beverage industries require units that will be resistant to wash down with caustic chemicals that enhance sanitation efforts by having external surfaces that are totally smooth. Properly configured C-face mount units lend themselves particularly well to these requirements.

Another popular version of the friction design is a combination clutch/brake unit as depicted in Figure 5.5.

This particular configuration is commonly used on stamping presses and other metal forming equipment. The design allows for single-point actuation from the clutch position to brake position in a rapid movement. This prevents overlap between the clutch and brake as only a single function can be actuated at a time. This design can be pneumatically or hydraulically actuated. To enhance rapid responsiveness, rotating components are kept to a minimal inertia and weight.

Multiple plate units (sometimes called multiple disks), such as the one shown in Fig.5.6, will produce more torque in a given diameter than a single disk design, providing a solution for applications where tight space is a consideration. While multiple plate designs may provide a higher torque capacity than single plate designs, the lack of external ven-

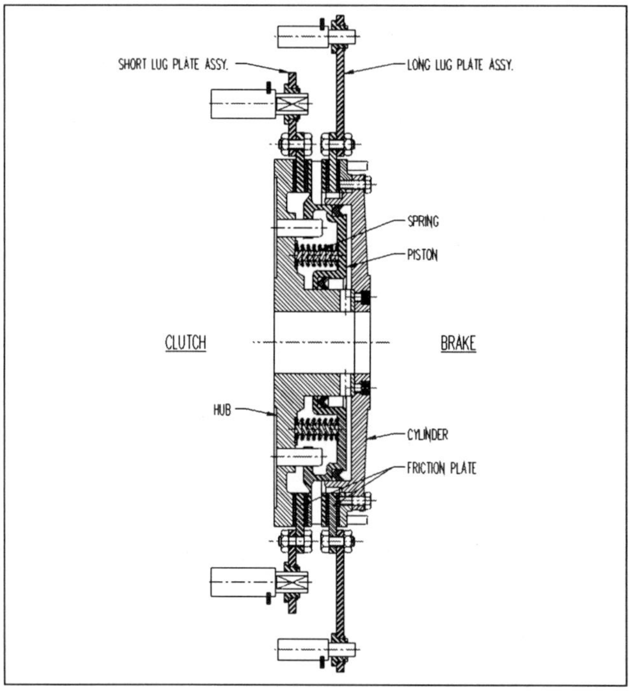

Figure 5-5: Hydraulic combination Clutch/brake for Stamping Press

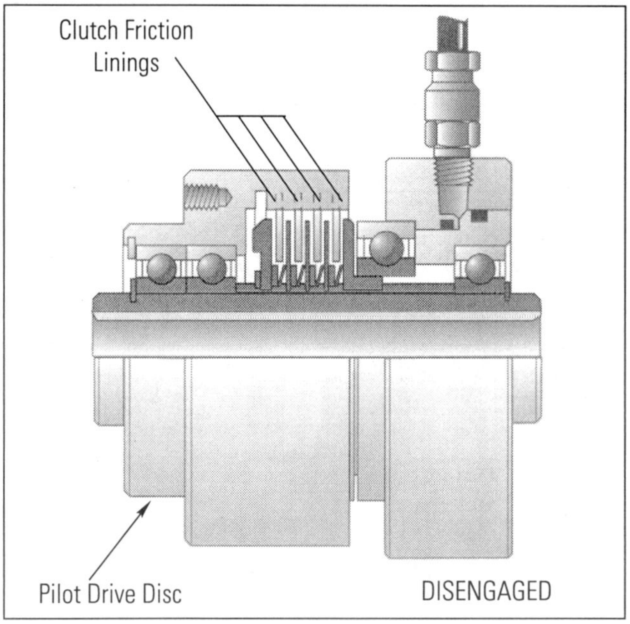

Figure 5-6: Multiple-plate clutch

tilation can mean that their thermal capacity is less than single disk designs.

Caliper units are an additional type of friction clutches that lend themselves to a range of applications. The unit shown in Figure 5.7 shows a multiple friction module device intended for continuous slip duty such as that needed for tension control of webs of paper, metal, or strand of wire during unwind processing. Typically a unit like this will be provided with a well-ventilated friction disk to allow for beneficial heat dissipation. The ability of caliper units to have multiple actuator modules and a range of disk diameters allows them to meet a wide range of application needs for both unwind or simple braking requirements.

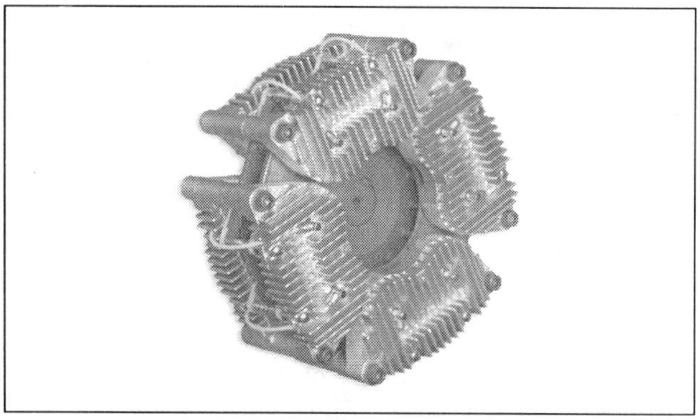

Figure 5-7: Multiple friction module device

A discussion of friction clutches and brakes would not be complete without a discussion of drum style units (see Fig. 5.8). Drum units are either contracting types (shoes move inward against the drum's outer surface) or expanding types (shoes move outward to engage the inside surface). Actuation can be mechanical, pneumatic, or hydraulic.

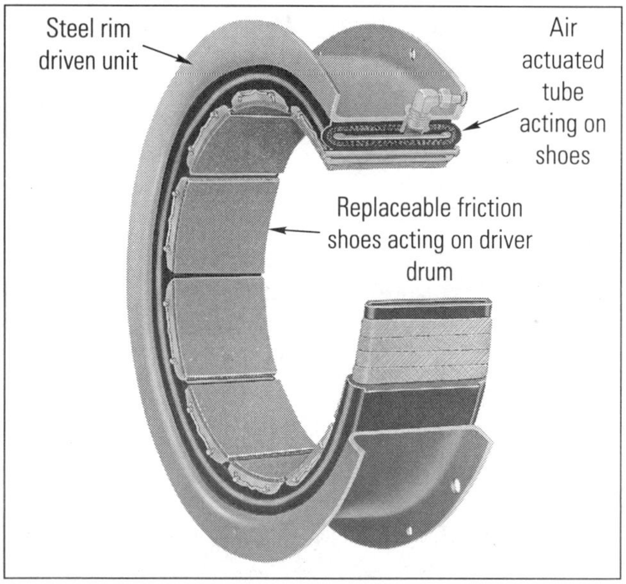

Figure 5-8: Drum clutch

Shoe brakes are similar to drum brakes but function only as a brake. A typical brake may be spring, pneumatically, or hydraulically applied with the shoe engaging around a drum that is mounted to the shaft to be stopped (see Fig. 5-9).

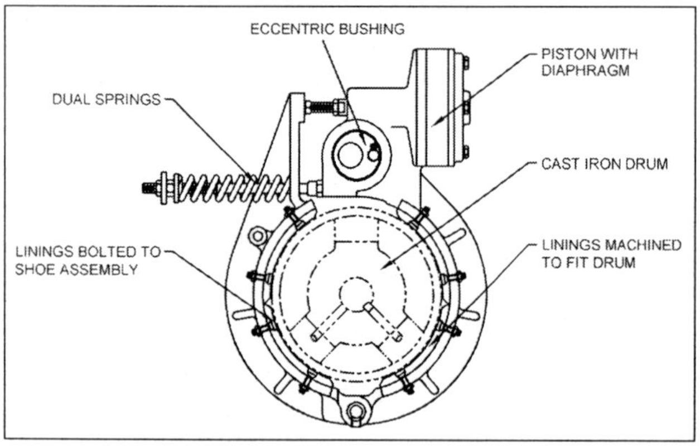

Figure 5-9: Shoe Brake

An increasingly popular design of friction brake are spring-set/power-released designs. These can be pneumatically or electrically released. When no power or air pressure is applied to the unit, springs engage and hold the load stationary. When air or power is applied, the unit releases. In this way, the unit engages in a power or pressure failure so that loads are safely controlled. Spring-set brakes are commonly adapted to servo motor applications where the motor does

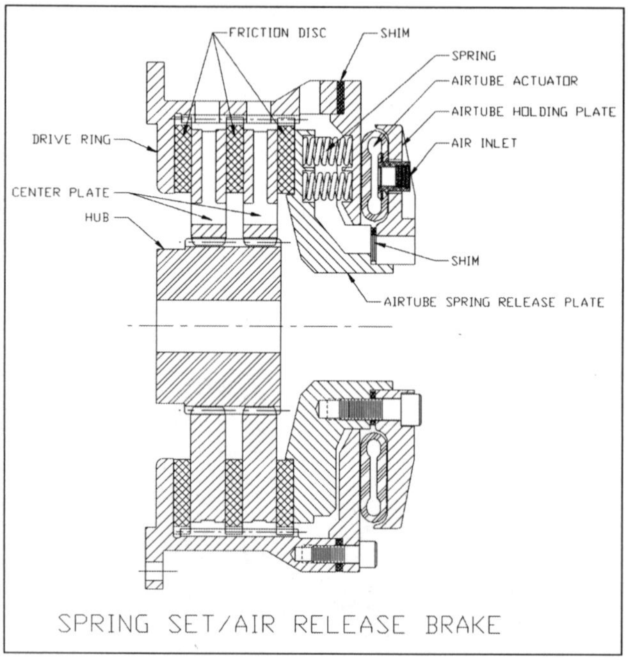

Figure 5-10: Spring-set, air release brake

the positioning and stopping and the brake provides positional holding of the load.

An alternative to the spring-set brake is the permanent magnet electrically released brake. In this case, the brake is released by applying power to an internal electromagnetic coil that cancels the magnetic effect of the permanent magnets. Like the spring-set designs, this unit engages when the power is off.

Magnetic and Electromagnetic Designs

These non-friction clutches and brakes employ magnetic attraction to transfer torque from input to output. They are commonly used in applications that require variable slip or soft engagement. Generally, friction designs will provide greater torque transfer for a given diameter so that magnetic and electromagnetic designs are larger in diameter than friction units. Eddy current, hysteresis, and magnetic clutches and brakes are used in light- to medium-duty applications such as wire and film tensioning, labeling, and printing.

Magnetic particle designs (see Fig. 5-11) are engaged using electromagnetic force as an actuating medium. The space between the input and output members is filled with dry iron particles. When power is applied, magnetic flux lines span the space, and particles line up to form a bond between input and output members. The degree of particle bonding, which is directly proportional to current, determines the level of torque transfer. Magnetic particle clutches and brakes are commonly used in tension and positioning applications in which speed changes continually.

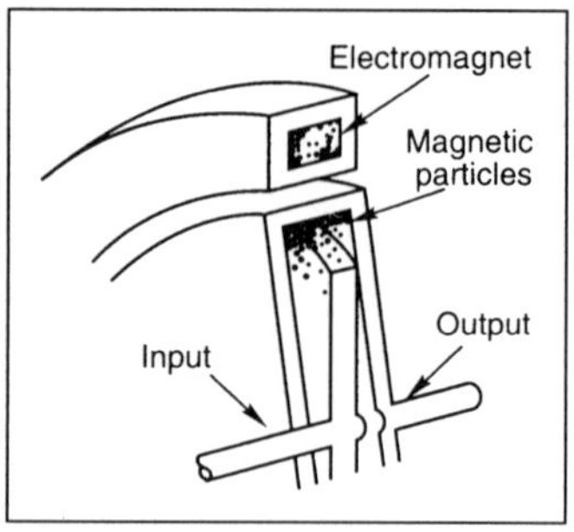

Figure 5-11: Magnetic-particle clutch

When the coil of an eddy current clutch is energized with DC current, the input rotor is magnetized. The magnetic flux of the input rotor then causes small currents called eddy currents to flow in the non-ferrous output rotor. These eddy currents create their own magnetic flux that reacts with the input rotor to transmit torque. The output torque is proportional to the DC coil current and slip speed. For eddy current clutches and brakes to work, there must be relative motion between the two magnetic surfaces. This means that the clutch's output speed will not match input speed and that an eddy current brake will not hold when the brake rotor is at rest.

Eddy current brakes have a fixed output member and a rotating brake rotor. If the brake rotor is spinning when the field coil is energized, eddy currents created in the brake rotor produce a magnetic field that opposes the rotation.

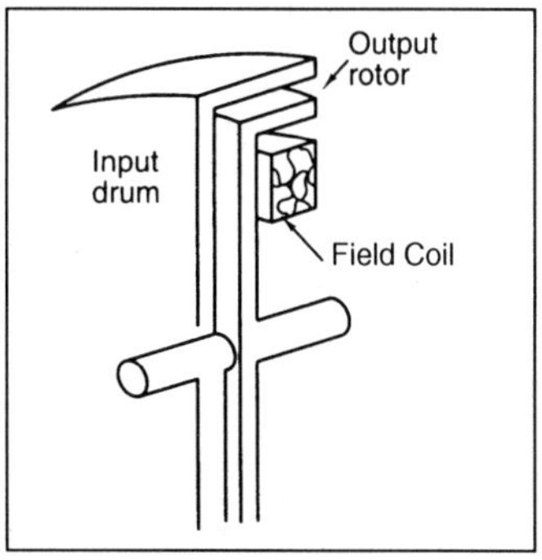

Figure 5-12:
Eddy current clutch

Hysteresis clutches are designed similarly to eddy current clutches, except that the output rotor of the clutch is made of ferrous material. When a current is applied to the actuator coil, the input rotor becomes magnetized. The magnetic field of the input rotor then causes the output rotor to magnetize, rotating the output rotor through interaction of the fields.

In a hysteresis brake, the input field assembly is held stationary causing the output rotor to come to a stop to line up with the stationary field assembly when the actuator coil is energized.

Precise control is a valuable, widely exploited feature of hysteresis brakes. They suffer very little wear and excel in fractional horsepower applications.

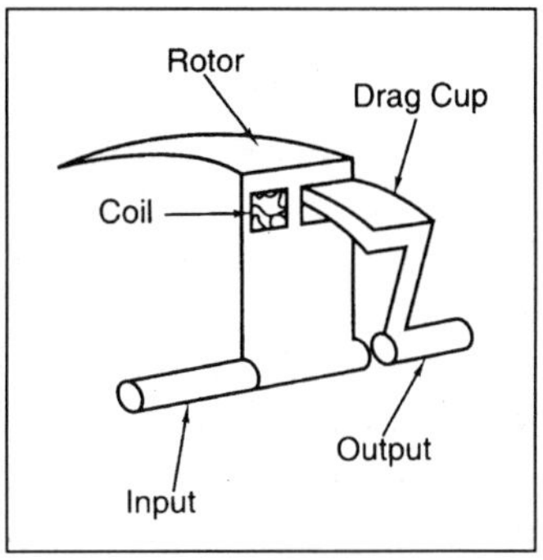

Figure 5-13:
Hysteresis clutch

High strength magnets allow for the manufacture of permanent magnet clutches and brakes that require no input power to operate. A key feature of these designs is their consistent torque level regardless of unit speed. They can be used as a clutch or brake depending on the mounting configuration. Internal magnets are user adjusted to provide the torque necessary to cause the output shaft to rotate (in a clutch) or cause the rotating shaft to come to a stop (in a brake). A recent development in the use of these products has been their adaptation for use in the tightening of caps onto bottles of soda, water, and similar screw-on cap products.

Their consistent torque output ensures consistent quality of cap tightness and seal.

Mechanical Lock-up Interfaces

Direct mechanical connection between input and output members is one of the oldest and most basic of clutching methods. In mechanical lock-up clutches, members engage by direct contact. Their controllability and simple design provide solutions in a variety of applications where exact no-slip engagement and cycling is required.

Actuation of mechanical units can be by any conventional method including manual or remote mechanical engagement. Self-actuating types may use centrifugal force, wrapping or wedging action to engage input and output. Rotational speed, speed differential and direction of rotation may trigger engagement. Other types of engagement for mechanical lock up may be pneumatic or hydraulic pressure to push the mechanism together or electromagnetism to pull surfaces together.

Jaw Clutches

Square-jaw clutches (see Fig. 5-14) use square teeth that lock into matching recesses in facing members. They provide instant, positive engagement in either direction of rotation. However, because they cannot slip, safe engagement is possible only when shafts are stationary or moving at the same speed in the same direction. Running engagement is only possible up to about 10 rpm.

Spiral-jaw clutches (see Fig. 5-15) are designed with ramped surfaces on one side of the teeth. This allows the clutches

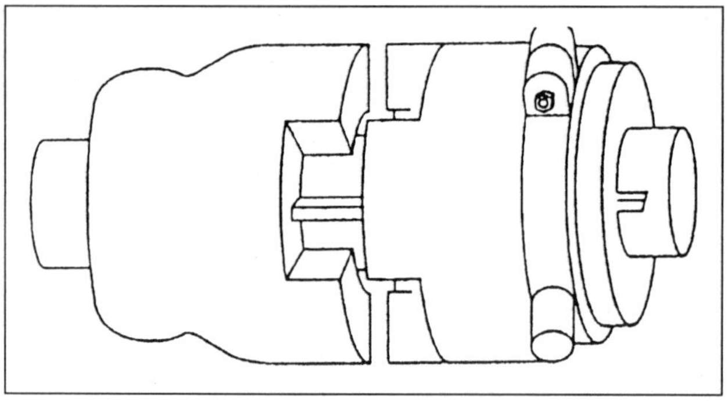

Figure 5-14: Square-jaw clutch

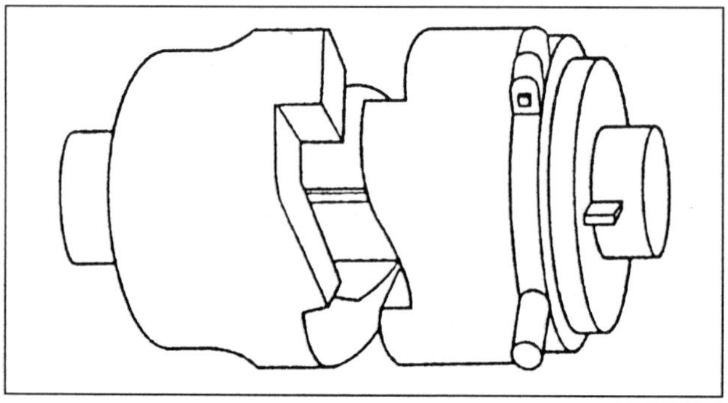

Figure 5-15: Spiral-jaw clutch

to be engaged at speeds up to 150 rpm but does limit the clutch to being able to drive in one direction only.

Multiple Tooth Clutches

Multiple tooth clutches are a refinement of the jaw clutch principle. They provide the advantages of mechanical lock up while allowing electromagnetic, hydraulic, or pneumatic engagement. For any given size, they will offer significantly higher torque-carrying capacity than friction units in the same size. Engagement is between large numbers of tapered, gear-like teeth (see Fig. 5-16). Similar to jaw clutches, tooth clutches require a low or zero speed engagement.

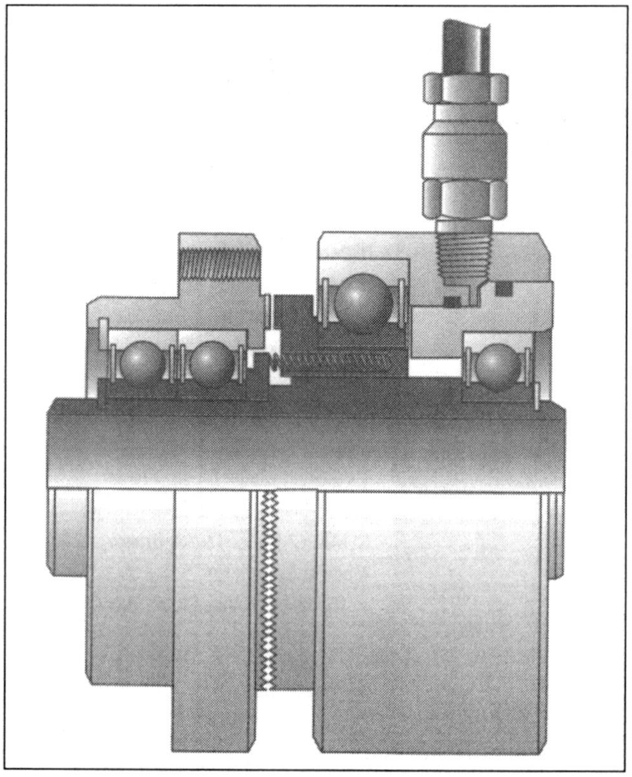

Figure 5-16: Multiple tooth clutch

Wrap-Spring Clutches

Wrap-spring clutches and clutch/brakes link input and output members with a coiled spring (see Fig. 5-17). Rotation in the direction of drive tightens or wraps the spring and transmits torque while rotation in the opposite direction will not transmit torque. Similar to a sprag clutch (to be discussed shortly) wrap-spring clutches can drive in one direction and freewheel in the other.

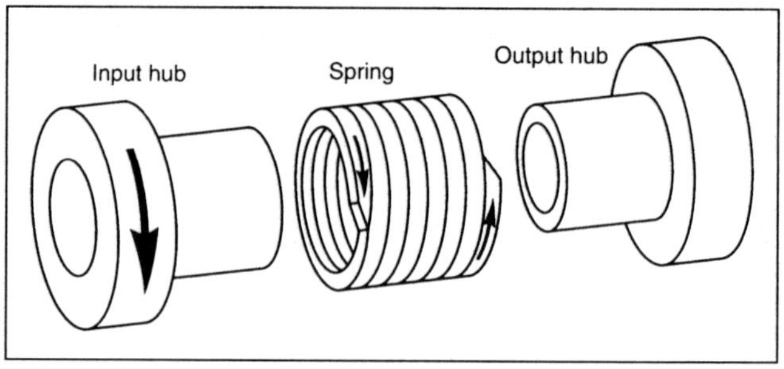

Figure 5-17: Wrap-spring clutch

If one member is anchored, a wrap-spring unit serves as a backstop or brake, allowing free rotation in one direction but preventing rotation in the other. When combined together, a combination clutch and brake can provide accurate indexing in applications with repeating movement. In such units, an external actuator (commonly an electric solenoid) allows for control of the start-stop function.

Centrifugal forces on the springs limit speeds of wrap-spring designs as the unit size increases. They are commonly used in copiers, packaging systems, collators, and die cutting systems.

Sprag Clutches

Another overrunning clutch type is the sprag clutch. These use cylindrical inner and outer races with sprags filling the space between. Sprags are sized, shaped, and mounted in a manner that causes them to wedge between the two races transmitting torque from one to the other when rotated in the drive direction (see Fig. 5-18). No outside control is needed for sprag clutches. In one direction they will drive, and when rotation is reversed, or if the output rotates faster than the input, they will slip or overrun.

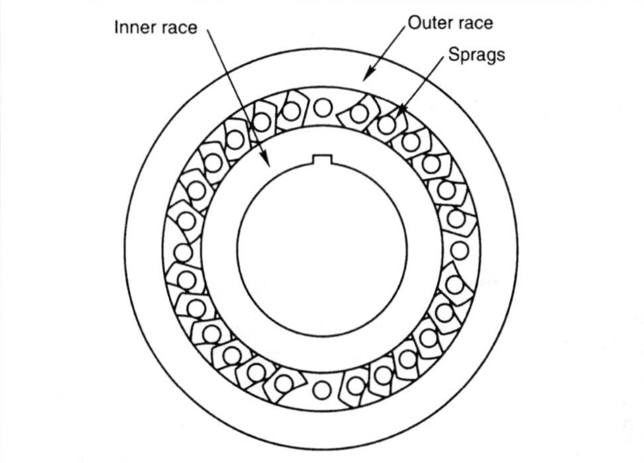

Figure 5-18: Sprag clutch

Ramp and Roller Clutches

The ramp and roller clutch provides performance similar to the sprag clutch, but with a different internal mechanism. In one direction, the roller moves along the ramp and wedges between input and output. In the other direction, the roller moves back on the ramp into an open space, disconnecting input and output. Therefore, it provides the same overrunning function as a sprag clutch (see Fig. 5-19).

Both sprag and ramp-roller designs can be used as a braking mechanism. If the outer race is mounted stationary, the unit will freewheel in one direction and lock in the other. This is often called a backstopping function, as it stops loads from moving backward.

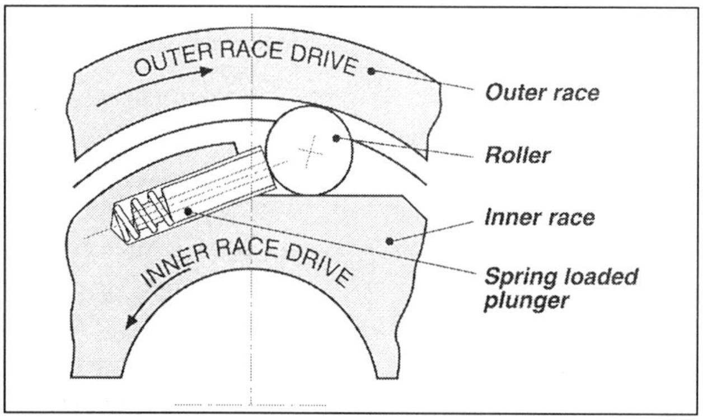

Figure 5-19: Ramp and roller clutch

In addition to backstopping, both sprag and ramp-roller designs are used in overrunning applications where the input and output are at two different speeds. Back-up drives or creep drives are common overrunning applications. In Fig. 5-20, a dual drive system is shown with clutches to the left and right of a blower. When the left side motor is running, that clutch drives and the other overruns, disengaging the engine and gearbox on that side. When the right side engine is running, that clutches runs and the other unit overruns.

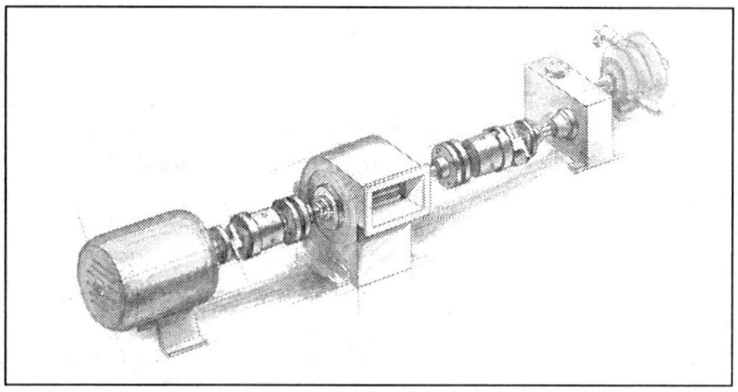

Figure 5-20: Typical dual drive arrangement using overrunning clutches

For indexing functions, when the input member is rotated and then reversed, the output member will provide an indexing start/stop function.

Centrifugal Clutches

Centrifugal clutches are an additional example of a fully mechanical design. These units consist of an input hub with friction material shoes held in place. As the unit rotates, centrifugal force pushes the shoes against the outer member or drum. Greater speed causes greater engagement and thus greater torque transmission (see Fig. 5-21). Torque rise in centrifugal units is gradual so that the effect is to accelerate the load more gradually than is the case for sprag or ramp-roller units where engagement is immediate.

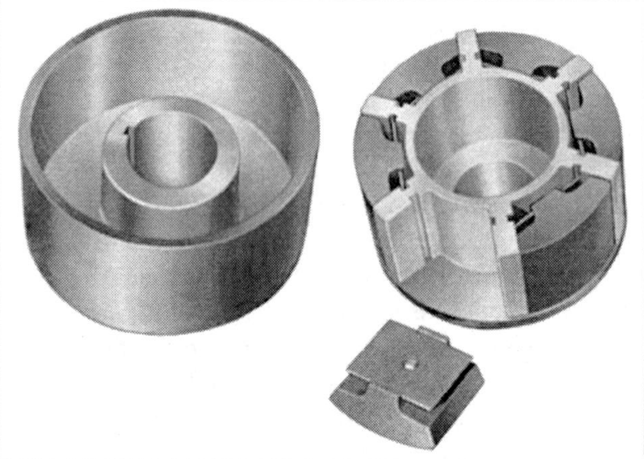

Figure 5-21: Centrifugal clutche

Centrifugal clutches are commonly used on high inertia loads such as large bulk conveyors, mixers, or pumps where it is best to allow the motor to accelerate unloaded and introduce the load more gradually.

Actuation Methods

Clutches and brakes are generally available in a broad range of ratings. Fractional HP up to around 50 HP is common for industrial clutches with engine-powered applications exceeding 2,000 HP. Unit size and torque are key selection criteria, but method of actuation will be an important consideration as well.

Mechanical Actuation

Manual actuation is used where an operator will mechanically control engagement. Actuation, whether by hand or by foot, is commonly accomplished by rods, cables, levers, or cams. Compound linkages are used on larger equipment. This method is usually practical only when the actuating mechanism can be located near the clutch or brake. The major operating advantage of mechanical actuation is the feel of engagement afforded the operator. Tactile feedback from a pedal or lever helps the operator gauge the amount of braking force applied or the degree of clutch slip obtained.

A disadvantage of this method is the need for an operator. Since actuation depends upon human strength, actuation force is limited to around 75 pounds. This limits not just clamping force and torque, but response time and cycle rate as well. For this reason, manually actuated clutches and brakes are commonly restricted to vehicular applications, lawn and garden implements and small industrial applications such as hoists and cranes.

Electric Actuation

Electrically actuated clutches and brakes offer the potential for high cycle rates up to 300+ per minute in low inertia applications. Electric actuation may be considered where remote control is desired or where signal following is desirable. Electric controls are generally simple to install and actuate.

Electromagnetic Actuation

Electromagnetically actuated clutches and brakes are controlled by a wire coil that is embedded into a steel shell. When DC voltage is applied, magnetism is created that will cause the two friction surfaces to clamp together where the friction transmits input to output.

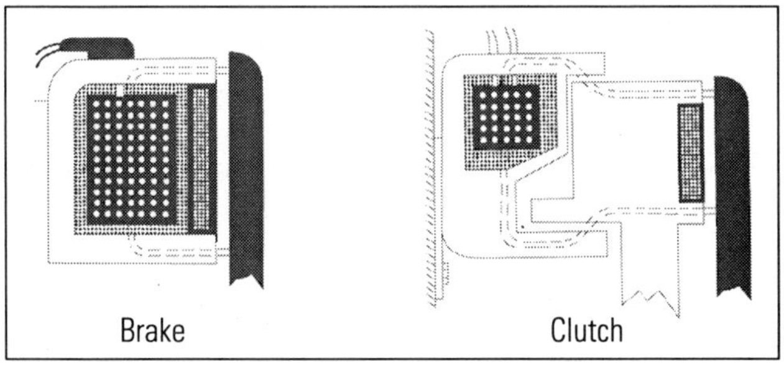

Figure 5-22: Electromagnetically actuated brake and clutch

In the case of an electrically engaged tooth clutch, the magnetism pulls together the two tooth sections where the teeth transmit the torque from input to output.

Pneumatic Actuation

Pneumatic actuation is widely used in industrial applications, especially in higher capacity systems. Compressed air systems are common in most factories.

Air pressure to 100 PSI either acting on pistons or inflating tubes engage or disengage the working surfaces. Some units are air-engaged while others are air-released. Combination clutch/brakes often use air to engage the clutch, while using spring pressure to engage the brake when air pressure is not present.

Pneumatically engaged, water-cooled brakes are used in applications such as steel or paper mill roll unwind stands where high torque and high heat dissipation are required. They are also commonly used as cable tensioners for large ocean sites such as oil drill platform mooring cables. Water is circulated through a set of valves and pathways to provide added cooling. These units should not be confused with hydraulically actuated designs. In these cases, the water is a cooling agent, not an engaging agent.

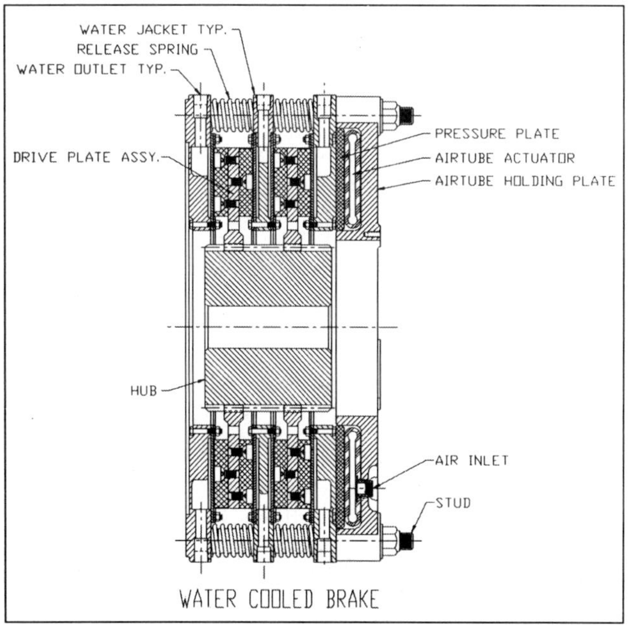

Figure 5-23: Water cooled pneumatic brake

Hydraulic Actuation

Hydraulically actuated units operate similar to pneumatic ones. Instead of using air pressure, hydraulic units use oil pressure to cause engagement of the friction plates. Hydraulic clutches and brakes generate higher torque per diameter than any other type. They are common in industrial and construction equipment.

Hydraulic units offer low response times and very smooth engagement. To assure rapid response, fluid lines are kept relatively large in diameter and short in length. Remote locations or special control requirements may prescribe additional fluid-control devices to maintain pressure. As in electric types, multiple friction elements may be immersed in oil for cooling.

When dealing with electric, pneumatic or hydraulic designs attention should be paid to the supporting equipment needed to control the clutch or brake. Electro-magnetic designs may require DC voltage and therefore a power supply or control to convert power to from AC to DC.

Appropriate switching will be needed to guide power to the proper unit at the proper time.

Pneumatic designs will require a properly plumbed pressure regulator, line filters, lubricators, control and exhaust valves to control units. Hydraulic units will also require pressure regulation, proper valves, and plumbing to work correctly. These additional requirements for each type of unit should be evaluated to ensure they are appropriate for the application and location.

Self-actuating Clutches

Some types of clutches and brakes are self-actuating. That is, they require no external device to cause them to engage. The sprag and ramp-roller clutches are examples of this type of clutch. As soon as rotation begins in the direction of drive, the unit will be engaged. Centrifugal clutches will begin to engage once the input has crossed over a preset speed, but the engagement will be gradual rather than immediate.

Common Application Notes

The basic purpose of all clutches is for the prime mover, whether electric motor, gas, or diesel engine, to continue running while the load is engaged and disengaged. When combined with a brake, the load can be repeatedly started and stopped while the motor/engine continues to run. An example of this function might be an automated conveyor in a warehouse distribution center that will start and stop a belt conveyor 15–20 times per minute as it sorts product. A different application would be a large pneumatic clutch/brake controlling a large stamping press. The clutch/brake ensures that the press cycles only when material is present and when it is safe to do so. Otherwise, the spring-set brake holds the ram in place.

In mobile equipment such as farm combines or road salt spreaders, the clutch allows the implement to be engaged when needed, but shut off when not needed, thus removing load from the engine. In such applications, a brake is not needed.

Conversely, there will be some applications such as large inclined bulk conveyors where a clutch is not needed but a holdback brake is needed so that the large load on the conveyor cannot backdrive if the drive train fails.

Selection Criteria

To choose the best possible clutch or brake for a given application, many factors must be considered. Specifying a clutch or brake on horsepower alone may yield a half-dozen potential candidates. Adding criteria such as cycle rate, availability of actuating method and preferred mounting will all have the effect of focusing choices. Data charts and section guides within manufacturers' literature and websites will help with selection for a given application.

The following questions can serve as a checklist to determine if all factors have been considered:

- How much HP and torque is required in the application?
- What is the maximum speed for the system?
- What type of engagement is needed? (rapid engagement, soft engagement)
- Are there cycle rate and/or thermal constraints to consider?
- How will the unit be actuated (remote, manual, automatic)?
- Are there environmental or contamination concerns?
- Is the site accessible for maintenance?
- What service life is required?
- What are the physical mounting considerations? (bore size, mounting surfaces)
- What added equipment will be needed for the possible actuating methods (e.g., power supply for electric, filter/regulator for pneumatic)?

Chapter 6

Conveyors and Components

Chapter Objectives

When you have finished Chapter 6, "Conveyors and Components," you should be able to do the following:

1. Summarize typical components employed on material and unit handling equipment.
2. Describe the important characteristics of the more widely used material handling equipment.
3. Differentiate between unit and bulk materials handling equipment.
4. Describe several varieties and styles of floor conveyors.
5. Differentiate the characteristics between gravity-powered and motor-powered roller conveyors.
6. Identify the characteristics of and applications for the following conveyors: (1) accumulation, (2) vertical reciprocating, (3) overhead conveying, (4) trolley overhead, (5) power and free, and (6) open and enclosed track.
7. Describe how monorail conveyors differ from other overhead conveyors.
8. Describe characteristics and applications for hoist and winches, motorized trolleys, and other overhead rail systems.
9. Explain how components must be selected for severe service applications such as in metals and food processing.
10. Describe characteristics and components used for bulk belt conveyors, screw conveyors, bucket elevators, pneumatic conveyors, and chip conveyors.
11. Identify the power transmission components typically used for palletizers, sortation conveyors, small package sorting, powered mechanical storage systems, automated storage and retrieval systems, and packaging machinery.
12. List considerations of special requirements that certain customers may demand.
13. Identify the organizations that set industry standards.

Introduction

A conveyor is a mechanical apparatus for moving articles or bulk materials from place to place. It is only one of the many items of equipment that relates to the movement, storage, control, and protection of materials, goods, and products throughout the process of manufacturing, distribution, consumption, and disposal that are classified as material handling equipment.

The history of conveyors can be traced back to the third century B.C. when Archimedes designed a screw conveyor for removing water from ships and irrigating farm land, but it wasn't until the latter half of the 1700s that the first belt conveyor was used to load grain onto ships. These early belt conveyors had wooden framework with belts of leather, canvas, or rubber that traveled over a wooden bed. The first patent for the roller conveyor was received in 1908, but it wasn't until 1919 when they were first used in the automobile industry that they began to prosper.

Although many of these early conveyor designs were rudimentary, they were successful enough to provide engineers with the incentive to consider conveyors as an economical and efficient way to move large quantities of material from one location to another. Today, conveyors have continually been improved to operate more efficiently and require less maintenance.

There are generally two broad categories of conveyor equipment distinguished primarily by characteristics of the material handled: unit (separated) and bulk (unseparated).

Unit item handling equipment transports individual items, large or small, as individual items usually not jumbled together. Examples include pharmaceuticals, watch components, auto body parts, TV picture tubes, printed circuit boards, packaged goods, barrels, and loaded pallets. Unit handling is the largest segment of handling equipment. A broad sampling of unit handling equipment is shown in Fig. 6-1.

Bulk handling equipment transports a homogeneous stream of material. The equipment handles chemicals, ores, sand, gravel, coal, and similar products. In some cases, material is handled in the open, as on bulk belt conveyors, and in other cases enclosed, as in pneumatic tubes and screw conveyors.

Unit Handling Equipment

Unit handling equipment includes a wide range of mechanical systems and components that convey, shift, orient, lift, lower, and turn items of various sizes and shapes. The items may be as large as locomotives or earth-moving equipment or as small as CD jewel box casings.

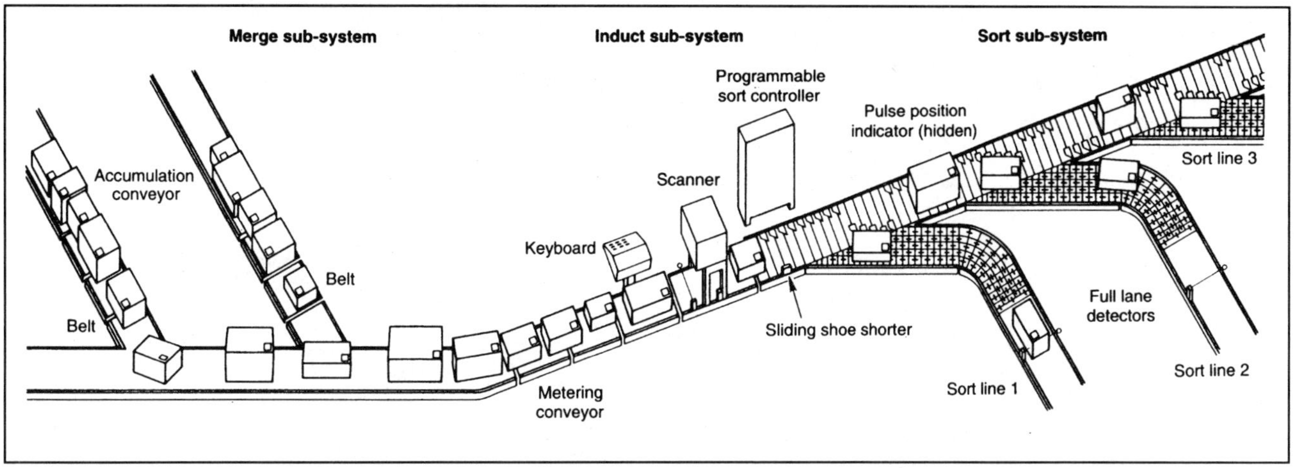

Figure 6-1: A group of conveyor types joined as one conveyor system. Types shown are belt, wheel conveyor, sliding shoe sorter, metering belts, and roller conveyor. The functions performed are conveying, accumulation, metering, merging, scanning, and sorting. This single system uses sensors for full lane detection, metering, or spacing systems, keyboard data entry, bar code scanning, position indication, and photoelectric detection. All are under the direction of a programmable sort controller.

Floor Conveyors

Unit handling conveyors are differentiated by a number of factors, the most basic being whether they are non-powered (gravity) or powered (live). They also are classified by mounting configuration and whether their loads can accumulate or not. Based on mounting configuration, floor conveyors are categorized as either in-floor or on-floor types. In-floor conveyors are mounted flush with the floor and have their drive power located below the floor, while on-floor types can be any conveyor that rests on the floor.

In-floor conveyors are available in a variety of styles and configurations. A towline conveyor, as shown in Fig. 6-2, is an example of an in-floor conveyor. It consists of a channel recessed in the surface of the floor with a chain running in it. The chain, in some cases, is a horizontal loop powering carriers continuously. In others, a vertical loop powers them for a distance and then returns unexposed below the floor. The chain is sometimes greased to minimize friction between it and the channel, or it may ride on spaced pairs of bearings that support it in the channel.

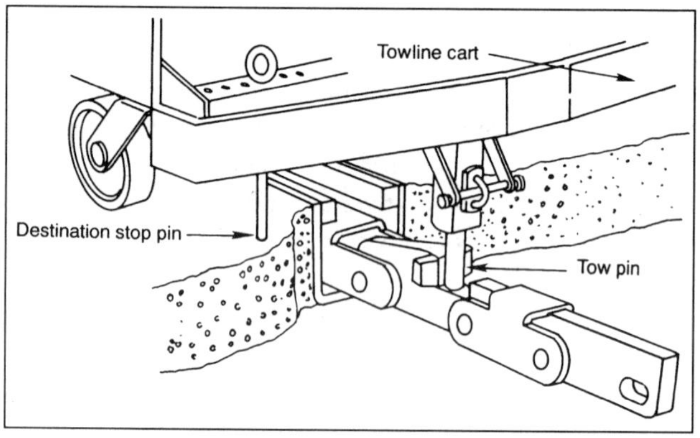

Figure 6-2: A low-profile or shallow in-floor towline conveyor. Stop pin selects spur, transfer, or stop point disengaging the tow pin.

The chain-in-channel is the active part of the conveyor. It pulls wheeled trucks or rolling platforms along on the surface of the floor. A pin interconnects the trucks or platforms with the chain. The pin can generally be engaged or disengaged manually, mechanically, or electro-mechanically.

Other in-floor conveyor types employ various flat-top conveyor chains that ride in channels in the floor. The hinged chains can be as wide as needed to support the conveyed items. Two or more can be used in tandem spaced some distance apart. This type is used on assembly lines for mobile equipment and to transport heavy finished coils of hot metal.

On-floor conveyors include roller conveyors, belt conveyors, and many other types and varieties.

Roller Conveyors

Falling into one of two categories, roller conveyors are either gravity (non-powered) or live roller (powered). In either case, the basic roller conveyor structure consists of side rails punched with spaced holes along their length. Rollers, assembled from steel or plastic tubes with bearings in each end, ride on shafts that fit between the rails. The shafts and the inner race of the bearing are usually stationary. The rollers either rotate on the bearings or are fixed to the shaft, and the shaft ends are mounted in flange or pillow block bearings. The rollers may be free running (see Fig. 6-3 [A]) or driven in any of a variety of ways. Power may be applied continuously or intermittently with devices employed to apply and remove drive power to the rollers.

Drive methods include direct drives, where every roller has a drive; roller-to-roller drives, where one roller is connected to the next by chain and sprockets, gears, or belts; sliding

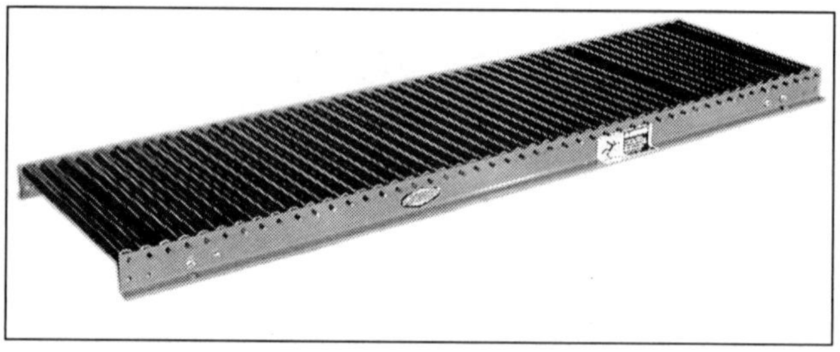

Figure 6-3 (A): Gravity Roller Conveyor

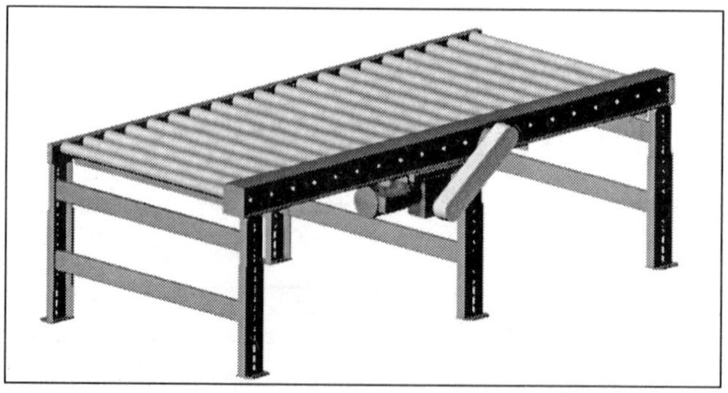

Figure 6-3 (B): Chain Driven Roller Conveyor

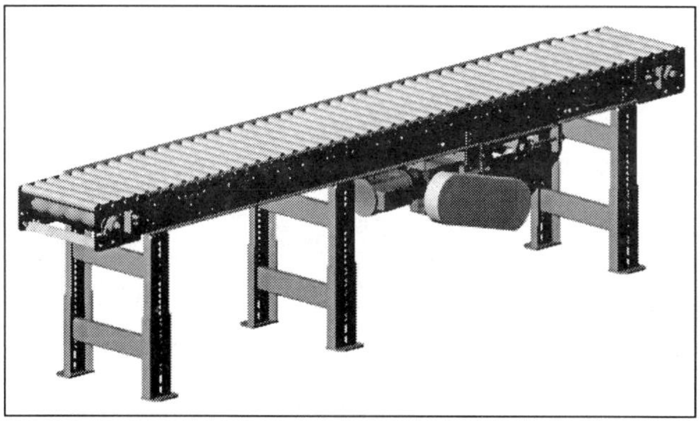

Figure 6-3 (C): Belt Driven Roller Conveyor

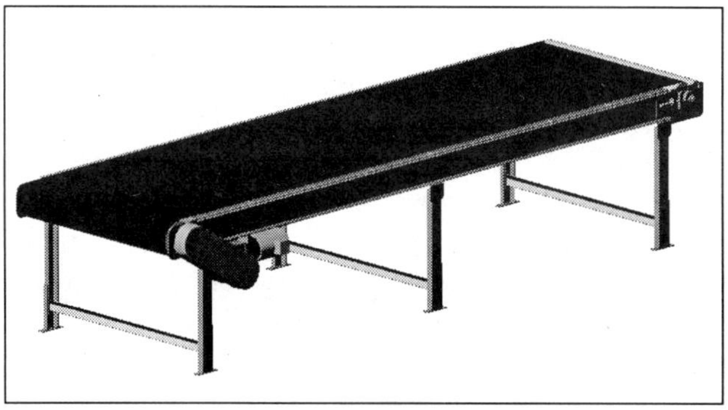

Figure 6-3 (D): Slider Bed Belt Conveyor

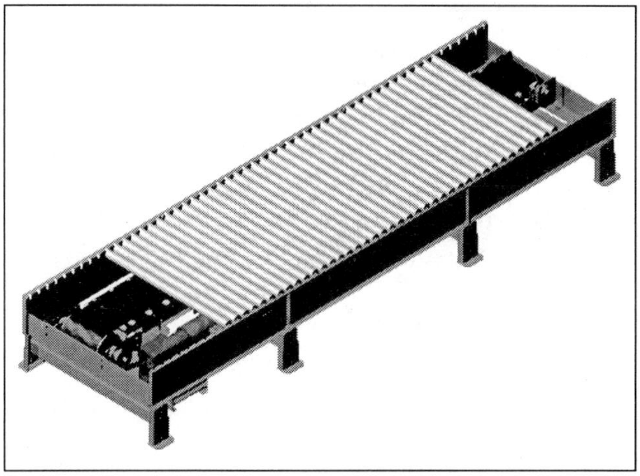

Figure 6-3 (E): Live Roller Conveyor

contact drives (see Fig. 6-3 [E]), where a belt runs below the rollers contacting them to provide the driving force; and self-contained motor-in-roller drives, where each roller has an internal drive.

When the rollers form the top conveying surface, the conveyor is called a roller conveyor, but when the belt rides on top of the rollers with the product riding on the belt, the conveyor is called a belt conveyor. The latter type is covered in greater depth later in this chapter.

When a belt runs beneath the rollers and contacts them to drive or power the rollers, the conveyor is designated a belt-driven live roller conveyor (see Fig. 6-3 [C]).

Chain-driven roller conveyors (see Fig. 6-3 [B]) may employ a double sprocket on each roller with power transmitted between rollers by short loops of roller chain, or there may be one sprocket per roller with a continuous chain powering all.

In some light-to-medium-duty applications, such as handling boxes in a warehouse, very long conveyors are driven by a single line shaft running beneath the rollers the length of the conveyor bed (see Fig. 6-4 [B]). The line shaft is supported by pillow blocks. Universal joints carry the line shaft around curves and spur conveyors and diverters, where it powers other actions. The line shaft powers the rollers through short belt loops. The belts run from a groove in the roller down and around plastic pulleys on the line shaft. The groove can be anywhere along the roller. To drive stub rollers or spurs, the belts may be applied between rollers rather than between the line shaft and the roller.
A roller conveyor may have only one drive per section. This can be a motor, speed reducer, pulleys and belts, or chains and sprockets. It may also be a gearmotor or shaft-mounted gear reducer and motor plus appropriate coupling devices.

On heavy-duty roller conveyors, such as runout conveyors used for handling coils, billets, and slabs in steel mills, there may be one drive or gearmotor per roller, and these may be electronically synchronized or servo controlled.

In some cases, power is selectively and intermittently applied to individual rollers or groups of rollers to stop and start the flow of product or to accumulate product close together in a line. Some of these configurations are called accumulation conveyors (see Fig. 6-4). Drive methods for the action include mechanical, a combination of air and mechanical, and electrical or electronic techniques,

Some roller conveyors are now being replaced by modular plastic chains. There are several retrofit kits available on the market to replace roller systems with these chains.

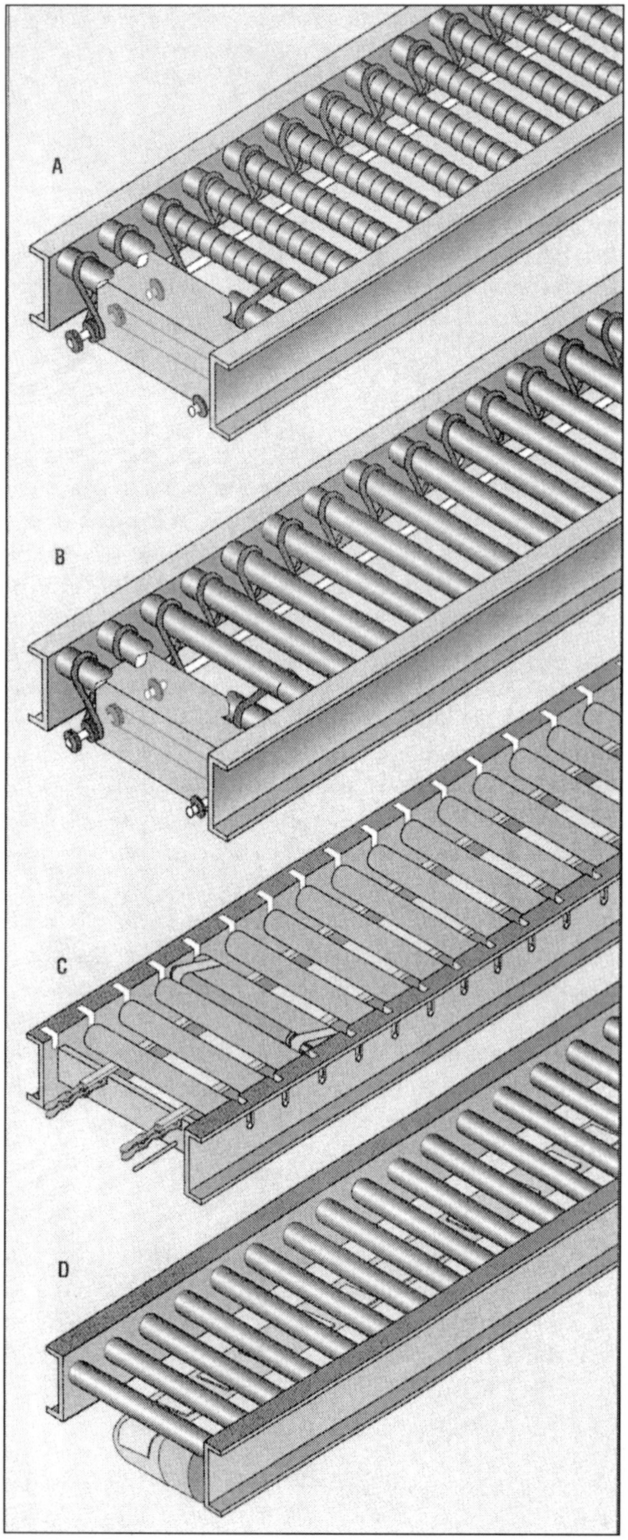

Figure 6-4 A: Methods of driving roller conveyors: (A) Free rollers on driven shafts; (B) Line shaft drive; (C) Tube-on-chain friction drive; (D) Padded or flat belt drive.

NOTE: For technical specifications on roller conveyors discussed in this section, refer to the following ANSI/CEMA Standards.

- 401 "Roller Conveyors – Non Powered."
- 403 "Belt Driven Live Roller Conveyors."
- 404 "Chain Driven Live Roller Conveyors."
- 406 "Line shaft Driven Live Roller Conveyors."

Commonly Used PT Components

Live roller conveyors employ a wide range of components in a variety of sizes and layout configurations. Bearings are in abundance and of every type and style used. These range from simple unground bearings (used on inexpensive rollers) to precision double-row ball bearings, tapered roller bearings, bushings, and pillow and flange block bearings.

Motors may be foot mounted and use a coupling for attachments to a gear reducer or attached directly to a gear reducer with a C-face mounting. The gear reducer may be an in-line or right angle drive. The gear reducer may be connected to the load using belts, roller chain, or flexible couplings, or directly mounted using a shaft mount reducer.

In elaborate multi-section conveyor systems, many drives are adjustable speed. Heavy start-up torque—changes in load, drag, and friction—is automatically compensated for by controlling the speed of the drive motor electronic frequency drive. The electronic frequency speed drive can also be easily programmed to match the speed of down-line equipment.

Pneumatic or electric brakes, clutches, and retarders may be used to control conveyor motion, product deflectors, and diverters.

Control and sensor systems for conveyors may be elaborate and under computer or programmable controller direction.

Belt Conveyors

A belt conveyor consists of a head and tail pulley with a belt stretched between them. Between the head and tail pulleys is a bed of polished steel or other low-friction material or spaced free-running rollers. The belt either slides on

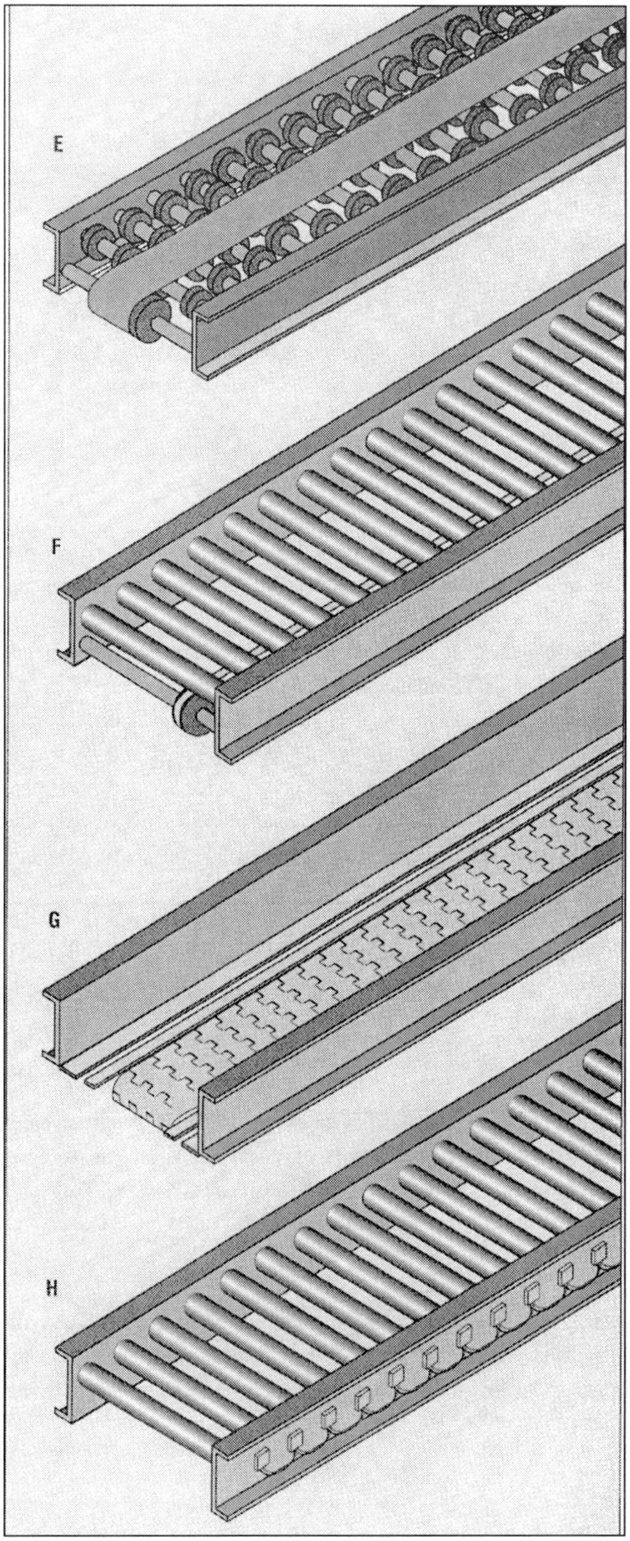

Figure 6-4 B: Methods of driving roller conveyors: (E) Flat belt between rollers; (F) Chain, cabled, or v-belt drive; (G) Chain flight between rails; (H) Individual motorized rollers.

the steel surface (slider bed) or runs on the rollers (roller bed). The conveying surface may also be made up of various kinds of belts running side by side. These may be round elastomeric belts, v-belts, round spring steel belts, or narrow flat belts.

Belts that slip on a solid polished surface instead of on rollers are called slider-bed belt conveyors (see Fig. 6-3 [D]). Slider-bed belts tend to slow down as weight increases due to friction between the back of the belt and the bed. Roller-bed conveyors are more popular than slider-bed types because they can carry greater weights with less friction.

Belt conveyors are most often used to move material up or down a grade. Here, the friction between the product and the belt keeps material in place. Belt conveyors sometimes include curved sections, which can be either slider or roller bed types.

NOTE: For technical specifications on belt conveyors discussed in this section, refer to the following ANSI/CEMA Standards 402 "Belt Conveyors" and 406 "Slat Conveyors."

Belts

The significant difference between roller and belt conveyors is the belt itself. There is a great variety of belt types and materials.

Conveyor belts may be made from rubber, PVC, woven fabric, steel, or woven wire. In some cases, it is thin and flexible as paper; in others, it is as stiff as 1/8-inch plywood. The material may have to withstand the heat of bakery ovens, the abrasive action of shot-blast tunnels, or the solvent spray of parts washers. High impact may also be encountered where material is dropped onto the belt. Belts also have to withstand continuous pull without stretching significantly.

Hinged belts consist of metal or plastic plates attached together to form a flexible belt that will bend around drive and return pulleys (see Fig. 6-4 [G]).

Stainless steel belts are applied where precisely controlled surface friction is called for. Flexible rubber and composition belts are generally purchased in bulk, cut to size, and joined by various lacing and splicing techniques.

Plastic modular chains are also another type of belt system. These use several modules bricked together to create a custom-configured chain. These types of belts offer low friction to minimize product damage, reduce energy consumption, and optimize product handling.

Accumulation Conveyors

Where products hitting or colliding with each other cannot be tolerated, or where this action must be light or gentle, either low- or zero-pressure live roller conveyors are used.

An example of non-contact accumulation is a line of automobiles at a traffic light. An example of contact accumulation is a pileup at the same light.

In low-impact accumulation, the rollers drive product into contact with each other. When the resistance is detected, drive power is removed.

In zero-pressure or zero-impact accumulation, the first item reaches a permanent stop. The upstream conveyor then removes power to rollers a few at a time as the items approach each other, minimizing pressure between them. Variations of this technique accumulate product but prevent contact between products. Accumulation conveyors handle loads from the size of CD jewel box casings to coils or stacks of sheet metal. Products have different handling characteristics and are handled in various volumes. These conveyors are configured for each situation. Some systems employ mechanical techniques while others use electrical or electronic. Stopping as well as starting considerations govern sensor and control design.

Vertical Conveyors

Some vertical conveyors transfer product in only one direction. Others can receive, carry, and off-load product in either direction. (None are built to elevator standards, nor are they intended to carry passengers.) The lifts can be built to any capacity and platform size.

Vertical conveyors employ a wide range of power transmission components including shafts, drums, motors, gearboxes, seals, bearings, wire rope, sensors, controls, and interlock switches.

Special Unit Handling Applications

Special conveyor applications are those that require components to operate under unusual conditions of water, dust, dirt, or temperature; to start up under full load; or to take extreme shock loads. The metals and food and beverage processing industries have many of these applications.

Metals Processing

The primary metals industries use conveyors to transport material between compression rollers while transforming material from red-hot billets to sheet form. The mechanical and electrical components are subjected to high heat, dirt, and shock loads.

Motors, gears, and bearings employed here often differ from standard ones in non-obvious ways. Motors, for example, may use double or labyrinth seals, and bearings often are double-row types using special high-melting-point grease.

Some drives spend their lives covered with dirt and debris. The result of poor cooling calls for motors with Class H or higher insulation. Couplings may be totally enclosed, grease purged to suit the application, and designed to take high-impact shocks.

Coil baskets on rod and wire extrusion lines, for instance, and the banding equipment that finishes the coil, operate in a high-temperature environment. Here, every rotating and sliding component and lubricant must be customized to the situation.

Remember that parts are cool when they are replaced or repaired. This is the opposite of how they operate, so operating conditions must be addressed before a replacement is selected.

Food and Beverage Processing

When considering conveying applications in the food and beverage industry, it is important to determine the specific needs of each application. The type of product it is handling as well as the environment it will be in will determine what type of conveying system and what type of materials to use in that system (see Fig. 6-5 and 6-6). Some items to take into consideration may be:

- Temperature.
- Chemicals (whether they are in the product or being used to clean the conveyor).
- Outdoors (UV light).
- Water or air flow requirements.
- Electrostatic environments.
- Noise concerns.
- Product sensitivity/product damage.
- Food applications – FDA approval.

Figure 6-5: FlatTop Beverage Conveyor.

For example, power transmission applications are subject to temperature extremes and extremely wet and dry conditions. Handling equipment moves products through fillers, warmers, ovens, water sprays, microwaves, nuclear radiation, and steam baths.

In many instances, the mechanical components must meet USDA or FDA approval for direct contact with the product. Regular washdown operations require careful consideration of the materials used in all conveyor belts and components.

In frozen food plants, handling equipment is subjected to a variety of process area conditions such as form, fill, and seal packaging lines, followed by sub-zero temperatures in blast freezing processes and holding freezers. Components that function perfectly at the normal range of ambient temperatures will definitely perform differently under sub-zero conditions.

Ovens use a variety of conveyors ranging from metal mesh belts to high-temperature composition belts. Often, the bearings for the rollers that support or drive the belt are placed outside the oven but still in a high-heat area. The bearings or pillow blocks may be standard; the lubricant and seals are not.

Autoclaves and pressure cookers usually don't incorporate power transmission components internally, but some special handling racks and carts that are loaded into them may. Drying and cooling lines, as well as wet pre-processing conveyors, also use components that either operates continuously in water or in close proximity to it.

In meat, fish, and poultry processes, steam cleaning is a standard requirement. In some cases, motors and drives must be epoxy coated or may require stainless steel or plastic enclosures. Electrical connection boxes require waterproof seals, and some electrical components will require special NEMA enclosures.

Food processes are difficult to shut down and, in the case of sub-zero conditions, difficult to maintain. Newly installed components must be normalized before they can be brought on line in sub-zero conditions. Properly specified motors seem to operate forever in freezer applications because their number one killer (heat) is not a problem.

Figure 6-6: FlatTop Food Conveyor.

Overhead Conveying Equipment

Overhead conveyors are just as common as floor conveyors. The various available types are segregated mostly by capacity. Overhead structures support both the conveyor weight and the loads. Overhead conveyors are often a single-rail design. This means that the rail must carry the total weight of the conveyor and the load.

Overhead Trolley Conveyors

I-beam rails for simple overhead trolley conveyors are suspended overhead. Trolleys with a bearing-mounted wheel on each side of the I-beam are interconnected and powered by chain links or cable. Loads are suspended from the carriers.

Overhead conveyor loops may be long and travel in up, down, and level planes. Some of these conveyors make turns by traveling around a wheel and others by traveling a curved track. They can be driven by the wheels at the turns or by a caterpillar drive.

A caterpillar drive consists of a strand of drive chain that has engagement spurs (sometimes called drive dogs) around the outside (see Fig. 6-7). The chain engages the overhead conveyor chain and drives it continuously. Take-ups and adjusters in the conveyor system remove the slack in the chain to keep the movement smooth. Sometimes, as in the case of wheel drives, the drives become the slack adjusters.

Overhead conveyors are used extensively in automobile manufacturing and fabrication shops because they leave the floor clear for other operations.

A variation, the so-called overhead towline, consists of trolleys and chains much like a common overhead conveyor. But instead of pendant suspension from the conveyor, towline carts are attached by a wagon-handle-like tow gear. These conveyors often pull hundreds of towline carts. The carts use heavy-duty casters or axle-mounted wheels with tapered roller or straight roller bearings and seals.

Power and Free Conveyors

P&F conveyors (see Fig. 6-8) consist of two overhead rails mounted in close proximity to each other. One rail is the power rail, with a chain driving continuously; the other is the free rail holding independent trolleys. The trolleys on the free rail run only when engaged with the power rail chain.

When a trolley, driven by the power rail, contacts another trolley or switch, it automatically disengages from the power chain and accumulates or travels a separate path. When a trolley disengages from another, it automatically re-engages the power rail.

The trolley rail can employ many configurations of load-holding fixtures, which are suspended from one or more free trolleys.

Multiple trolleys sometimes share awkward or heavy loads. And two conveyors often operate in parallel to handle large loads such as automobiles.

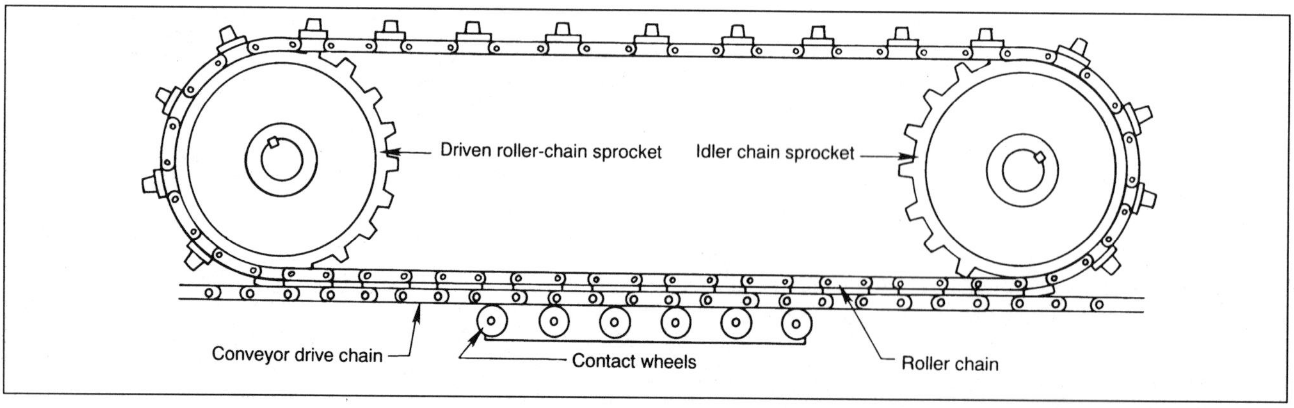

Figure 6-7: A caterpillar drive used to drive overhead and in-floor conveyor chains for tow line conveyors, overhead power & free conveyors, and overhead trolley conveyors.

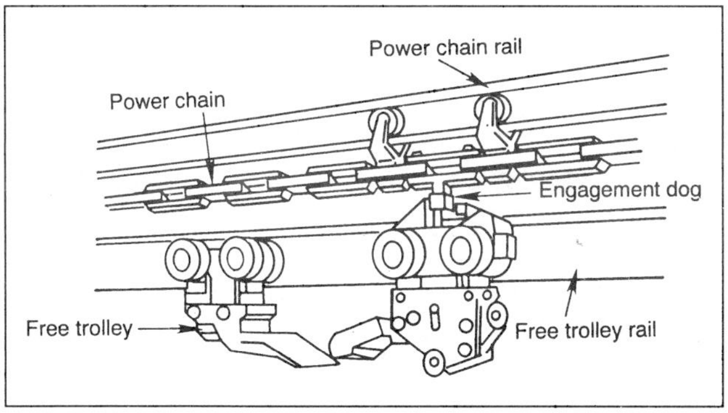

Figure 6-8: Power & free conveyor showing powered (top) and non-powered (bottom) rail with trolleys.

Complex systems of overhead power and free conveyors are found in the auto plants where large components are built and accumulated, awaiting release to assembly or other operations. These systems can be extensive with multiple levels, lift and drop sections, and every variation of horizontal and inclined conveying accessory. They are used primarily to buffer and accumulate product overhead out of the way of floor-based equipment.

Open- and Enclosed-Track Overhead Conveyors

Two types of construction are used for most overhead conveyors: open and enclosed track. Both are used in washing, painting, and other processes. They differ from power and free conveyors in that they are continuously moving and are lighter duty with a lower capacity per trolley. Both the chain and transport trolleys are enclosed in a single track.

Open-track conveyors consist of a track or channel beam rail on or between which trolleys travel linked by a chain or cable. Each trolley runs on open, sealed, or fitting-lubricated bearings.

Some of these conveyors use spot or spray lubrication systems to automatically lubricate the trolley bearings. Enclosed track overhead conveyors have a track that largely surrounds the chain and trolleys with a bottom slot for the trolley load fixture. These are often lubricated by brushes or spray lubrication.

NOTE: For technical specifications on overhead conveyors discussed in this section, refer to the following CEMA Standard: 601 "Overhead Trolley Chain Conveyors."

PT Components

Power and free conveyors employ a wide variety of power trolleys, free trolleys, and chain of various types. The chain is driven at multiple points by synchronized drives. Many are large horsepower applications. Multiple drives are also used. Variable speed drive controls synchronize the drives to equalize chain pull.

This type of conveyor uses many varieties of ball and roller bearings, seals, motors, gearboxes, lubrication systems, chains, and sprockets. Couplings, bushings, brakes, and clutches will also be found along with shafts and keys.

Straight overhead conveyors use most of the same components plus cable, clutches, backstops, and safety stops.

Monorail Conveyors

Differing from other overhead conveyors, monorail conveyors employ I-beam rather than rolled or fabricated tracks.

The rail carries a chain and trolleys. Loads are suspended from the trolleys, and the trolleys are spaced to suit the loads.

Automated Electrified Monorails

So-called AEMs are a type of monorail conveyor where each carrier is powered independently of the others. Each trolley rides on an I-beam rail or an extruded aluminum rail, picking up power from a bus system in or adjacent to the rail.

Each powered trolley consists of a gearmotor and drive wheel, which contacts the rail, driving the trolley. The powered trolley may have follow-on or attached non-powered trolleys to share the load (see Fig. 6-9).

The trolleys may do more than transport. They sometimes include hoists that lift and lower the load into dips, baths, or processes. They may also interface with floor processes, including floor-running conveyors, by synchronizing their speed and transferring loads.

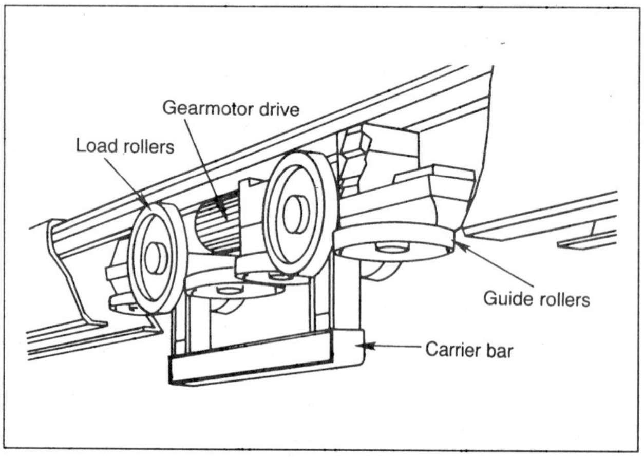

Figure 6-9: An automated electrified monorail (AEM) in a totally enclosed track showing drive and guide system.

AEM carriers have built-in travel, direction, and speed controls including diagnostic circuits. They require only the power and communication conductors in the rail itself.

Carousel Conveyors

Sometimes arranged in an ellipse, carousel conveyors are suspended from floor-standing frames. They are generally light duty and often found in dry cleaner shops to hold garments. The trolley carriers in some versions support suspended wire shelf sections. The shelves are usually used for dynamic inventory storage but may also be used for work-in-process and hot test of small electrical components.

Gearmotors drive the conveyors through chains and sprockets. There is usually a soft-start or fluid coupling and brakes in the drive line. These storage conveyors are built for repetitive start/stop operation.

Other Overhead Equipment

Heavy items are often moved over entire bays or sections of a plant by three-axis overhead equipment such as overhead bridge or monorail cranes.

Overhead bridge cranes ride on rails mounted along the side of the bay. The bridge carries one or more hoists and auxiliary hoists.

These cranes can be controlled by an operator in a cab suspended from the bridge, by an operator or rigger walking with the load at floor level, using a pendant control, or by an operator using radio-remote control.

A monorail crane setup is similar, but here a single cross beam carries the hoist. These are usually controlled by an operator on the floor.

The hoisting medium is wire rope and pulley block hooks, but there may be one or more hooks on each hoist and more than one hoist on each monorail crane.

Hoist capacity for bridge and monorail cranes varies by type. Bridge cranes can handle up to 100 tons. Monorail cranes can handle up to 20 tons. Most cranes use hoist units incorporating specially designed gearing. In some cases, drum brakes, wire rope drums, and controls on the larger cranes are specially built for the load and the industry. For many cranes, only exact replacement parts are acceptable.

A wide range of below-the-hook grabs and lifters is used in various industries. These may be quite elaborate with mechanical drives and controls to open tongs, tilt coils, grab bundles, rotate the hook, and many other special actions. They are used primarily in areas that are serviced exclusively by cranes as opposed to floor-running equipment.

Hoists and Winches

Hoists are used in a large number of manufacturing plants. They serve as lifts in machine loading, in assembly, and for handling of one-at-a-time lifts. Hoist capacity ranges to about 10 tons, although some are larger.

Hoists are suspended from a fixed overhead mount, hook, or a trolley running on an overhead rail. The rails may be part of a transport system or simple jibs cantilevered from a building column.

Packaged hoists are portable pre-assembled units. They may be manually, electrically or pneumatically powered, and use wire rope, link chain, or roller chain as the lifting medium. Package hoists are usually an assembly of proprietary components; only exact replacement parts are satisfactory.

Winches are most often floor mounted and used for various lifting and pulling operations. They consist of motors, gearing, cable drums, brakes, and a variety of pulleys, tackle blocks, drum switches, and pendant controls. Their open construction exposes many common power transmission components.

Motorized Trolleys

In some cases, hoists move along a beam or monorail under load driven by a motorized trolley. The trolley consists of a framework with wheels two on either side of an I-beam or rail. One or both sets of wheels are powered by a gearmotor. The motor may also incorporate a built-in brake. The trolley has a link or suspension point where any type of hoist or hook device can be attached. A push-button pendant controls both the hoist and the trolley direction and speed.

Trolley, Beam Lift Sections, and Spurs

Overhead rail systems used in, for instance, primary metals manufacturing often travel from one bay to another and from one level to another. This requires special structures to move entire rail sections, loaded or unloaded, as well as elevate and lower them.

These rail sections may also rotate on large wheel-type mounts to align the rail in another direction.

The wheel mountings are often a concentration of power transmission components. The central pivot point is a bearing-mounted shaft that can be 5 in. in diameter, and the wheel may be driven by a large sprocket and gear drive.

Monorail carriers or trolleys are generally moved vertically with an elevating lift section. Such sections ride on vertical guide rails aligning the horizontal load rail with one or more levels during the lift. Special hoists or shaft, sprocket, and chain arrangements are used to lift and lower the elevating section. Precise position often calls for servo controls or shot-pin alignment systems.

The hoist mechanisms usually use chain but may also use wire rope. The mechanisms include a variety of standard power transmission components such as pillow blocks, mounted bearings, shafting, sprockets, drives, brakes, drums, clutches, controls, and motors.

Bulk Handling Equipment

Handling bulk materials differs markedly from handling discrete items. Bulk materials are usually dirty, dusty, corrosive, and often lumpy, whether wet or dry.

All types of agricultural products are handled in bulk. These include sugar, grains, fruits, vegetables, and farm chemicals of all types. Bulk materials also include coal, dirt, rock, gravel, and both processed and unprocessed ores.

Bulk materials are generally moved by conveyors that were especially designed to suit the characteristics of the material. Texture and fragility of bulk materials play a big part in the design of handling equipment.

There is a wide range of conveyor designs available for the handling of bulk materials.

The troughed belt is the common method for handling ore and minerals mined from underground or open-face mines. By changing belt width, thickness, and speed of travel, various materials may be moved at different rates depending on the user needs.

The CEMA book titled "Belt Conveyors for Bulk Materials" is considered by many to be the belt conveyor industry basic handbook. While written for the experienced engineer, it provides detail subject areas critical to selecting bulk handling belt conveyors and is a valuable resource for all who must design or approve projects involving the handling of bulk materials with belt conveyor systems.

Apron conveyors with steel pans, either formed or hinged, are used in handling large, sharp, or abrasive materials that would damage rubber belts. Like troughed conveyors, they can be of different widths, chain speeds, and thicknesses to meet application requirements.

Powdered bulk materials are handled by various types of equipment, including troughed belts, vibratory conveyors, screw conveyors, pneumatic conveyors, and drag chain or en-masse conveyors. Granular and similar bulk products are handled in similar fashion and may also be moved vertically with bucket elevators.

Plastic modular chains are also used to convey bulk materials. The modular chains may include pusher flights and integrated side rails to contain bulk product such as snack food, vegetables, etc.

In most applications, moving a material from one point to another is the primary consideration. In specialty applications such as pharmaceutical and chemical industries, other factors come into play, for example, mixing and blending, influencing the conveyor selected.

Figure 6-8 (A): Troughed belt conveyor

Bulk Belt Conveyors

Trough belt conveyors are used for a variety of applications, a common application being the movement of rock and minerals in the mining industry (see Fig. 6-10 [A]). Conveyors may move the material from the mine to the rock crusher, and then from the crusher to a shaker screen, which is a device that sorts the rock into different sizes. From there, troughed belt conveyors move the material into piles of like material.

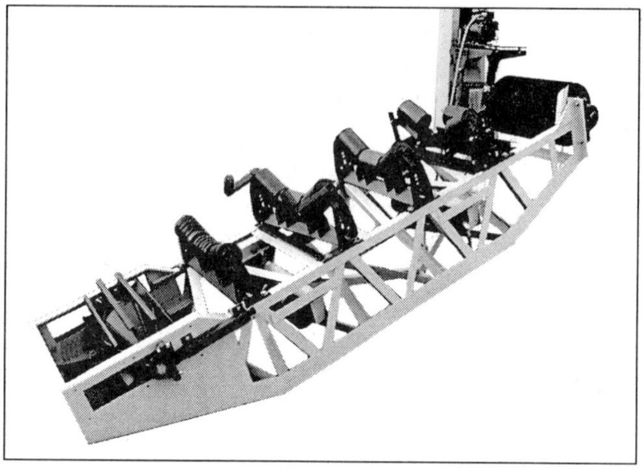

Figure 6-8 (B): Typical Troughing Conveyor

Trough belt conveyors (see Fig. 6-10 [B]) are constructed of a steel framework with a flat or the slightly crowned drum (head) pulley (see Fig. 6-10 [C]) mounted to the drive end.

The other end of the conveyor system will have a tail (wing) pulley (see Fig. 6-10 [D]) attached. The "wings" vibrate the belt and dislodge material adhering to the belt.

Mounted to the top of the framework are replaceable troughing idlers (see Fig. 6-10 [E]). They are made of a frame supporting a horizontal roller in the middle and a roller on each side, which slant up either at 20 degrees, 35 degrees, or 45 degrees to keep material on the belt. As the belt passes over the head pulley, it returns in the opposite direction beneath the conveyor supported by flat return idlers (see Fig. 6-10 [F]).

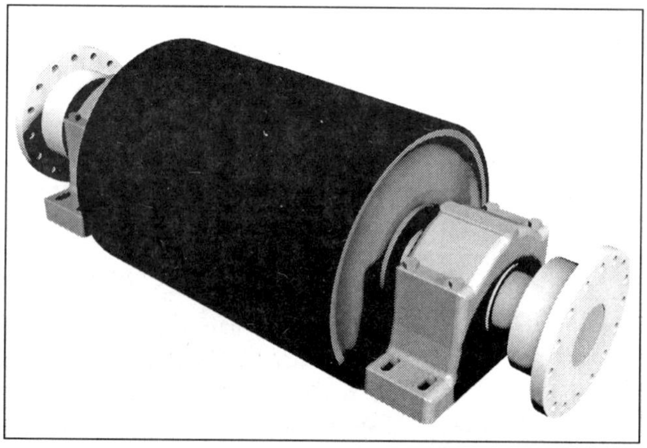

Figure 6-8 (C): Head Pulley Assembly

Normally, these conveyors are driven by electric motors driving a gear reducer. Accessories to these conveyors include belt edge guide rollers, belt tensioning devices, and belt cleaners. Fasteners are sometimes used to join or repair belts, or they can be vulcanized and spliced endless.

NOTE: For technical specifications on Bulk Handling Equipment discussed in this section, refer to the following CEMA Standards and Publications: Belt Conveyors for

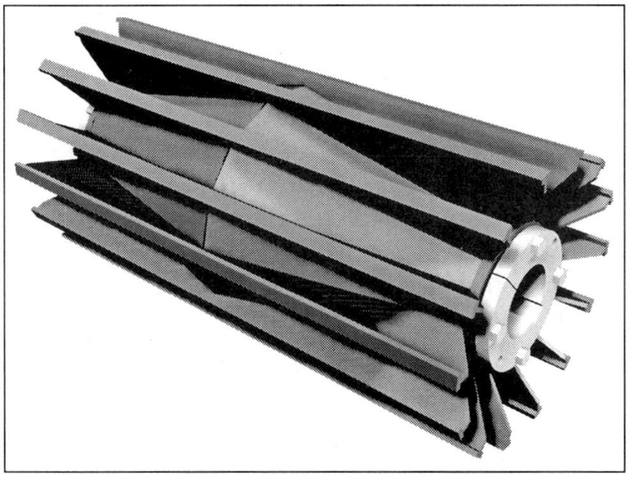

Figure 6-8 (D): Winged Tail Pulley

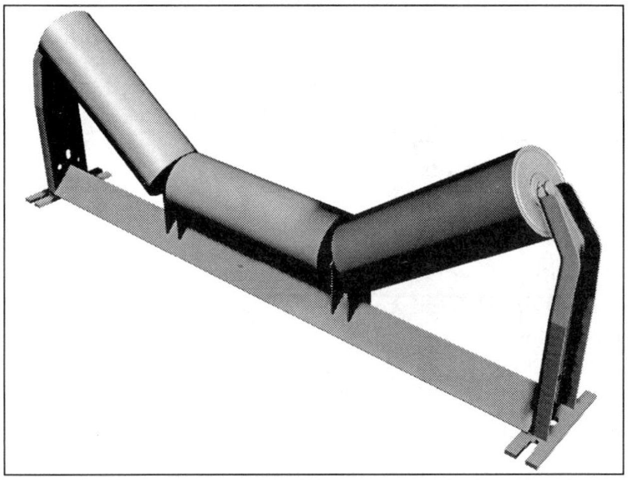

Figure 6-8 (E): Troughing Idler

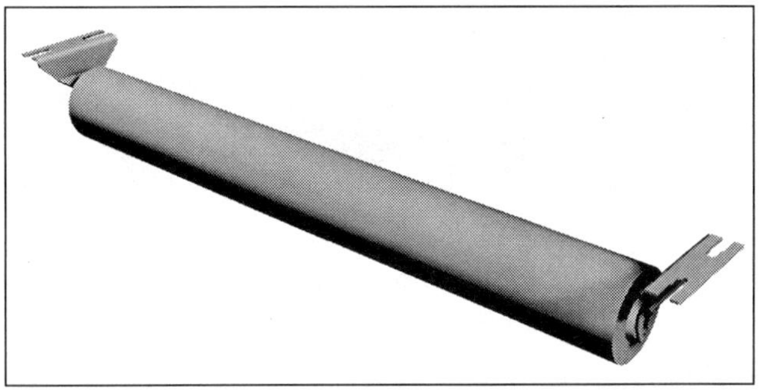

Figure 6-8 (F): Flat Returning Idler

Bulk Materials; 300 "Screw Conveyor Dimensional Standards;" 350 "Screw Conveyors."

Screw Conveyors

Screw conveyors are used to transfer bulk materials in and between processes, and sometimes simultaneously blend and meter the product. The screw rotates inside a U-shaped trough or tubular housing and moves the product forward with each revolution. Screw conveyors are used extensively in many industries including food, grain handling, chemical processing, municipal waste treatment, and minerals processing.

Traditional screw conveyors (see Fig. 6-11 [A]) are available in specific CEMA diameters up to 24 in. Larger and nonstandard diameters can be custom-designed based on the particular application. Screw conveyors are modular in design and multiple sections may be joined together to provide the appropriate length required for a precise application.

Figure 6-11 (A): Screw Conveyor with Product

Extending the conveyor length through modularity is accomplished by utilizing "hanger" or internal bearings. Internal bearings are required at each conveyor section to support the screw in the trough or housing and to prevent excessive deflection over the length of the screw. These internal bearings utilize bushing or journal type bearings. Typical bearing materials are UHMW plastic, oil-impregnated hardwood, cast iron, ceramic, or bronze depending on the compatibility of the bearing material to the product

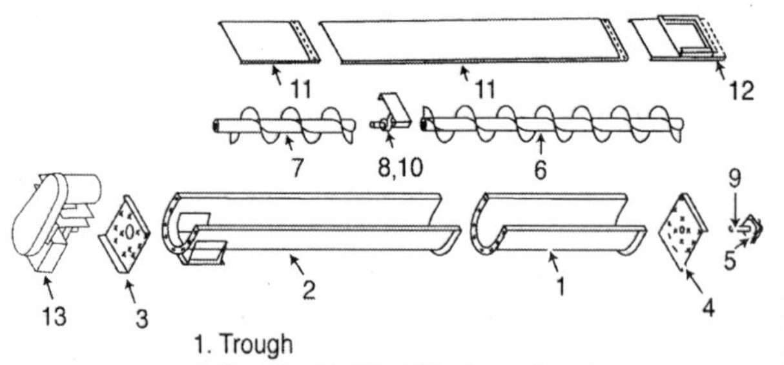

1. Trough
2. Trough with Fitted Discharge Spout
3. End Plate for Screw Conveyor Drive
4. Trough End Plate with Ball Bearing
5. Seal Plate
6. Screw
7. Screw with Bare Pipe at Discharge End
8. Misc. Coupling Bolts & Various Assembly Bolts
9. Tail Shaft
10. Hanger with Bearing and Coupling Shaft
11. Flanged Covers with Buttstrap
12. Inlet
13. Screw Conveyor Drive, Motor Mount, V-Belt Drive and Guard.

Figure 6-11 (B): Exploded View of Typical Screw Conveyor

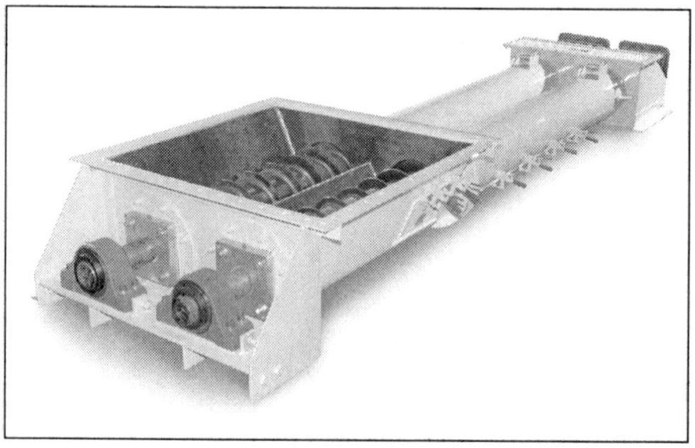

Figure 6-11 (C): Twin Screw Feeder

being conveyed. Ball or roller bearings are typically not used for internal bearings because of the severe contamination presented by the product being conveyed within the trough.

Screw conveyors are used in many applications. For example, in grain handling storage silos, the entire bottom of the silo may contain a row of open conveying screws. The screws form a "live bottom" for metering the grain from the storage silo. The "live bottom" design prevents bridging or voids in the stored product, which might prevent flow.

Screw conveyors are chosen for applications that take advantage of their special features such as total sealing, blending in transit, and the fact that they can convey vertically. While the screw conveyor is basically simple in design, the selection and application is much more complicated and should be referred to an experienced designer. The CEMA Standard No. 350 (ANSI/CEMA 350-2003) "Screw Conveyors" provides acceptable engineering and application practices. Various types of bearings, motors, drives, v-belts, chain, gate operators, shafting, and couplings are applied by the designer to complete the screw conveyor selection process.

Screw Conveyor Flighting

Screw conveyor flighting can be manufactured using two different methods. One method is by forming a flat bar into a continuous helix. Flighting formed by this method is called helicoid flighting. Helicoid flighting provides a continuous one-piece construction with a work-hardened, smooth finish. Another method is by cutting out disks from flat plate and forming each disk into a helix of approximately one revolution. Flighting formed by this method is called sectional flighting. A continuous helix is made by joining a number of sectional flights together.

Special Flight Types

In addition to transporting bulk materials from one location to another, screw conveyors can also simultaneously mix various materials together by utilizing special modifications to the flighting.

Cut screw flights, Fig. 6-12, have special notches for moderate mixing of bulk materials, and they can also be used for breaking up lumps.

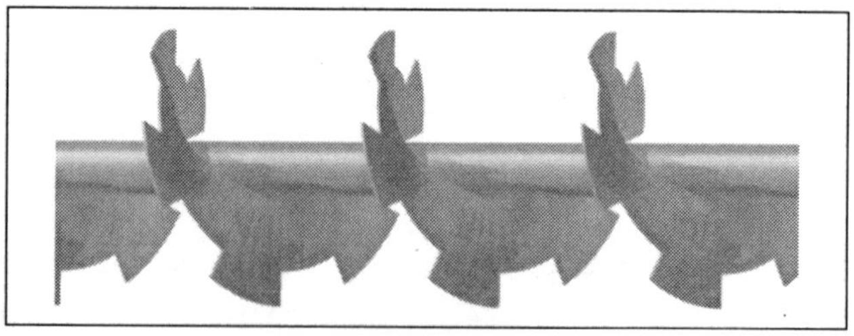

Figure 6-12: Cut Screw Flights

Cut and folded screw flights, Fig. 6-13, have a tab formed or welded on the cut screw flight for lifting and aggressively mixing bulk materials. The folded tab improves aeration and agitation of bulk materials.

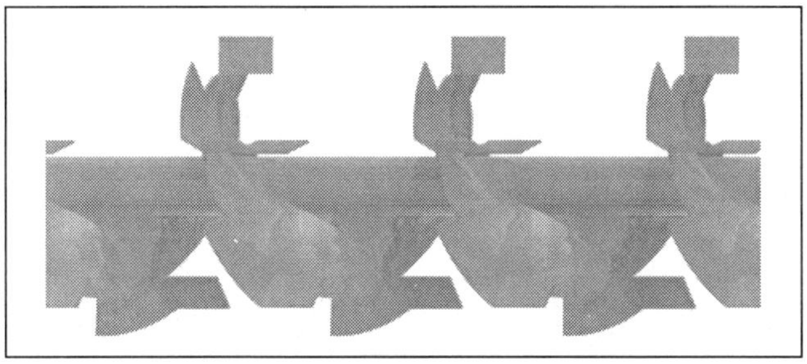

Figure 6-13: Cut and Folded Screw Flights

Regular screw flighting with mixing paddles, Fig. 6-14, have fixed or adjustable paddles that can be located every 90 deg. around the diameter of the center pipe. The paddles are used to retard the flow of material and provide moderate mixing. Mixing action can be controlled by adjusting the angle of the paddles on the center pipe.

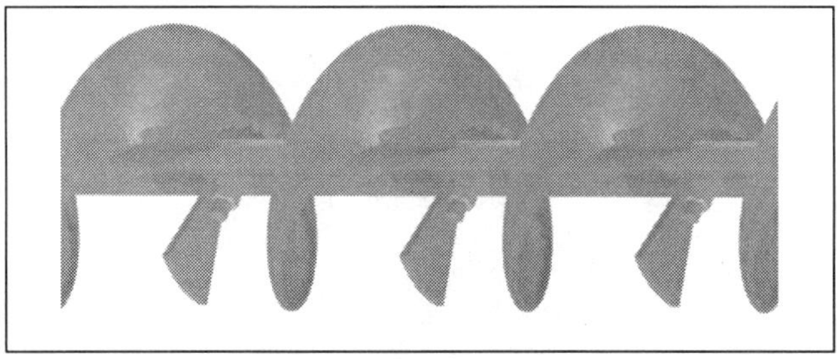

Figure 6-14: Regular Screw Flights with Mixing Paddles

Ribbon-screw flighting (see Fig. 6-15) consists of helical ribbon flighting attached to the center pipe using round bars or integral legs that are part of the ribbon design. Ribbon screws are used for mixing sticky or viscous materials or where the materials tend to adhere to the flighting in the corner where the flight meets the center pipe.

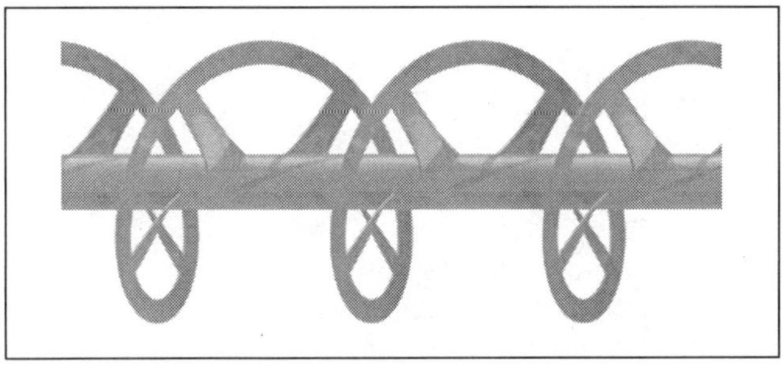

Figure 6-15: Ribbon Screw Flights

Inclined and Vertical Screw Conveyors

In addition to horizontal conveying, the versatile screw conveyor can be designed to operate at inclines up to 45 deg. and even vertically. Modified designs are available that improve the conveying efficiency of inclined screw conveyors as the vertical angle increases. For example, shortening the screw flight pitch enhances the carrying angle of the flight on which the material is conveyed. Conveying a bulk material up an incline using a screw conveyor is typically a cost-effective solution and requires a minimal amount of space.

Nevertheless, several key factors must be considered when designing inclined screw conveyors. As the angle of incline of a screw conveyor increases, the available capacity decreases. The screw conveyor flighting becomes less efficient at moving the bulk material forward as the degree of incline increases. The bulk material slips off the flighting and falls back down the incline of the conveyor. This reduction in capacity causes material turbulence and tumbling. The solution requires additional speed and horsepower to overcome the fallback of material and to overcome gravity.

Several design rules have been developed to overcome many of the issues associated with inclined screw conveyors. These design rules are:

1. Limit the use of standard screw conveyor components to inclines of 15 deg. or less.
2. Use close clearance between trough and screw. The close clearance will limit the amount of material fallback.
3. Increase the speed of the screw conveyor. The increased speed will aid in pushing the material up the incline.
4. Shorten the pitch of the screw flight. As previously mentioned, the screw becomes more efficient as the pitch is shortened. Screw speed must be increased to compensate for the shorter screw flight pitch.
5. Eliminate intermediate hanger bearings. Hanger bearings can be an obstruction to the flow of material and should be eliminated.
6. Use tubular housings. Standard U-troughs allow material to fall backwards over the top of the rotating screw. Tubular housings contain the material and increase conveying efficiency.

Vertical screw conveyors offer many advantages for conveying bulk materials in a vertical path. They require less space and are very compact when compared to bucket elevators or other elevating conveyors. Figure 6-16 shows a typical vertical screw conveyor consisting of a conveyor screw rotating in a vertical tubular housing with an inlet on the lower end and a discharge at the upper end. The drive is typically located at the top.

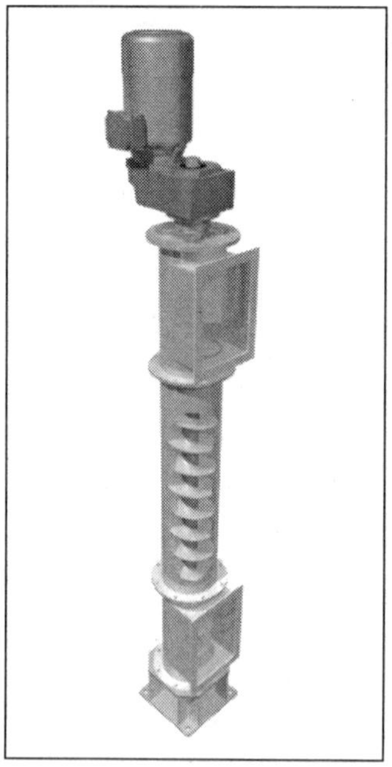

Figure 6-16 : Typical Vertical Screw Conveyor

The method of feeding the vertical screw conveyor is important to the success of the operation. A horizontal screw feeder is required to provide a controlled, uniform flow rate of material to the vertical screw conveyor. The speed of the vertical screw conveyor is constant. Changes in material flow rate are made by varying the speed of the horizontal screw feeder. If the horizontal screw feeder is turned off, the vertical screw conveyor will not completely empty itself. Some material will be left in the vertical screw conveyor. The amount of material left in the vertical screw conveyor depends on the characteristics of the material.

Bucket Elevators

One of the most efficient ways to vertically lift bulk materials is with a bucket elevator (see Fig. 6-17 [A]). A bucket elevator consists of a series of buckets attached to a belt or chain and driven with pulleys or sprockets located at the top and bottom of the unit. The buckets are loaded as they move past an inlet point located in the boot or bottom section. After loading, each bucket is raised vertically up and over the head pulley or sprocket, and the bulk material is ejected or dropped through a discharge chute in the head or top section. The buckets (see Fig. 6-17 [B]) can be made of malleable iron, nylon, fabricated steel, or other materials to suit the application. A bucket elevator is totally enclosed using intermediate housings or casing sections to provide dust-tight conveying. A wide variety of bucket elevator designs are available based on the characteristics of the bulk material and the process requirements.

Types of Bucket Elevators

Centrifugal Bucket Elevators – Bulk materials are loaded into each bucket by digging materials out of the boot section. The bulk materials are raised vertically and then discharged by centrifugal force over the head pulley or sprocket. Centrifugal bucket elevators operate at comparatively high speeds and are designed for bulk materials such as limestone, sand, cement, or various grain products. Centrifugal bucket elevators are great for bulk materials of relatively larger lump size. Buckets for centrifugal bucket elevators are reinforced at the edges and are designed for digging materials out of the boot section. The bottom of the bucket is rounded so bulk materials easily discharge.

Figure 6-17 (A) Cement Bucket Elevator

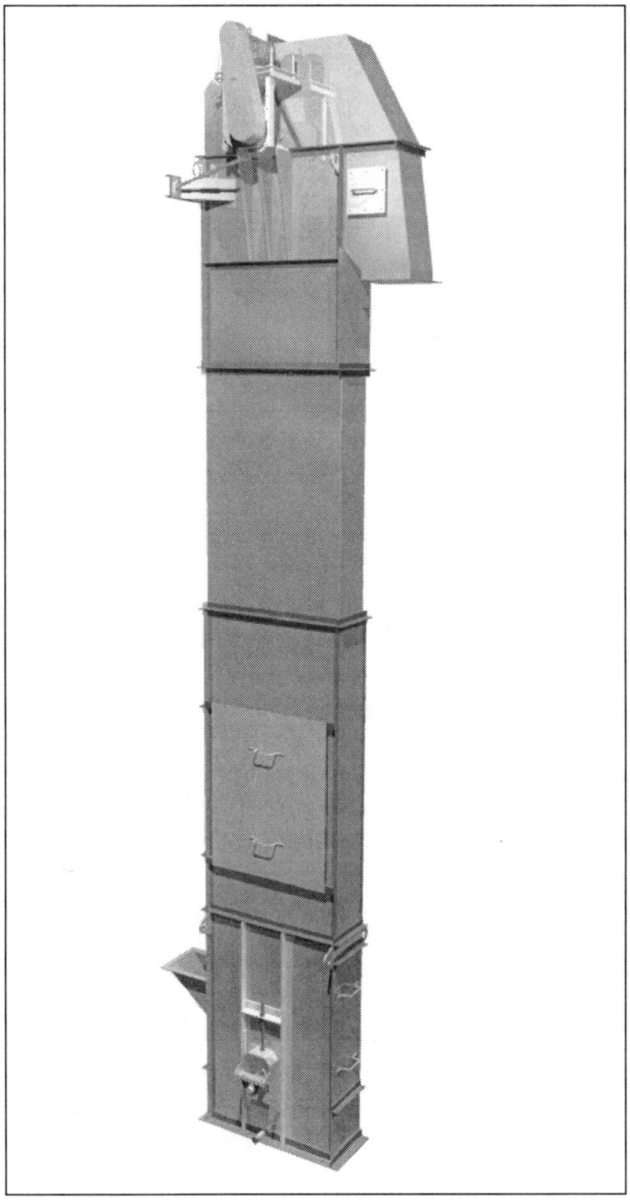

Figure 6-17 (B) Bucket Elevator

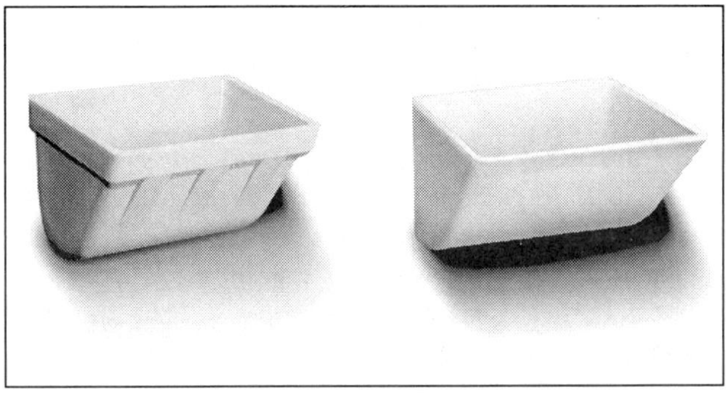

Figure 6-17 (C): Typ. Elevator Buckets

Continuous Bucket Elevators – Bulk materials are loaded directly into each bucket, and digging materials out of the boot section is not desired. The bulk materials are raised vertically and then discharged gently by gravity over the head pulley or sprocket. The buckets are spaced very closely together on continuous bucket elevators when compared to centrifugal bucket elevators. The shape and spacing of the buckets allow for the bulk materials to discharge from a bucket to the front of the preceding bucket, using the preceding bucket as a chute to gently discharge the materials. Continuous bucket elevators operate at much lower speeds than centrifugal bucket elevators and are primarily used for bulk materials that are fragile or abrasive, such as urea prills, carbon black, coal, or soda ash. Buckets for continuous bucket elevators are v-shaped and designed to gently discharge bulk materials.

Belt or Chain – Centrifugal or continuous bucket elevators can be driven by either belt or chain. The choice to use belt or chain depends on the characteristics of the bulk material being elevated. Belt type bucket elevators should be used for corrosive or abrasive bulk materials such as powdered chemicals or sand. These abrasive materials can easily get into the joints of a chain type elevator and cause rapid wear. Chain type bucket elevators are great for dense bulk materials such as cement or limestone. Also, chain type bucket elevators are required for elevating hot materials above about 232 deg. C/450 deg. F. Most belts are not designed to operate at high temperatures.

Bucket elevators are very versatile and can handle a wide variety of bulk materials at very high capacities. Bucket elevators are designed to be very low maintenance with a minimum number of moving parts. They are completely enclosed, dust and weather tight so many bucket elevators are located outdoors. Bucket elevators have many advantages when compared to other types of equipment used for vertically elevating bulk materials.

Pneumatic Conveyors

An air stream, generated by a pneumatic conveyor, can transport loose material a considerable distance either horizontally or vertically. Material is fed into the stream by a so-called rotary lock. Rotary locks separate the air stream from the input and output of the conveyor. A rotary lock typically consists of a wheel with pockets in it. Material is fed into the pockets on the outside. The pocket or wheel rotates to the pressure side, dropping the material into the stream. Material is fed out in a similar manner.

Serviceable components found in these systems include the belt, motor drives for the high-volume blowers, rotary locks, exhaust fans, and diverters for directing product flow.

Magnetic Chip Conveyor

Chips and turnings are a byproduct of machine tools. They are removed from machines, then treated and sometimes broken up to make them suitable for sale.

Magnetic chip conveyors, one type of bulk feeder conveyors (see Fig. 6-18 [A]) have a belt of magnets inside stainless steel housing.

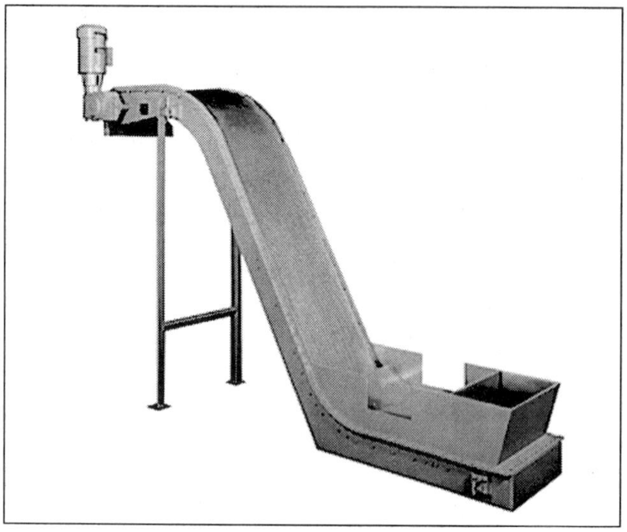

Figure 6-18 (A): Magnetic Chip Conveyor

The foot of the conveyor rests in the machine sump. Chips are attracted to the magnets traveling behind the stainless steel bed. Clumps of chips are thus pulled up the stainless steel surface. Coolant from the chips drains back into the sump. At the top of the conveyor, the belt pulls the magnets away from the chips, dropping them into a container or a take-away conveyor.

Other styles of chip conveyors use hinged plate belts (see Fig. 6-18 [B]), vibrating conveyors (see Fig. 6-18 [C]), or chain flights with special surfaces that serve the same purpose.

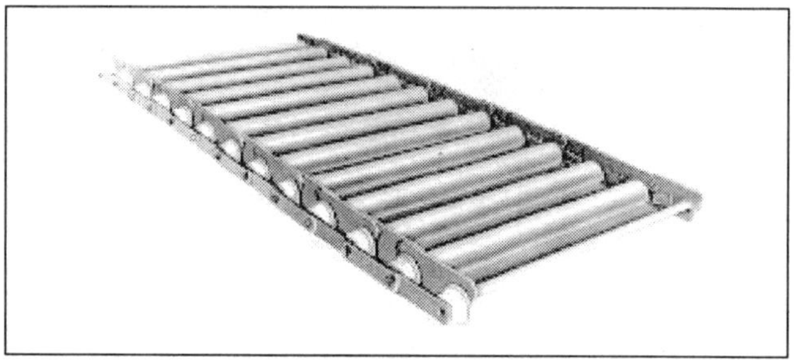

Figure 6-18 (B): Typ. Hinged Plate Belt

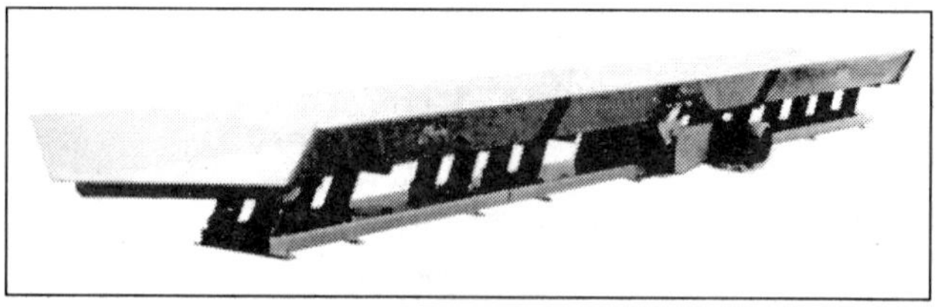

Figure 6-18 (C): Typ. Vibrating Conveyor

In other cases, chips are carried under the floor in a trough by a reciprocating harpoon conveyor or a continuously moving steel belt. The collected chips first go into a centrifugal chip ringer and then to a crusher that turns them into a uniform size. Both the chips and the coolant have value; some have more than others.

All components are subject to abrasive and dirty conditions, even when sealed from contact with the chips.

All standard conveyor components are in play here. They include shafts, pulleys, drives, motors, seals, bearings, roller chain, and sprockets.

Special Concentrations of PT Components

Roller and belt conveyors generally have a drive at one end and some bearings along the way. But there are some other material handling assemblies and accessory sub-systems that are highly concentrated with shafts, gears, mechanisms, actions, and other power transmission parts. Here are a few of them.

Palletizers

Accepting a flow of cartons or boxes fed to it by a conveyor system, the palletizer assembles the product in a patterned layer on a roller assembly platen. Each completed layer is slipped off the assembly platen onto a pallet or onto a previous layer. The layer pattern is changed or alternated so that each successive layer interlocks with the one below. An elevator lowers the pallet in steps until it is full. The load is then ejected over an output roller conveyor or picked up by a fork lift truck.

Up to 12 separate drives may be used on some palletizers, powering pallet elevators, diverters, pallet or separator sheet dispensers, case orienters, roller platens, layer strippers, and other devices. And there may be numerous hydraulic and pneumatic devices.

All functions are timed, usually by a programmable controller that is fed information by sensors, photo cells, upstream conveyors, and production machines. The position movements of palletizers are usually handled by chains, sprockets, synchronous shafts, couplings, cams and eccentrics, clutches and brakes, and linear motion devices.

Sortation Conveyors

Warehouses often pick items for numerous customers simultaneously and then sort them by customer later in the cycle. The primary system that does this is a sortation conveyor (see Fig. 6-1).

Sortation conveyors vary widely in design, depending on the weight, size, fragility, and volumes of items handled.

Heavy items in a low-speed system may simply travel down a powered roller conveyor line with numerous spur lines (powered or non-powered) on either side. Pop-up rollers, pneumatic diverters, or other devices push selected items onto the spurs.

High-speed conveyor sorters (in the range of 500 sorts per minute) take fast-running conveyors full of boxes and merge them into a single lane. The flow is fed to fast start/stop belts that put space between the boxes. These packages are fed single file onto a wide roller conveyor where they may be merged with other induction systems performing the same function. All lanes are then merged into a single file line.

An analog code is assigned to each package as it travels over the spacing belts; from that point on, the system knows where every package is.

Sliding shoes traveling across the conveyor between the rollers engage each package to guide it off onto spur conveyors or sort points.

Small Package Sorting

Small parcels or items are often sorted with a tip-tray or cross-belt sorter. A tip-tray sorter is a loop of fast-traveling, gull-wing-shaped trays on special carriages. Synchronous input belts feed packages onto the trays at high speed. At the correct destination or sort point, a tray tips the product off to the right or left into a chute or slide where the parcels accumulate.

Cross-belt sorters are an endless chain of short belts connected to each other side by side. This chain of cross belts is then powered around a large elliptical track. Each cross-belt conveyor has its own motor and drive.

Chutes, or destination slides, are arranged on both sides of the track. As empty cross belts pass an induction point, packages are placed on them. Farther down the track, when a sort destination point is reached, the cross belt powers the item off into the chute.

Each cross belt is a conveyor. Each has a drive, a motor, bearings, and a power source. And, there may be other powered equipment downstream in the run-out or accumulation system at each chute or sort point, depending on how elaborate the system is.

The cross-belt loop forms another conveyor that also uses a drive and controls. The cross-belt loop can be circular, elliptical, or other shapes to suit the application.

These sorters are concentrations of stepping motors, synchronous and adjustable-speed drives, timing belt drives, and sophisticated controls. They also employ special drives, chains, belts, pulleys, sprockets, cams, and controls.

Powered Mechanical Storage Systems

A good portion of material handling equipment involves storage using automated mechanisms, including horizontal and vertical carousels and a variety of automated storage and retrieval systems. Horizontal carousels were described earlier under the heading of conveyors. As automatic storage structures, they also employ insertion and extraction devices that load and unload the shelves of the carousel.

Some specialized horizontal carousels are hundreds of feet long, many stories high, and may hold a million pounds of material. They use only a few drives because the loads are in constant motion; loading and unloading are done on the fly.

Vertical carousels are made up of two loops of chain traveling over large sprockets on shafts mounted some distance apart (see Fig. 6-19). Trays or shelves are supported between the chain and sprocket arrangements. Controls rotate the shelves inside a sheet-metal housing, exposing one shelf at a time to an access opening for product picking.

Vertical carousels can be configured to handle a range of products from carpet rolls to structural steel, whatever can be suspended between the chains and sprockets.

Power transmission components used here include very large pillow-block bearings, chains, sprockets, speed-reducers or gear-reducers, motors, brakes, clutches, and a variety of interactive sensors.

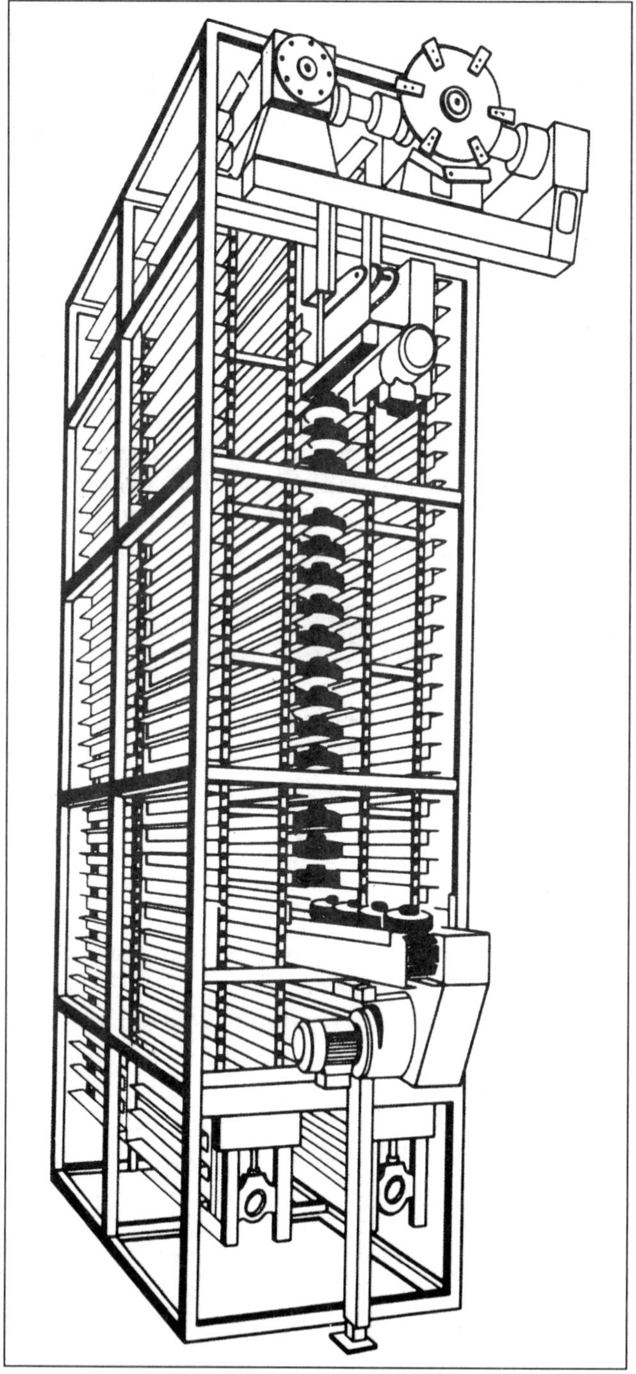

Figure 6-19: A cutaway of a vertical carousel system showing internal components.

Automated Storage and Retrieval Systems

The storage/retrieval machine works in three axes. They travel down an aisle, lift to the proper height, and then deposit or extract a load from a bin. Travel and lift can be accomplished simultaneously.

Storage and retrieval mechanisms may be stationary or travel up and down an aisle. Traveling automated storage/retrieval systems (AS/RS) place loads in openings in a rack or other storage structure.

Storage/retrieval machines employ numerous power transmission components including gearmotors, chains, sprockets, bearings, rollers, sensors, microprocessor-based controls, optical encoders, resolvers, and limit switches. Each manufacturer has developed methods of powering the machines and indexing them precisely.

Small-load systems (see Fig. 6-20) typically handling up to 700 lb per container are called mini-load AS/RS.

These may include a complex front end consisting of conveyors or pickup and drop-off stations that may also have extensive conveying and transfer devices.

Large-load AS/RS (see Fig. 6-21) handle pallet loads or containers typically weighing up to 3,000 lb; some handle loads up to 30,000-lb rocket booster engines.

There are many variations of unit load AS/RS. These can have elaborate front ends. The front end is the load pickup and drop-off station. These stations can be simple in-and-out shuttle conveyors that place the load in a position to be picked up by the storage machine, or they may be continuous loops. The loops circulate loads to be stored or ones that have been picked from storage.

Devices that raise and lower loads, transfer them at right angles to the travel path, tip containers up for easy access, and other movements are powered by drives, actuators, and escapements of many kinds. Gears, sprockets, pulleys, gearmotors, air cylinders, bladders, bearings, shafts, cams, and movable stops are part of most designs.

These highly complex automated storage machines are directed by microprocessors and programmable controllers. Every motor uses a motor starter, and many drives are adjustable-speed types.

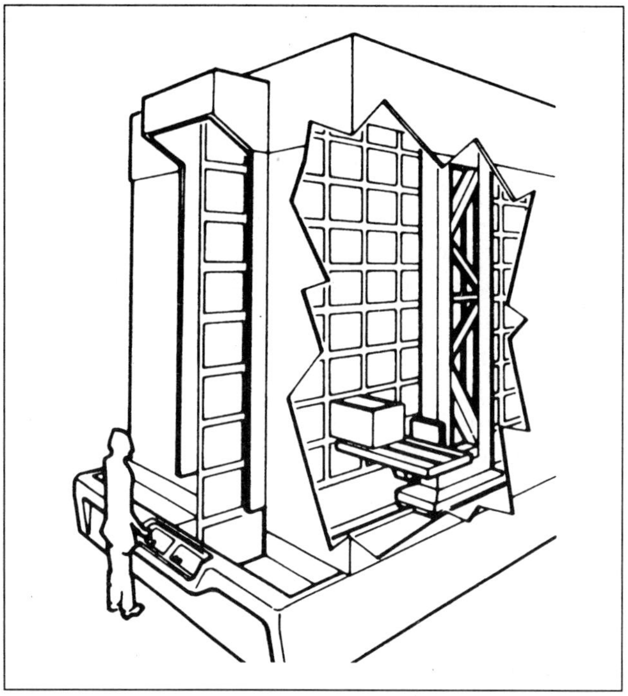

Figure 6-20: Configuration of a miniload Automated Storage/ Retrieval System.

Figure 6-21: Man-ride automated storage and retrieval system.

Packaging Machinery

While packaging machines are not strictly material handling, they are one of the largest users of power transmission components. Manufacturers producing consumer products are the biggest users of packaging machines.

Typically, there are machines for primary packages and others that place and wrap a block of primary packages into secondary or shipping packages.

Packaging machines are compact with dedicated sections that open cartons, insert product, and close and wrap the product. Preceding these, there are filler, labeler, capper, and sealer machines.

All these machines employ a wide range of drives and motions from Geneva movements to progressive pitch screws, cams, and levers. Most machines are almost watch-like in their precision and complexity, and all provide for removing lash or looseness in every movement.

Packaging machines use a lot of open gears and drives that employ unusual components such as precision ladder or bead chain and timing belts, timing screws, and pulleys. Many fractional horsepower drives and servo motors are also used.

Farther downstream in packaging systems are automatic strapping and case-closing machines as well as stretch wrappers and baggers for unitizing pallet loads of products ranging from pails and drums to boxes.

Special Considerations

No matter what the power transmission situation appears to be on the surface, it is best to ask if a customer has any special requirements that must be kept in mind. Industries evolve handling their material in special ways for very good reasons. This thinking may be totally transparent to those outside of the business.

Gunpowder manufacturers, for example, pay a premium to ensure that no part in a system can rub against another and make a spark. Motors are all sealed and explosion-proof. Brass, bronze, plastic, and wood are materials of choice.

Food and pharmaceutical businesses rely heavily on conveyor components made of stainless steel and plastics to prevent corrosion from wash-down with hot water, caustic chemicals, and steam. Wash down gearmotors may be coated with white epoxy paints, be totally enclosed, and have sealed electrical connection boxes.

Painted components are usually finished with non-toxic paints or epoxies, and many are sealed in some way. Some use special lubricants. For sanitation, all components in non-wash down areas may still be sealed to exclude vermin.

Pharmaceutical manufacturing takes place under almost cleanroom conditions, but there are degrees of cleanliness depending upon the manufacturing stage.

Standards

As you become familiar with the conveyor designs used in various industries, you will notice similarities of designs within those industries. The automotive industry uses guidelines specified by the Joint Industry Council (JIC) or the newer Automotive Industry Action Group (AIAG). Other industries have watchdog groups setting standards that provide for employee and product protection.

Primary metals manufacturers have Mill Standards or Mill Specifications. The government has a wide range of federal specifications or military standards relating to equipment that it buys for both military and military supplier industries. Fungus proofing, for example, is a special requirement for equipment used in the tropics.

There may be separate standards for each of the armed services plus GSA (General Services Administration) and merchant marine. Never attempt to supply a look-alike aircraft component.

Industry Associations are very active in developing standards for their industries. The Conveyor Equipment Manufacturers Association (CEMA) maintains size and dimensional standards for most conveying equipment described in this chapter. Many of those standards also address proper horsepower and other power transmission engineering calculations. They also have included in these standards, or have developed stand-alone, Safety Best Practices Recommendations for proper and safe operation of their equipment. CEMA, the American Society of Mechanical Engineers (ASME), and the American Chain Association (ACA) also maintain standards for proper chain application and design. Many of these standards are American National Standards Institute (ANSI) Standards. Other ANSI Standards should be consulted depending on the project components involved.

Be aware that these standards exist and that they evolved for specific reasons. Some of the reasons are the result of a long and expensive experience; others are based on user requirements for total reliability, minimum downtime, or to suit the operating environment.

Customization

In automotive manufacturing, for example, many drives must meet JIC standards, but each choice carries impacts if it fails. In one case, a drive will have a foot-mounted motor coupled to a foot-mounted speed reducer that powers a machine through chains and sprockets or belts and pulleys (see Fig. 6-22). Here, a failed component can be quickly replaced with a minimum of downtime. Where components can be quickly separated from one another and are easily removed, a two-wheel hand truck can often handle the spare. In another case, a unitized direct-coupled or shaft-mounted speed reducer could be used, but that requires that an entire duplicate unit must be stocked as a replacement. To replace this component, an overhead crane might be needed.

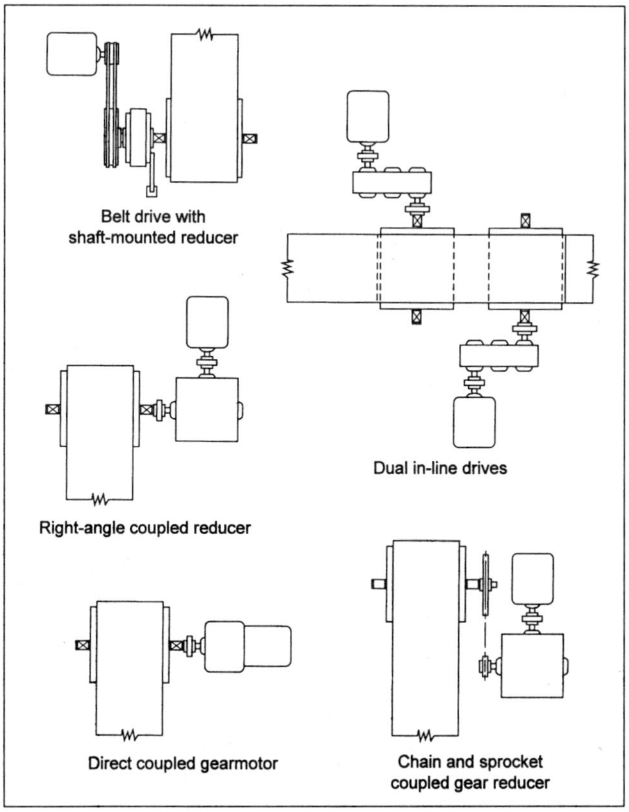

Figure 6-22: Drive line configurations showing shaft-mounted reducers, in-line drives, coupled motor/reducer drives, chain and sprocket coupled reducer, and direct coupled in-line gearmotor.

In another example, a flexible coupling or U-Joint might freeze to the shaft and be difficult to remove; chain or belt connections and split sprockets, on the other hand, are more easily removable.

There are also times when components are standardized in an application for unusual reasons. This might result in some that are far larger than the application calls for or are of higher quality than needed. The reasoning here is that the higher quality encourages rebuilding rather than scrapping the failed component. Uniform sizes also reduce the number and variety of spares required.

Automotive plants employ all types of handling equipment and, depending upon the type of manufacturing plant, use almost every type of conveyor. While much of their equipment is standard, automotive plants require many special functions and specially designed handling devices.

The reason automotive systems are customized is in the volumes of product they produce. Most systems for auto assembly, for example, are built for a specific model year. The equipment must often take punishing three-shift continuous service with little or no maintenance. This may mean high-quality bearings, automatic lubrication systems, and sometimes redundant drives.

On the whole, power transmission equipment distributors will find that the automotive industry uses more proprietary components in its handling systems than almost any other single industry. However, disassembly of the proprietary components reveals many common replacement parts assembled in unique arrangements.

Additional Information

Addresses and phone numbers for associations are provided in the Association Appendix.

Conveyor Equipment Manufacturers Association (CEMA)

American Society of Mechanical Engineers (ASME)

American National Standards Institute (ANSI)

Chapter 7

Couplings and U-Joints

Chapter Objectives

When you have finished Chapter 7, "Couplings and U-Joints," you should be able to do the following:

1. Differentiate between rigid, mechanically flexible, elastomeric element, metallic element, and universal joint shaft couplings.
2. Describe the characteristics of flange, sleeve, and ribbed couplings.
3. Explain the functions and features of flexible couplings.
4. Compare the characteristics and features of gear, chain, and grid couplings.
5. Describe how metal-beam, spring, slider block, and metal-bellow couplings function.
6. Differentiate between the functions and features of elastomeric couplings loaded in compression and those loaded in shear.
7. Understand why most flexible couplings have two flexible elements.
8. Identify the features of block-and-pin and bearing-and-cross type U-Joints.
9. Identify the features of block type, split eye, solid eye, and yoke type U-Joints.
10. Describe the features and applications of some of the miscellaneous type couplings.
11. Identify the source for coupling and U-Joint standards.
12. List the application requirements when selecting a coupling.

Introduction

A shaft coupling is used to connect two shafts that turn in the same direction on the same centerline. The major types of shaft couplings are rigid, flexible, universal joint, and miscellaneous. Rigid couplings are used to connect shafts that are precisely aligned. Flexible couplings and U-Joints accommodate varying degrees of misalignment between shafts. The wide range of coupling types allows customers the ability to select couplings that will provide such performance features as vibration dampening, zero backlash, environmental tolerance, and electrical isolation. Such features can address concerns or mitigate forces within machine operations in ways that extend machine system life and ensure accurate machine operation. As might be expected, not all coupling types will perform equally in a given application, and selection of the proper coupling type can require significant evaluation of the application goals.

Rigid Couplings

Where two shafts are precisely aligned, rigid couplings are used to connect the shafts. These couplings, the simplest type, provide a fixed union between two shafts where torsional flexibility is not required, shaft alignment is maintained, and proper bearing support is provided. Shaft misalignment can cause excessive wear or failure of the shaft, coupling, or bearings.

To reduce shaft deflection, shaft support bearings should be located as close to the coupling as possible. Large couplings and those running at high speed may require balancing to reduce vibration.

The basic types of rigid couplings are flange, sleeve, and ribbed. While these coupling designs provide no dampening or misalignment tolerance, they are often inexpensive and simple to install.

Flange Couplings

Flange couplings are used to connect shafts of the same or different diameters. These devices consist of two flanged hubs, bored and keyed to fit their respective shafts and joined by bolts. Some types (see Fig. 7-1) have removable bushings for easier replacement. Their ease of installation and removal, plus minimum maintenance and wear, are the chief advantages of this type of coupling.

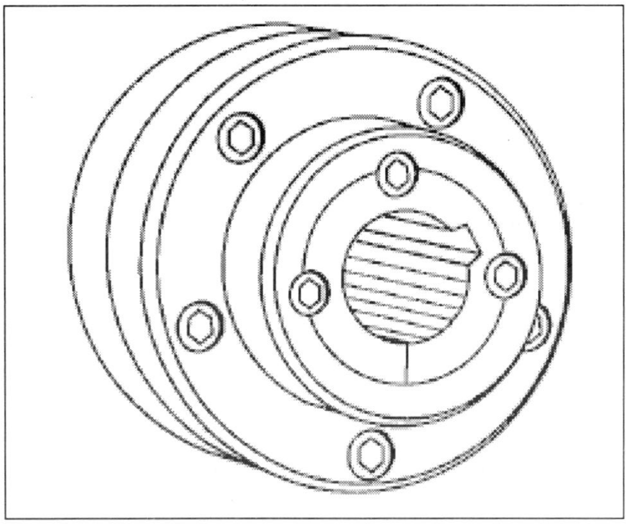

Figure 7-1: Rigid flange coupling with replacable bushing

Usually made from carbon steel, flanged couplings are available in shaft sizes to over 8 in. diameter. Designed for heavy-duty applications, these couplings handle up to 1,500,000 lb-in. torque and speeds to 10,000 rpm.

Sleeve Couplings

A sleeve coupling (see Fig. 7-2), is made from one piece of material. It is factory bored to accommodate standard shafting. Keyways are broached, and setscrews are installed. Shaft keys transmit torque from the shaft to the coupling hub, and setscrews keep the coupling from sliding along the shaft. A shaft key is not normally required for 1/2 in. diam-

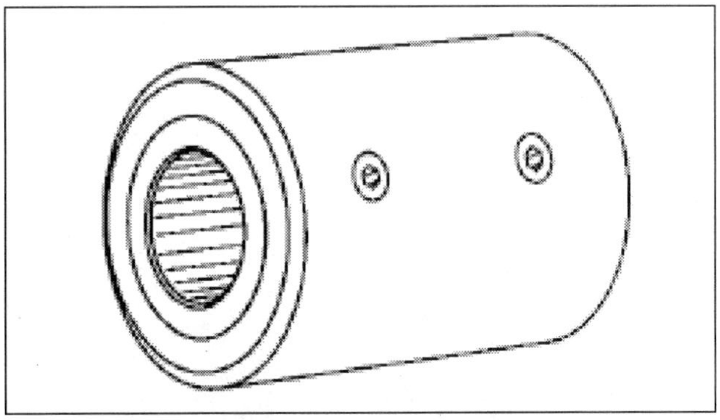

Figure 7-2: Sleeve coupling attached to shaft with locking setscrews

eter or smaller. Sleeve couplings may be provided with two different diameters, but an easier solution may be to use a sleeve coupling with detachable hubs.

Designed for light- to medium-duty applications, sleeve couplings have torque ratings ranging from 100 to 4,000 lb-in. for speeds up to 6,200 rpm. Bore sizes typically range from 1/4 to 1 3/8 in.

Ribbed Couplings

For heavy-duty services, a ribbed coupling may be used. It is made in two pieces and is used to connect shafts of the same diameter. The two halves of the coupling clamp onto the shaft and are held together by bolts (see Fig. 7-3). The coupling is keyed to both shafts. These couplings permit quick and easy installation and removal. They exhibit no wear and require no maintenance. Available in shaft sizes from 1 to 7 in., ribbed couplings are rated to transmit the same load as the steel shafts that they connect.

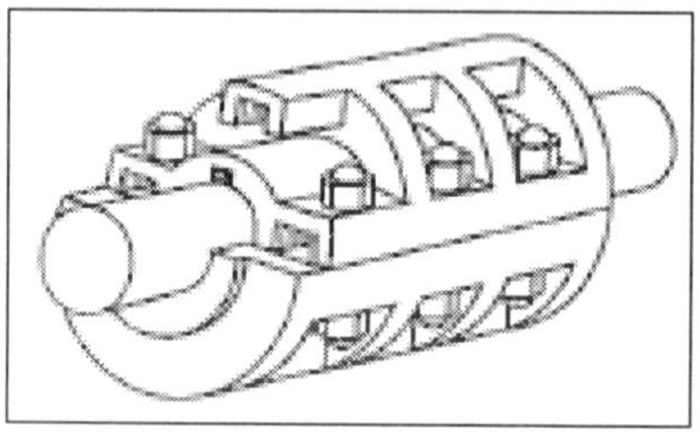

Figure 7-3: Ribbed coupling is split axially to enable easy shaft asembly

Flexible Couplings

As with rigid couplings, one function of a flexible coupling is to connect a driving shaft with an adjacent in-line driven shaft and transmit the required torque or power. But flexible couplings also flex while rotating to accommodate misalignment between shafts. There are two types of shaft misalignment: angular and parallel (see Fig. 7-4 [A] and [B]).

Proper shaft alignment means that both types of misalignment must be reduced to within the equipment manufacturer's requirements to assure satisfactory operation of the equipment.

Depending on type, flexible couplings can accommodate from a 1/4 degree of misalignment up to 4 degrees (in fact, some special couplings can handle 15 degrees of misalignment). But most equipment cannot handle the reactionary loads the couplings generate when misaligned, so equipment is aligned to a much tighter tolerance. Low-speed general-purpose couplings are aligned to 0.001 in/in (or roughly 3.5 seconds). For high-performance applications, they are aligned to even tighter numbers, 0.0005 in/in (1.75 minutes).

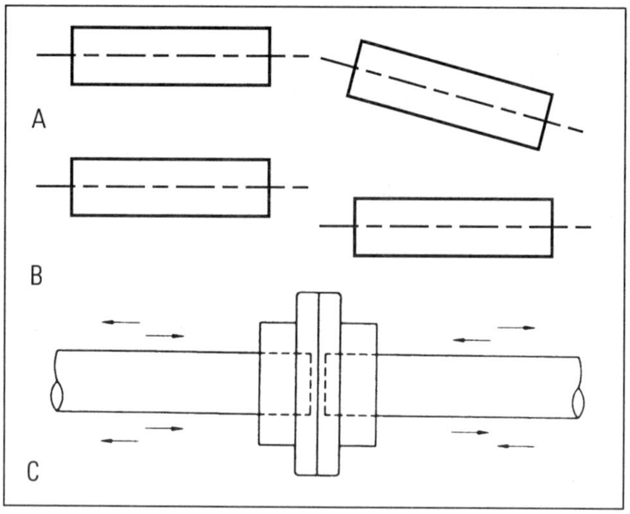

Figure 7-4: Types of shaft misalignment. (A) Angular offset misalignment, (B) Parallel, (C) Axial movement

Maximum parallel misalignment can range from a few hundredths of an inch to over an inch, again depending on the type of coupling. Couplings that have two flexible elements can accommodate more parallel offset than elastomeric couplings with a single flex element.

Most flexible couplings accommodate axial movement of the shafts (see Fig. 7-4 [C]). This movement is called end float.

Manufacturers typically rate a coupling for its maximum misalignment capability. Misalignment that is greater than what is recommended causes excessive flexing. With lubricated and elastomeric couplings, this can create heat and reduce the service life of the coupling or cause high restoring forces in the coupling, which can reduce the service life of the equipment. To minimize this effect, the user should align connected shafts as closely as possible when installing a flexible coupling.

Flexible shaft couplings provide flexing by one or more of the following means:

1. Mechanical motion either sliding or rolling.

2. Flexing motion of elastomers, thin metallic elements, or composite material.

3. Compressing or stretching of resilient or composite materials.

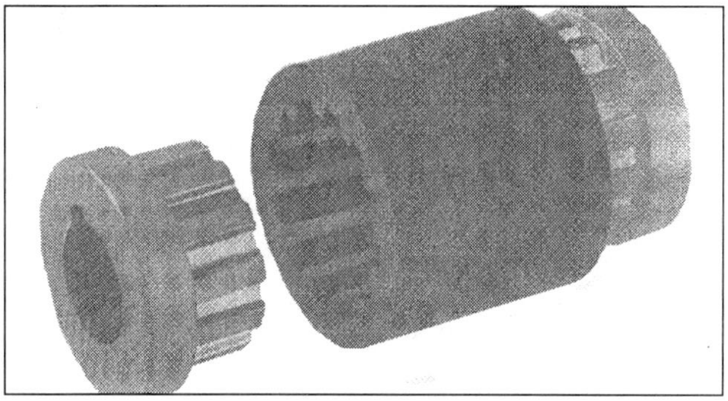

Figure 7-5: Miniature nonlubricated gear coupling

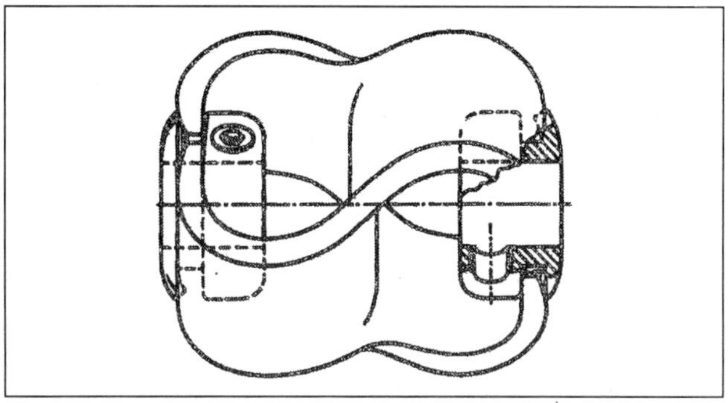

Figure 7-6: Miniature urthane ribbon elastomeric

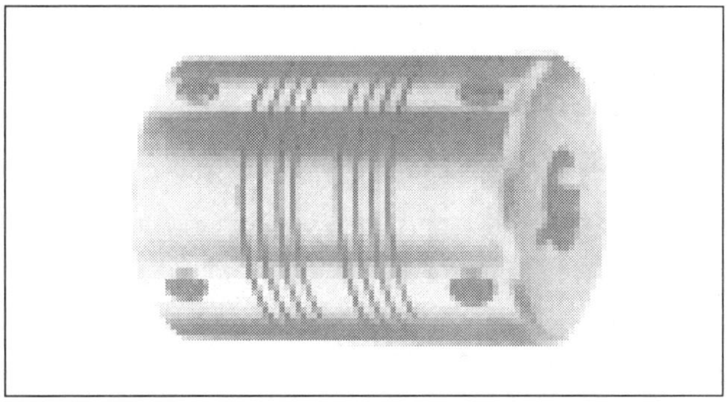

Figure 7-7: Miniature metallic beam coupling

Figure 7-8: Miniature disc coupling

Other Functions of Flexible Couplings

Besides the basic function, flexible couplings are sometimes required to do the following:

1. Dampen vibration and reduce peak or shock loads.
2. Protect equipment from overload.
3. Measure output torque of driven equipment.
4. Insulate the driver from the driven equipment.
5. Position a rotor of a motor or generator.
6. Tune a system out of a torsional critical condition.

The following are the three disciplines for the application of flexible couplings:

1. The Miniature discipline, which covers couplings used for office machines, servomechanisms, instrumentation, light machinery, and so on. Examples include the miniature non-lubricated gear coupling (see Fig. 7-5), miniature urethane ribbon elastomeric coupling (see Fig. 7-6), miniature metallic beam coupling (see Fig. 7-7), and miniature disc coupling (see Fig. 7-8).

2. The General Purpose Industrial discipline, which covers couplings used in the steel industry, the petrochemical industry, utilities, machinery, and so on. Many of these couplings are considered "off the shelf" or modifications of it.

3. The Special Purpose Industrial discipline, which covers couplings that are specifically designed for a particular application or piece of equipment; they are generally not "off the shelf."

It should be noted that any given type of coupling may fulfill application requirements in all three disciplines.

There are four basic types of couplings:

1. Mechanically Flexible Couplings
2. Elastomeric Element Couplings
3. Metallic Element Couplings
4. Miscellaneous Couplings

The mechanically flexible types generally obtain their flexibility from a clearance between components that allows rolling or sliding of mating parts, or from both. They require lubrication unless one moving part is made of a material that supplies its own lubrication (i.e., a nylon gear coupling).

The elastomeric element types obtain their flexibility from shear or compression of a resilient polymer material (rubber, plastic, etc.).

The metallic element types obtain their flexibility from the flexing of thin metallic discs or diaphragms.

Miscellaneous coupling types obtain their flexibility from a combination of mechanisms previously described, the use of composite materials, or through a unique mechanism.

General Purpose Couplings

General-purpose couplings can be classified according to size. Less than 100 HP is classified as small. Between 100 and 1,000 HP is characterized as medium and usually, HP greater than 1,000 is considered critical and therefore will be covered in the special-purpose coupling section. If a small piece of equipment (i.e., pump) shuts down, it usually does not affect the plant operation. This equipment uses a coupling type that allows the flexible element to be easily inspected and replaced, often considered a throw-away item. The couplings are very flexible and require very simple alignment techniques, calipers, scales, and perhaps if one is sophisticated, a dial indicator. A failure from over-torque or over-misalignment is usually due to the flex element, and little or no damage occurs to other components. A few examples of these coupling types on the market are grid, disc, and elastomeric. Also found on this equipment are small gear couplings; some will have the outer sleeve made of nylon or plastic and therefore require no lubrication.

Medium-size equipment (100–1,000 HP) is not normally critical to the operation of the plant but is problematic and costly if constant maintenance and downtime are required. They use grid, gear, disc, and elastomeric type couplings. The type of coupling to use is usually based on what comes with the equipment. Each equipment manufacturer bases its selection on past price, experience, and a preference to work with one coupling manufacturer over another. Over time, maintenance departments eventually develop a preference based on pump duty, type of maintenance performed, as well as personal preference. This group of couplings usually fails because of lack of proper installation, lubrication, and alignment.

There are various types of general-purpose couplings, including gear, chain, grid, U-Joint, compression donut, block, jaw, urethane tire, urethane dual diaphragm, corded tire, shear donut, disc, and special purpose.

Mechanically Flexible General Purpose Type Couplings

The mechanically flexible types generally obtain their flexibility from a clearance between components that allows rolling or sliding of mating parts, or from both. They require lubrication unless one moving part is made of a material that supplies its own lubrication (i.e., a nylon gear coupling).

Gear Couplings (Mechanically Flexible Type)

Available in a wide range of sizes, gear couplings are capable of transmitting high torque at high speed. Each coupling consists of two essentially identical hubs with external gear teeth. Most commonly, the hubs are joined by flanged sleeves with internal gear teeth, and the flanges are bolted together (see Fig. 7-9). There is also a one-piece continuous-sleeve version where hubs are connected through the internal teeth of the sleeve. The one-piece sleeve is sometimes made from nylon, and this design requires no lubrication, but all metal sleeve types require lubrication.

Gear couplings accommodate misalignment with axial clearance between gear teeth, called backlash. End float is accommodated by providing axial clearance between hubs. Some have crowned teeth, which permit increased angular misalignment and relieve edge loading between mating teeth. These are available in bore sizes up to about 40 in.

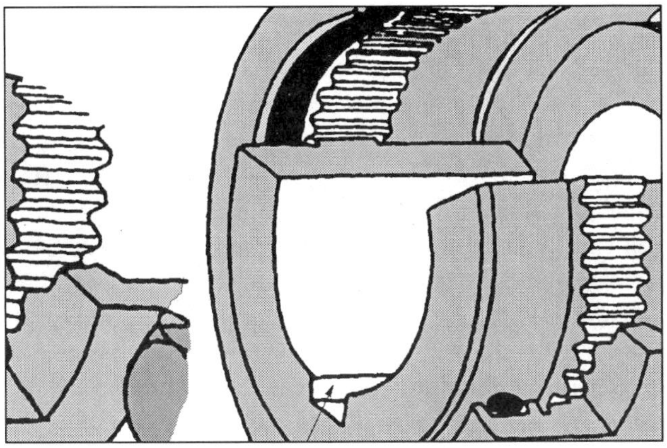

Figure 7-9: Gear tooth coupling

General-purpose gear couplings are usually made from carbon steel. Those designed for more severe applications (higher torque, speed, or misalignment) are typically hardened to reduce wear, or they are made from stronger materials to increase load capacity.

In applications where coupled equipment is precisely aligned, load is distributed among all of the coupling teeth. However, for misaligned equipment, the load is not equally distributed. Thus, misalignment reduces the load rating of a gear coupling, increases wear, and reduces the coupling life. It is common to find gear couplings in metals-industry applications such as steel or wire mills and paper mills where the high torque-to-size ratio and speed capabilities provide value.

Gear type couplings consist of two hubs with external teeth that engage internal teeth on a two- or one-piece sleeve. The teeth may be straight or curved (crowned). For applications requiring over 1/2 degree, curved teeth may be better. These couplings obtain their flexibility due to the looseness (backlash) between the mating teeth. Gear couplings are used for medium- and large-equipment applications and are probably the most power-dense type available.

They require periodic lubrication, every 1–2 years depending on duty and type of lubrication. If properly maintained (good lubrication and reasonable alignment), these coupling have a service life of 3–5 years and in many cases 5–10 years.

Some gear couplings have a sleeve that is made of plastic (nylon, high molecular plastic), and these do not require lubrication. These have much lower torque capacity than the all-steel couplings and are used mainly on small pumps.

Chain Couplings (Mechanically Flexible Type)

Chain type couplings (see Fig. 7-10) are very similar to gear couplings. They are usually composed of all metal (hardened sprockets and chains), but some use plastic chain (see Fig. 7-11). They have two hubs with sprockets rather than teeth. The sprockets are connected by a length of double-strand roller chain. Most of the couplings are open and are lubricated by "brushing" grease on the chain. Sometimes a cover is used to help keep the lubrication in. These couplings do not transmit as much power (per the same outside diameter) as gear couplings but are usually less costly. Chain couplings are used for low-speed applications on medium- and small-equipment applications.

These couplings accommodate shaft misalignment by means of clearances between the chain and sprocket teeth and clearances within the chain. Chain couplings are used in applications requiring low to moderate torque and speeds.

Figure 7-10: Chain coupling - steel

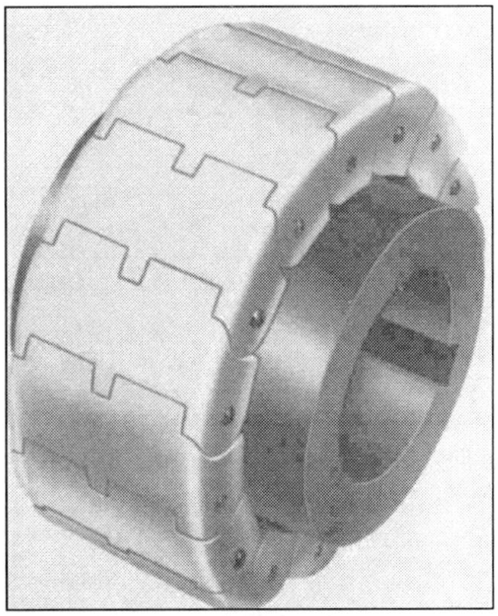

Figure 7-11: Chain coupling - nylon

The chain is easily removed and replaced without disturbing the driving or driven shaft. Covers are used for chain couplings where the rotating speed is capable of throwing off the lubricant or where the atmosphere is corrosive or abrasive. The cover rotates with the coupling.

There are three types of chain couplings: double roller, silent, and plastic. Double-roller-chain couplings, the most common, are suitable for moderate speeds and loads. With this type, double-strand roller chain is wrapped around two sprockets mounted face to face on shafts. A connecting link attaches the ends of the chain together. Hardened bushings and sprocket teeth reduce wear. These couplings permit up to 2 deg. angular misalignment, 0.015 in. parallel misalignment, and 0.25 in. end play, depending on coupling size.

Silent-chain couplings (see Fig. 7-12), are used for heavy-duty drives requiring high torque. They are rated for 1/2 deg. misalignment and 0.125 to 0.50 in. end play, depending on the chain construction. These couplings operate at speeds up to 5,000 rpm, depending on their size.

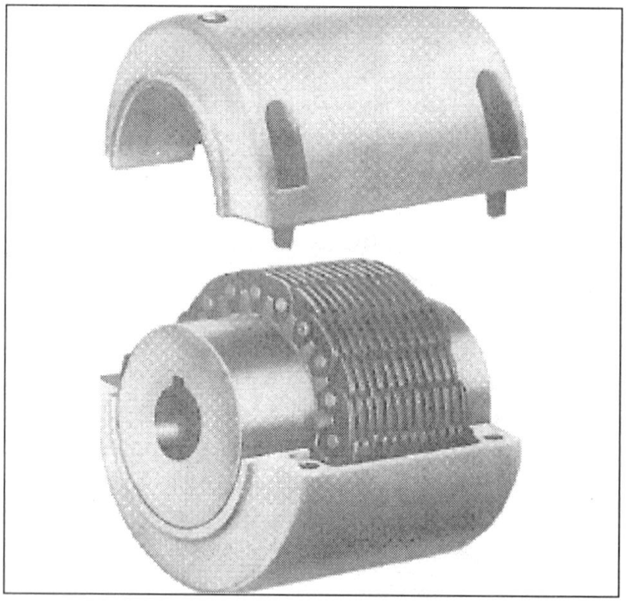

Figure 7-12: Silent chain coupling

Plastic-chain couplings generally use nylon chain, which requires no lubrication. They are suitable for light loads in applications where lubricants are undesirable because of possible contamination, such as in the textile and food-processing industries. These couplings are also suitable for corrosive atmospheres. They are quieter in operation than metal-chain couplings and require no protective cover.

Grid Couplings (Mechanically Flexible Type)

Grid-type couplings are also very similar to gear couplings (see Fig. 7-13). They consist of two hubs with multiple slots through which a steel strip or grid weaves back and forth. Usually composed of all metal, they have some degree of resilience. Parallel and angular misalignments and end float are accommodated by sliding movement of the grid within the lubricated slots. These slots are generally tapered to allow space for the grid to flex. This configuration provides torsional flexibility under different loads, vibration dampening, and the cushioning of shock loads.

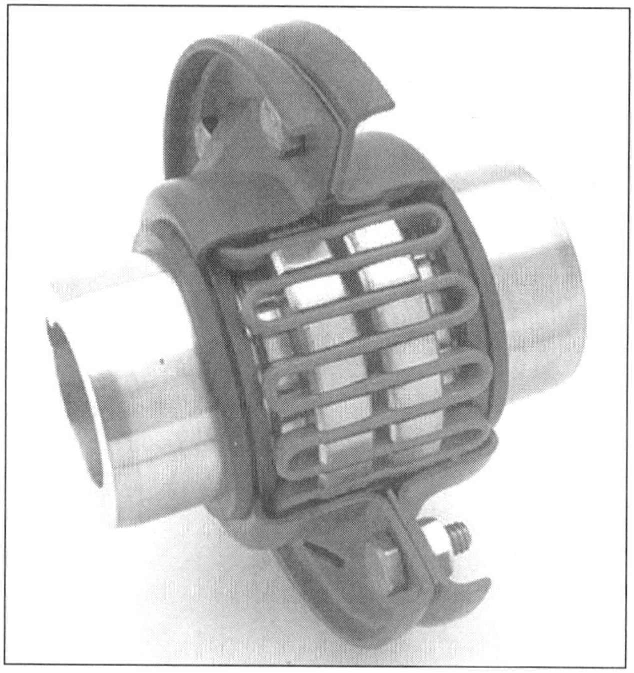

Figure 7-13: Grid coupling-vertical split cover

Grid couplings typically allow 1/3 deg. angular misalignment up to 0.040 in., and end float to 3/16 in. Bore sizes range up to 20 in., with the smaller sizes accommodating speeds up to 6,000 rpm. These couplings can dampen vibration and reduce peak or shock loads by 10–30 percent. Covers are normally provided to retain lubricant, prevent contamination, and act as a safety device.

The covers are either vertically split (see Fig. 7-13) or horizontally split, (see Fig. 7-14). These couplings do not transmit as much power (per the same outside diameter) as gear couplings but are usually less costly. Grid couplings are used for medium- and small-equipment applications. Like the gear- and chain-type couplings, the grid designs require lubrication.

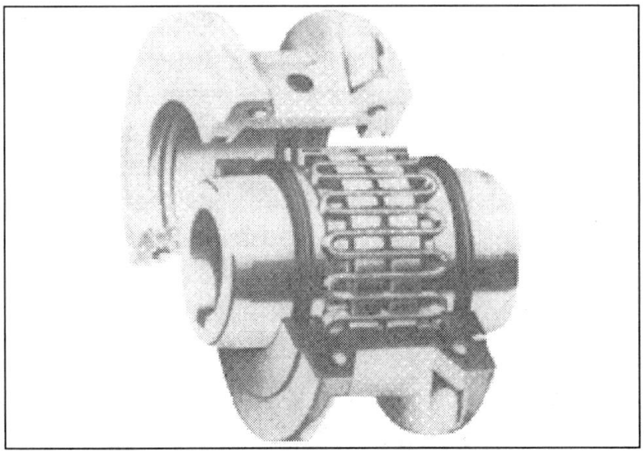

Figure 7-14: Grid coupling- horizontal split cover

These couplings are commonly used on pumps and high-capacity bulk conveyor applications.

U-Joints (Mechanically Flexible Type)

They can accommodate end movement by using a telescoping shaft (square shafting or splines). U-Joints function by a sliding motion between two "flanges" that are fork-shaped (a yoke) and having a hole (eye) radially through the eye that is connected by a cross. This allows flexibility by sliding, and if one gets sophisticated, it has bearings and seals between the yoke and the cross (most industrial U-Joints use bearings), and now we have the modern day U-Joint.

More commonly known as U-Joints, universal joints allow transmission of power at larger angles than is possible with a flexible coupling, generally used in applications where high misalignment must be accommodated (generally in excess of 1 degree and up to 30 degrees). U-Joints are used in a wide variety of power transmission applications including automobile and truck drive shafts, farm machinery, and industrial equipment (see Fig. 7-15).

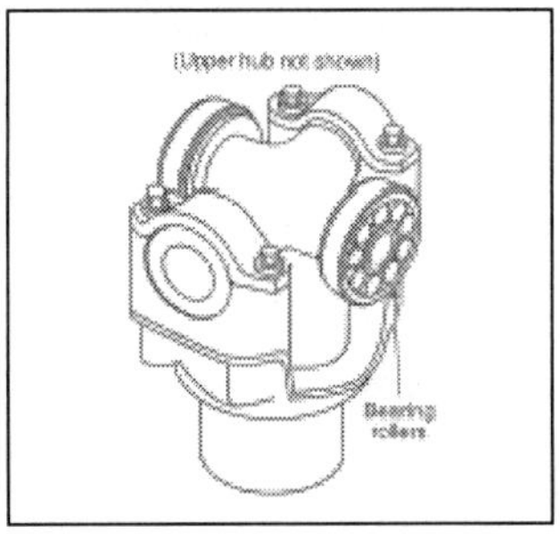

Figure 7-15: Vehicle U-Joint

There are many varieties of U-Joints, some of which are very complicated. The simplest category, called Cardan U-Joints, are either block-and-pin or bearing-and-cross types.

Block-and-Pin Joints

Available in both single and double configurations (see Fig. 7-16), these U-Joints are most commonly used in industrial applications. A single U-Joint consists of two shafts with U-shaped and drilled ends, a center block that fits between the two ends, and two pins that secure the block to the shaft ends via the drilled holes.

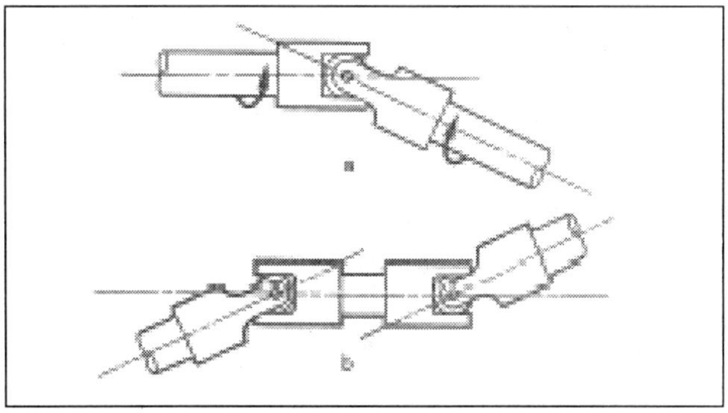

Figure 7-16: Block and pin U-Joint included both single (A) and double (B), versions. Double U-Joint opterates at near-constant velocity

A double U-Joint has a center member connected on each end with separate joints just as if two singles were connected back to back. Though the single U-Joint handles angular misalignment, it doesn't compensate for parallel misalignment, a job best handled by double U-Joints.

Where changes in misalignment or shaft end movement occur during operation, a telescoping splined shaft provides the necessary adjustment in shaft length.

Bearing-and-Cross Joints

In this design, two U-shaped hubs are joined by a cross-shaped piece. Needle bearings at each end of the cross-shaped piece fit into the hub arms (see Fig. 7-17).

Universal joints are torsionally rigid but are capable of up to 15 degrees of angular misalignment. However, when a U-Joint is operating at an angle, a new problem must be considered.

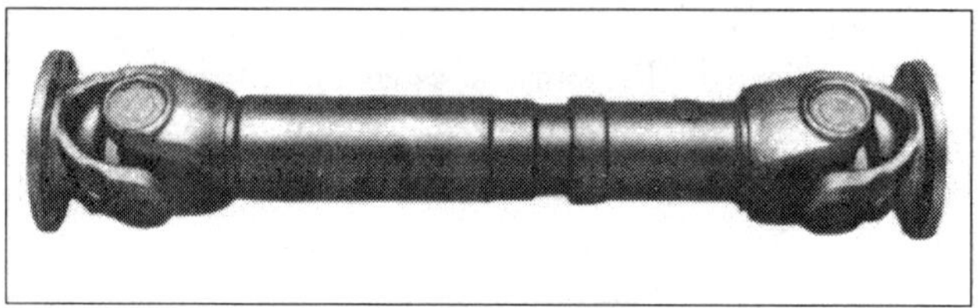

Figure 7-17: Yoke type U-Joint

As the input shaft rotates at a constant rpm, there is no acceleration or deceleration of the shaft as it makes each revolution. It has a constant velocity. When a U-Joint is used, the output shaft will rotate at the same rpm as the input shaft. However, as that output shaft rotates, it speeds up through 90 degrees of rotation and then slows down during the next 90 degrees of rotation; it continues this process as long as the shafts are turning. These changes in velocity create vibrations that limit single U-Joint applications to very low speeds. Using U-Joints in pairs solves this problem.

Elastomeric Flexible Element Couplings

The elastomeric element types obtain their flexibility from stretching or compressing a resilient material (rubber, plastic, etc.). Several design approaches are employed in these couplings. They all incorporate a flexible material in compression or in shear between two hubs. The material may be mechanically interlocked, bolted, clamped, or molded to the hubs. Those loaded in compression transmit high torque, whereas the shear type deflects more and allows for greater misalignment and torsional damping.

Elastomeric couplings generally offer better shock and vibration damping than other coupling types but are subject to operating temperature limitations, which depends on the elastomeric compound used. It is common to find elastomeric couplings used with pumps where the vibration dampening characteristic of the coupling serves to protect shafts and keyways of both pumps and motors.

The basic types of elastomeric couplings that are loaded in compression include jaw, block, and donut types. Elastomeric couplings that are loaded in shear include rubber-tire, bonded, and donut-shear types.

Bonded-Rubber Shear Couplings (Elastomeric Element Type)

Bonded-rubber shear couplings have a resilient elastomeric member bonded between two hubs (see Fig. 7-18), which generally provides a more constant torsional stiffness than other designs. The rubber compounds are hard and nearly incompressible, which limits their shock-absorbing effect. Designed for low-frequency isolation, these couplings range from those used in instruments to a 4 in. bore.

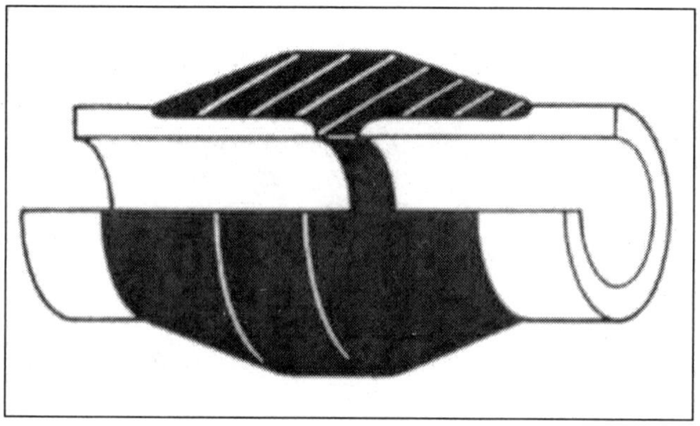

Figure 7-18: Bonded Rubber in shear

Urethane Tire Couplings (Elastomeric Element Type)

Common on small-type equipment, urethane tire couplings (see Fig. 7-19), have their elastomeric element in shear and are made of urethane, polyurethane, or polyether materials. The tires are split to enable easy assembly without removing the hubs. These couplings offer a high degree of flexibil-

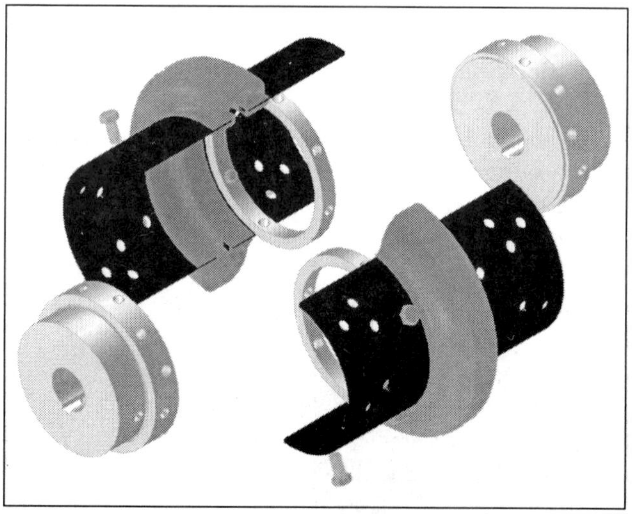

Figure 7-19: Urethane tire coupling

ity and can commonly function with greater misalignment than similar-sized couplings. When the coupling fails, usually only the elastomeric element is replaced. The type of material in the element will impact unit tolerance of misalignment and of environmental factors such as moisture or chemical exposure.

Urethane Dual Flex Diaphragm (Elastomeric Element Type)

Urethane dual flex diaphragm coupling is unique to elastomeric couplings (see Fig. 7-20 and 7-21). Made of urethane bonded at the inside diameter (ID) to a composite tube, this composite spacer makes the center section light enough so any unbalanced force that tries to eccentric the center tube is insignificant and therefore allows it to run at balance levels of AGMA 9. This coupling, because of the dual flexible element, has good offset capacity that increases as the shaft separations increase. Also, because the dual flex element is at the end, the reactionary force on the equipment is lower than any of the single flex elastomeric couplings when they are used as a spacer coupling. The element is easily removed by unbolting at the ridges and dropping out the center section. Dual flex urethane couplings offer a high degree of flexibility.

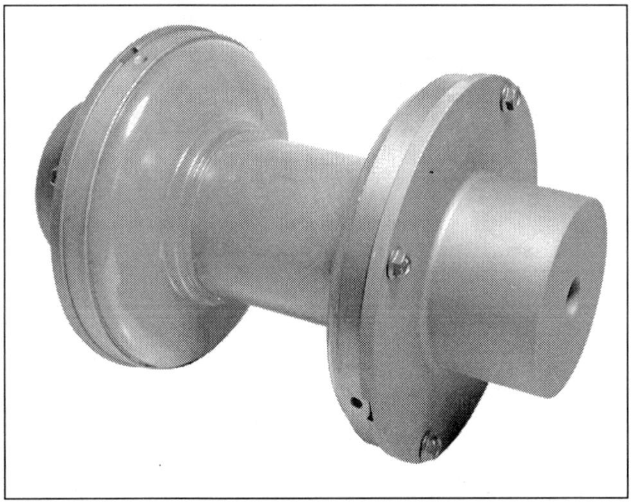

Figure 7-20: Urethane Dual Flex Diaphragm

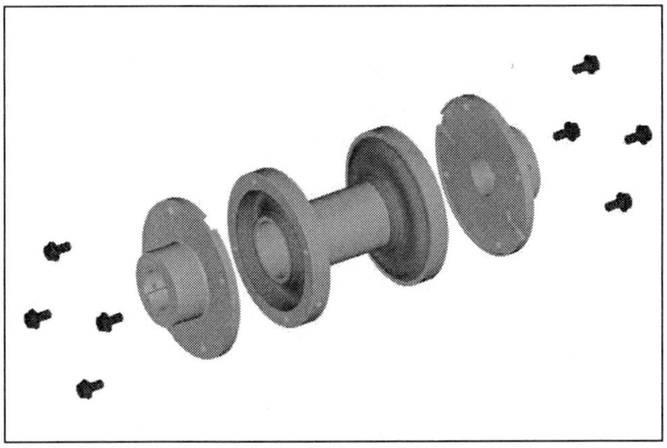

Figure 7-21: Urethane Dual Flex Diaphragm-exploded view

Corded Tire Couplings (Elastomeric Element Type)

Corded tire couplings also have their elastomeric elements in shear. They use a reinforced element (similar to belted auto tires). Because of the reinforcement in the element, the torque capacity is greater per outside diameter than the urethane tire type. Many small- and some medium-size equipment applications use these couplings.

This shear-type coupling is basically a flexible rubber tire with tension-member cords that carry the load. The tire is secured to the flanges, which are secured to the shafts (see Fig. 7-22). Internal reinforcement and external clamping of the tire enhances the torque capacity and stiffness. Rubber-tire couplings are often used to reduce the transmission of shock loads to the surrounding equipment in drives subject to jamming. They also are used with motor generator sets to reduce vibration transmittal from the generator back to the motor. Available in shaft sizes up to 8 in., these couplings enable transmission of a wide range of horsepower at speeds up to 9,000 rpm. They typically accommodate up to 4 degrees of angular misalignment, 1/8 in. parallel misalignment, and 5/16 in. end float.

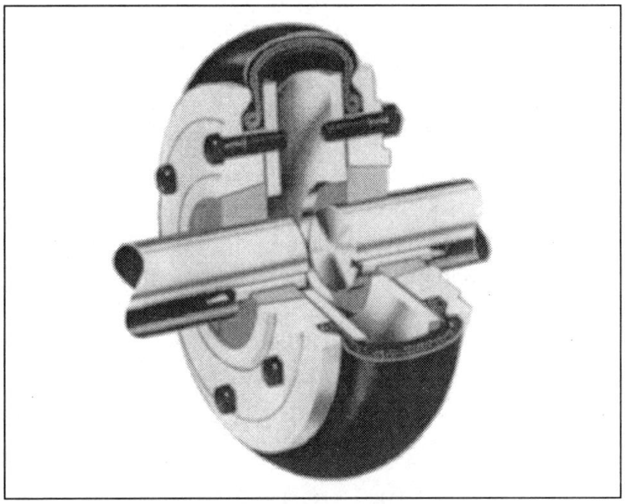

Figure 7-22: Corded tire coupling

Elastomeric Donut Couplings-Unclamped (Elastomeric Element Type)

Another type of coupling that is common on small- to moderate-size equipment, elastomeric donut couplings have their elastomeric elements in shear. They provide low torsional stiffness and low reactionary forces. Oversizing can lead to premature failure by wearing of the engaging teeth rather than material failure flexing.

This shear-type coupling uses a rubber element with internal and external gear teeth that engage with mating teeth in the two shaft hubs (see Fig. 7-23). The rubber elements may be one piece, one piece split, or two pieces split with a steel locking ring that fits into a groove in the rubber. All three types have high torsional flexibility. Because of the flexibility of the element, radial loads imposed on equipment bearing due to parallel misalignment are minimized.

Figure 7-23: Elastomeric donut coupling (unclamped)

In addition to accommodating 1 degree shaft misalignment, these couplings also dampen shock and vibration. Optional elastomeric compounds increase the load-carrying capacity, which in turn generally increases torsional and lateral stiffness. This type of coupling is generally used in electric motor applications.

Compression Donut Couplings (Elastomeric Element Type)

Compression-donut couplings (see Fig. 7-24) have a pre-compressed elastic element. Screws force the donut to a smaller diameter. All legs of the donut are in compression before the load is applied. Medium and some small equipment incorporate these couplings.

Low torsional and lateral stiffness are inherent in this design. These couplings typically accept 3 deg. angular misalignment and 0.080 in. parallel offset.

Figure 7-24: Compression donut coupling

Block Couplings (Elastomeric Element Type)

Block couplings (see Fig. 7-25) use rubber in compression. The rubber blocks are installed in cavities formed by internal sleeve blades, an external hub blade, and two end plates. The blocks are usually precompressed into cavities between blades on the sleeve and hub. This type is unique among couplings due to its fail-safe feature. If the elastomeric fails, the coupling may run for some time on the metal blades. These couplings can provide up to 1/4 deg. of misalignment and parallel offset capabilities of 1/64 to 1/32 in. Sometimes medium and small equipment will use these couplings because, if properly aligned, they require no maintenance (except for replacement of blocks every 3–5 years).

Figure 7-25: Block coupling

Jaw Couplings (Elastomeric Element Type)

Jaw couplings have their elastomers in compression. Jaw couplings have two hubs with projecting jaws that mate with a one-piece elastomeric spider insert (see Fig. 7-26). For high-power applications, individual load cushions may be substituted for the spider. The elastomeric elements can be supplied in various types and hardnesses to provide suitable load-carrying capacities and torsional characteristics. Both torsional stiffness and torque capacity vary with the number of jaws and their width and shape. The flex element can be one piece or split to facilitate replacement. Jaw couplings also have a fail-safe feature. Flex elements are made of many types of elastomeric materials, such as rubber or urethane. The properties (that is, hardness, resilience, etc.) can be varied to suit required loads. These couplings are used primarily to accommodate misalignment and transmit power. Small- and medium-size equipment employ these couplings. Use caution, as sometimes they are used to absorb energy and dampen loads, particularly when equipment is engine driven. Also, they are used at times because these couplings are torsionally soft and are used to tune a system out of a torsional critical mode. Jaw couplings are widely used in both fractional and integral-horsepower service with capacities up to 170,000 lb-in. torque. Misalignment capabilities typically range up to 1.5 deg. angular and 0.086 in. parallel.

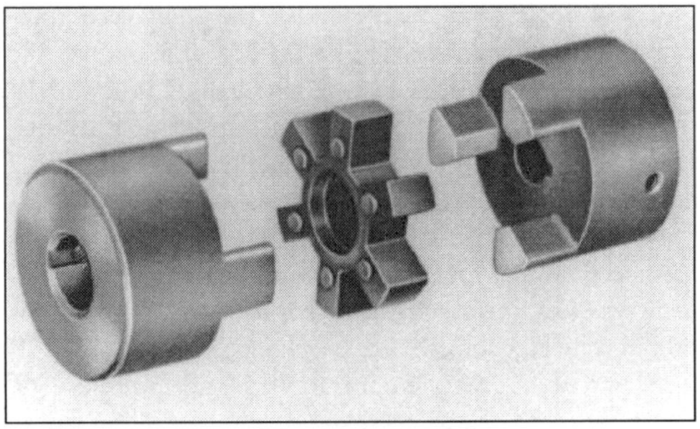

Figure 7-26: Jaw coupling

Pin and Bushing Couplings (Elastomeric Element Type)

Pin and bushing couplings (see Fig. 7-27) are used primarily to accommodate misalignment and transmit power. Relative movement is accommodated by compression of steel-bonded bushings. Small- and medium-size equipment employs these couplings. They are used to absorb energy and dampen loads, particularly when equipment is engine driven.

Figure 7-27: Pin and bushing coupling

Sliding-Block Couplings (Elastomeric Element Type)

Sliding-block couplings, generally known as Oldham's couplings, have two hubs with jaw-shaped flanges. These are positioned at right angles to each other to engage opposing surfaces of a square, sliding-center member (see Fig. 7-28). The center member may require light lubrication, depending on the material used, but others are oil-impregnated or made from polyacetal elastomers, which do not require lubrication.

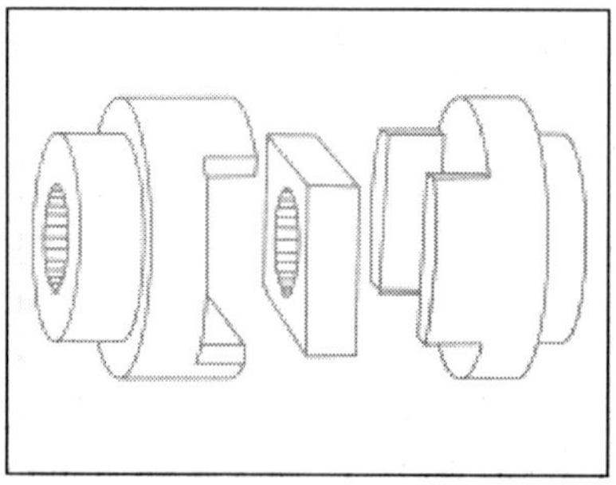

Figure 7-28: Sliding block coupling

This type of coupling transmits torque through an intermediate floating member and compensates for all three types of misalignments by the sliding action between the closely fitted center member and the adjacent driving and driven jaw flanges. These couplings are used in small-size and sometimesmedium-size equipment applications.

Metallic Element Type Couplings

The metallic-element types obtain their flexibility from the flexing of thin metallic discs or diaphragms. Disc types are commonly used for general-purpose application. Diaphragm couplings are mainly used in the special purpose application area and will be covered in that section.

Disc Couplings (Metallic Element Type)

The disc coupling transmits torque by a simple tensile force between alternating driving and driven bolts on a common bolt circle. Misalignment is accomplished from the flexibility that comes from the length of material between the bolts.

There are several types of metal-disc couplings, all of which contain a series of thin, flexible steel discs bolted between coupling flanges (see Fig. 7-29). Made from spring steel, these discs flex to permit angular or parallel misalignment plus end float while providing torsional rigidity. Because there is no sliding between parts, these couplings do not require lubrication. They are used over a wide range of horsepower and speed on equipment such as pumps, fans, blowers, compressors, and motor generator sets. This type of coupling has become very common for gas compression in natural gas and refinery applications due to its ability to be configured to meet API 610 and 671 requirements. It is used for general-purpose-type disc couplings with bore sizes ranging up to about 16 in.

Figure 7-29: Disc Coupling

The flexible discs come in different shapes including square, octagonal, scalloped, circular, and are made from linked segments, (see Figs. 7-30 [A] to 7-30 [D]). One type consists of separate links that are connected at the ends for easy disassembly and replacement.

Metal-disc couplings are usually used with a spacer and have two flexible elements. Disc couplings are sometimes used as a single flexible element and then only accommodate angular misalignment up to 1/2 deg. The double-disc version, however, is the most frequently used in industrial applications. This type accommodates greater angular misalignment (1 deg.) or a parallel misalignment within 1/2 deg. Some special metallic discs can handle up to 2 deg. of angular and 1 deg. of offset misalignment. Disc couplings have been around for years, but with the use of Finite Element Analysis, this type can and has been enhanced for optimum characteristics. These couplings are composed of all metal and do not require lubrication. The discs are usually continuous and of various shapes (see Figures 7-30 [A] to 7-30 [D]: circular, square, scalloped, but can be individual links. Most disc couplings use multiple thin discs rather

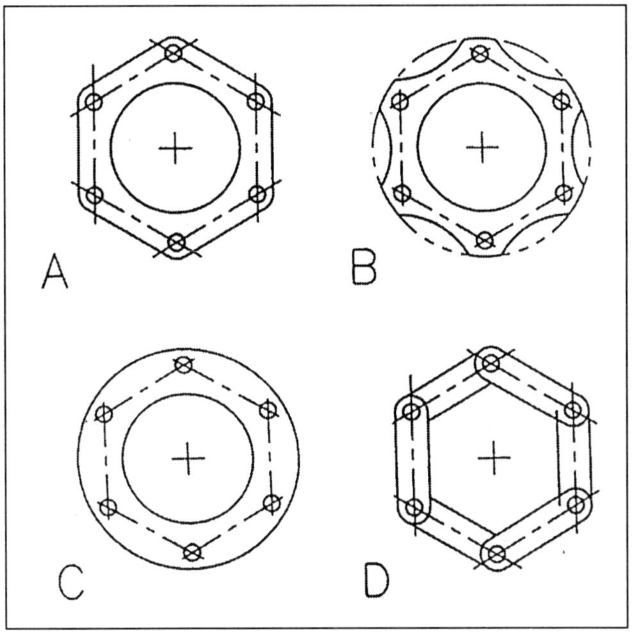

Figure 7-30 (A): Octagonal disc coupling
Figure 7-30 (B): Scalloped disc coupling
Figure 7-30 (C): Circular disc coupling
Figure 7-30 (D): Link disc coupling

than one thick disc/link because stresses from misalignment are proportional to T3 vs. Nt3. These couplings are used in medium-size equipment applications. If the misalignment is beyond 1/2 deg. during operation, the flexible element will probably fail in fatigue.

Miscellaneous General Purpose Coupling Types

Miscellaneous coupling types obtain their flexibility from a combination of mechanisms previously described or through a unique mechanism.

Metallic Beam Couplings (Miscellaneous Type)

Metallic-beam couplings transmit torque through a spiral cut beam. A beam coupling, also known as a helical coupling, is a one-piece metal beam cylinder into which a helical or spiral slot is cut to form a spring-like flexing portion. Beam couplings are considered miniature couplings and are available for shaft applications from 1/8 in. to 1/2 in. (see Fig. 7-31). Helical couplings transmit rotary motion with constant torque.

The beams are shaped to allow torque to be transmitted in either direction. Misalignment is accommodated by deflection of the curved beam. They can accommodate misalignment up to 1/2 deg. These couplings are used in small-size equipment applications where simplicity of installation and zero backlash are beneficial.

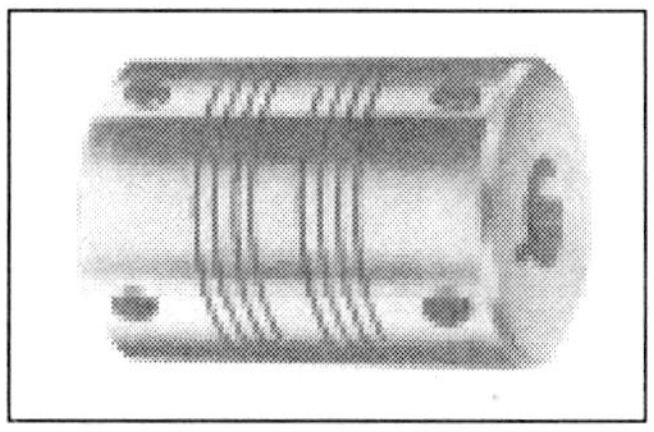

Figure 7-31: Metallic beam coupling

Metal Bellows Couplings (Miscellaneous Type)

Metal-bellows couplings consist of two hubs connected by a flexible bellows section (see Fig. 7-32) and typically accommodate 5 to 10 deg. angular misalignment in light-duty industrial applications, such as motor-to-encoder connections.

Available in shaft sizes to 1 in., metal-bellows couplings are capable of operating at speeds greater than 10,000 rpm. Because they exhibit high torsional stiffness and negligible backlash, these couplings are often used with encoders and resolvers and in servomotors, robots, instruments, and machine tools.

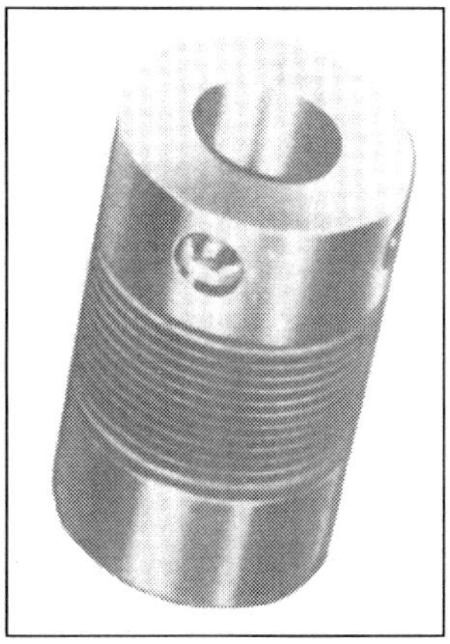

Figure 7-32: Metal-bellows

Tangential Spring Couplings (Miscellaneous Type)

Similar in appearance to the metal-beam device, tangential spring couplings (see Fig. 7-33), consist of one or more

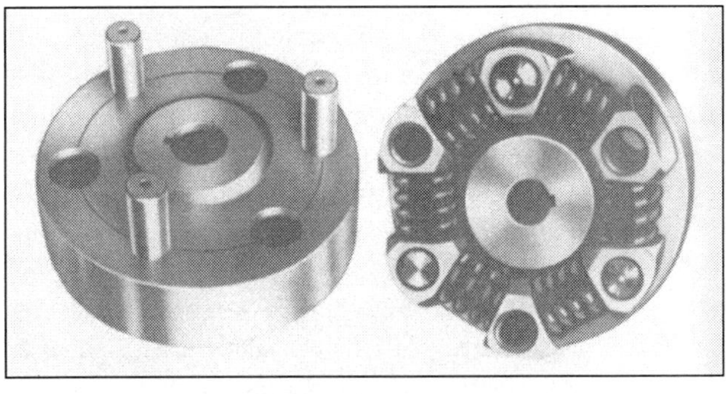

Figure 7-33: Tangential spring coupling

square-wire springs that are concentrically wound and connected to hubs at each end of the spring. Available in both miniature and standard sizes, spring couplings allow approximately 4 deg. angular and up to 1/8 in. parallel misalignment. The hubs may be straight-bored or straight-bushed with bore sizes ranging normally up to 1 1/2 in.

Special-Purpose Couplings

There are two categories, "lubricated" and "non-lubricated." Much of the older equipment still in use today uses lubricated couplings, whereas most new equipment or upgraded equipment will use couplings with low maintenance and require infrequent lubrication, therefore non-lubricated couplings. Generally, the types are gear coupling (mechanically flexible couplings) that require oil or grease lubrication and flexible element types that require no lubrication. The gear couplings category has many subgroups to allow for the many different designs; however, they are only variations. These couplings are usually made of alloy steel and have surface-hardened teeth (usually nitrided). U-Joints also require lubrication, but they have positive type seals that do not flex under misalignment. There are two primary types of non-lubricated couplings (metallic-element couplings). These are used for special-purpose applications: the diaphragm type and the disc type.

A coupling moves from general purpose to special purpose once it is applied to very critical equipment within the production or process system. Thus, a pump coupling on a spared, redundant system in almost any process plant— from refinery to ammonia plant— is likely not a special purpose unit. On the other hand, a 67,000 HP unspared boiler feed pump in a base station or the compressor train in the same ammonia plant as mentioned previously is certainly special purpose and critical to trouble-free operation. A turbine generator set of 40,200 HP at 3,600 rpm (606,615 in/lb.) is also critical. Most gas turbine-driven generators on peaking or cogeneration systems are special-purpose machines that require special purpose couplings. It is easy to identify the end points or obvious situations that require a special-purpose coupling. They are applied to a special purpose or special machine. Those machines can be identified as expensive, high powered, and high speed. Those machines are driven by synchronous motors, gas turbines, or steam turbines at 3,600 rpm or more. The horsepower is usually in excess of 1,000. Usually, due to expense, they are not spared. Also, although these machines are high powered, they are also sensitive to nearly everything in their environment. That is, forces or moments that would seem insignificant to high-powered mill machinery become life-threatening to sensitive machines. As a result of that sensitivity and the speed and power, coupling criteria for the machines take on an entirely different perspective.

When the critical application is found in a refinery or refinery-related setting, the coupling comes under the API 671 specification. That specification has definite requirements for coupling construction as well as coupling selection. For example, the specification calls out certain service factors and certain torque selection variables. A gear coupling selected for the continuous operating torque might have a service factor as high as 1.75. If selected by motor size rather than driven equipment output, it could be as low as 1.2. Transitory torque may also be used for coupling selection.

This is not intended to be inconsistent, but to encourage a dialog between the equipment designer and the coupling manufacturer. That dialog is necessary. Too much coupling can cause operational problems and high cost, just as too little coupling could result in a failure.

High-Performance Gear Couplings (Mechanically Flexible Type)

High-performance gear couplings are available in several configurations. These couplings are usually made from alloy steels and operate at speeds often in excess of normal motor speeds. These couplings are available in sealed-lubrication and continuous-lubrication types. There are four basic styles. Most of these couplings conform to the requirements of API 671.

The sealed lubricated gear coupling (see Fig. 7-34) was adapted and modified during the 1950s and 60s to carry more power at higher speeds. The major problem with gear couplings is lubrication. Grease separates under centrifugal force and seals may leak as a result. Continuously lubricated designs (see Fig. 7-35) using the coupled machine's bearing oil are more common for this type of application. There are several variations of designs including couplings with external teeth on or with the external teeth on the spacer in the "marine" type (see Fig. 7-36). This design could be used with integral flanges on the machine shafts such as on ship propulsion turbines and gears, hence the name marine type. Reduced moment versions were also readily used. The coupling itself tends to act as a centrifuge and separates grease particles out of oil. This produces sludge (see Fig. 7-37), which causes the coupling to lockup and the equipment

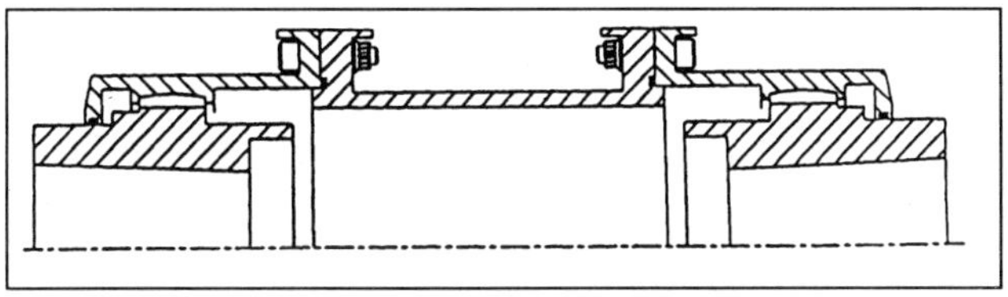

Figure 7-34: High speed gear coupling (sealed lubed)

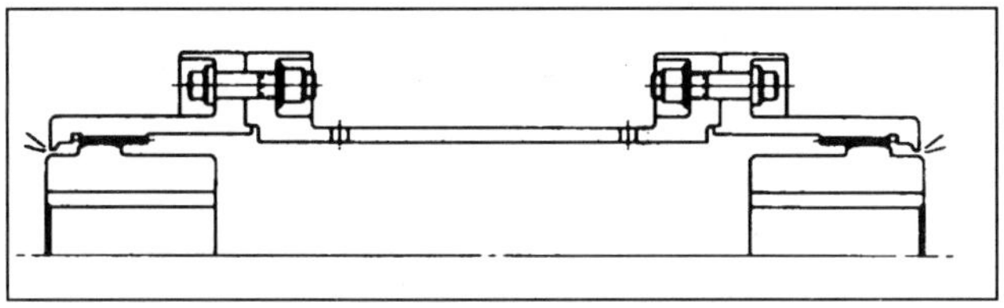

Figure 7-35: High speed gear coupling (continuously lubed)

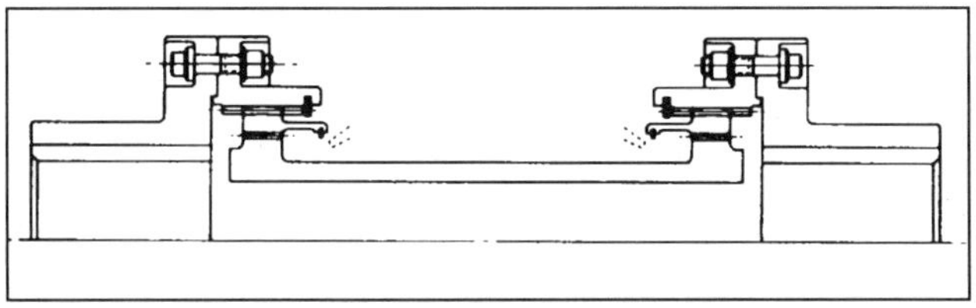

Figure 7-36: High speed gear coupling (continuously lubed marine style)

to vibrate. The most common mode of failure for a gear coupling is wear because of lubrication problems. If proper lubrication is maintained, these couplings will operate successfully for years.

Figure 7-37: Sludged gear coupling

High-Angle Gear Couplings/Gear Spindles (Mechanically Flexible Type)

High-angle gear couplings (gear spindles) are used on heavy-duty, high-torque applications such as metal rolling mill main drives. The gear teeth of these spindles are usually made of alloy steels and are surface hardened (induction hardened, nitrited, or carburized). Deciding which material and heat treatment is best depends on the specific application.

Bonded-Shear Couplings (Elastomeric Element Type)

Bonded-shear couplings (see Fig. 7-38) are used in special applications requiring specific torsional stiffness to tune the system (such as diesel drives). This type of coupling can be very large (over 60 in.). Bonded-shear couplings tend to be 30–50 percent larger than elastomeric compression type in order to handle the same torque. They are much more flexible (2–3 times) than compression-type elastomeric types. If the elastomeric fails, the coupling totally disengages. These couplings can provide up to 1/2 deg. of misalignment and parallel offset capabilities of 1/32 to 1/16 in.

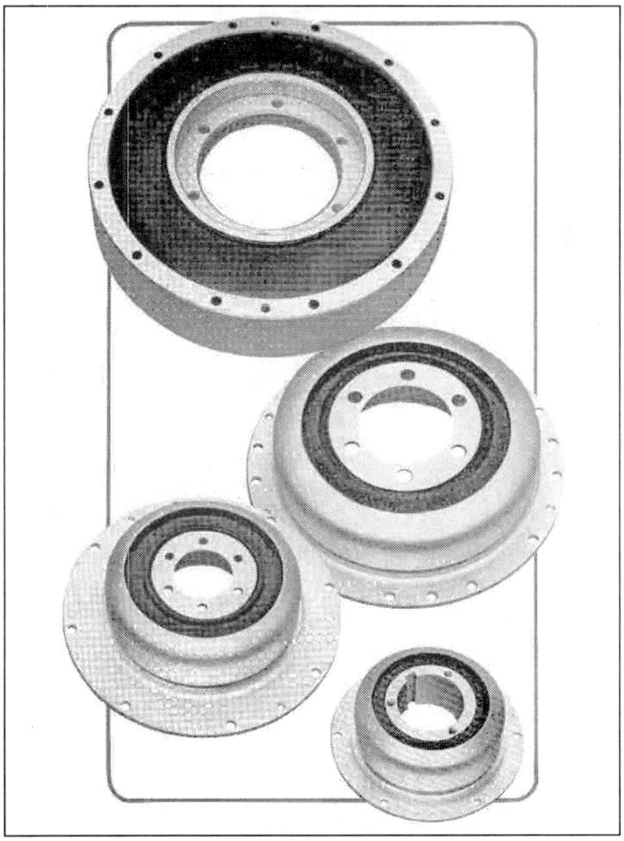

Figure 7-38: Bonded shear coupling

Elastomeric Block Couplings (Elastomeric Element Type)

Elastomeric block couplings (see Fig. 7-39) are used on large critical equipment (such as synchronous motor-driven equipment) and used for special-purpose applications to reduce vibratory torques or because it is torsionally soft, tune a system. Block couplings use rubber in compression. The rubber blocks are installed in cavities formed by internal sleeve blades, an external hub blade, and two end plates. This type is unique among couplings due to its fail-safe feature. If the elastomeric fails, the coupling may run for some time on the metal blades. These couplings can provide up to 1/4 deg. of misalignment and parallel offset capabilities of 1/64 to 1/32 in. These couplings are often used in combination with another type of special-purpose coupling on the gear box driven end, such as a gear, disc, or diaphragm. The elastomeric is usually on the low-speed end of a train and usually mounted on the motor.

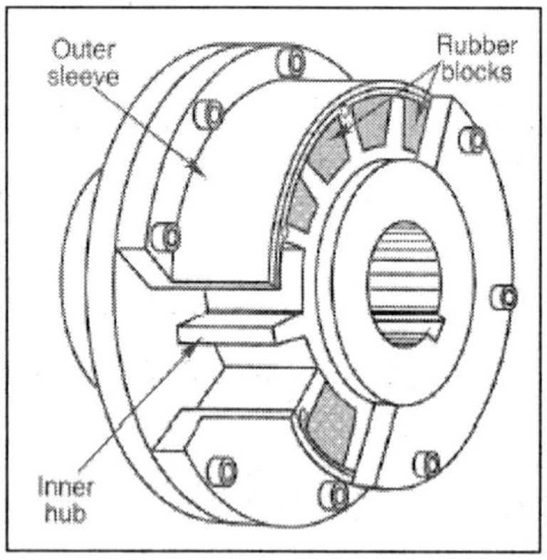

Figure 7-39: Block coupling

Disc Couplings (Metallic Element Type)

The disc coupling is available in a number of forms (see Fig. 7-40), all having driving and driven bolts on the same bolt circle. The amount of misalignment that each type can handle depends on the length of material between bolts. Torque is transmitted by driving bolts pulling driven bolts with disc material. More bolts provide greater capacity but reduce the coupling flexibility. For special-purpose applications, the discs are provided as a pack. Some disc packs are factory installed, while others are supplied with pilot rings for installation in the field. They are available in many styles, including close coupled, reduced moment (see Fig. 7-41), and marine (see Fig. 7-42).

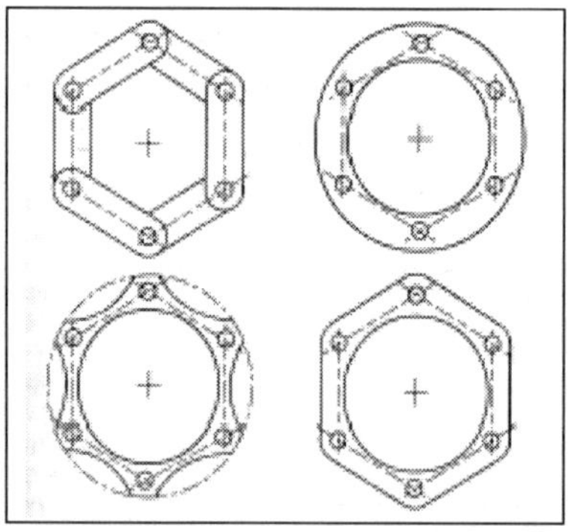

Figure 7-40: Various disc shapes

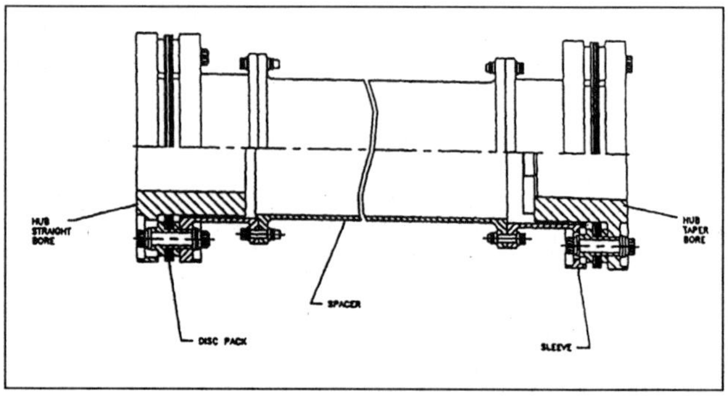

Figure 7-41: Reduced moment disc coupling (scalloped)

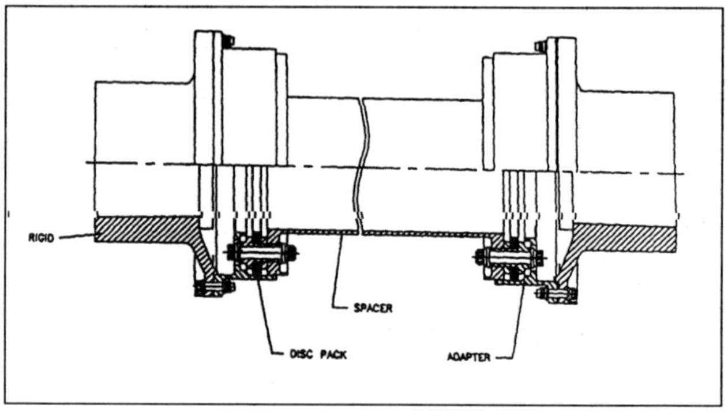

Figure 7-42: Marine style disc coupling (scalloped)

Disc Coupling Operating Principle

One style of coupling used to replace a gear coupling on special-purpose machinery is disc coupling. The principle of operation is that torque is transmitted through a flexible element by tensile loading between alternate bolts that are on a common bolt circle. One of the alternate bolts was the load transmitter, and the other was the load receiver. They are fastened to opposite sides of the torque path.

The misalignment is accommodated by the flexing of the element between adjacent bolts (see Fig. 7-43). The element must be thin to be flexible. Stacks of elements provide parallel load paths, and the diameter of the bolt circle is an indicator of the amount of torque to be carried. The amount of misalignment is related to the chord length be-

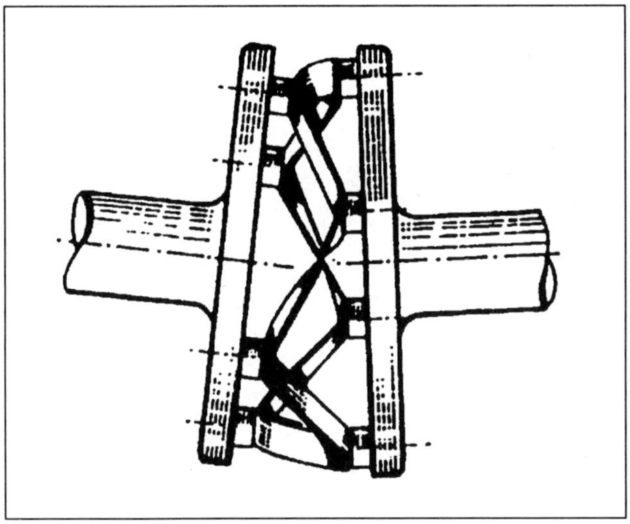

Figure 7-43: Disc Coupling

tween bolts and the thickness of the disc. The thin element had both tensile and bending imposed on the element so that a fatigue life analysis was required. Once the fatigue life was determined, an infinite life coupling could be designed to keep loading below that fatigue limit. It took some time to develop high fatigue strength materials and advances in material load analysis, such as finite element, were necessary to speed the development of reliable couplings. The disc coupling is non-lubricated and designed for infinite life. The amount of misalignment available is a function of the bending that can be accommodated while under tension. The unit will allow axial movement, as seen with thermal growth; however, that is not unlimited. Axial capability and angular capability are interrelated. An important feature of disc couplings, and couplings that work in a similar manner, is the low reactionary load that is transmitted to the machinery that it couples. That low load is relative to the high load of gear couplings. Disc couplings can be initially more expensive than gear couplings but are less expensive in total. The key element in the disc coupling is the disc pack.

Disc Coupling Design Considerations

When we apply the couplings to critical services, we have to make concessions and modifications to accommodate high speed and high torque. High speed means balance, while high torque means strength. The combination of misalignment, speed, and torque means fatigue resistance.

As is the case with other types of couplings, additional considerations are needed for the special-purpose application. In order to save space and reduce cost, the coupling designer will select disc packs to closely match the application. In some cases, a disc will be designed specifically for the equipment. That design will involve the bolt circle diameter, number of bolts, size of bolts, and number of discs needed. Once the disc pack unit is designed and built at the factory, the pieces should not be disassembled. Piloted disc packs or factory-assembled disc packs help to ensure against fatigue failure.

Balancing of the disc coupling is no different from balancing of any other rotating part. API 671 is very specific about the balance requirements of special purpose couplings. Both high-speed and low-speed categories are addressed. First, the design is made symmetrical, and then it is manufactured to tight tolerance on concentricity and squareness. Finally, the components are balanced. The disc coupling can also be easily balanced as an assembly because the flexing pieces are not a loose fit.

Designing for strength is a function of the disc pack materials and the shape of the disc at critical points, such as the bolt holes. The high-performance discs are made from cold-rolled stainless steel (generally 300 series). Special discs are made of monel, inconel, PH stainless, and other special materials. Sometimes discs are coated to minimize the effects of fretting at high angles. Corrosion, if it is a factor, is controlled by material selection. Bending, which comes from misalignment, is controlled by geometry, thickness, number of bolts, and fatigue strength of the design.

API 671 covers the strength issue by specifying a fatigue factor of safety using the proportional increase method with the Modified Goodman Diagram or the Constant Life Curves. Those references are used with material fatigue strength and ultimate strength. It is an issue best left to the coupling designer, but one for which the designer needs to have complete application information. Axial movement, axial thrust, and maximum allowable angular misalignment are important at this juncture.

The criterion for coupling selection is again based on torque requirements versus torque capabilities. For the disc coupling, it is important to understand and know that the misalignment requirements intermesh with the torque requirements. The angular misalignment and axial displacement both distort or bend the elements. With each revolution of the coupling, the bending from misalignment is reversed or flexed. That bending is the source of fatigue loading. The coupling manufacturer will help select the coupling so that the effects of the bending are within the coupling capabilities.

Diaphragm Couplings (Metallic Element Type)

The diaphragm coupling comes in two basic forms: a single-tapered profile that is either welded (see Fig. 7-44) or one-piece (See Fig. 7-45) and a multiple-modified profile either straight with cutouts (see Fig. 7-46) or contoured (see Fig. 7-47). Torque path is through the diaphragm member in the radial direction—from the outer diameter to the inner diameter. The Diaphragm couplings are used in most special-purpose applications and are available in many shapes and styles, including marine (see Fig. 7-48) and reduced moment (see Fig. 7-49).

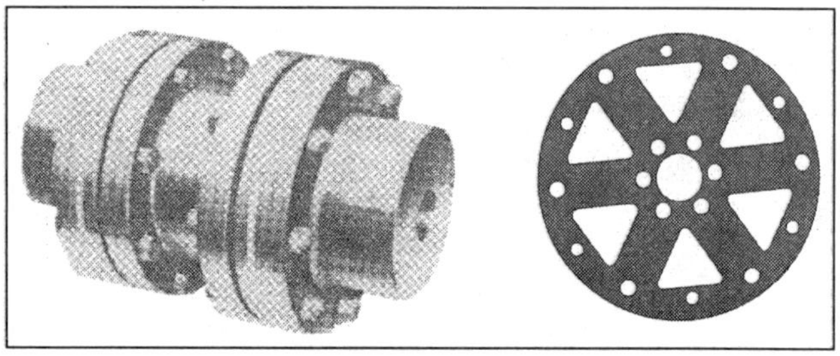

Figure 7-46: Multiple straight diaphragm coupling

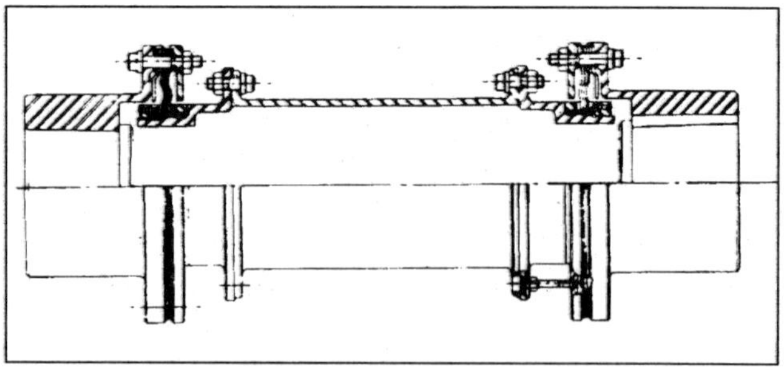

Figure 7-47: Multiple convoluted diaphragm coupling

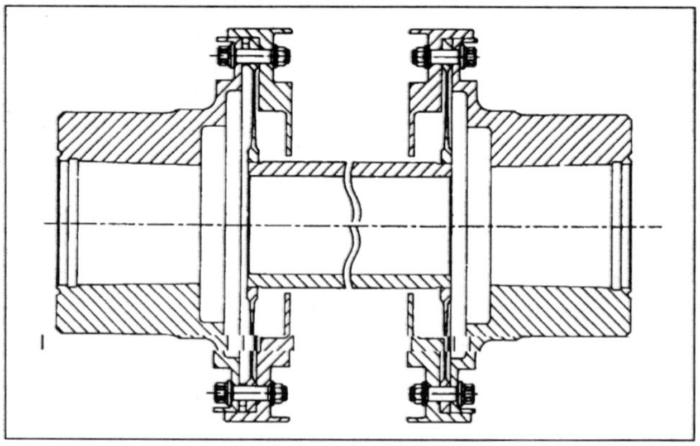

Figure 7-44: Taper diaphragm coupling (welded)

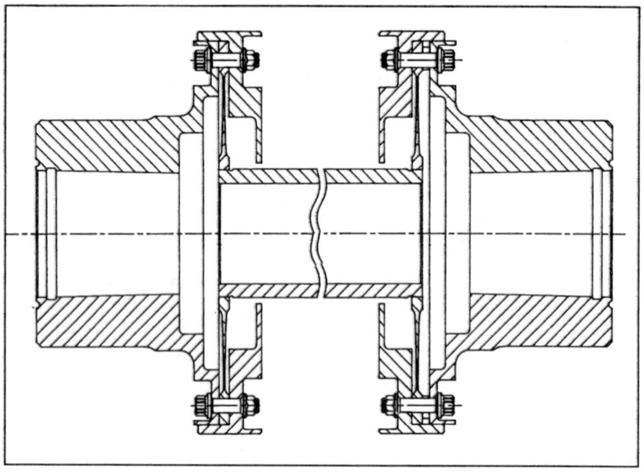

Figure 7-48: Marine style Diaphragm

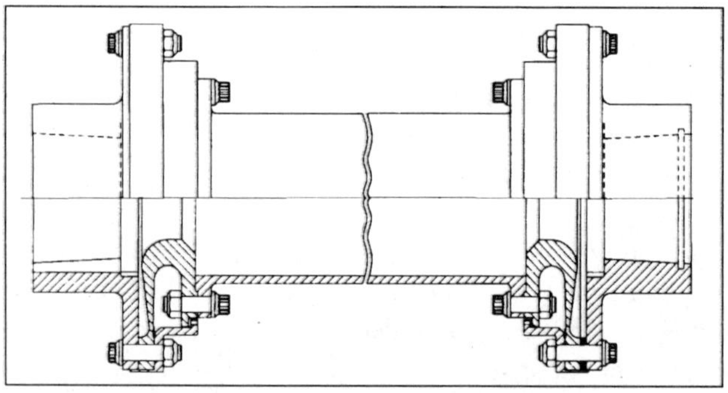

Figure 7-45: One piece diaphragm coupling

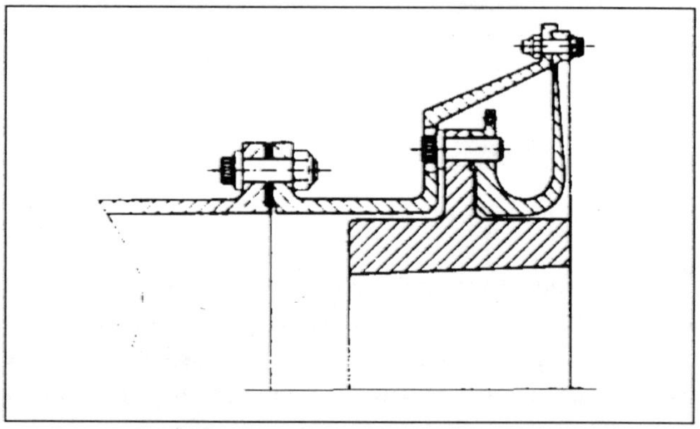

Figure 7-49: Reduced moment Diaphragm

Diaphragm Coupling Operating Principles

Diaphragm types can be classified as couplings that utilize a single plate or a series of plates or diaphragms for the flexible members. The torque transmission path through the diaphragm members is in the radial direction—from the outer diameter to the inner diameter, or vice-versa. Load from operating torque is seen as a shear stress on the diaphragm member(s).

There are two basic types of diaphragm couplings:

1. Single (see Fig. 7-50)
2. Multiple (see Fig. 7-51)

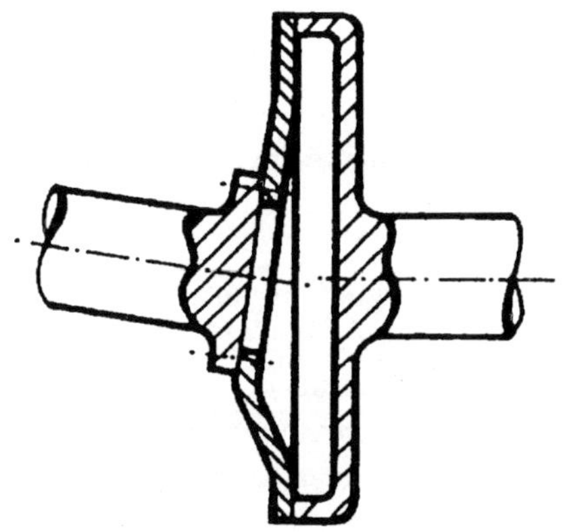

Figure 7-50: Single Diaphragm

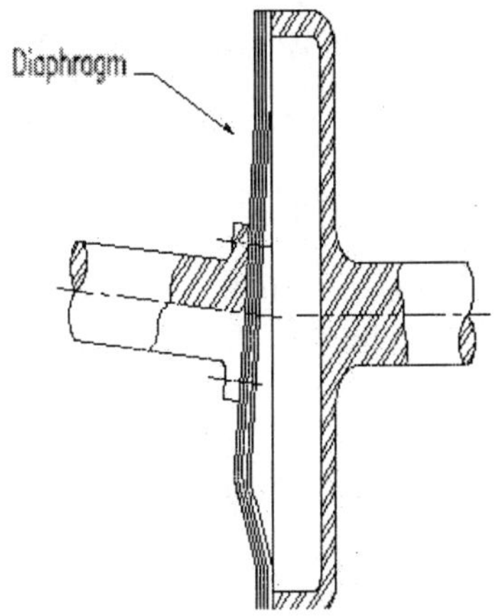

Figure 7-51: Multiple Diaphragm

They come in various profile shapes: contoured tapered, convoluted wavy, and flat-profile modified with spokes or a series of cutouts.

All shapes have some type of profile modification that helps reduce size, increase flexibility, and control stress concentrations. A contoured diaphragm coupling typically uses a single diaphragm plate for the flexible member; the plate has a contoured or wavy profile, which usually has a variable thickness from OD to ID to provide an optimum stress condition. A convoluted and flat-profile diaphragm coupling typically uses multiple diaphragm plates that have a wavy profile or other modified profile. All types of diaphragm couplings attach the flexible member to other components with bolts, splines, or welds, and both transmit torque in the same manner. Diaphragms are made of high-strength materials; some are corrosion resistant (15-5/17-4 PH), and others use 4,300 steel and coat the diaphragms for corrosion protection. Some diaphragm couplings are shot-peened to reduce the residual stresses that are imposed during the manufacturing process.

Diaphragm Coupling Design Principles

Diaphragm couplings use a single contoured plate for the flexing members: the plate is relatively thin and called a diaphragm. Each diaphragm can be deformed much like an oil can lid (prior to plastic). This deflection of the outer diameter relative to the inner diameter is what occurs when the diaphragm is subject to misalignment (see Fig. 7-50).

Angular misalignment twists the outer diameter relative to the inner diameter, and produces a complex shape on the diaphragm where it must stretch one way at one point and then stretch the other way at 180 deg. In between these points, the diaphragm is subject to a combination of stretching and twisting; axial displacement attempts to stretch the diaphragm, which results in a combination of elongation and bending of the diaphragm profile.

Multiple diaphragms accommodate misalignment somewhat differently. They use multiple thin plates (see Fig. 7-51) that are made to be wavy from OD to ID. They react similarly to the contoured diaphragm under misalignment except that they unfold the wavy profile of the plates instead of stretching the diaphragm, or they have holes or cutouts to make them more flexible.

Contoured Diaphragm Coupling

The contoured diaphragm coupling has as its flexible element—a thin profiled diaphragm machined from a solid disk of heat-treated, vacuum-melted alloy steel. This diaphragm is contoured so that it has a nearly uniform torsional shear stress throughout the profile, which is therefore thicker at the hub, or ID, and thinner near the rim, or OD (see Fig. 7-52). The purpose of contouring the profile is to keep the diaphragm as thin as possible, consistent with the transmitted torque. This keeps the misalignment bending and axial bending stresses as low as possible for a given torque capacity.

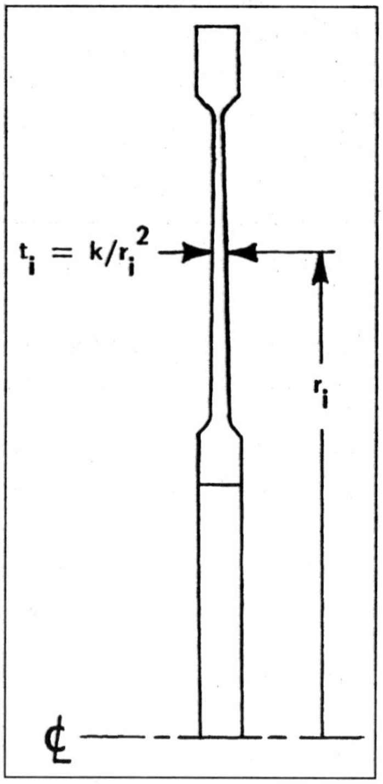

Figure 7-52: Typical Shape of Contoured Diaphragm

The thickness of a diaphragm can be changed to permit a tradeoff between torque capacity and flexibility. A thicker diaphragm has greater torque capacity but is not as flexible, and vice versa. Smooth fillet junctions are provided between the flexing portion and the rigid integral rims and hubs, which connect to the rest of the coupling.

In one configuration, the diaphragm hub is electron-beam welded to the spacer tube (see Fig. 7-53) in a permanent connection.

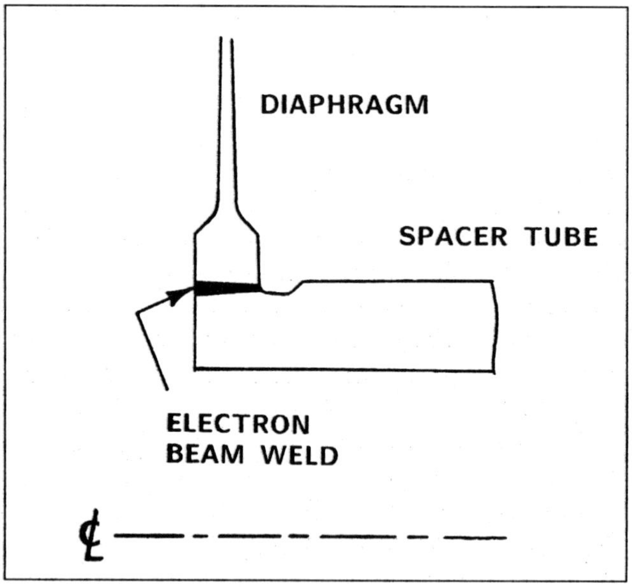

Figure 7-53: Electron beam welded diaphragm

In another configuration, the diaphragm incorporates an integrally machined flange (see Fig. 7-45).

Whereas one design may minimize the number of mechanical connections in the coupling, the other allows flex element replacement. The most often used marine style, coupling configuration (see Fig. 7-48) allows the mounting hub bore to vary considerably without affecting the diaphragm diameter. Thus, the size of the coupling is being chosen to fit the torque and misalignment requirements rather than being dictated by the connected machine shaft size.

A piloted guard configuration incorporates diametrical-locating pilots, which meet the API Standard 671 and enhance the balance repeatability of the coupling. This makes using the technique of only balancing the coupling components (for component interchangeability) more practical.

For those special applications requiring a reduced moment coupling, the contoured diaphragm coupling is made with the diaphragms machined from forgings with integral hubs. Or, the diaphragm is reduced in ratio and inverted so it sits over the hub, or it is splined onto the hub (see Fig. 7-54, 7-55, and 7-56).

This configuration shifts the flexible center closer to the machinery bearings to reduce the overhung moment from the weight by moving the coupling center of gravity (CG) toward the machine bearings.

Multiple Convoluted Diaphragm Coupling

The multiple convoluted diaphragm coupling uses a stainless steel diaphragm pack (see Fig. 7-47). The pack consists of several thin, separated, convoluted diaphragms.

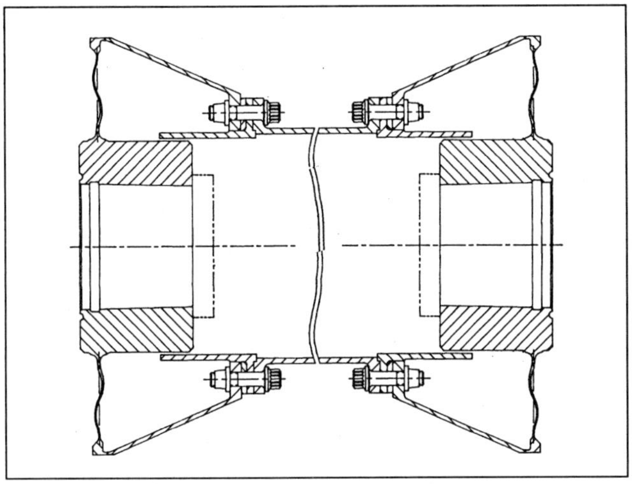

Figure 7-54: Reduced moment coutoured

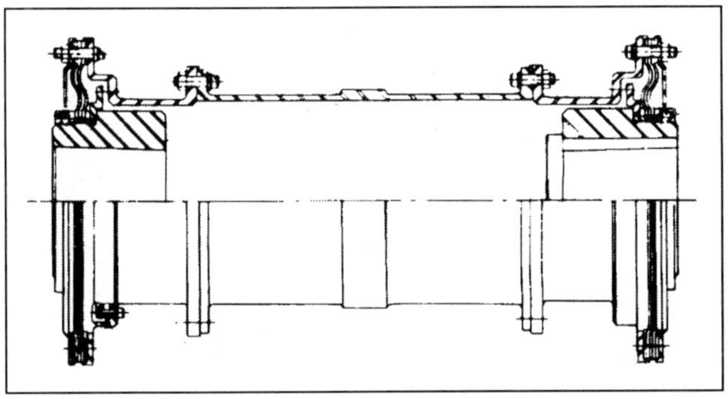

Figure 7-55: Reduced Moment Multiple Convoluted diaphragm

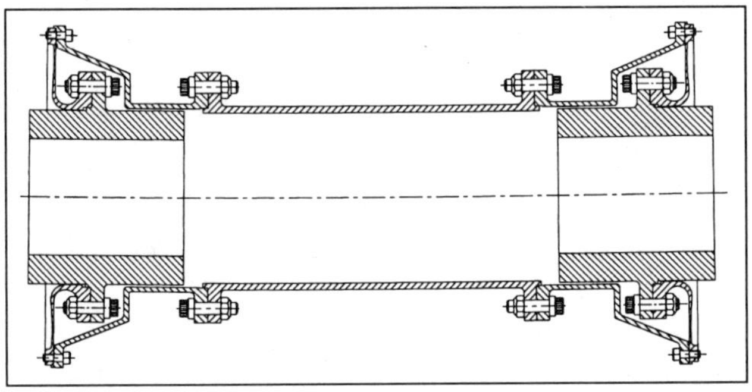

Figure 7-56: Reduced moment one piece contoured diaphragm

The design comes in two standard styles of diaphragms: a large OD/ID ratio and a reduced OD/ID ratio. The diaphragm pack consists of several thin convoluted diaphragms, separated at the OD and ID. The diaphragms are attached at the ID by a fine pitch spline to an adapter that is clamped tight by a nut.

Multiple Flat Diaphragm Couplings

The multiple flat diaphragm coupling (see Fig. 7-46) consists of a series of thin plates assembled as packs with welds, rivets, etc. Torque is transmitted from the hub to the pack by splines, bolts, or friction fits. There are three basic styles of multiple flat diaphragm couplings.

Miscellaneous Special Purpose Coupling Types

Other unique styles of couplings have developed to solve difficult industrial problems. Offset coupling is unique in its flexibility, and there are several types that have unique features and applications. Two types of magnetic couplings, three types of fluid couplings, and two types of special-composite couplings each have advantages over traditional couplings. Couplings can even have torque-measuring capabilities incorporated into them.

Offset Coupling/Schmidt Coupling (Miscellaneous Type)

Offset couplings (see Fig. 7-57) are designed to handle large parallel shaft offsets up to 18 in. and torques to 1 million in/lbs of torque. The coupling is used on steel mill equipment, conveyors, etc.

Figure 7-57: Offset coupling (Schmidt coupling)

Magnetic Type Couplings

Two types of magnetic couplings are available: the magnet-to-magnet type and a new eddy current type. Both offer the advantages of no lube, no contact, large installation tolerances, and long life. A magnet-to-magnet coupling (see Fig. 7-58) transmits torque without physical contact. Power is transmitted through air as a result of the magnetic

Magnetic coupling: Contains three primary elements: a driving and a driven member, both containing permanent magnets, and a nonmagnetic barrier. The coupling develops torque through interaction of the two sets of magnets. Torque ratings run from a few oz-in. to over 500 lb-ft.

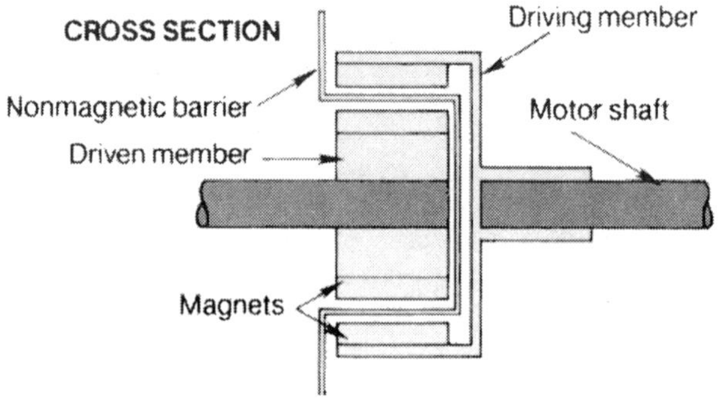

Figure 7-58: Magnet to Magnet coupling

fields interacting with each other, opposite poles attracting, and like poles repelling. This type of coupling provides the advantage of fully synchronous motion between the driver and driven equipment under operating conditions. The disadvantage is the vibration induced in the drive train when the coupling is overloaded. When the two sets of magnets pass by one another during overload conditions, the magnetic fields jump from one to the next, creating torsional vibration. This also occurs at startup, as the driven equipment resembles a stalled condition until the inertia is overcome and the speeds of driver and driven match.

Eddy Current Couplings

This type of coupling (see Fig. 7-59) eliminates the overload and startup vibration problems. High-energy permanent magnets induce an eddy current flow in a nearby conductor when relative motion is present, the same way the rotating magnet moves the needle inside a speedometer.

When the equipment is started, the relative motion between the two components creates a drag on the driven equipment and smoothly accelerates the load. Under normal operating conditions, the driver rotates slightly faster than the driven equipment. This is necessary to produce the relative motion that creates the eddy currents. A significant advantage provided by this type of coupling is the reaction to overload conditions. As torque increases, the relative motion between the two components increases, producing more torque until a peak torque is reached. When the torque load exceeds the capacity of the coupling, the driven side of coupling smoothly slows down. If the load stalls, the driver side continues to rotate, unloading the driver. The result is a smooth overload condition that does not induce vibration. The same phenomenon absorbs vibration produced in the drive train by the driver or driven equipment. A disadvantage of this type of coupling is the small loss in rpm from driver to driven, which is frequently less than 5 percent, but it can be less than 2 percent.

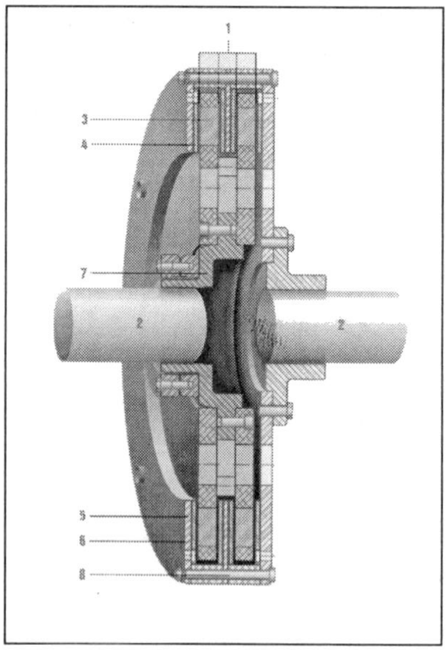

Figure 7-59: Eddy Current Coupling

Fluid Couplings

Fluid couplings are available with three basic types of fill: liquid, silicone, and shot filled.

Liquid-Filled Couplings

Liquid-filled fluid couplings (see Fig. 7-60) allow the motor to start under light load conditions, provide torque overload protection, and absorb shock loads. Optional reservoirs delay full engagement and provide extended start time, further reducing motor inrush currents. Although these devices are called couplings, they cannot accommodate misalignment. To reduce stress and strain on the connected equipment shafts, conventional couplings are used to accommodate misalignment.

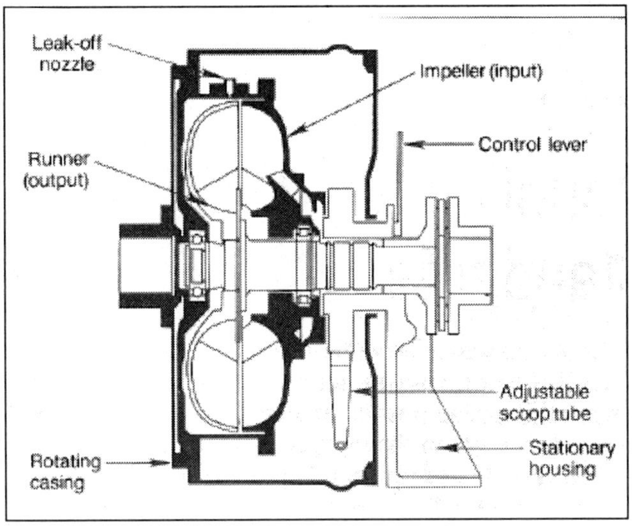

Figure 7-60: Liquid filled fluid coupling

Silicone-Filled Fluid Couplings

Silicone-filled fluid couplings (see Fig. 7-61) transmit torque through the natural shear resistance of silicone fluid between the input and output members. Silicone-filled fluid couplings provide similar soft-start and overload protection as other fluid-filled couplings, but with a greatly reduced risk of fire.

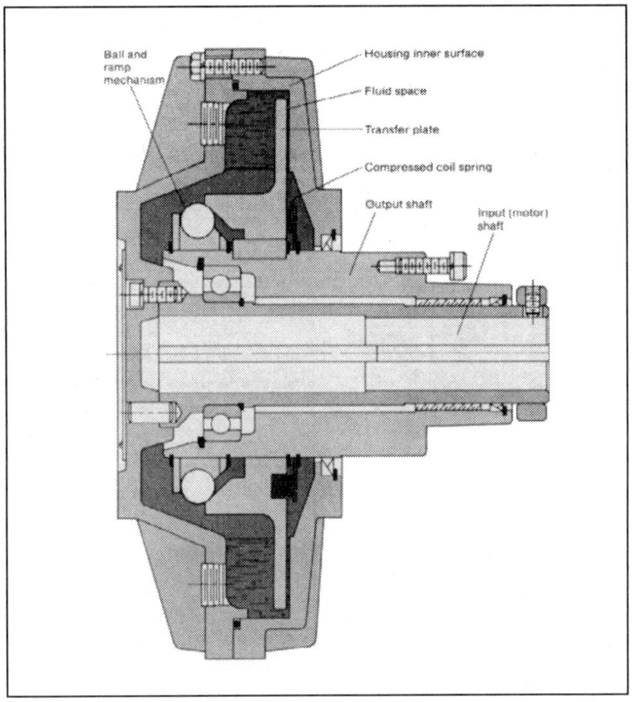

Figure 7-61: Silicone filled fluid coupling

Shot-Filled Fluid Couplings

Shot-filled fluid couplings (see Fig. 7-62) are dry fluid couplings. The hardened steel balls inside are distributed inside the driver housing by centrifugal force. The shot locks together, clamping on the driven rotor and transmitting power without slip.

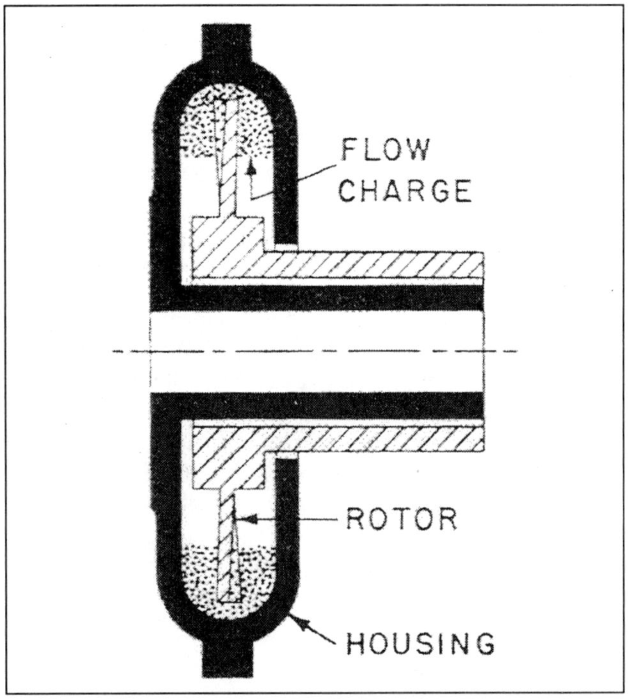

Figure 7-62: Shot filled coupling

Composite Couplings

Composite couplings (see Fig. 7-63) transmit torque without lubrication the same as a disc coupling, but they provide additional advantages of higher strength-to-weight ratio, very low thermal expansion, and higher critical speeds. Composite flex elements utilize high strength graphite fibers wound around stainless steel bushings to eliminate the fretting failure associated with disc couplings. The bonded elements are then coated with urethane to ease handling and improve appearance.

Size for size, composite couplings do not have the torque capacity of traditional disc couplings and have lower maximum speeds. As with disc couplings, the misalignment of a composite coupling is limited by the angular capacity of

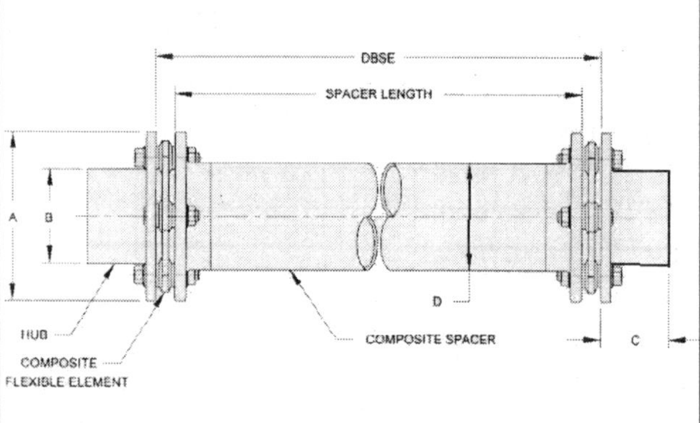

Figure 7-63: Composite Coupling-Cooling tower

each flex element. Typical misalignment limits are 1 deg. per flex element.

Windmill applications have several unique requirements: large misalignment (1/2–1 deg.), cyclic torque because of changes in the wind (cyclic torque can be of 10–15 percent), the need for a unique flexible element that can last 5–10 years without replacement due to it being in a place difficult to service—100 feet in the air and in a tight Nutall structure. These elements are not only made of composite but are unique in shape (see Fig. 7-64) as they are cantilever links.

Figure 7-64: Windmill composite caniliver disc coupling

Also, besides composite flexible elements, there is an advantage in some applications to make the spacers of various composite materials. This allows for the elimination of support bearing on long drives (cooling tower couplings and marine propulsion shafting). Also, composites can allow for a better solution to help torsionally and laterally tune a system.

Coupling and U-Joint Standards

The American Gear Manufacturers Association (AGMA) publishes standards for flexible couplings that cover nomenclature, unbalance classification, lubrication, bores and keyways, keyless fits, and dimensions for standard gear-coupling flanges.

- AGMA 9000 – Flexible Couplings—Potential Unbalance Classification
- AGMA 9001 – Flexible Coupling—Lubrication
- AGMA 9002 – Bores and Keyways for Flexible Couplings (Inch Series)
- AGMA 9003 – Flexible Couplings—Keyless Fits
- AGMA 9004 – Flexible Couplings—Mass Elastic Properties and Other Characteristics
- AGMA 9008 – Flexible Couplings—Gear Type—Flange Dimensions, Inches
- AGMA 9009 – Flexible Couplings—Nomenclature for Flexible Couplings
- AGMA 922-A98-A98 – (Information Sheet) Load Classification and Service Factors for Flexible Couplings
- The Mechanical Power Transmission Association (MPTA) has developed standard terms for elastomeric couplings.
- MPTA-C2C-2002 Elastomeric Coupling Alignment

The American Petroleum Institute has two specifications that are for couplings or include requirements for couplings. API 671 Special-Purpose Couplings for Petroleum, Chemical, and Gas Industry Services is now in its third edition and is over 25 years old.

The ISO equivalent to the API relevant API standard is ISO 10441 – Flexible couplings for mechanical power transmission – Special Purpose applications. This committee also has written ISO 14691 – Petroleum and Natural Gas Industries – Flexible Couplings for Mechanical Power Transmission—General Purpose Applications.

Table 7-1

GENERAL PURPOSE COUPLINGS	COUPLING FUNCTIONAL CAPACITIES				
	MAX. CONTINUOUS TORQUE (IN-LBS)	MAX BORE (IN)	ANGULAR MISALIGNMENT (DEGREE)	PARALLEL OFFSET (IN/IN) or (IN X OD) or (IN)	AXIAL TRAVEL (IN)
MECHANICALLY FLEXIBLE TYPE					
GEAR TYPE					
STRAIGHT TOOTH	54,000,000	45	1/2	.009 IN/IN	1/8-1
CROWNED TOOTH	54,000,000	45	1 1/2	0.026 IN/IN	1/8-1
CHAIN COUPLING					
STEEL	1,350,000	10.5	1-2	0.004 IN/IN	1/4
NYLON	50,000	8	1-2	.009 IN/IN	1/2
GRID TYPE					
VERTICAL SPLIT COVER	4,000,000	10	1/3	0.004 IN/IN	3/16
HORIZONTAL SPLIT COVER	4,000,000	20	1/3	0.004 IN/IN	3/16
U-JOINT TYPE					
VEHICLE	30,000	6	15	.25 IN/IN	12[a]
MECHANISM	30,000	2.5	35	.57 IN/IN	NA
YOKE	135,000	6.5	20	.34 IN/IN	12[a]
ELASTOMERIC ELEMENT TYPE					
SHEAR TYPE					
URETHANE TIRE	175,000	8	4	3/16 IN	1/8
CORDED TIRE	450,000	11	4	1/8 IN	1/4
SHEAR DONUT	75,000	6	1	1/4 IN	5/16
COMPRESSION TYPE					
COMPRESSION DONUT	26,500	5.5	1	0.086	1/16
BLOCK	20,000,000	34	1/4	0.0003 X OD	1/32
JAW	550,000	10	1/4	0.0003 X OD	1/16
PIN & BUSHING	1,000,000	12	1/8	0.0003 X OD	1/16
METALLIC ELEMENT TYPE					
DISC TYPE					
CIRCULAR DISC	4,000,000	15.5	1/2	.009 IN/IN	3/8
SQUARE DISC	4,000,000	15.5	1/2	.009 IN/IN	3/8
SCALLOPED DISC	4,000,000	15.5	1/2	.009 IN/IN	3/8
LINK DISC	250,000	7.5	1/2	.009 IN/IN	1/4
MISCELLANEOUS TYPE					
PIN & BUSHING	150,000	6	1/8	0.001 x OD	1/8
METALLIC BEAM	20,000	3	1/2	.009 IN/IN	1/4
SLIDING BLOCK	20,000	3.5	1	0.001 x OD	1/8

[a] telescoping shaft is used to accommodate axial travel NA = NOT APPLICABLE

Table 7-2

SPECIAL PURPOSE COUPLINGS	COUPLING FUNCTIONAL CAPACITIES				
	MAX. CONTINUOUS TORQUE (IN-LBS)	MAX BORE (IN)	ANGULAR MISALIGNMENT (DEGREE)	PARALLEL OFFSET (IN/IN) OT (1N X OD) OR (IN)	AXIAL TRAVEL (IN)
MECHANICALLY FLEXIBLE TYPE					
GEAR TYPE					
HIGH ANGLE (GEAR SPINDLE)	85,000,000	26	6	0.1 IN/IN	12
HIGH SPEED (SEALED LUBED)	10,000,000	18	1/2	0.009 IN/IN	1
HIGH SPEED (CONTINUOUSLY LUBED)	10,000,000	18	1/2	0.009 IN/IN	1
HIGH SPEED (CONTINUOUSLY LUBED MARINE STYLE)	10,000,000	18	1/2	0.009 IN/IN	1
U-JOINT TYPE					
CONSTANT VELOCITY	56,000	4	20	.34 IN/IN	NA
YOKE	70,000,000	30	15	.25 IN/IN	12[a]
BLOCK	70,000,000	30	15	.25 IN/IN	12[a]
ELASTOMERIC ELEMENT TYPE					
SHEAR TYPE					
BONDED SHEAR	350,000	16	1/2	0.0005 X OD	1/16
COMPRESSION TYPE					
BLOCK	20,000,000	34	1/4	0.0003 X OD	1/32
METALLIC ELEMENT TYPE					
DISC TYPE					
REDUCED MOMENT (SCALLOPED)	3,500,000	9	1/3	.006 IN/IN	.375
MARINE STYLE (SCALLOPED)	3,500,000	18	1/3	.006 IN/IN	.5
FRAME	100,000	6	1	0.018	.25
DIAPHRAGM TYPE					
TAPER	6,000,000	20	1/2	0.009 IN/IN	1
ONE PIECE	6,000,000	20	1/2	0.009 IN/IN	1
MULTIPLE STRAIGHT	6,000,000	20	1/2	0.009 IN/IN	1
MULTIPLE CONVOLUTED	6,000,000	20	1/2	0.009 IN/IN	1
MISCELLANEOUS TYPE					
OFFSET (SCHMIDT)	1,000,000	8	NA	18^0	NA
TANGENTIAL SPRING	15,000,000	20	1/8	0.002 IN/IN	NA

[a] telescoping shaft is used to accommodate axial travel NA= NOT APPLICABLE

Table 7-3

GENERAL PURPOSE COUPLINGS	COUPLING EVALUATION CHARACTERISTICS: AXIAL FORCES	TORQUE CAPACITY TO OD	HIGH SPEED CAPACITY	INHERENT BALANCE	BENDING MOMENT	FLEXIBLE ELEMENTS EASILY REPLACED	TORSIONAL STIFFNESS	DAMPING	BACKLASH	LUBRICATION	EASY OF ASSEMBLY	RELATIVE COST	LIFE OF FLEXIBLE ELEMENT
MECHANICALLY FLEXIBLE TYPE													
GEAR TYPE													
STRAIGHT TOOTH	M-H	H	H	G	H	N	H	N-L	M-H	Y	E	M	3-5
CROWNED TOOTH	M-H	H	H	G	H	N	H	N-L	M-H	Y	E	M	3-5
CHAIN COUPLING													
STEEL	M-H	M	L	F	H	Y	M	N	H	Y	E	L-M	1-3
NYLON	L-M	M	L	F	H	Y	M	N	H	Y	E	L-M	1-3
GRID TYPE													
VERTICAL SPLIT COVER	M	M	M	G	M	Y	L-M	L-M	M	Y	E	L-M	2-3
HORIZONTAL SPLIT COVER	M	M	M	G	M	Y	L-M	L-M	M	Y	E	L-M	2-3
U-JOINT TYPE													
VEHICLE	H	L	L	F	H	N	L-M	N	N (a)	Y	G	L	3-5
MECHANISM	H	L	NA	F	H	N	L-M	N	N (a)	Y	G	L	1-3
YOKE	H	M	L	F	H	N	L-M	N	N (a)	Y	G	L-M	1-3
ELASTOMERIC ELEMENT TYPE													
COMPRESSION TYPE													
COMPRESSION DONUT	L-M	L	L	F-G	L	Y	L	M-H	N	N	E	L	3-5
BLOCK	M	M-H	L	F-G	L-M	Y	L-M	M-H	N-L	N	G	M-H	3-5
JAW	M	M	L	F-G	L-M	Y	L-M	M-H	N-L	N	E	L-M	3-5
PIN & BUSHING	M	M	L	F-G	M	Y	M	L-M	N-L	N	G	M	3-5
SHEAR TYPE													
URETHANE TIRE	S	L	L	F	L	Y	L	H	N	N	G	L	2-3
CORDED TIRE	S	L	L	F	L	Y	L	M-H	N	N	G	L-M	3-5
SHEAR DONUT	S	L	L	F	L	Y	L	H	N-L	N	G	L	2-3
METALLIC ELEMENT TYPE													
DISC TYPE													
CIRCULAR DISC	L-M	M-H	H	E	L-M	Y	M	N-L	N	N	G	M-H	4-8
SQUARE DISC	L-M	M-H	H	E	L-M	Y	M	N-L	N	N	G	M-H	4-8
SCALLOPED DISC	L-M	M-H	H	E	L-M	Y	M	N-L	N	N	G	M-H	4-8
LINK DISC	L	M-H	H	G	L-M	Y	M	N-L	N	N	F	M-H	4-8
MISCELLANEOUS TYPE													
PIN & BUSHING	H	M	L	F	H	Y	M	N-L	L	N	F	M	1-3
METALLIC BEAM	L-M	L	M	G	L	Y	M	N	N	N	G	M-H	4-8
SLIDING BLOCK	H	L	L	F	H	Y	M	N	M	Y	G	M	1-3

a) No backlash unless telescoping shaft is used to accommodate axial travel

M = Medium, S = Small; H = High, L = Low, NA = Not Applicable; E = Excellent, G = Good, F = Fair; Y = Yes, N = No/None, ? = Maybe

Table 7-4

SPECIAL PURPOSE COUPLINGS	COUPLING EVALUATION CHARACTERISTICS: AXIAL FORCES	TORQUE CAPACITY TO OD	HIGH SPEED CAPACITY	INHERENT BALANCE	BENDING MOMENT	FLEXIBLE ELEMENTS EASILY REPLACED	TORSIONAL STIFFNESS	DAMPING	BACKLASH	LUBRICATION	EASY OF ASSEMBLY	RELATIVE COST	LIFE OF FLEXIBLE ELEMENT
MECHANICALLY FLEXIBLE TYPE													
GEAR TYPE													
HIGH ANGLE (GEAR SPINDLE)	M-H	H	H	G	H	N	H	N-L	M-H	Y	E	M	3-5
HIGH SPEED GEAR (SEALED LUBED)	M-H	H	H	G	H	N	H	N-L	M-H	Y	E	M	3-5
HIGH SPEED GEAR (CONTINUOUSLY LUBED)	M-H	H	H	G	H	N	H	N-L	M-H	Y	E	M	3-5
HIGH SPEED GEAR (CONTINUOUSLY LUBED MARINE STYLE)	M-H	H	H	G	H	N	H	N-L	M-H	Y	E	M	3-5
U-JOINT TYPE													
CONSTANT VELOCITY	H	L	L	F	H	N	L-M	N-L	Na	Y	E	L	3-5
YOKE	H	H	NA	F	H	N	L-M	N-L	Na	Y	E	L	1-3
BLOCK	H	H	L	F	H	N	L-M	N-L	Na	Y	E	L-M	1-3
ELASTOMERIC ELEMENT TYPE													
SHEAR TYPE													
BONDED SHEAR	L	L	L	F	L	Y	L	H	N	N	E	L	3-5
COMPRESSION TYPE													
BLOCK	M	M-H	L	F-G	L-M	Y	M	M-H	N-L	N	G	M-H	3-5
METALLIC ELEMENT TYPE													
DISC TYPE													
REDUCED MOMENT (SCALLOPED)	L-M	M-H	H	E	L-M	Y	M	N-L	N	N	G	M-H	4-8
MARINE STYLE (SCALLOPED)	L-M	M-H	H	E	L-M	Y	M	N-L	N	N	G	M-H	4-8
FRAME	L-M	M-H	H	E	L-M	?	M	N-L	N	N	G	M-H	4-8
DIAPHRAGM TYPE													
TAPERED	L-M	M-H	H	E	L-M	?	M-H	N-L	N	N	G	M-H	4-8
ONE PIECE	L-M	M-H	H	E	L-M	?	M-H	N-L	N	N	G	M-H	4-8
MULTIPLE STRAIGHT	L-M	M-H	H	E	L-M	?	M-H	N-L	N	N	G	M-H	4-8
MULTIPLE CONVOLUTED	L-M	M-H	H	E	L-M	?	M-H	N-L	N	N	G	M-H	4-8
MISCELLANEOUS TYPE													
OFFSET (SCHMIDT)	H	M	L	F	H	Y	M	L	L	N	F	M	1-3
TANGENTIAL SPRING	L-M	L	M	G	L	Y	M	L-M	N	N	G	M-H	4-8

a) No backlash unless telescoping shaft is used to accomodate axial travel

M = MEDIUM, S = SMALL; H = HIGH, L = LOW, NA= NOT APPLICANLE; E = EXCELLENT, G = GOOD, F = FAIR; Y = YES, N = NO/NONE, ? = MAYBE

API 610 Centrifugal Pumps for Petroleum, Petrochemical and Natural Gas Industries has a section of the requirements for couplings for this type of pump. ISO 13709 is its equivalent and has the same requirements for flexible couplings. ISO has two standards for couplings: Special Purpose Couplings and Standard Couplings.

AGMA has recently become the writing body for ISO and is in the process of writing several specifications for couplings, including metric bores and keyways, coupling balance requirements, metric mass elastic data, and conversion of several of its specifications to metric.

Coupling ratings can be confusing because there is no industry standard rating system. Manufacturers' ratings are based variously on yield strength, endurance strength, and life. They also may require application of service factors. To ensure that ratings are properly applied, use the selection procedure recommended by the manufacturer.

Design calculations for parameters such as misalignment, torque and loads, allowable stress, and coupling physical characteristics can be found in manufacturers' design guides and textbooks.

Application and Selection

The coupling selector (equipment designer or system designer) must decide what coupling or U-Joint is best for the system. The designer must review the possible candidates for a flexible coupling or U-Joint and make a selection. The person responsible for the selection of couplings should build a file of the most recent coupling and U-Joint catalogs. This file should be reviewed at regular intervals because designs, models, materials, and ratings are constantly updated and improved.

Couplings and U-Joints are usually selected based on their capacities and characteristics. The two most important capabilities are torque and speed. One should consider the capacities of the couplings by reviewing the application requirements and determining which couplings meet the requirements (see Table 7-1) for general-purpose couplings and special-purpose couplings (see Table 7-2). Finally, the designer must also look at the application and its history and then tradeoff features and characteristics of available couplings and pick the best one suited for the equipment and operating condition (see Table 7-3) for general-purpose couplings and special-purpose couplings (see Table 7-4).

When selecting a coupling, consider all the application requirements: speed, horsepower, torque, misalignment, shaft sizes, space limitations, and environmental conditions. In addition, determine the type of prime mover and driven machine, their interface requirements, and potential vibration conditions.

At any given horsepower, the torque applied to a coupling will change as the speed of the coupling changes. For example, a coupling between a 1,800 rpm motor and a gear reducer will see less torque than the coupling between the same gear reducer and a conveyor drive pulley rotating at 18 rpm. Therefore, when sizing a coupling, horsepower, torque, and speed are equally important factors. Coupling selection specifications and requirements are available in coupling manufacturer's catalogs.

Additional Information

(Addresses and phone numbers for associations are given in the Associations section.)

Some of this chapter was taken from a short course that was first presented at the 27th Turbomachinery Symposium and also for "Couplings and Joints—Design, Selection, and Application," Jon R. Mancuso, Marcel Dekker Inc., New York, 2nd edition 1999.

American Gear Manufacturers Association-AGMA, Alexandria, Va. (http://www.agma.org)

Mechanical Power Transmission Association-MPTA, Naples, FL. (http://www.mpta.org)

American Petroleum Institute-ISO Washington DC (http://api-ep.api.org)

International Organization for Standardization-ISO, Specifications are available through (http://www.iso.org)

Chapter 8

Gears

Chapter Objectives

When you have finished Chapter 8, "Gears," you should be able to do the following:

1. Understand the purpose of gears and drives.
2. Identify five gear design categories and their orientation types.
3. Describe and compare various gear types, including: spur helical, herringbone, straight bevel, spiral bevel, cylindrical worm, double-enveloping worm, cycloidial, and hypoid.
4. Define the most common gear terms.
5. Explain the application requirements when selecting gears.
6. List the important specifications needed when ordering gears.
7. List five causes for gear tooth failure.
8. Name the associations that standardize gear classification.
9. Describe and compare the basic types of enclosed gear drives.
10. Explain the function and importance of seals and breathers for enclosed gears.
11. Describe the purpose and the lubrication essential for gear life.
12. Explain gear rating standards.
13. Explain the major factors for selecting and installing gear drives.

Introduction

A **gear** is a rotating machine part having cut teeth, or cogs, which mesh with another toothed part in order to transmit torque. Two or more gears working in tandem are called a *transmission* and can produce a mechanical advantage through a gear ratio and thus may be considered a simple machine. Geared devices can change the speed, torque, and direction of a power source. The most common situation is for a gear to mesh with another gear; however, a gear can also mesh with a non-rotating toothed part, called a rack, thereby producing translation instead of rotation.

The gears in a transmission are analogous to the wheels in a pulley. An advantage of gears is that the teeth of a gear prevent slipping.

When two gears of an unequal number of teeth are combined, a mechanical advantage is produced, with both the rotational speeds and the torques of the two gears differing in a simple relationship.

Where called upon, gears can change both speed and direction of rotation. Gears can be mounted on shafts, and their centerlines can be parallel or at any angle relative to each other and in one or more planes. Gears are supplied either unassembled, which is referred to as open gearing, or assembled as part of an enclosed gearbox or speed reducer.

Open Gears

Gears are grouped into five design categories: spur, helical, bevel, hypoid, and worm. They are also classified according to the orientation of the shafts on which they are mounted, either in parallel or at an angle. Generally, the shaft orientation, efficiency, and speed determine which type should be used for a specific application. Table 8-1 compares the different gear types.

Parallel Shaft Gearing

Spur Gears

Spur gears, or straight-cut gears, are the simplest type of gear. They consist of a cylinder or disk with the teeth projecting radially, and although they are not straight-sided in form, the edge of each tooth is straight and aligned parallel to the axis of rotation. These gears can be meshed together correctly only if they are fitted to parallel shafts. Figure 8-1 shows a typical spur gear. When a pair of gears rotate, the teeth mesh with a combined sliding and rolling motion. To avoid excessive wear of the sliding motion and forces, proper lubrication is important.

Figure 8-1:
Spur gears have straight teeth and are parallel to the shaft axis

Because the teeth are parallel to the shaft axis, spur gears produce reaction loads only radially

TABLE 8-1 – Comparison of gear types				
			Maximum pitch line Velocity, fpm	
Gear type	Approximate range of efficiency, %	Range of reduction ratio	High-precision	Commercial
Spur	97-99	1:1-10:1	20,000	4,000
Helical or double helical	97-99	1:1-9:1	40,000	5,000
Straight bevel	97-99	1:1-10:1	10,000	1,000
Spiral bevel	97-99	1:1-10:1	25,000	5,000
Cylindrical worm	50-90	3:1-100:1	10,000	5,000
Double-enveloping worm	50-98	3:1-100:1	10,000	4,000
Hypoid	90-98	1:1-10:1	10,000	4,000

to the shaft. They do not produce axial thrust loads, as do some other types of gears.

Spur gears are applied in moderate speed applications such as mill drives, hoisting equipment, and general machinery. Spur gears at high speeds can be very noisy and exhibit higher-wear performance. Spur gears are the most widely used gear type due to their low cost, lack of end thrust, and low maintenance.

When two gears have the same number of teeth, the speed of one matches the other, but the two rotate in opposite directions. When two meshing gears have different numbers of teeth, the gear with the least number of teeth is called the pinion and the larger is called the gear. In this case, the two gears rotate in opposite directions as before, but now the gears rotate at different speeds. The speed ratio equals the number of teeth in the larger gear divided by the number of teeth in the smaller pinion gear.

If an idler or an intermediate gear is inserted between the driving pinion and driven gears (see Fig. 8-2), the pinion and gear rotate in the same direction. This configuration does not affect the speed ratio, which is still equal to the number of teeth in the driven gear divided by the number of teeth in the pinion.

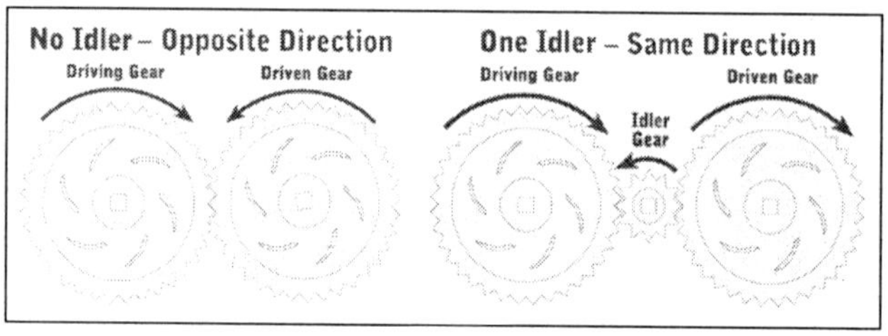

Figure 8-2:
Using Idler Gear to Change Gear Rotation

Many types of spur gears are available. Pinions come with or without a hub, and gears are available with solid, webbed, or spoked bodies.

Tooth Forms

The shape of a spur gear's teeth is based on an involute tooth form (see Fig. 8-3), which produces a rolling contact rather than a sliding contact between mating teeth. Common tooth forms for spur gears are defined in terms of pressure angle (the angle of contact between the teeth) and diametrical pitch (the number of teeth given per inch

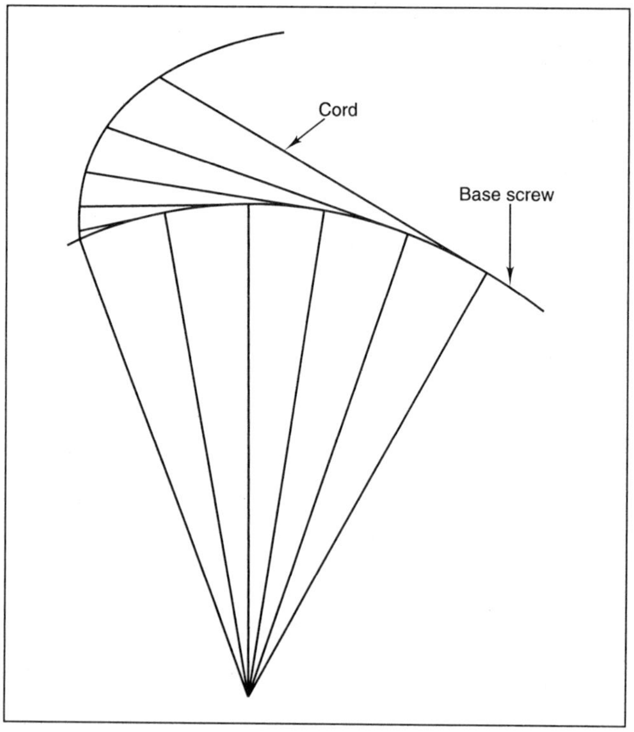

Figure 8-3: Involute form is generated by unwrapping a cord from a circle

of a gear's pitch diameter). Diametrical pitch can be either coarse (one to 19 diametral pitch) or fine (20 diametral pitch or higher).

The American Gear Manufacturers Association (AGMA) has established standards for gear teeth. These standards include the following:

- 14 ½ degree course pitch (former standard – not recommended for new designs)
- 20 degree coarse pitch
- 20 degree fine pitch

The low pressure angle, 14 ½ degrees, provides smooth, quiet meshing between mating teeth. However, for gears that have fewer than 32 hobbed teeth, each tooth is undercut at its base, thereby weakening it. Thus, this tooth form is no longer an AGMA standard, although it is still used for replacement gears and special design. For teeth formed by other methods, this undercutting does not apply.

The larger 20 degree pressure angle produces stronger teeth that have higher load capacities. These larger angles also minimize undercutting of hobbed teeth. Thus, undercutting shows up in the 20 degree tooth form only when the number of teeth is fewer than 18. Of major importance is the fact that 14 ½ degree pressure angles cannot be used with gears of 20 degree pressure angles and vice versa.

Helical Gears

Helical gears are capable of carrying more load than spur gears. They have teeth that are oriented at an angle, when compared to the shaft's axis. This is called the helix angle (see Fig. 8-4). This arrangement provides an overlapping tooth engagement, which results in smoother, quieter operation. Helical gears are quieter with less vibration and run at higher speeds than spur gears.

Figure 8-4: Helical gears have teeth across the face at an angle

Helix angles range from a few degrees to about 45 degrees. However, tooth bending-load capacity drops off at large angles, generally those above 20 degrees. The teeth in mating helical gears that are operating on parallel shafts must have the same helix angle, but must be oriented in opposite directions such that a right-handed gear meshes with a left-handed pinion. Unlike spur gears, helical gears require bearings designed to take thrust loads.

Double Helical Gears

A double helical gear is a variation of the helical gear. A double helical gear has two sets of teeth on one gear with opposing tooth angles (see Fig. 8-5). A groove separates the two sets, which have identical helix angles and tooth pitches.

Figure 8-5: Double helical gear sets use two pairs of opposed gear teeth to eliminate thrust load

Unlike single helical types, these double helical gears exhibit no side thrusts because the thrust of one side cancels out the opposing side. For this reason, the teeth can be cut at a greater helix angle, providing greater tooth overlap and smoother operation. Double helical gears are particularly effective when shock and vibration are present. Double helical gears are also used for applications that require high speed and high ratio in a single stage (one pair of gears).

Herringbone Gears

Similar to double helical gears, herringbone gears have two sets of teeth that are joined in the middle (see Fig. 8-6). The continuous tooth form allows herringbone gears of commercial quality to operate at pitch line velocities up to 5,000 feet per minute. Herringbone gear sets are also well suited for shock and vibration applications or applications where a high single reduction is needed. In general, the terminology for helical gears is the same as that for spur gears. In some instances, the design formulas are different because they depend on the helix angle.

Figure 8-6: Herringbone gears have opposite teeth joined in the middle

Internal Gears

Internal gears have teeth on the inside (see Fig. 8-7). These gears are available with either spur or helical-type teeth. In either case, an internal gear meshes with a pinion that has external teeth.

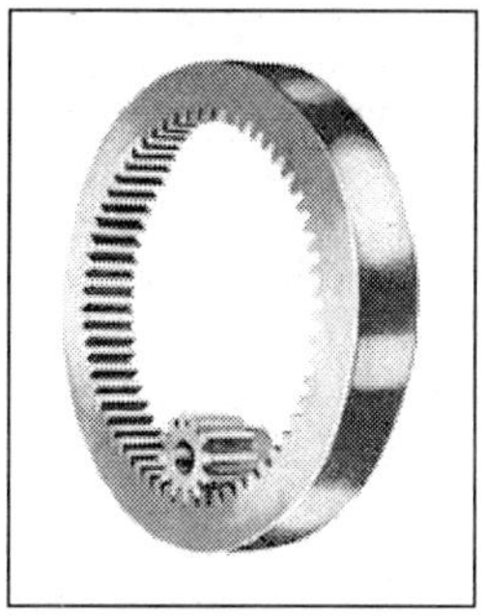

Figure 8-7: Internal spur gear set

Internal gears allow a closer center distance with their mating pinions than do external gears of the same size. This allows for a more compact design and also provides a protective guard over the meshing gear teeth. Internal gears and pinions rotate in the same direction; as a result, an idler gear may not be required.

Although more compact, internal gears cannot be used in all cases. The number of teeth in the pinion and the gear cannot be equal because of interference between the tips of the mating teeth. As a general rule, the difference in tooth count must be at least 8 teeth for a 20 degree stub-tooth form and 10 teeth for a 20 degree full-depth form.

Internal gears are typically used in planetary gear systems (see Fig. 8-8). In this example, the sun gear, or center pinion,

Spur Gear Terminology

A few of the most common terms are defined here to provide a working knowledge of gear terms. Generally, the basic terminology for spur gears applies to all the other gear types as well. More complex gear types call for additional, rather than different terminology.

Addendum – Height of a tooth above the pitch circle.

Backlash – The distance by which a tooth space exceeds the thickness of a mating tooth at the pitch circles. Also described as the play between mating teeth.

Center distance – Distance between the centerlines of mating gears.

Circular pitch – Distance between corresponding points on adjacent teeth measured along the pitch circle.

Circular thickness – Thickness of a tooth measured as an arc along the pitch circle.

Clearance – Radial distance between the top of a tooth and the bottom of the mating tooth space.

Dedendum – Depth of a tooth space below the pitch circle.

Diametral pitch – Ratio of the number of teeth to the pitch diameter, in inches.

Face width – Length of a tooth in an axial plane.

Fillet radius – Also known as the root radius, this is the radius of the fillet at the base of the tooth.

Flank of tooth – Surface between the pitch circle and the bottom of the tooth space.

Gear ratio – The ratio of the number of teeth in the larger gear to that of the smaller gear in a mating pair.

Module – Ratio of the number of teeth to the pitch diameter, in millimeters.

Outside diameter – Distance from the top of one tooth to the top of the diametrically opposite tooth.

Pitch circle – A circle defined by the number of teeth and the diametral pitch.

Pitch diameter – The diameter of the pitch circle.

Pressure angle – The angle between the normal to the tooth profile at the pitch circle and the tangent to the pitch circle at the point.

Root diameter – Distance from the bottom of one tooth to the bottom of a diametrically opposite tooth.

Tip relief – A modification of the tooth profile whereby a small amount of material is removed at the tip of the tooth.

Tooth profile – One side of a tooth as viewed in a cross section of the tooth. Also can be described as the shape of a gear tooth, usually based on the involute of a circle.

Tooth surface – The side of a gear tooth.

Undercut – Condition where the tooth generating process removes material from the lower portion of the tooth profile to prevent interference with the tip of the mating tooth.

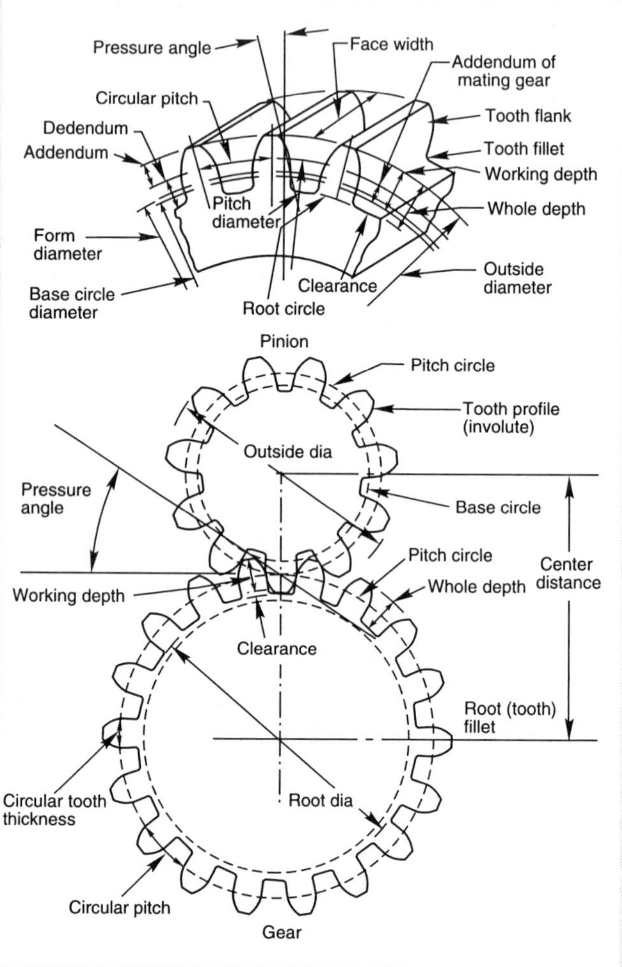

Whole depth – Depth of a tooth space from the outside diameter to the root diameter.

Working depth – The depth to which a tooth extends into the tooth space of a mating gear. Also described a the depth of engagement between two meshing gear teeth.

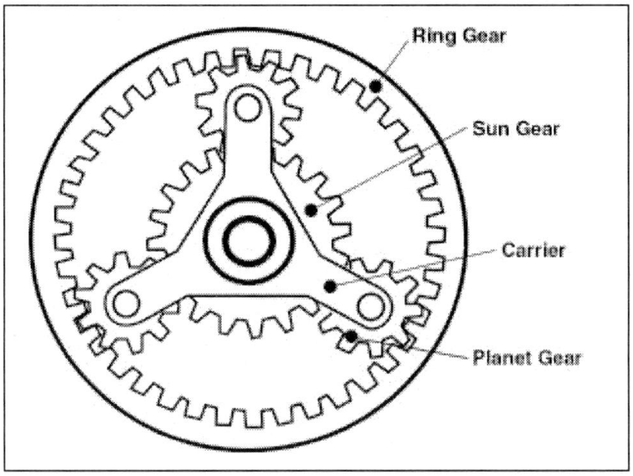

Figure 8-8: Planetary gear system consists of a sun, planets, and internal ring gear

rotates in a clockwise direction, causing the planet or idler gears to rotate counterclockwise. The fixed internal or ring gear causes the planet gears to travel as a group in the same direction as the sun gear but at a lower speed. The system can also be arranged so the planetary group is fixed and the ring gear is free to move and connect to the output shaft.

Gear Racks

A **rack and pinion** is shown in Fig. 8-9. A rack and pinion is a type of linear actuator comprised of a pair of gears that converts rotational motion into linear motion. A circular gear called "the pinion" engages teeth on a linear gear bar called "the rack." Rotational motion applied to the pinion causes the rack to move, thereby translating the rotational motion of the pinion into the linear motion of the rack. The rack and pinion is suited for oscillating or reversing applications. It is used extensively in machine tools, lift trucks, power shovels, and other heavy machinery where rotary motion of the pinion drives the straight-line motion of a reciprocating part.

Figure 8-9: Rack and pinion for linear motion

Helical and herringbone racks are sometimes used, but the most common rack and pinion system uses 14 ½ degree or 20 degree spur gears. Out of these three types, the helical requires provision for thrust loads at the pinion. Herringbone rack and pinion systems are sometimes selected for heavy-duty, precision mechanisms.

Gears with Angled Shafts

Helical gears

When a design requires shafts to transmit power at different angles, a crossed-axis helical gear can be used. As shown in Fig. 8-10, these shafts transmit motion at an angle to each other. These gears operate with a sliding action and as a result are used for light to moderate loads. Care must be taken when selecting the proper lubricant and materials to minimize wear, heat, and seizing.

Figure 8-10: Cross axis helical gears

Helical gears with 45 degree helix angles are widely used. When 45 degree helix gears have the same number of teeth and pitch, they can be positioned either on parallel shafts or on shafts that are perpendicular to each other.

Worm gears

Used to transmit power through right angles on nonintersecting shafts, a worm-gear set, Fig. 8-11, consists of a cylindrical worm that meshes with a larger gear, often called a wheel. The worm, which has a screw-type thread, requires several revolutions to drive the wheel through a single revo-

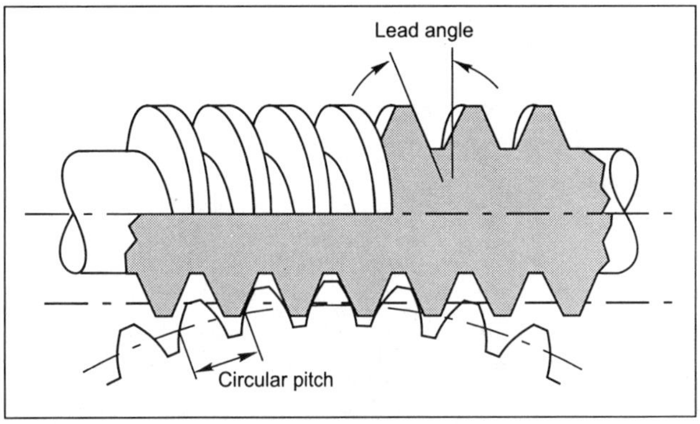

Figure 8-11: Cylindrical worm gear set

lution. In this way, a wide range of speed-reduction ratios (up to 100:1) can be obtained in a limited space.

Worm threads are available in different forms. Some have a single thread; others have more. The number of threads on a worm can be determined by observing the number of thread starts at the worm end. One, two, three, or four starts are common. Starts may be either right or left hand. The direction of the helix angle is the same for both the worm and its mating gear.

Worm gears are suitable for applications where shock loads are encountered. The screw action is quiet and produces a constant output speed free of pulsations.

Generally, the worm thread slides along the wheel teeth, rather than rolls, as it drives the wheel. For this reason, worm gears exhibit lower efficiency than other gear types.

Worm gears also produce a thrust load, the level of which depends on the lead angle, along the axis of rotation. Mounting assemblies and bearings should be designed to accept this thrust load.

The worm almost always drives the wheel, not vice versa, though some sets can drive in either direction. When the worm has a lead angle (helix angle) less than 5 deg, the wheel can't drive the worm in reverse, a condition called self-locking. Such gears can be used where reversibility of direction is not permitted. However, vibration can cause a self-locking set to slip, allowing the wheel to temporarily drive the worm. Applications where this action is unacceptable call for a brake on the worm.

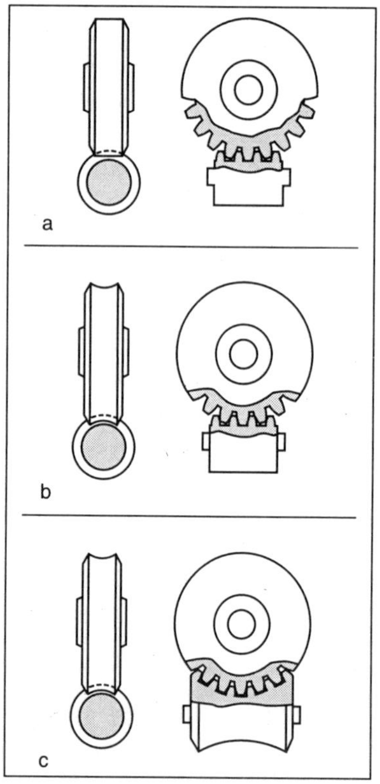

Figure 8-12: Three types of worm gears include cylindrical (a), single-enveloping (b), and double-enveloping (c)

There are three types of worm gear configurations, Fig. 8-12, the most basic and least common of which has a cylindrical worm meshing with a helical gear, which is also cylindrical. Such gears provide high speed-reduction ratios. But, because of a small tooth contact area, these gears are suitable only for light loads.

In a second, more common version, called single-enveloping, the gear has throated (curved) teeth that wrap partially around the worm to achieve greater tooth contact area, higher load capacity, and better shock resistance.

A third version, called double-enveloping, provides higher load capacity than either cylindrical or single-enveloping types. In addition to curved gear teeth, this type also has a concave or hourglass-shaped worm that wraps partially around the gear, further increasing the number of teeth in engagement, thus providing the highest load capacity and shock resistance of the three types.

There are wide variations of worm gearing designs, each predicated on the desired efficiency. Gears generally come with a 14 1/2, 20, 22 1/2, or 25 deg pressure angle. As the pitch increases (smaller teeth), it is necessary to increase the pressure angle on multiple thread worms to avoid excessive undercutting.

Bevel gears

As with worm and helical gears, bevel gears exhibit certain geometric and strength characteristics that can't be obtained with spur gears. Bevel gears generally transmit power between shafts with axes that intersect at 90 deg, but are sometimes used on shafts oriented at other angles. These gears have teeth cut on an angular or conical surface, and typically have a 20 deg pressure angle.

Bevel gears produce a thrust load on both shafts, and this should be considered in designing the shaft mounting assemblies.

There are four basic types of bevel gears: straight, Zerol®, spiral, and skew tooth.

Straight bevel gears — These are the simplest and most common bevel gears. Straight bevel gear teeth are cut on a truncated conical surface, which, if extended, would come to a common point on the axis of the supporting shaft, Fig. 8-13. The teeth are tapered in thickness, with the outer part being the widest. Straight bevel gears of commercial quality are generally capable of pitch line velocities up to 1,000

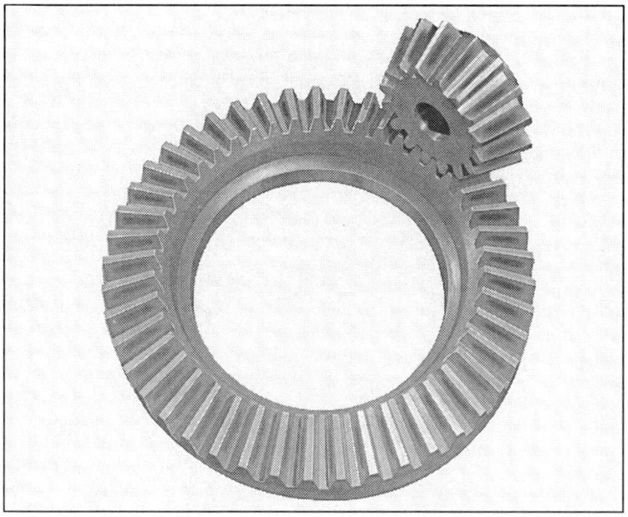

Figure 8-13: Straight-tooth bevel gears

fpm, and are the most economical gears for transmitting power at right angles.

When shafts are at right angles and both shafts turn at the same speed, the two mating gears are usually alike and are called miter gears.

Though most bevel gears operate at right (90 deg) angles, other angles are also available, both acute (less than 90 deg) and obtuse (greater than 90 deg). When the pitch cone angle of a gear is 90 deg (or less) from the shaft, it is called a crown gear. When the angle is greater than 90 deg, it is called an internal bevel gear.

Two methods are typically used to cut bevel gear teeth:

- Uniform addendum system. Here, the addendums of gear and pinion are the same and all tooth contours converge to the same center point. Because miter gears have a 1:1 ratio and converge at a 45 deg angle, they are typically generated in this system.
- Gleason Coniflex® system. This method generates teeth that are slightly crowned in the lengthwise direction to prevent load concentration on the ends of the teeth. Gleason bevel gears usually have a 20 deg pressure-angle tooth form, but they may also be cut with a 14 1/2 degree pressure angle.

Because the tooth contours of the two mating gears do not converge to a common center point, most cutting formulas depend on the diametral pitch and the pitch angle, plus the number of teeth on both gear and pinion.

Spiral bevel gears — These gears are also cut on conical surfaces. But, the teeth are curved and oriented at an oblique angle to the shaft centerline, Fig. 8-14. This characteristic is usually defined in terms of the spiral angle, which is typically 35 deg.

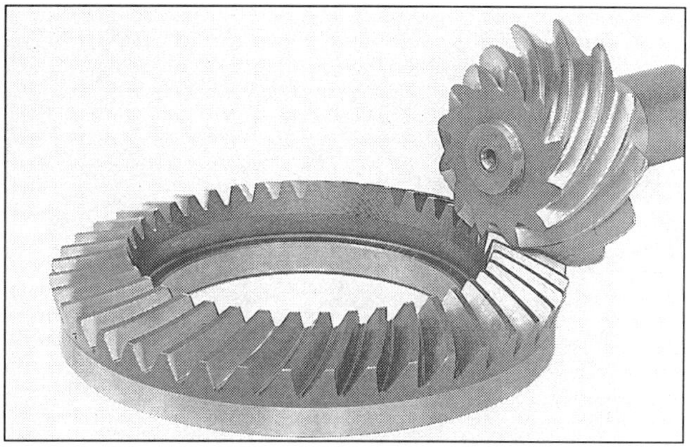

Figure 8-14: Spiral bevel gears have curved teeth for smoother operation

The curved teeth of spiral bevel gears have a considerable amount of overlap, which ensures that more than one pair of teeth is in contact at all times. Though more complex, spiral bevel gears have a higher load-carrying capacity and they run more smoothly and quietly than straight bevel gears. And they are generally capable of pitch-line velocities up to 5,000 fpm.

Zerol® bevel gears — Similar to spiral types, Zerol® bevel gears also have curved teeth, Fig. 8-15. But, these gears generally have zero-deg spiral angles.

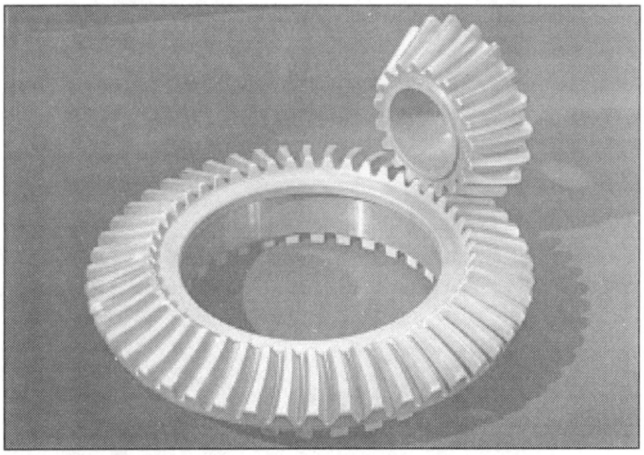

Figure 8-15: Zerol® bevel gears

Zerol® bevel gears are also smoother and quieter than straight bevel types due to their tooth curvature and slight overlap of teeth. Their zero-deg spiral angle produces the same thrust load as straight bevel gears and the two types of gears are interchangeable without changing thrust bearings.

Skew tooth gears — The skew tooth gear, Fig. 8-16, is similar to the spiral gear except that the teeth are not curved. They are cut straight but at an angle to the shaft centerline. This improves load capacity when compared to a straight bevel gear. These gears are used primarily in large (over 30-in. pitch diameter) sizes.

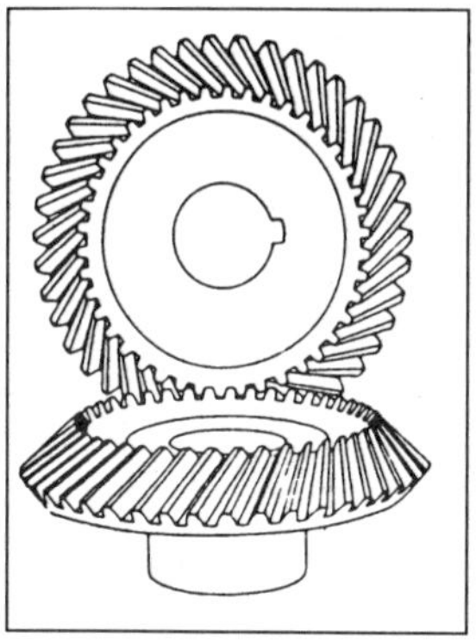

Figure 8-16: Skew tooth (bevel) gears

Hypoid gears

Though hypoid gears, Fig. 8-17, resemble spiral bevel gears, they operate on shafts with centerlines that do not intersect. Here, the pinion is positioned in a different plane than the gear. This arrangement allows each shaft to be supported by bearings on both ends. The pitch surfaces of both gears and pinions are hyperboloids of revolution, hence the shortened term, hypoid.

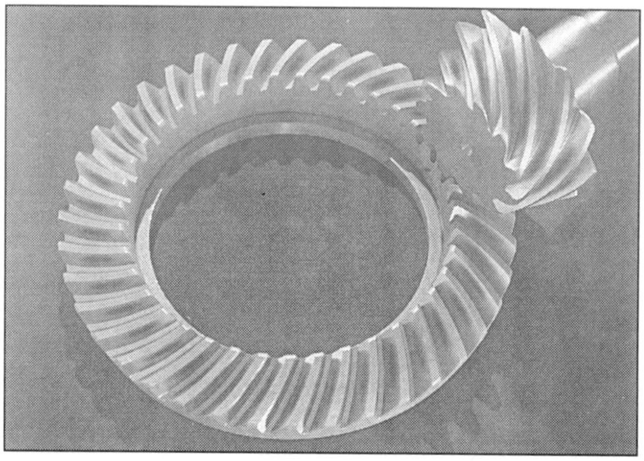

Figure 8-17: Hypoid gear shaft axes do not intersect. Therefore, both shafts can be supported at both ends

Tooth action is a combination of rolling and sliding. Because the number of teeth in a gear and pinion is not proportional to the pitch diameter, it is possible to make larger, sturdy pinions without enlarging the driven gears.

Hypoid gears are stronger, and they operate smoother and quieter than spiral gears due to their higher contact ratio. And, they generally tolerate higher shock load. Though high contact pressures and sliding speeds require special lubricants, these gears are used extensively in automobile rear-wheel-drive differentials.

Face gears

In a face gear, Fig. 8-18, the axis of each tooth lies in a plane that is perpendicular to the shaft axis. The mating pinion is either a spur or helical gear. The pinion and face gear are usually oriented at 90 deg to each other. As with straight bevel gears, face gear teeth are tapered, but these teeth are thicker near the inside diameter. Face gears have a smaller load capacity than bevel gears, thus, they are used mostly to transmit motion and light loads. However, recent efforts are being made to adapt face gear designs to high-power applications such as helicopter transmissions.

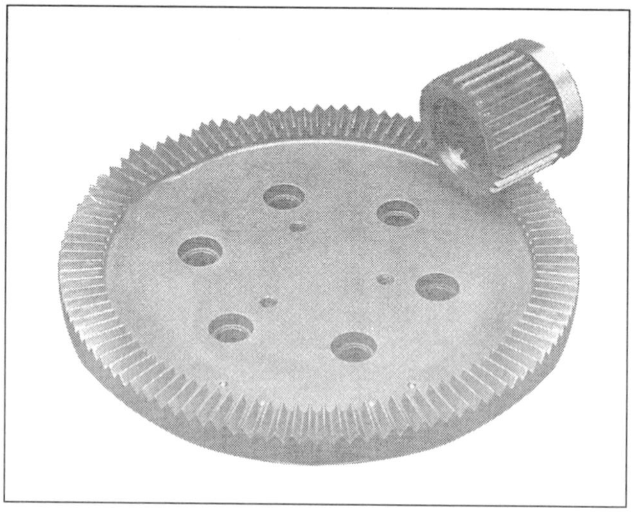

Figure 8-18: Face gear and pinion

Traction drives

A cousin to the gear family, traction drives transmit power by means of mating rollers, Fig. 8-19, which may be considered gears with infinite numbers of teeth. The rollers can be cones, cylinders, discs, rings, or spheres. The speed ratio of a traction drive varies with geometry factors such as the radius of rotation of the driver, the distance between rollers, and their orientation with respect to each other.

Traction drives come in both dry and lubricated types. With dry traction drives, slippage between driving and driven members is prevented by a spring-loaded system.

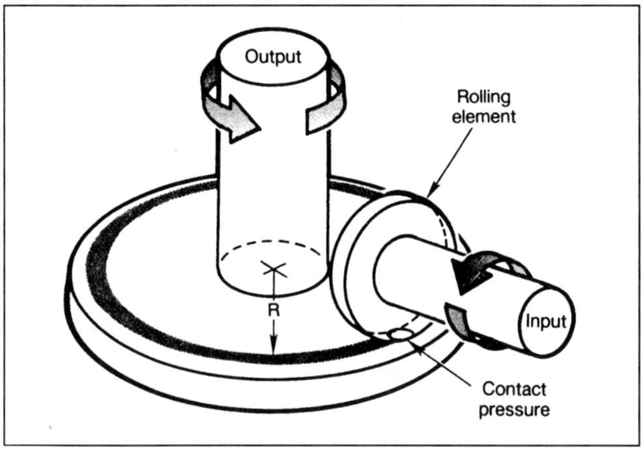

Figure 8-19: Traction drive transfers power from the input to output shaft with a set of mating rollers

Other traction drives use synthetic fluid lubricants. Under pressure, viscosity of the lubricant increases so it behaves like a plastic film, enabling the drive to transmit high power without slip.

Traction drives are used in place of belts, gears, or chain drives where space is limited. Moreover, they are quieter than most other mechanical drives because there is no backlash or shock loading due to engagement. Other advantages include high speed reduction or multiplication, low vibration, and high efficiency.

Backlash

The amount of play between the meshing teeth of two mating gears, called backlash, is an important factor in gear operation. A small amount of backlash must be built into the drive to allow the gears to perform properly, and the amount required depends on the application.

Backlash provides clearance for proper lubricant penetration and prevents binding. It also compensates for conflicting tolerances, misalignment, and thermal expansion. Insufficient backlash causes early failure because of overloading, overheating, and tooth surface fatigue.

Backlash is more specifically defined as the amount by which the width of a tooth space exceeds the thickness of the mating tooth at the operating pitch circle. In a set of meshing gears, the amount of backlash depends on the thickness of the teeth and the distance between gear centerlines.

Some drive applications, such as instruments, machine tools, and robots, require precise positioning with minimum backlash. Such applications may require special gear designs that limit backlash, hardened-and-ground precision gears, or traction drives that transmit torque through compressively loaded rollers.

Materials

Gears are made from a wide variety of materials, both ferrous and nonferrous. Material selection is based on the application requirements, especially the design loads, but also including speed, service conditions, and cost.

Ferrous materials

Gears are generally made from ferrous materials such as carbon and alloy steels (both forged and cast), cast iron, and ductile iron. Some are made from sintered powder metal.

Cast iron is often used for spur, bevel, and worm gears, generally in conjunction with steel pinions or worms. Cast iron wears well, but has poor bending strength.

Carbon steels are generally satisfactory for gears subjected to uniform or moderate shock loads. Alloy steels are used for high load or high-speed applications or where longer life is required. Hardened alloy steels are used to improve wear resistance, improve durability, or reduce size where space is limited.

Reference material and selection data are usually available from both gear and steel manufacturers.

Heat treatment

Manufacturers typically harden the teeth of metal gears to improve wear resistance and strength, which enables a higher gear rating. Hardening is done by one of two basic methods: either through-hardening or surface (case) hardening. Through-hardening imparts essentially the same hardness all the way through a tooth, whereas case-hardening penetrates only to a shallow depth. Hardened gears are usually made from alloy steel with 0.30 to 0.55 percent carbon for through-hardened gears and 0.10 to 0.55 percent carbon for surface-hardened gears. The selection of through or surface hardening depends on application factors, such as torque, reduction ratio, operating environment, loads, lubrication, alignment, and cooling, as well as cost.

Surface-hardened gearing, because of high hardness and residual compressive stress in the case, has higher strength. Thus, it can be rated higher or made smaller for the same rating. Heat treatment and finishing generally cost more for surface-hardened gears compared to through-hardened gears.

Nonferrous materials

These materials include various alloys of copper, aluminum, and zinc, plus nonmetallic materials such as plastics, laminates, and fibers. Bronzes (copper alloys) account for most of the nonferrous gears. They are widely used in worm gears because they resist wear, reduce friction, and improve efficiency. The worm is generally steel and the worm wheel, in most cases, is either brass or bronze.

There are four basic types of gear bronzes: phosphor, manganese, aluminum, and silicon bronze. In worm gearing, bronze withstands high sliding velocity in contact with a steel worm better than other materials.

When weight, nonmagnetic requirements, and corrosion resistance are factors, aluminum alloys or nonmetallic materials are generally specified. Frequently, nonmetallic gears are driven by steel pinions. When loads are light and the parts small, it is often possible to make both gear and pinion of the same material.

Nonmetallic materials

Lightly loaded gears, especially those that transmit motion rather than power, are often made from nonmetallic materials. The most common of these is plastic, which is light, exhibits a low coefficient of friction, operates quietly, and is low cost. Many plastic gears have inherent lubricity and thus require little or no lubrication. Also, plastic gears are well suited to corrosive environments that damage metal gears. But, plastic gears have limitations such as lower strength and higher deflection under load. Moreover, they are sensitive to temperature and humidity changes. The most commonly used gear plastics are acetal, nylon, and Delrin®.

Gear Specifications

Standard gears are generally specified and ordered from suppliers' catalogs. When nonstandard gears are required, the purchaser should provide information on the operating requirements, as typically called for on the supplier's inquiry or specification form.

When ordering gears, the diametral pitch, tooth size, pitch diameter, gear size, pressure angle, tooth shape, and shaft size should be specified. At this time, it should also be determined if the gears are to be metric. In many areas of the world the metric system is the standard. In these cases the module gear system is used. The module is the ratio of the pitch diameter in millimeters, to the number of teeth. Diametral pitch is the ratio of the pitch diameter in inches, to the number of teeth. Two gears will not engage properly unless the diametral pitch (or module) and pressure angle are the same for both. For worm gear sets, both members should be obtained from the same manufacturer to ensure proper tooth mesh.

Gear Tooth Failures

Gears fail in numerous ways, typically through tooth pitting, scoring, fatigue, or breakage. Failures generally fall into one of the following categories:

- Wear.
- Surface fatigue.
- Plastic flow.
- Breakage.
- Process-related failures.

Wear

Wear is a more-or-less gradual removal of metal from the contacting surfaces of the meshing teeth. Factors that contribute to wear include load, inadequate lubrication (allows metal-to-metal contact), abrasive particles in the lubricant, and corrosion. Wear caused by high localized stresses combined with localized failure of the oil film is called scoring.

Surface fatigue

Also known as pitting, surface fatigue consists of the removal of small pieces of metal, leaving cavities or pits in the surface. Caused by repeated loads that produce stresses above the endurance limit of the material, it usually progresses over a long period of time. A severe form, in which large pits occur over a considerable area, is called spalling.

Plastic flow

Heavy load, in combination with the rolling and sliding action of meshing teeth, causes the contacting surfaces of teeth to yield and deform, a condition called plastic flow.

Breakage

Breakage is the fracture of an entire tooth or a substantial part of a tooth due to overload or repeated overstressing of the material. Fractures generally occur due to high bending stresses in the tooth root or fillet radius, sometimes accentuated by cracks or notches. A gear is typically useless after a fracture.

Process-related failures

This category includes such failure modes as grinding burns or checks (fine cracks), tool marks or nicks, and quenching cracks due to heat treatment. Such defects can ultimately lead to catostrophic failures.

For more information on gear failures, see standard ANSI/AGMA 1010.E95, appearance of gear teeth-terminology of wear and failure.

Gear Classification Standards

Of the gears described herein, spur, bevel, worm, and standard helical gears in popular sizes are generally carried as stock items by most gear manufacturers and suppliers. Other gears are usually made to customers' specifications.

More sophisticated types of gears, such as those that are ground, shaved, or lapped, and those for precision, instrument, or aircraft applications, are generally manufactured by specialty gear houses and are not classified as standard gears.

The American Gear Manufacturers Association has adopted gear classification standards for loose (unassembled) gears including ANSI/AGMA 915-1-A02, 915-2-A05, 2015-1-A01, and 2015-2-A06 for spur and helical gears, and AGMA 2009-B01 and 2011-A98 for bevel, hypoid, and worm gears, and racks.

The AGMA classification system is based on a multidigit classification number that includes a quality number, tooth thickness tolerance code, and a material and hardness code. For example, in the number Q8 A-HA-14, the Q8 is the quality number, A is the tooth thickness tolerance code, HA is the material designation, and 14 is the heat treatment and hardness designation.

The quality number, which indicates the accuracy level of tooth element tolerances, ranges from 3 to 15 in order of increasing precision. This number must be included in all classification numbers, whereas the other codes are optional depending on the service application.

Enclosed Gears

Enclosed gears are open gears contained in a housing. The housing normally supports the bearings and shafts, and holds lubricants for the transmission of torque. In addition, the enclosure protects the gear components from the environment. Enclosed gears, also known as gearboxes, are available in a wide range of load capacities and speed ratios to transmit power from a driving source. In most cases, the source of power is either a hydraulic or electric motor.

Enclosed drives are used to increase or reduce speed. As a result, the torque output from the enclosed drive will be the inverse of the function. If the enclosed drive is a speed reducer, then the torque output will increase; if the enclosed drive increases speed, then the torque output will decrease.

Basic Types

Worm, helical, and bevel gears are the most prevalent gears used in small- and medium-sized gear enclosures. Worm gear units are low in cost but also low in efficiency when compared to the higher cost helical gears of equal ratio. Helical-gear drives offer higher efficiency, with more torque capacity in a larger size. The size and weight difference becomes larger when there are multiple reduction ratios.

Worm Gear Speed Reducers

A single reduction speed reducer can typically achieve up to a 100:1 reduction ratio in a small package. Known as right angle drives, the basic configuration consists of two members: a cylindrical worm with screw threads and a worm or worm wheel.

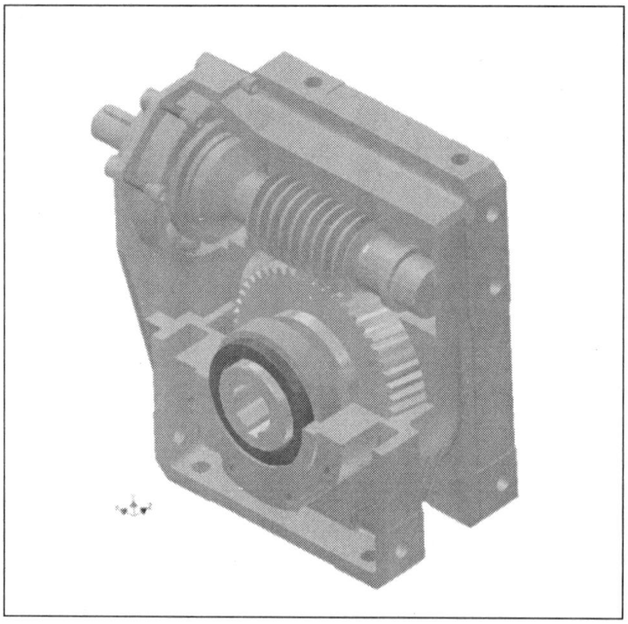

Figure 8-20: Worm gear speed reducer has perpendicular nonintersecting shafts; several revolutions of the worm pull the wheel through one revolution

With a single start worm, the worm gear advances only one tooth of the gear for each 360-degree turn of the worm. Therefore, regardless of the worm's size, the gear ratio is the *"size of the worm gear to 1."* Given a single start worm, a 20 tooth worm gear will reduce the speed by a ratio of 20:1. The design of worm gear drives allows for a continuous sliding contact between the gear teeth and the worm thread during rotation. This design is extremely quiet and helps eliminate impact between gears.

While the worm gear drive has many benefits, the main disadvantage is that the efficiency of the worm gear tends to be lower than other gear types. The efficiency of a typical worm gear in respect to its ratio is shown in Fig. 8-21. The graph demonstrates that the efficiency decreases as the ratio increases. The efficiency of a worm gear speed reducer also depends on the input speed, lead angle, material, design, and output torque.

Higher Reduction Ratios

Higher reduction ratios can be generated using double and triple reduction ratios. For example, a double reduction worm gear reducer, when both worm gear sets have a 50:1 ratio, provides a 2,500:1 (50 x 50) combined speed reduction ratio.

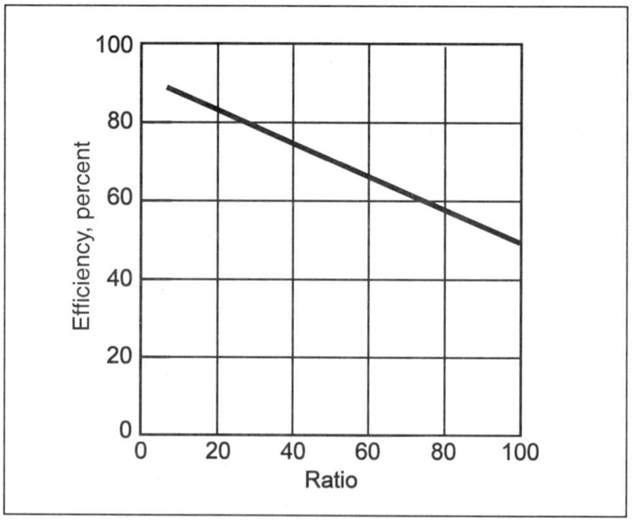

Figure 8-21: Efficiency of worm gears.

Another common method of providing multiple reductions is to create multiple stages by combining worm and helical gear reduction units. This method will produce high reduction ratios while maintaining high overall efficiency.

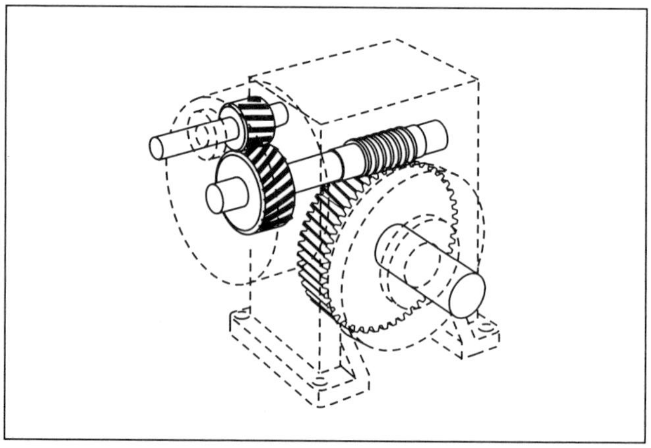

Figure 8-22: Double reduction helical and worm gear set.

Helical-Gear Speed Reducers

Helical reducers provide high-efficiency speed reduction through 1, 2, 3, or even 4 sets of gears. Power is transmitted from a high-speed pinion to a slower-speed gear through rolling, sliding contact action between their respective gear teeth. Helical gears usually operate with their shafts parallel to each other. The two most common types are the concentric, in which the input and output shafts are in line, and the parallel shaft, in which the input and output shafts are offset, Fig. 8-23.

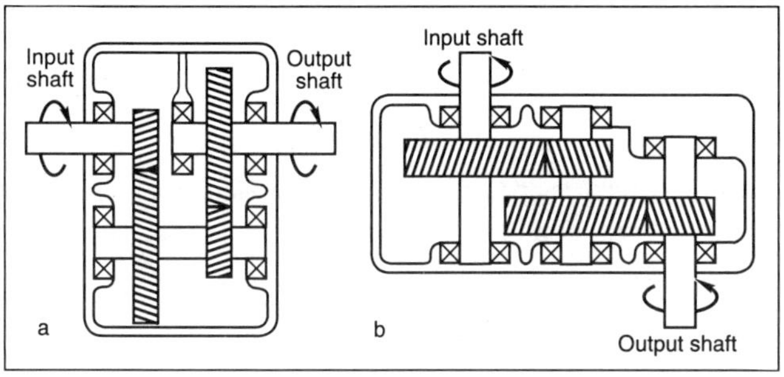

Figure 8-23: Helical gear speed reducers with concentric (a), and offset (b), shafts.

Single-stage helical-gear reducers are generally used only for gear ratios up to about 8:1. Thus, where lower speeds and higher gear ratios are required, double, triple, or even quadruple gear reduction stages are used.

When multiple stages of reduction are used, the total gear ratio of the reducer is obtained by multiplying the individual ratios of each reduction stage by each other. For example, a triple-reduction gear reducer made up of a primary stage ratio of 4:1, a secondary stage of 7:1, and a final stage of 3:1, has an overall gear ratio of 4 x 7 x 3 = 84:1.

Helical gears are very efficient, the total efficiency of a single pair ranging from 97 to 99 percent, including bearings and seals. For example, a triple-reduction speed reducer with an average efficiency of 98.5 percent per gearset has an overall efficiency of 0.985 x 0.985 x 0.985 = 0.956 or 95.6 percent.

This can be compared to a worm-gear reducer of similar ratio which would typically have an efficiency of about 60 percent. Because of this difference in efficiency, a helical-gear speed reducer operates at lower cost than a worm-gear speed reducer for the same amount of output torque.

Cycloidal Disc Speed Reducers

Cycloidal gearing is different from conventional gearing in that these units operate without a high speed pinion or gear teeth, and all components of cycloidal gearing operate in compression rather than shear. This is due to the design of the cycloidal discs which roll within a ring gear housing, similar to a planetary design. The main components are the eccentric cam, the internally flanged output shaft, the cycloidal discs and the ring gear housing, Fig. 8-24. The cycloidal drive operates simply by the action of the eccentric cam mounted to the input shaft. The eccentric cam rotates inside the bore of the cycloid disc, forcing the cycloidal disc to roll inside the ring gear housing. These type drives provide high ratios while delivering high efficiencies. Furthermore, two thirds of the reduction components within cycloidal gearing maintain contact at all times, whereas only two or three teeth in conventional gearing will share the load. In this respect, cycloidal reducers can withstand shock loads well beyond conventional ratings. Cycloidal gearing presents a more torque dense design when compared to other gearing, therefore smaller units have higher thermal limitations and consequently can be grease lubricated. Overall, the cycloidal design has made its way through the ranks of the power transmission industry, and today is regarded as one of the highest quality gearing designs on the market.

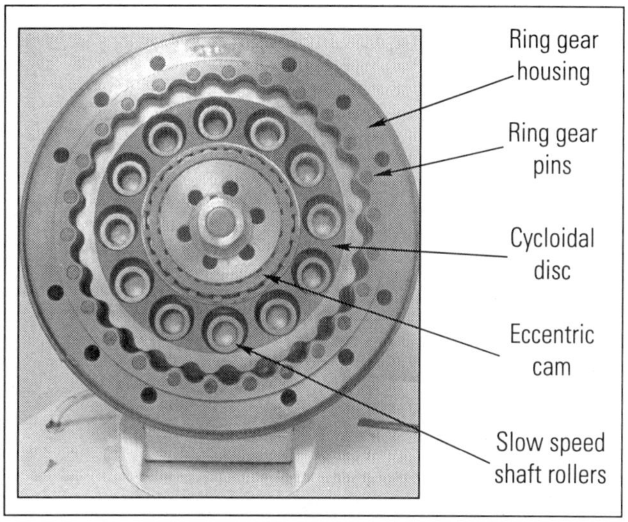

Figure 8-24: Cycloidal Input Assembly

Mounting

Speed reducers are grouped in three categories for mounting purposes: base-mounted, shaft-mounted, and gearmotor. Base-mounted reducers, which have feet for bolting to a stationary structure, are the most common, Fig. 8-25. Here, the prime mover is generally mounted on the same structure as the speed reducer.

Shaft-mounted reducers, Fig. 8-26, have a hollow output shaft that slips over the driven shaft. Free rotation of the shaft housing is prevented by a reaction member such as a torque arm or flange attached to the stationary part of the machine.

A gearmotor, Fig. 8-27, combines an enclosed gearset with a motor. A motorized reducer, which performs similar functions, resembles a gearmotor except that it is driven by a separate NEMA C-face motor, mounted on the input flange of the motorized reducer, Fig. 8-28. Generally, the distinction depends on whether the motor is an integral part (as in a gearmotor) or a modular part (as in a motorized reducer-motor combination).

Seals and Breathers

Seals are used between the gear housing and both input and output shafts to retain oil and exclude dirt. The most commonly used type, the radial lip seal, consists of a metal casing that fits into the housing bore, and an elastomeric sealing lip that presses against the shaft, Fig. 8-29. Lip seals are generally suitable for operation at speeds up to 3,600 rpm. Some types are spring-loaded to better maintain the sealing pressure throughout the life of the seal. Others are suitable for high-pressure applications. Because elastomeric oil seals are a wearing member of the speed reducer, they should be periodically inspected and replaced as necessary.

Labyrinth seals are used for high-speed applications. These seals consist of a stationary housing with a series of rings that limit leakage by maintaining a close radial clearance with the shaft. Though they permit some leakage, labyrinth

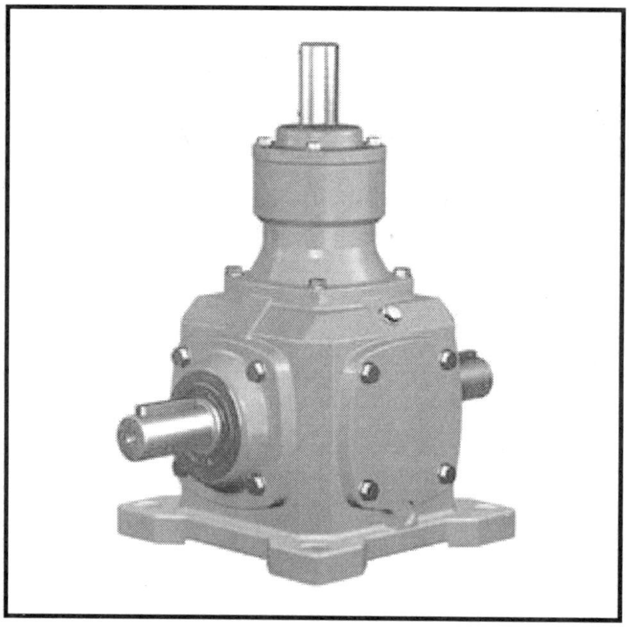

Figure 8-25: Base-mounted speed reducer.

Figure 8-26: Shaft-mounted speed reducer. Torque arm attached to bottom prevents housing rotation.

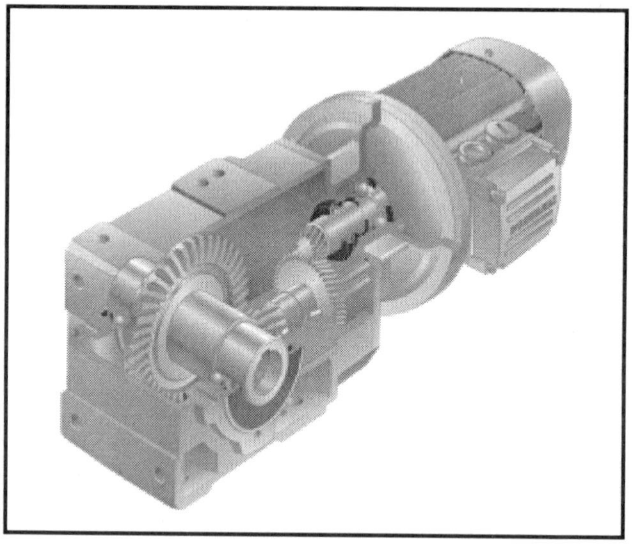

Figure 8-27: Gearmotor.

Figure 8-28: C-face input flange on a speed reducer accepts motor mounting.

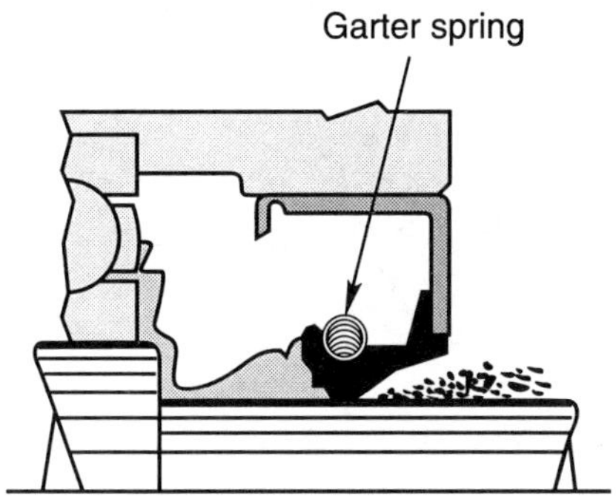

Figure 8-29: Radial lip seal.

seals are simple and reliable. And, because they are noncontacting, they introduce no friction.

A breather is a plug with a hole that is mounted in the gear housing to permit air flow between the inside of the gearbox and the outside atmosphere. This action prevents internal pressure from forcing oil out around shaft apertures or destroying the seal.

Lubrication

Proper lubrication is essential to gear life. Its main purpose is to minimize tooth wear and prevent tooth failure. But, lubrication also reduces friction, protects against corrosion, and prevents heat buildup.

Enclosed gears are generally lubricated with oil. The most common types of oil are rust and oxidation inhibiting, extreme pressure, compounded, and synthetic. Rust and oxidation-inhibiting oil is normally used for moderate loads, whereas extreme pressure types are required for high loads. Compounded gear oils are generally used in worm-gear drives. Synthetic lubricants are suitable for severe operating conditions such as temperature extremes.

Other lubricant types include grease and solid film. Grease can be used for specific types of speed reducers. These would include: Worm, Planetary, Cycloidal, and Hypoidal.

A lubrication system for enclosed gears should have an oil-tight case, a sump filled to a specified level with lubricant, and a means for distributing the lubricant to the gear-tooth surfaces and the shaft bearings.

Two methods are widely used for distributing oil to gears, a splash system and a circulating system. Enclosed gears are usually lubricated by splash systems in which one gear dips into an oil bath and transfers the lubricant to the mating gear teeth as the gears rotate. This method is generally suitable for gears that operate at pitch line velocities up to 5,000 fpm.

Circulating systems are often found in larger gearboxes or those where gears operate at higher speeds. These systems, which pump the lubricant from a reservoir to the gear set, deliver a fixed amount of lubricant in a stream or spray onto the gears. Frequently, the stream is directed to the point of contact between meshing teeth. After passing between the gears, the oil is returned to the reservoir to be recirculated. And, circulating systems often provide filtering and cooling.

In some cases, splash or circulating systems also distribute oil to the bearings where one type of lubricant is suitable for both gears and bearings. Other gearboxes (particularly vertical units) have grease-lubricated bearings.

Proper lubrication is especially important for open gearing because of exposure to contaminants and other environmental conditions that induce wear, corrosion, and early failure.

Several methods are used to lubricate open gears. For slow-speed, low-load applications, a lubricant can be applied with a brush, paddle, or a drip-feed system. Drip-feed or splash systems are used for pitch-line velocities up to about 2,500 fpm. Splash pans or oil troughs are often employed for pitch-line velocities up to 3,500 fpm.

Information on lubricants, lubrication systems, and application methods is available from the manufacturers, many of which provide engineering assistance.

Gear Rating Standards

Ratings for enclosed speed reducers depend on a number of factors, primarily the mechanical and thermal capacity of the individual gears. Additional factors considered in the ratings include the design of other components such as housings, shafts, bearings, keys, and fasteners.

The AGMA has established standards for the rating of both individual gears and enclosed drives. For example, individual gears are rated according to ANSI/AGMA 2001-D04 for spur and helical gears and ANSI/AGMA 2003-B97 for bevel gears. In addition, ANSI/AGMA 6013-A016 establishes ratings for enclosed drives that contain spur, helical, herringbone, and bevel gears.

The rating, or load-carrying capacity, of an individual gear is based on its ability to resist two types of failure: pitting in the tooth surface caused by stress at the point of contact between mating teeth, and fatigue cracks at the root of the tooth caused by bending stress. The pitting resistance and bending strength of the gear teeth are calculated by fundamental rating formulas that contain factors related to geometry, dynamic loads, application, size, load distribution, allowable stress, fatigue life, reliability, and temperature.

Ratings for enclosed drives are based in part on the previously calculated pitting resistance and bending strength of the gears used in the speed reducers. In addition to these mechanical capacity requirements, the overall drive rating considers application factors, thermal capacity (ability to operate without overheating), and the overall power rating of drive components other than gears, which includes the housing, bearings, and shafts.

ANSI/AGMA standard 6013-A016 also gives recommendations on lubrication and installation.

Two standards apply to the rating of gearmotors: ANSI/AGMA 6009-A00 provides a method for rating the pitting resistance and bending strength of spur, helical, herringbone, and bevel gears used for gearmotors. ANSI/AGMA 6034-B92 covers the design and rating of worm-gear speed reducers and gearmotors.

International standards for gears include those of the International Organization for Standardization (ISO), and the Deutsches Institut Fur Normung (DIN).

Application and Selection

Gear drives can be selected and installed with the aid of standards and practices developed by the AGMA. The major selection factors include: shaft orientation, speed ratio, design type, nature of load, gear rating, environment, mounting position, and lubrication.

Shaft orientation

The first consideration in selecting a gear type is the required orientation of input and output shafts: parallel, right angle, intersecting, nonintersecting, skewed, and concentric.

Speed ratio

The ratio of input to output speed is a significant factor both in selecting the type of gearing and determining whether it should be single or multistage. The user should also consider the required efficiency in making this selection. Gears can be custom designed to meet specific speed requirements or standard ratios may be selected from manufacturers' catalogs.

Design type

The application should be evaluated to determine if open gears are sufficient or if an enclosed speed reducer is required. Generally, an enclosed drive with built-in oil lubrication is best, but grease-lubricated open gears can be used in clean environments.

Nature of load

Theoretically, gears and bearings operating at stresses below the endurance limit and lubricated properly will last indefinitely, provided the loads are within design specifications. Thus, determining the nature of the load is important to selecting gears for long life and reliable service. Applying a service factor to horsepower and torque requirements will help to assure correct design specifications.

Gears may be subjected to occasional, intermittent, or continuous-service loads. The life of the geared system is the period of operating time or cycles during which the gears transmit the required load.

Determining the nature of the load involves consideration of the maximum horsepower of the prime mover, inertia of drive components, overhung load on the shaft produced by other drive components (chain sprocket or belt pulley), and speed limitations of the gears.

Gear rating

All components of an enclosed gear are usually rated by established methods, with the rating for the entire system (gears, bearings, shafts) determined by the lowest-rated part. There are two types of ratings: mechanical and thermal. The mechanical rating is based on the strength of the various components or the load compatible with the required life of gears or bearings for the specific application. The thermal rating specifies the power that can be transmitted without exceeding a specified rise in operating temperature.

Environment

The type of gear drive selected may need to compensate for adverse environmental conditions such as dust, heat, wide temperature variations, moisture, and chemicals. Each of these can adversely affect gears, bearings, seals, or lubricant.

Mounting position

Most gear drives are designed to operate in either a horizontal or vertical position only. In some applications, however, the unit may have to operate in an inclined position. This can affect the oil level, air vent position, and drain hole locations in the gear unit selected.

Lubrication

Reliability of a lubrication system is important because failure to supply lubricant to the gears and bearings can cause their damage or failure. Therefore, the type of lubricating system should be chosen carefully.

The user should know the temperature range in which the unit will operate. If temperatures vary widely, the oil viscosity should be changed to suit the conditions.

Additional Information

Addresses and phone numbers for associations are provided in the Association appendix.

AGMA 2009-B01 and 2011-A98, "Handbook-Gear Classification, Materials, and Measuring Methods for Unassembled Spur and Helical Gears."

American Gear Manufacturers Association (AGMA), Alexandria, Va. Some of the more important AGMA publications include:

ANSI/AGMA 915-1-A02, 915-2-A05, 2015-1-A01 and 2015-2-A06, "Gear Classification and Inspection Handbook—Tolerances and Measuring Methods for Unassembled Spur and Helical Gears."

ANSI/AGMA 2001-D04, "Fundamentals Rating Factors and Calculation Methods for Involute Spur and Helical Gear Teeth."

ANSI/AGMA 2003-B97, "Rating the Pitting Resistance and Bending Strength of Generated Straight Bevel Gear Teeth."

ANSI/AGMA 6013-A016, "Standard for Spur, Helical, Herringbone, and Bevel-Enclosed Drives."

Dudley's Gear Handbook, 2nd Edition, Edited by Dennis P. Townsend, McGraw-Hill Inc., New York, 1992.

Fundamentals of Gear Design, Raymond J. Drago, Butterworth Publishers, Boston 1988.

Gear Drive Systems, Peter Lynwander, Marcel Dekker Inc., New York. 1983.

Mechanical Components Handbook, Robert O. Parmley, P.E. Editor in Chief, McGraw-Hill Inc., New York, 1985.

Plastic Gearing, Clifford E. Adams Marcel Dekker Inc., New York, 1986.

Chapter 9

Hydraulics & Pneumatics

Chapter Objectives

When you have finished Chapter 9, "Hydraulics and Pneumatics," you should be able to do the following:

1. Describe the basic principle behind fluid power.
2. Explain the advantage of fluid power.
3. Provide examples of fluid power applications.
4. Describe the function of a hydraulic pump.
5. Differentiate between the features and applications for gear, vane, and piston pumps.
6. Explain the importance of pressure, flow, speed, and efficiency rating factors.
7. Describe the purpose and features of proportional and servo valves.
8. Explain the concept of flow capacity.
9. Describe the features and principles of linear and rotary actuators.
10. Describe the types of fluid conductors, connectors, and filters.

Introduction

Fluid power, as the name implies, involves using a fluid to do useful work. Fluid power encompasses two areas: hydraulics when the working fluid is a liquid (usually oil), and pneumatics when the fluid is gas. The basic principle behind fluid power is that a fluid under pressure produces an outward force on all surfaces in contact with the fluid. When fluid pressure acts on a movable surface, such as the piston in a cylinder, that force can be harnessed to do work (see Fig. 9-1).

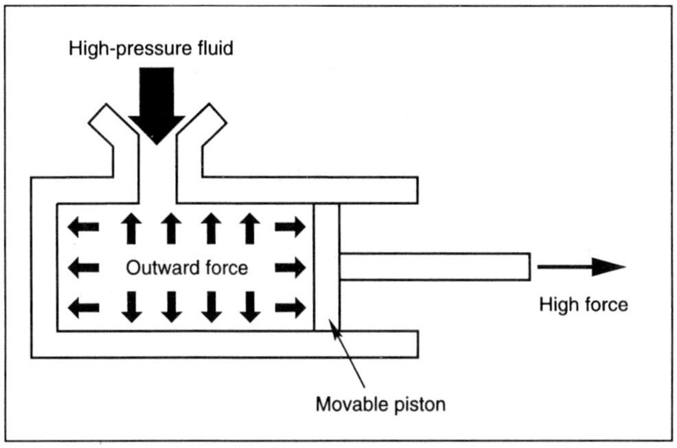

Figure 9-1: High-pressure fluid acting on a movable surface can be harnessed to produce force and motion.

One advantage fluid power holds over other technologies is that it can produce a lot of force in a small package. For example, a hydraulic cylinder small enough to be held in one hand can produce enough force to lift a small car.

Fluid power is typically the best way to produce high-force linear motion. Electric motors are usually preferred for rotating applications over fluid motors and rotary fluid actuators. Exceptions are where extremely high torques are required. Examples include the drive wheels of a large excavator, or where safety concerns prevent the use of electricity, such as under water. Hydraulic motors also have the advantages of torque limiting, compact size per horsepower, and ease of speed control.

The technique is used in a wide variety of applications. Aircrafts, for example, employ hydraulics to control aerodynamic surfaces because a small, lightweight actuator can produce high forces. Construction equipment uses hydraulic devices because they are rugged and extremely powerful. Pneumatics, however, is preferred for industrial applications, like stamping presses or pick-and-place robots, because the systems are inexpensive, fast acting, safe, and clean.

Fluid-power mechanisms are also recognized as being extremely durable. Both hydraulic and pneumatic equipment stand up to shock and vibrations, dust and dirt, heat and cold, and require little maintenance, so they are often used where harsh conditions would quickly destroy other systems. The schematic for a basic fluid-power system is shown in Fig. 9-2. Here is a closer look at the critical components in such a system.

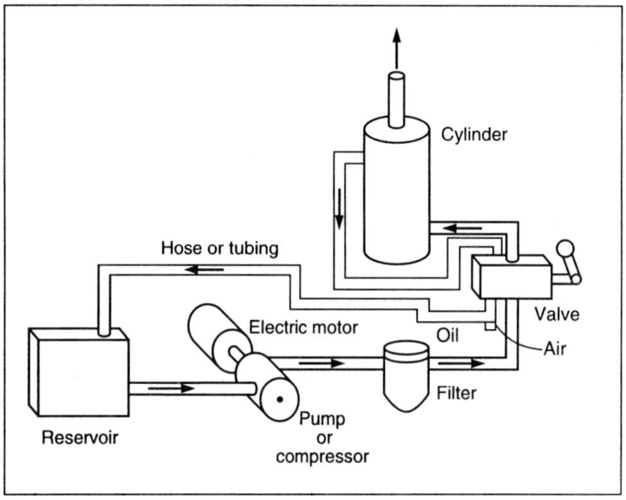

Figure 9-2: A typical fluid-power system consists of a pump or compressor to propel the fluid, valves that control where the fluid goes, and actuators that do the work. Hose or tubing carries the fluid from one place to another, and filters keep the fluid clean.

Hydraulic Pumps

A pump is a device that converts mechanical force and motion into hydraulic fluid power. Normally driven by an electric motor or internal-combustion engine, the pump is the heart, or power source, of a hydraulic system. A wide variety of pumps are available that deliver from less than one to many hundreds of gallons per minute (gpm). Output pressures typically range from 500 to about 6,500 psi, although some specially designed pumps produce pressures exceeding 50,000 psi. Pumps are available in three basic designs: gear, vane, and piston.

Gear Pumps

Compact and inexpensive gear pumps feature few moving parts. Two types are widely used: gear-on-gear and gear-within-gear.

Gear-on-Gear Pumps

These pumps (see Fig. 9-3) consist of two gears, usually of equal size, that mesh with each other inside a housing. The driving gear, connected to the drive shaft, drives the second gear. As the gears rotate, liquid is drawn into the housing, compressed between gear teeth and housing, and delivered at pressure to the discharge port. Both spur and helical gears are used, but spur gears are the most common. Helical gears release segments of fluid more evenly, minimizing ripple.

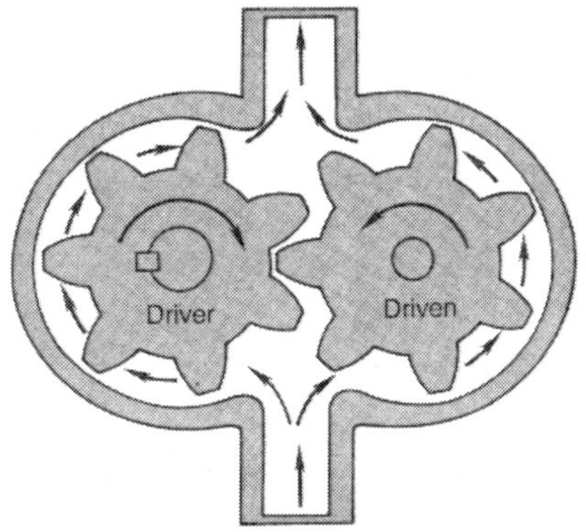

Figure 9-3: In gear pumps, the gears rotate within the housing and sweep fluid from the inlet to the outlet.

Gear-within-Gear Pumps

This variety (see Fig. 9-4) consists of an externally toothed gear that rotates inside and drives a larger internally toothed gear. In one common configuration, the gerotor pump, the inner gear has one tooth less than the outer gear. Here, as the inner gear drives the outer one, the relative motion of the gear teeth creates sliding seal points. As the teeth unmesh, a vacuum forms in the pocket between the teeth, drawing oil from the inlet. When the teeth engage, the action forces the oil out the discharge port.

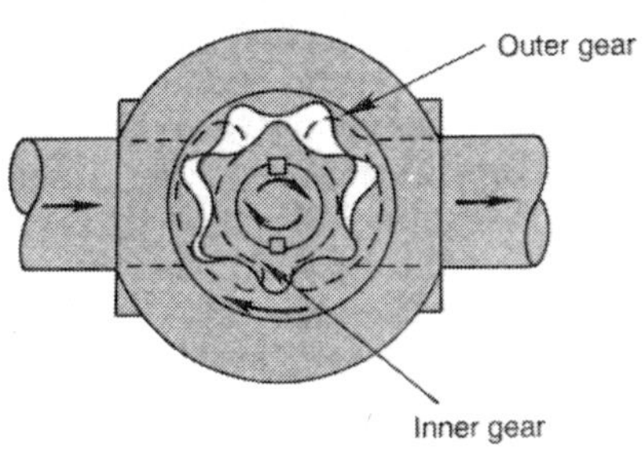

Figure 9-4: The inner gear has one tooth less than the outer gear in this gear-within-gear pump. Relative motion between the two causes pumping action.

Vane Pumps

A typical vane pump consists of a circular rotor mounted eccentrically in a circular cavity. As the rotor spins, vanes in the pump extend and retract to seal against the cavity surface, sometimes called a cam ring. Fluid is trapped between the vanes at the inlet, swept along by the vanes, and propelled through the outlet.

Vane pumps are efficient at speeds over about 600 rpm, the minimum rate where vane tips contact the cam ring. If below this speed, leakage is high, and efficiency is low. Vane length is adequate to accommodate appreciable wear. The pumps are more sensitive to contaminants than gear pumps but less so than piston pumps.

The basic vane pump is subject to high bearing loads because it is unbalanced. A high-pressure outlet area on one side of the rotor and a low-pressure inlet area on the other force the rotor and shaft down against the bearings. This drawback is overcome with the balanced vane pump (see Fig. 9-5). Here, the rotor is centered inside an elliptical cam ring, and there are two inlet and two outlet ports opposite each other. With this configuration, the two high-pressure zones are 180 deg. apart, balancing out the forces.

Piston Pumps

The most efficient of the three pump types, but also the most costly, piston pumps exhibit volumetric efficiencies up to 99 percent. They convert rotary motion of an input shaft to a reciprocating motion of one or more pistons. Fluid is drawn into and forced out of a chamber by the piston, with

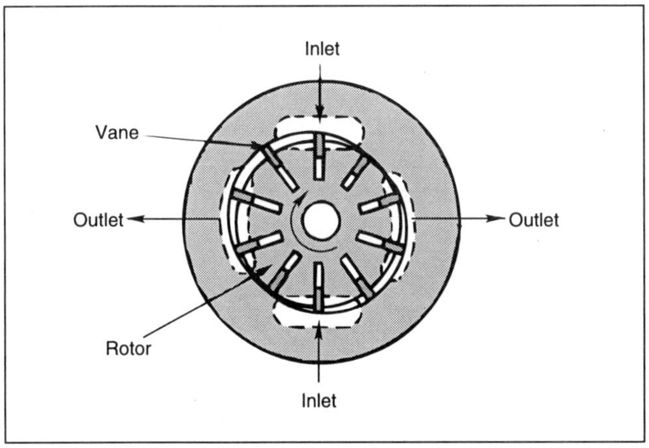

Figure 9-5: Two opposing high-pressure zones in this balanced-vane pump equalize forces on the pump shaft.

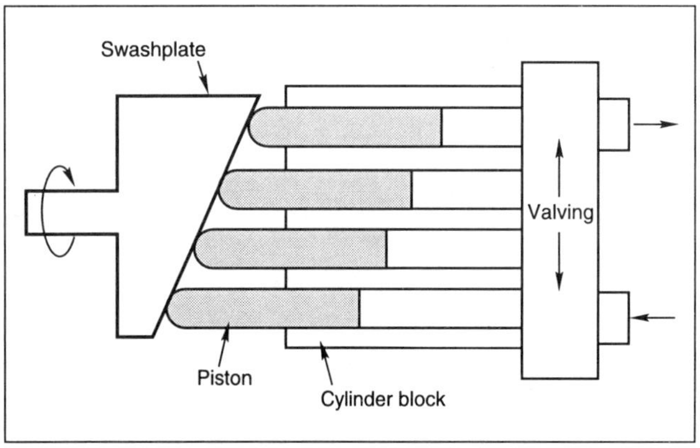

Figure 9-6: The pistons reciprocate as the swashplate rotates, taking in fluid while moving toward the thin part of the plate and expelling it while approaching the thick end. Valving controls flow direction.

valves controlling flow direction. The action is much like that of a piston and cylinder in an automobile engine or even a simple bicycle pump. Two types are available: axial-piston and radial-piston. The terms indicate the direction of piston motion in relation to the pump shaft.

Axial-Piston Pumps

These pumps (see Fig. 9-6) often use an angled cam or "swash plate" attached to the pump shaft. As the plate rotates, the pistons reciprocate, taking in fluid while moving toward the thin part of the plate and expelling it while approaching the thick end. An alternative way to reciprocate the pistons is to mount them at an angle in relation to the drive shaft and allow both the pistons and shaft to rotate. For this design, no swash plate is required because of the angled relationship of the pistons and shafts. This configuration is typically called a bent axis pump.

Radial-Piston Pumps

This design converts rotary shaft motion into a radial reciprocating motion of the pistons. One type is driven by a rotating cam that runs through the center of the pump, driving the pistons.

Variable Displacement Pumps

Hydraulic applications calling for high power (typically over 15 HP) and flow that varies over a wide range are often best served by variable-displacement pumps. These pumps allow control of output flow, usually between zero and maximum. Vane and piston pumps are usually available with this function. Vane pumps vary displacement by adjusting the position of the cam ring while piston pumps change axis angle or angle of the swash plate. Gear pumps typically change flow only by adjusting drive speed, which generally restricts them to fixed-displacement applications.

The main advantage of a variable displacement pump is low power consumption. They only use enough power to deliver the required flow—and no more. However, these pumps are generally not as efficient and cost more than fixed-displacement types. In recent years, newer technology in electric motor control has opened the path for fixed displacement pumps to be used as variable displacement pumps. Instead of changing the cam ring or the swash plate, it is now quite easy and inexpensive to change the speed of the electric motor using a variable frequency drive. This saves both energy and the cost of the pump.

In general, fixed displacement pumps should be used in applications where:

The duty cycle is on-off, and the pump can be unloaded completely when not in use.
Full flow from the pump is required under most operating conditions, even though the load may vary over a wide range.

Pressure Compensated Pumps

Only the piston and vane pumps can be manufactured to compensate for a pressure change during the pumping cycle. A pressure compensated pump is designed to hold the same pressure on the outlet during a back-stroke as during the power stroke. This is usually accomplished by using a swash plate to hold the pressure. Because less pressure is normally required on the return part of the stroke, system designers are using load sense controls. Using the pressure compensating control, they set it up for remote control. Then the load itself is used to tell the pump how much pressure is required. This produces major savings in energy in most systems.

Rating Factors

Pressure

The pressure rating is one of the major considerations in determining whether a pump can perform a job. The rating is generally limited by the capability of the pump to withstand pressure without undesirable increase in internal leakage and without damage to the pump parts.

Typically, maximum pressure is approximately 4,500 psi for external gear pumps and from 2,000 to 4,000 psi for vane pumps. Internal-gear units run somewhat lower, with maximums in the range of 1,500 to 2,500 psi. Piston pumps are rated to around 6,500 psi maximum, although some are suitable for pressures exceeding 8,000 psi. A few permit higher pressures for intermittent peak loads.

Flow

The second most important consideration in selecting a pump is how much fluid it can deliver. This is usually expressed in cubic inches or cubic centimeters per revolution. The flow rate in gpm is determined by multiplying cubic inches per revolution by the speed of the engine or electric motor and dividing by 231 cubic inches per gallon.

Speed

A third consideration is the speed rating, which may be a limitation imposed by the ability of the pump to fill without cavitating or by some mechanical factor. The permissible speed range and inlet pressure requirements for pumps are usually clearly defined.

Efficiency

Comparing efficiency is a good way to judge two otherwise similar pumps, but efficiency is defined in three ways:

Volumetric efficiency is the ratio of actual to theoretical delivery. The difference between actual and theoretical delivery is normally due to internal leakage necessary to lubricate the pump ("slippage") and other factors. Volumetric efficiency is typically very high, often in the mid to high 90s.

Overall efficiency is the ratio of hydraulic power output to mechanical power input.

Mechanical efficiency is the ratio of actual pressure to theoretical pressure. Mechanical losses are due principally to internal friction and fluid compression.

Pneumatic Compressors

A compressor is a device that moves specified volumes of gas against some working pressure. Compressors are similar in concept and hardware to hydraulic pumps, and selection considerations are much the same. There are important considerations in air compressors. Because the air gets compressed, the water trapped in the air releases and gets in the pneumatic components. Because of this, it is critical in large compressors to have a filter with water removal, a cooler to reduce the temperature of the air after it is compressed, and a dryer to remove as much water as possible. In addition, point-of-use dryers, typically with a dessicant material, are used to remove water that might foul the air valves.

Inside a plant or factory, numerous pneumatic systems are often powered by a single, large compressor. However, many applications call for individual, small compressors. These units are available in a wide variety of designs. The following are some of the most common.

Vane Compressors

Like their hydraulic counterparts, vane compressors (see Fig. 9-7) are inexpensive, operate at low cost, and demand little starting torque. They are compact, relatively vibration free, and generate little pulsation in the compressor output. The sliding vanes are closely fitted in the rotor slots and wear little during operation. Vane compressors are available in power ranges from 10 to 500 HP at pressures to 150 psi.

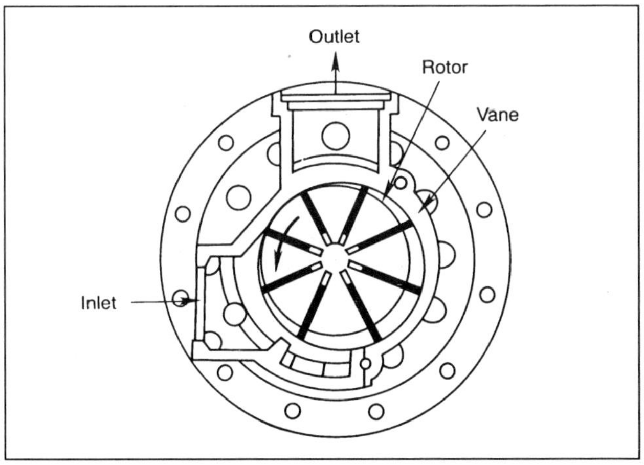

Figure 9-7: The vanes slide in and out of the slots as the compressor rotates, carrying air from inlet to outlet.

Reciprocating Compressors

This type of compressor (see Fig. 9-8) consists of pistons that compress the air in cylinders and valves that control its inlet and outflow. Sizes range from less than 1 to over 5,000 HP. Reciprocating compressors are efficient and useful for a wide range of operating conditions.

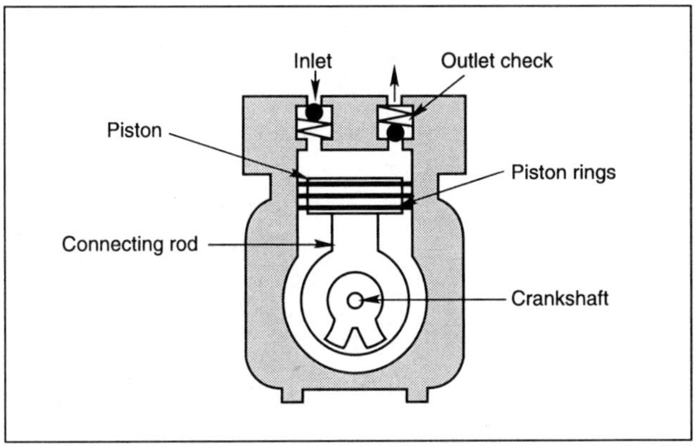

Figure 9-8: Check valves control flow direction as the reciprocating piston draws in and expels air.

Diaphragm Compressors

A modification of the reciprocating compressor, the piston in a diaphragm compressor (see Fig. 9-9) flexes a metal or elastomeric diaphragm. This motion compresses air trapped in the cylinder. The diaphragm provides a positive sealing barrier between the air and the compressor's internal parts, so they are often used where clean air is a must.

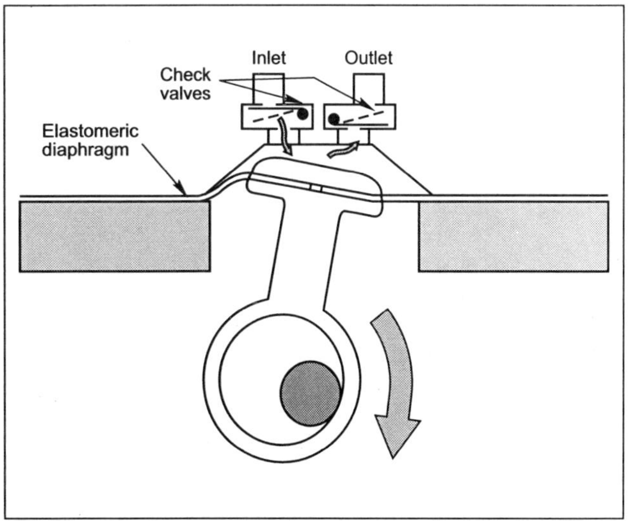

Figure 9-9: The reciprocating piston flexes a diaphragm which, in turn, compresses the trapped air.

Centrifugal Compressors

Best suited to moving large volumes of air, centrifugal compressors (see Fig. 9-10) operate at relatively low pressures. They consist of a high-speed rotating impeller, a diffuser section where air velocity decreases and pressure increases, and a collector section that further reduces velocity and increases pressure. The compressors typically employ multiple stages, supplying from less than 250 to well over 20,000 cfm.

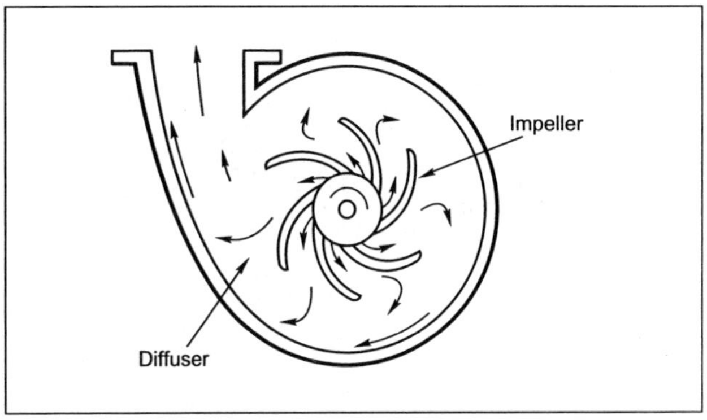

Figure 9-10: The high-speed rotating impeller in a centrifugal compressor forces air through a diffuser, where velocity decreases and pressure increases.

Sizing considerations for compressors are much the same as for hydraulic pumps, with pressure rating and flow of greatest concern. While compressors can be sized to provide only the required pressure and flow, 10 to 25 percent over-capacity is preferred. Inlet filters should be provided to protect the compressor, and outlet filters and dryers should condition the air and protect downstream components.

Valves

Hydraulic pumps and air compressors use valves to control or modulate the flow direction, fluid pressure, or the rate of fluid flow. Valve mechanisms are classified as a spool, poppet, slide, rotary, and diaphragm.

Spool Mechanisms

These devices (see Fig. 9-11) slide inside a sleeve, controlling flow between two ports. The device is used in both hydraulic and pneumatic valves. The device calls for a short stroke, exhibits low friction, requires little actuation force, and is suitable for high pressures. Because these valves depend on close clearances between spool and bore, low-leakage varieties may be relatively costly and sensitive to contaminants.

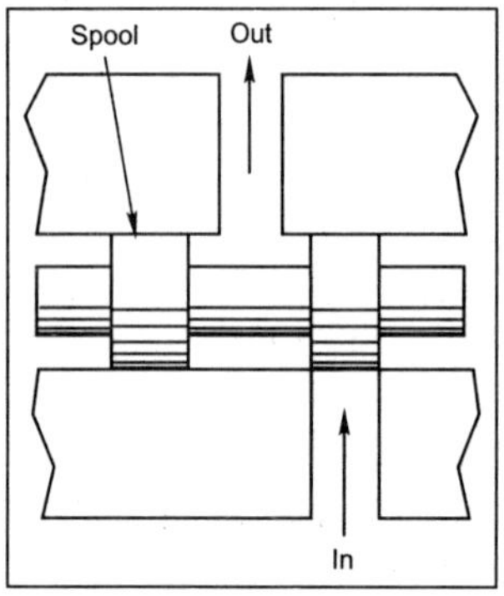

Figure 9-11: The spool sliding within the sleeve permits or blocks flow.

Poppet Mechanisms

Here (see Fig. 9-12), the relative position of the poppet with respect to a seat controls fluid flow. Cartridge valves consist of a poppet, sleeve, and spring that are contained within a machined manifold block. The valves typically are suitable for hefty flows with minimum pressure drop and are relatively insensitive to fluid-borne contaminants.

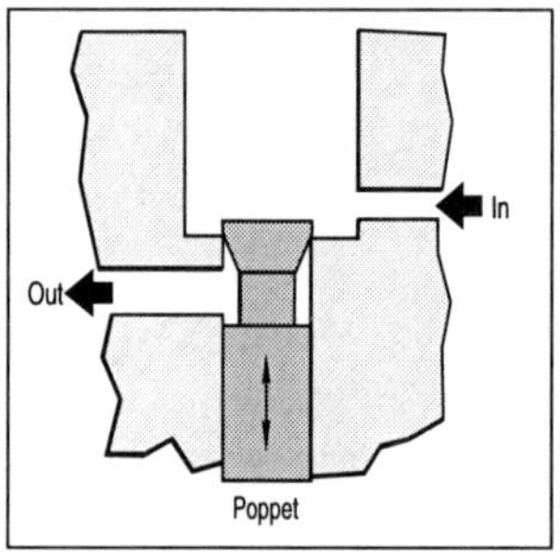

Figure 9-12: Poppets seal on the narrow contact area between poppet and seat.

Valves are actuated by various means. Some employ a manually operated lever, while others may be operated by cam or electric solenoid. Pneumatics and hydraulics can be used to operate yet other valves.

Electrically actuated valves are normally powered by small solenoids. The solenoid plunger moves the spool or poppet directly on small valves. On larger valves, the solenoid drives a miniature valve that, in turn, directs pressurized fluid to the spool or poppet of the main valve. These are called solenoid-controlled, pilot-operated valves. Position control is enhanced using proportional solenoids on the valves. These solenoids cause the spool or poppet to move only the distance that the solenoid commands it to move. By varying the voltage to the coil, the spool travel is modulated. This allows much greater speed control and positional accuracies which the machine operator can determine.

Pressure-Control Valves

These valves modulate pressure level in fluid-power circuits. Several types are available, categorized by function.

Relief Valves

Relief valves are used on hydraulic systems to keep the pressure in the lines from exceeding a level that would damage equipment or lines. A spring holds the relief valve closed until the pressure level is such that hydraulic force acting on the inlet port exceeds the spring pressure. When this pressure is exceeded, the valve ports the hydraulic fluid back to the fluid reservoir while maintaining pressure on the lines to meet the demands of the equipment being used. Pop-off valves are used in air compressor applications to protect air lines and system components. However, they are manually reset valves that open to the atmosphere only when the pressure is not being adequately controlled by the compressor pressure switch. Maximum pressure ratings may be found in manufacturer's catalogs.

Reducing Valves

Reducing valves are used to limit pressure levels of hydraulic system branch circuits. When the circuit reaches a predetermined pressure, the reducing valve restricts flow by moving the spool to a position such that there is a balance between the spring and the low-pressure side. When these valves are used in air systems, they are called regulators.

Sequence Valves

Used for machine sequencing, the valves sense pressures other than maximum. These normally closed valves permit flow between inlet and output ports when the pressure reaches preset levels. Typically, a certain minimum pressure must develop in one part of the circuit before fluid can pass through another part. Sequence valves normally come with a reverse check. The reverse check causes the system to sequence in one direction only. To sequence in the return direction, a second sequence valve needs to be used.

Unloading Valves

Unloading valves cause a pump to go off-line when a certain pressure is achieved while at the same time maintaining the pressure in the system. When the system pressure falls a set amount, the valve closes and boosts the system pressure back to the preset level and then unloads the pump again. Frequently, we will see accumulators in these systems. The accumulator is located on the pressurized side of the valve and allows the pump to stay unloaded for longer periods of time, making the operation much smoother.

Brake Valve

A brake valve is just like its name suggests. This valve is used for dynamic braking or stopping on a hydraulic motor. It consists of a relief valve where the back pressure to slow the motor can be set. It also has a pilot line, which causes it to completely open when the motor is commanded to run in the reverse direction.

Counterbalance Valve

The counterbalance valve, which is similar to the brake valve, is used in cylinder circuits. It is used to put a back pressure on the cylinder to prevent a load from falling uncontrolled. The valve back pressure is adjusted such that when no downward pressure is applied to the cylinder, the cylinder will not move or fall downward. This valve is considered an important safety valve in such machines as manlifts and presses where uncontrolled movement could cause serious injury to personnel on or near the machine. Typically, this valve should be hard plumbed to the bottom port of the cylinder, so that, in the event of a hose rupture, the cylinder does not fall. In the reverse direction, a pilot signal is sent that opens the valve completely to allow free movement of the cylinder back up.

Flow-Control Valves

Fluid flow is controlled by either throttling or diverting it. Throttling limits flow by reducing the size of an orifice and bypassing part of the flow around a circuit, so that an actuator receives only the portion needed to perform its task.

When the inlet flow of an actuator is controlled, the circuit is said to be a meter-in system. When the outlet flow of an actuator is controlled, it is called a meter-out circuit. When that part of the fluid being diverted to a reservoir or another part of the circuit is controlled, it is said to be a bleed-off system.

Noncompensated Flow Controls

These simple valves meter flow by restricting or throttling. When the amount of fluid passing through an orifice is restricted, there is a drop of pressure on the downstream side that is directly proportional to the amount of fluid restricted. A common type of noncompensated valve employs an adjustable needle valve. Needle valves create a restriction that meters flow (see Fig 9-15).

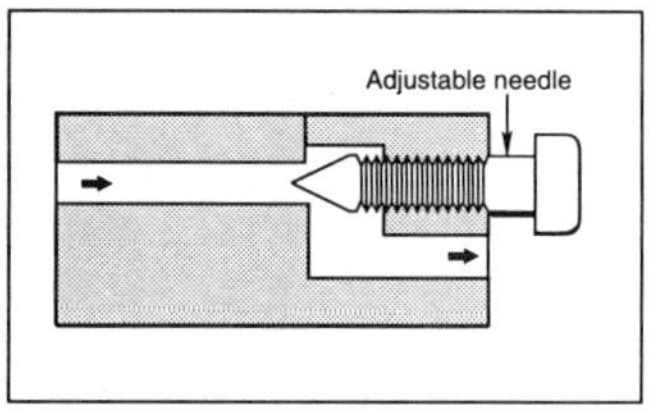

Figure 9-15: Needle valves create a restriction that meters flow.

Pressure-Compensated Flow Controls

To maintain nearly constant flow despite variations in circuit pressure, pressure-compensated valves incorporate a metering orifice like those in non-compensated units. Pressure drop across this orifice shifts a balanced spool against a control spring. This action maintains a constant pressure drop across the orifice which, in turn, maintains a constant flow.

Directional Control

Directional-control valves determine the fluid's flow path through a circuit. The valves are classified by ports, position, and way.

Ports

This is the number of plumbing connections to the valve. Thus, a three-port valve has one port each for connections to a pump, reservoir, and actuator.

Position

The number of stops the valve can make during operation is referred to as position. For example, a two-position valve can shift either in one direction or the other, like a light switch. Three-position valves have an intermediate position between the two end stops.

Way

This term refers to the possible flow paths through a valve. A check valve that permits fluid to flow in only one direction is a one-way valve. A two-way valve has two ports and allows flow in either direction. A common directional-control valve is shown in Fig. 9-16. A three-position, four-way valve can power an actuator to extend the rod, hold a load in an intermediate position, and power the retract stroke.

Proportional and Servo Valves

Proportional valves are solenoid valves that vary the flow in proportion to the amount of current that flows through the solenoid coil. Although fluid flow rate is nonlinear to current flow, they still provide an inexpensive method of controlling position, speed, or force of equipment requiring high-speed response at high flow rates.

Positioning accuracy can be improved by adding a positioning sensor to the end of the spool. Electronic feedback results in performance almost as good as that of a servo valve but at a lower cost.

Servo Valves

The operation of a servo valve is accomplished from a remote electronic controller. Sensors mounted on hydraulic or pneumatic actuators continuously monitor the position, speed, force, and acceleration of the actuator. These signals are transmitted to an electronic controller where they are compared with desired actuator movements that have been programmed into the controller. The controller then instantaneously opens or closes the valve so that the actuator follows the movements desired. This method of control, whereby information is constantly monitored, analyzed, and used to make adjustments, is called "closed-loop feedback."

Servo valves are precision valves that are capable of regulating a large pressure drop across the spool and are available in single and multiple-stage versions that offer a wide range of pressure and flow ratings. Multi-staging is employed in high-flow capacity valves. The first stage typically has an electromechanical actuator, such as a torque motor, force motor, or solenoid that controls a hydraulic metering valve. Sliding spools are generally used for the other stages, with first stage output driving the second stage spool.

Servo valves are precision valves that have high pressure drop across the spool to ensure precise metering and exacting control. They are used in applications requiring high load stiffness, good stability, precise positioning, good velocity and acceleration control, or predictable dynamic response.

Flow Capacity

Valve size can be related either to size of piping connections or to flow capacity. Valve flow capacity can be expressed as a nominal or maximum rating.

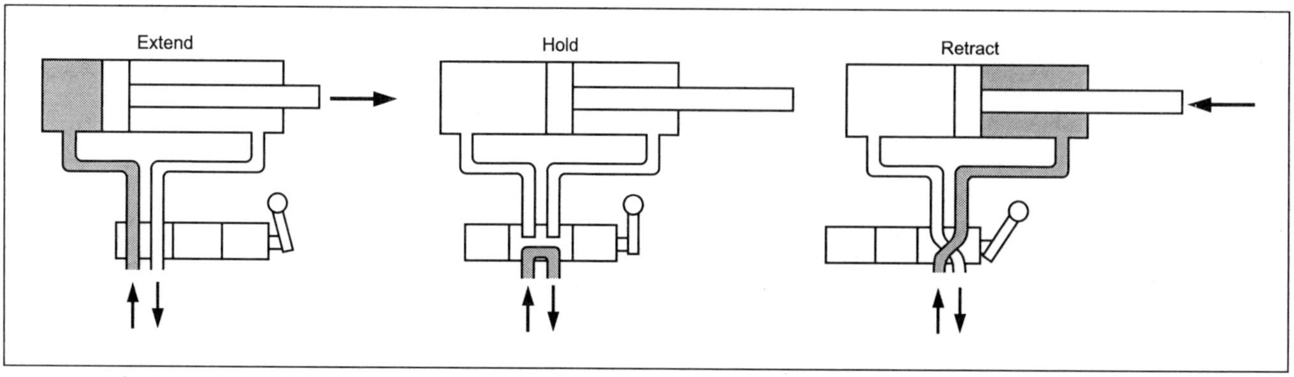

Figure 9-16: A three-position, four-way valve can power an actuator to extend the rod; hold a load in an intermediate position; and power the retract stroke. Pneumatic valves generally have 5 ports (2 exhaust).

Nominal Flow Rating

This rating refers to the amount of flow a valve can handle with relatively low pressure drop. The nominal rating quoted by valve makers is often the hydraulic flow that creates a 50 to 60 psi pressure drop through the valve.

Maximum Rating

The maximum flow a hydraulic valve can control without malfunction is referred to as the maximum rating. For example, a valve with a 30 gpm nominal rating might have a 50 gpm maximum rating. If subject to higher flow, available pilot (controlling) pressure may be insufficient to shift the valve spool mechanism. When a valve is used above its nominal limits but within its maximum limits, the pressure drop varies directly with the square of the flow. For example, a 50 gpm nominal valve is operating at 100 psi input and 50 output. In order to get a flow rate of 60 gpm, a 1.2 increase in flow rate, the input pressure would have to increase by 1.44 (1.2 squared) to 144 psi.

Size Ratings

Size ratings refer to the physical port size. A 3/4 inch valve will couple directly to a 3/4 inch pipe. Flow rates and pressure ratings are independent of this specification.

Flow Coefficient (Cv)

As a valve restricts the flow of a fluid, there is a proportional drop in pressure on the outlet or downstream side. When a valve is operating within its nominal flow rating, flow can be computed as a multiplier or coefficient of the pressure reduction. With this Cv data, valve manufacturers can create a flow rate table, whereby the user can enter required pressure readings on either side of a valve and read the flow rate directly from the chart. For example, if a user requires a 50 gpm flow rate at 100 psi and has an input pressure of 200 psi, the Cv chart would tell the user if the valve was big enough to pass 50 gpm with the required 100 psi pressure drop. If not, a larger valve would be selected. Slow valve response times and high actuator fill and exhaust times can induce errors into valve sizing calculations. A reliable approach is to build a prototype system.

Actuators and Motors

An actuator is a device for converting fluid energy into mechanical energy. Fluid-power actuators are available in a number of forms to provide specific types of action. Cylinders, by far the most common actuators, work through linear extension; motors impart continuous rotary motion to objects; rotary actuators turn an object through only a limited arc.

Generally, all types of actuators are available for either pneumatic or hydraulic operation. Often, the same cylinder can be used for either air or low-pressure oil operation. Air and hydraulic motors, though similar, are usually not interchangeable.

Cylinders

When fluid is pumped into a cylinder, the piston and rod are forced to move in or out against a load. Many different types are available, including the following:

Single-Acting Cylinders

The simplest type of cylinder (see Fig. 9-17) powers a stroke in only one direction. When the fluid is allowed to drain from the cylinder, some external force, such as gravity, must push the piston back to its starting position. An example is a hydraulic jack like a car lift.

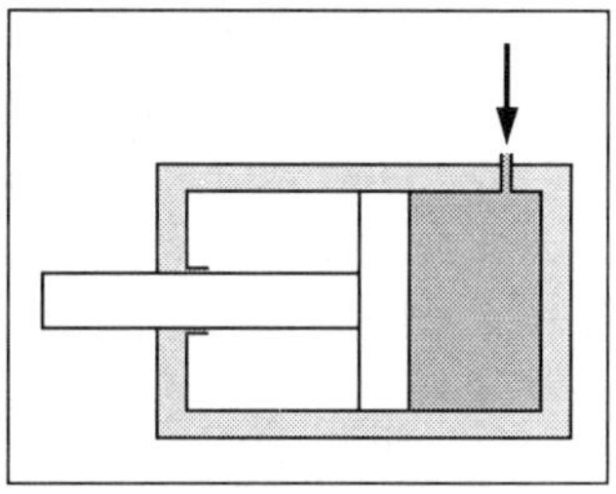

Figure 9-17: Single-acting cylinders provide power only to extend the rod.

Spring-Return Cylinders

Similar to the low-cost, single-acting types, spring-return cylinders (see Fig. 9-18) have a spring that repositions the piston to its starting point. This type is used in both hydraulic and pneumatic service. Cau-

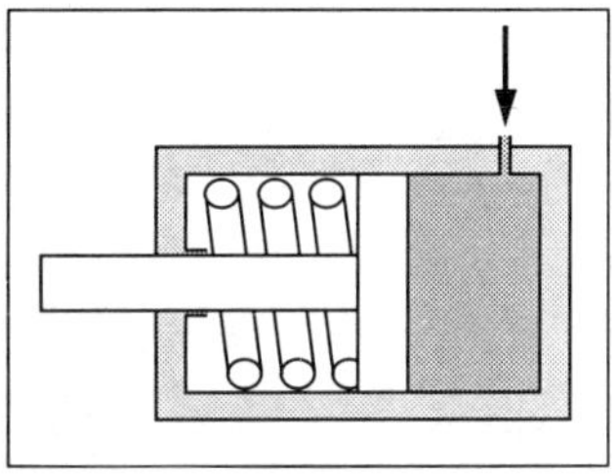

Figure 9-18: Spring force retracts the rod in a spring-return cylinder.

Sizing Cylinders

Important considerations when sizing a cylinder are how big and how fast it is. An oversized cylinder has a healthy extra margin of force that can override a bit of misalignment, binding, or overload. However, extra cylinder size boosts cost, increases weight, and retards actuation.

Force

Cylinder force is a function of piston size and the fluid pressure. Force (F) in lbs. is found from the relationship F = PA, where P is pressure, psi, and A is the piston area, inches^2. Area is determined from the piston diameter dp by A = $(\pi dp^2)/4$. Thus, a 1 inch diameter cylinder operating at 3,000 psi produces F = $(3{,}000)(\pi)(1^2)/4 = 2{,}356$ lbs. of force.

Note that double-acting cylinders produce more "push" force than "pull" force at the same input pressure. That is because pressure acts on the rod side of the piston when the cylinder retracts, and the rod covers part of the piston surface. Effective piston area Ae is found by subtracting rod area from the total piston area, or Ae = $\pi\ (dp^2\text{-}dr^2)/4$, where dr is the rod diameter, inches^2.

Therefore, if the 1 inch diameter cylinder mentioned previously has a 0.5 inch diameter rod, the cylinder produces F = $(3{,}000)(\pi)(1^2\text{-}0.5^2)/4 = 1{,}767$ lbs. of retraction force.

Speed

Velocity of a piston is determined jointly by its size and the flow volume into the cylinder. For steady-flow conditions, where flow Q in gpm and piston area A are known, velocity V can be calculated by V(ips) = Q(gpm)(231 in 3/gal)/A (inches^2)(60 seconds per minute). For example, if flow to a 1 inch diameter cylinder is 2 gpm, rod velocity is V = (2)(231)/(0.785)(60) = 9.8 ips.

Rod Size

The rod must be large enough to withstand the stresses imposed by load and cylinder. The rod diameter is a function of the load, the length of the rod, stroke, and the mounting of the cylinder. In long stroke applications with a relatively large pushing force, it is necessary to separate the rod end bearing from the bearing on the piston. This is accomplished by using a stop tube inside the cylinder to create separation. Generally, the manufacturer sizes the rod with a healthy margin of safety for the cylinder's maximum pressure rating.

tion should be used in pneumatic service because the compressibility of air can cause high actuator exhaust times. If the spring is heavy enough for speedy piston return, it may require too much force to compress. The cylinder must be about twice as long as the required stroke to include space for the spring. Cylinders may be spring loaded to the extended or retracted position and use hydraulic or pneumatic power to move in the opposite direction.

Double-Acting Cylinders

Double-acting cylinders (see Fig. 9-19) contain two fluid chambers so that pressure both extends and retracts the rod. This type of cylinder is the most common and can be used in nearly all types of applications. Effective working area of the rod side of the piston is less than that of the blind side, so double-acting cylinders retract faster than they extend and exert less force on the retraction stroke.

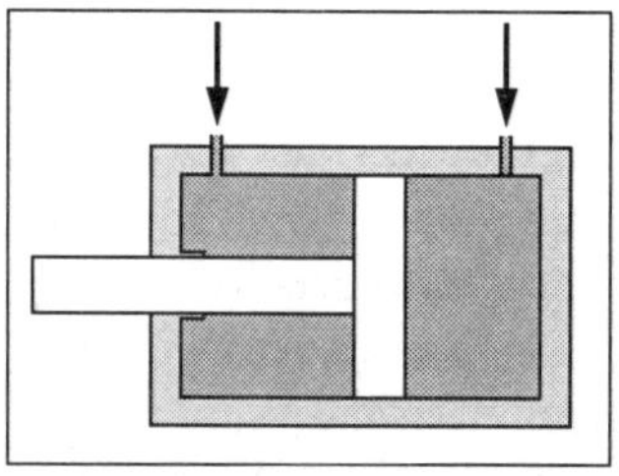

Figure 9-19: Fluid pressure can extend and retract the rod in double-acting cylinders.

Rodless Cylinders

As the name implies, rodless models (see Fig. 9-20) differ from their more conventional counterparts in that no piston rod extends from the cylinder body. Rather, an internal piston is connected either physically or by magnetic force to an external carriage. While the medium can be either hydraulic or pneumatic, it is far more common to see this type of actuator in pneumatic applications. Driving the piston hydraulically or pneumatically causes the carriage to traverse between the cylinder end caps. A major advantage of rodless cylinders is that they require considerably less mounting space—nearly 50 percent less in some applications—compared with rod-type models.

Servo Actuators

The coupling of electronic and fluid-power technology is becoming increasingly common, especially in electrohydraulic and electropneumatic actuators. Advances in trans-

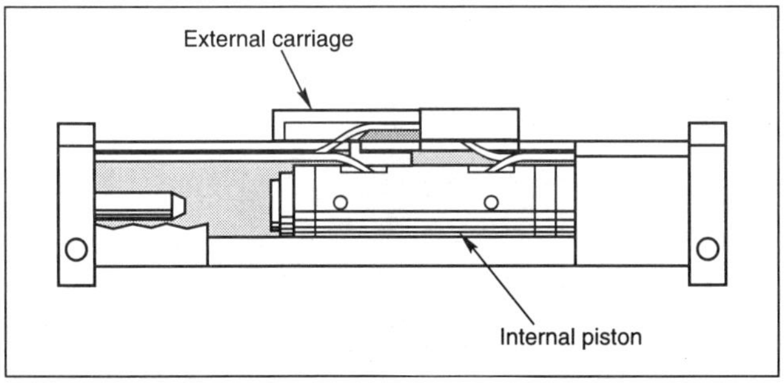

Figure 9-20: A carriage coupled to an internal piston travels between the end caps of a rodless cylinder.

ducers and electronic controls have resulted in cylinders that transmit high forces while allowing computer control of rod velocity, acceleration, and positioning accuracy to less than 0.001 inch.

The key to the operation of these actuators is precisely sensing cylinder rod position. One method uses a linear displacement transducer mounted inside a hollowed-out piston rod. The transducer measures position and sends that information to an electronic controller which, in turn, signals a servo or proportional valve to adjust flow and cylinder position accordingly. All this happens in a fraction of a second.

Rotary Actuators

Rotary actuators turn an output shaft through a fixed arc. They are compact, simple, and efficient. The actuators produce high torque instantaneously in either direction, occupy little space, and are simple to mount. Two popular designs are rack-and-pinion actuators and vane actuators.

Rack-and-Pinion Actuators

This type of actuator uses fluid pressure to drive a piston connected to a gear rack, which rotates a pinion (see Fig. 9-21). Standard units are available that rotate either 90, 180, or 360 deg. Some actuators have two parallel piston-rack units; this doubles output torque. Outputs exceeding 30 million lbs. per inch are available.

Vane Actuators

These units consist of a shaft mounted in a cylindrical housing with one or more vanes attached to the shaft (see Fig. 9-22). Applying fluid pressure to the vane rotates the shaft. An internal barrier between housing OD and shaft divides the interior volume into two chambers. For this reason,

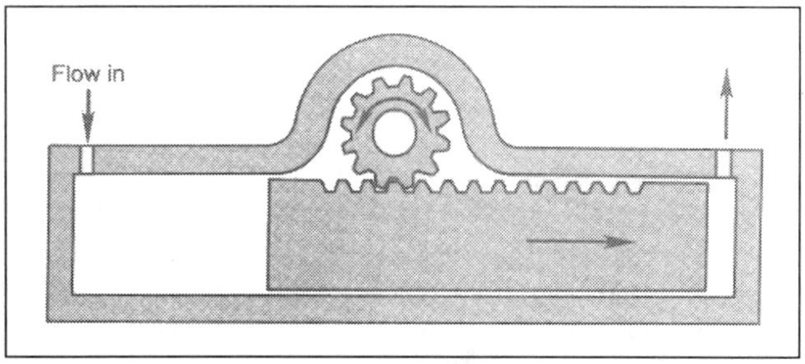

Figure 9-21: Fluid pressure slides the rack and rotates the pinion in this rotary actuator.

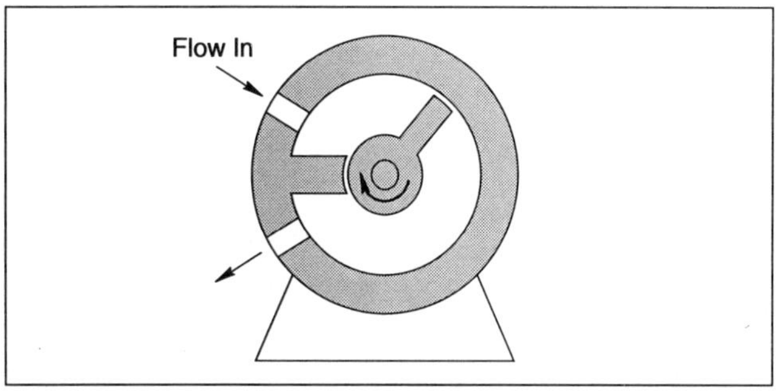

Figure 9-22: Fluid pressure in a single-vane actuator moves the vane and rotates the shaft.

single-vane actuators are normally limited to about 280 deg. of rotation and double-vane actuators to about 100 deg. Torque is directly proportional to both vane area and effective fluid pressure. Some vane actuators have torque outputs exceeding 500,000 lbs. per inch.

Fluid Motors

Hydraulic and pneumatic motors usually take a back seat to electric motors, but they are useful in specialized applications that include the following:

- Where very high torque is required.
- Where space or weight is limited.
- Where the motor is subject to stalling and holding loads.
- Where safety considerations prevent use of electricity.

Basic configuration of fluid motors is much the same as the corresponding hydraulic pumps. The difference is that flow causes shaft rotation in a fluid motor. Common designs include the following:

Axial-Piston Motors

These contain several pistons that are driven by high-pressure fluid.

Axial-piston motors have high volumetric efficiency at both high and low speeds. In hydraulic designs, typical maximum torque ratings are up to 20,000 lbs. per inch at pressures to 5,000 psi, with maximum speeds to approximately 4,500 rpm.

Radial-Piston Motors

Reciprocating pistons cause shaft rotation in radial-piston motors.

Radial-piston motors (see Fig. 9-23) are driven by high-pressure fluid. The pistons are arranged in a circle with their bases connected to a plate that is mounted off-center to the output shaft. This eccentric mounting causes the plate to act as a crankshaft. A valve that rotates with the shaft ports fluid to successive cylinders, so that the down stroke of each cylinder creates a smooth application of power. They can develop over one million lbs. per inch torque at pressures exceeding 5,000 psi. Speeds range from 0.1 to 2,000 rpm.

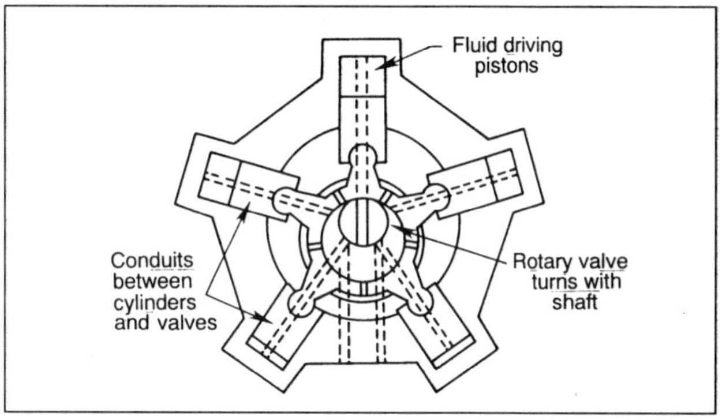

Figure 9-23: Reciprocating pistons cause shaft rotation in radial-piston motors.

Gear-on-Gear Motors

This design is one of the most common for hydraulic units. They consist of a pair of matched spur or helical gears enclosed in a case (see Fig. 9-24). These units typically develop maximum torques of about 6,000 lbs. per inch and speeds to 3,000 rpm.

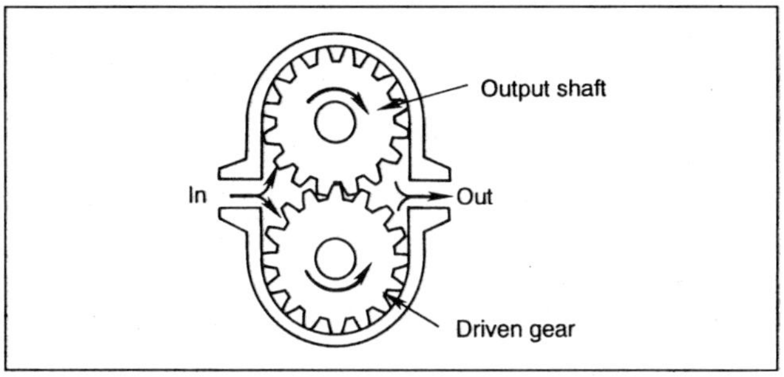

Figure 9-24: Fluid flow turns the gears and rotates the shaft in gear-on-gear motors.

Gear-within-Gear Motors

Often called gerotors, the motors are very compact for their displacement. An inner gear seals against an outer one to guard against fluid leakage. Tooth velocities and wear are low, and power density is high. Gear-within-gear motors deliver torques exceeding 1,500 lbs. per inch at speeds to over 5,000 rpm. A more common use of this type of motor is in the high torque/low speed application. Here the motor has an internal reduction using a "dog bone" drive shaft that moves in an orbit as the motor rotates. The intent here is to be at low speed, usually well below 1,000 rpm, but to give a lot of torque. From 2,000 to 5,000 lbs. per inch is very common but can run as high as 12,000 lbs. per inch.

Vane Motors

Used for both pneumatic and hydraulic operation, vane motors consist of a slotted rotor mounted eccentrically within a circuit cam ring. Vanes in the rotor slots are free to move in and out, often spring-loaded to the outward position. As air or fluid enters the motor, it applies force against the vane, turning the rotor and allowing the fluid to sweep from inlet to outlet ports. A typical rating is 4,000 lbs. per inch torque at 2,500 psi and 4,000 rpm top speed.

Hydrostatic Drives

Hydrostatic drives are widely recognized as an excellent means of power transmission where variable output speed is required. They offer fast response, maintain precise speed under varying loads, and allow infinitely variable speed control from zero to maximum.

A basic hydrostatic transmission is an entire hydraulic system. It contains pump, motor, and all controls in one package. Such a system combines the advantages of stepless adjustment of speed and torque with smooth, controllable acceleration.

A typical hydrostatic transmission (see Fig. 9-25) consists of a variable-displacement pump and either a fixed or variable-displacement motor. Adjusting pump flow varies hydraulic motor output. Various configurations produce either constant torque, constant power, or variable torque and power. Adjusting pump flow in a hydrostatic transmission varies hydraulic motor output speed.

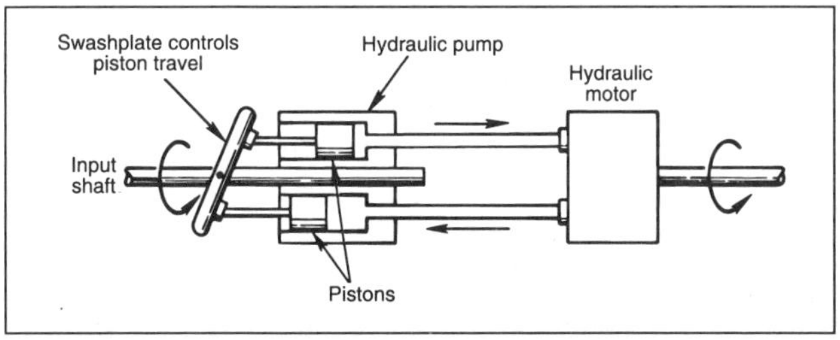

Figure 9-25: Adjusting pump flow in a hydrostatic transmission varies hydraulic motor output.

There are two types of hydrostatic transmissions—split and close coupled. A split transmission consists of a hydraulic motor mounted where torque is needed and the hydraulic pump, heat exchanger, filters, valves, and controls mounted on a remote reservoir. Hose or tubing connects the motor and power unit. This arrangement offers flexibility, the most efficient use of space, and the best weight distribution.

Close-coupled, or integrated, transmissions have a hydraulic pump and motor that share a common valve mechanism. This arrangement provides an extremely short oil-flow path, eliminating potential leak points. Housing provides a self-contained oil reservoir, structural support for the rotating elements, and heat dissipation. The assembly is usually bolted directly to a mechanical differential axle to form a hydrostatic transaxle. Close-coupled transmissions are typically found in light-duty applications where tight space constraints require compact units and high-volume production mandates easy assembly.

Unlike gear transmissions, hydrostatics have a continuous power curve without peaks and valleys, and they can increase available torque without shifting gears. But despite the superior performance of hydrostatics, a major drawback has been higher cost compared to their mechanical counterparts. Manufacturers, however, continue to boost performance levels, produce smaller and lighter packages, and offer advanced electronic controls. These factors now often make hydrostatics an economical choice.

Transmission Sizing

Hydrostatic transmission size is often based on corner horsepower. This is the product of the maximum speed required, even though these two conditions rarely occur simultaneously. Corner horsepower for vehicle propulsion is as follows:

$$Hc = FtV/3{,}600\eta$$

Where
Hc = corner horsepower, kW
Ft = maximum vehicle tractive force, N
V = maximum vehicle speed, kn/h
η = final drive efficiency, percent

Transmission corner horsepower, Ht, is the product of maximum output torque (generally at a specified maximum pressure) and maximum output speed:

$$Ht = T\tau N/9{,}550$$

Where
T = theoretical torque at maximum system pressure, N-m
τ = torque efficiency, percent
N = maximum transmission speed, rpm

Initial transmission selection is made by comparing the results of these calculations. Selection is refined by considering the effects of duty cycle, final-drive ratios that are available as a standard, rolling radius, prime-mover speed, and design life. Light-duty units (less than 20 HP) are used on equipment such as lawn tractors; medium-duty units (25 to 50 HP) are used on skid-steer loaders and similar vehicles; heavy duty transmissions (approximately 60 HP and higher) are used on large construction equipment.

Fluids and Conductors

Constructing a hydraulic or pneumatic system involves the design or selection of numerous components and the determination of how they will all interact. A point often overlooked is that both the fluid and the means for moving it from one location to another are critical.

Fluids

The functions of hydraulic fluid are rather basic: to transmit power efficiently and lubricate moving parts.

Petroleum-Based

These fluids are the most widely used for hydraulics because they are relatively inexpensive and perform well with little or no maintenance. They contain a base oil and additives that protect against rust, prevent wear, inhibit foaming, and lengthen life. Tough applications that require wide temperature range, extreme pressure protection, and long life call for premium-grade fluids.

Other fluids typically cost more than petroleum oils, restricting their use to special applications. Nonpetroleum fluids must also be evaluated for compatibility with the metals, seals, and elastomers in a system.

Fire-Resistant

These fluids are used when petroleum oils present a hazard, especially when a broken hydraulic line could spray fluid into an ignition source. Among the major types of fire-resistant fluids are phosphate esters, water glycols, invert emulsions, and high-water-content fluids. These tend to be more expensive than petroleum-based fluids and have some performance deficiencies.

Synthetic Oils

Generally consisting of esters or synthetic hydrocarbons, synthetic oils are premium hydraulic fluids. Qualities include wide temperature range, resistance to oxidation, and good lubricating properties. However, high price dictates that they be used only where absolutely necessary. They tend to be used where performance requirements are stringent, such as in systems that generate a lot of heat, where start-up temperatures are very low, or long life is a must.

Environmentally Acceptable (EA)

EA fluids are used where hydraulic oil spilled into water, wetlands, and other sensitive areas can be environmentally damaging. Today's vegetable-oil-based EA fluids are suited for most hydraulic applications with virtually any type of pump or valve.

Conductors

Transporting fluid from pump to valves, actuators, and motors is accomplished by a variety of means. Pipe and tubing are used where rigid lines are preferred. Pipe is generally used in permanent applications involving long straight runs with large amounts of fluid. Tubing bends easily and is preferred on applications requiring routing around obstacles or frequent disassembly. Steel tubing is the only type recommended for high-pressure hydraulic service. Copper, aluminum, and plastic tubing are generally suitable for low-pressure applications and pneumatic circuits.

A hose is widely used in applications where lines must flex and bend. Hose construction has been standardized by SAE International under SAE J517, better known as 100R-series hoses. Reinforcement, construction, and dimensions vary among these designs, providing different pressure ratings and other specific performance features. For example, SAE 100R1 is a rubber hose with one layer of braided-wire reinforcement between a rubber inner tube and outer cover. Working pressures range from 375 to 3,000 psi, depending on the hose ID, and temperature range is from -40 to 200°F (93.3° C). This hose is commonly used on lower-pressure hydraulic lines. Other hoses are available with more wire braids; 2 wire, 3 wire, 4 wire, and 6 wire braids are available. More wire braids allow higher pressures, which are required in today's market.

Sizes are designated in 16ths of an inch by using a "dash" equivalent to the numerator of the fraction. Thus, "—10" is 10/16 (or 5/8) inches in size. These dash sizes are marked on the hose.

Connectors

Connectors fill the gap between the hose or pipe and the mating port on a pump, valve, or actuator. The following are among the most common types.

Pipe Threads

The threads seal by metal-to-metal interference between fitting and housing (see Fig. 9-26). They are widely used, inexpensive, and suitable for low-pressure service. In recent years, there is a strong move away from using pipe threads in hydraulic service. Pipe threads are prone to leaks, and the fluid is considered hazardous to the environment. Because of this, the 37 deg. fitting and the SAE fitting described below are now used extensively in both mobile and industrial applications.

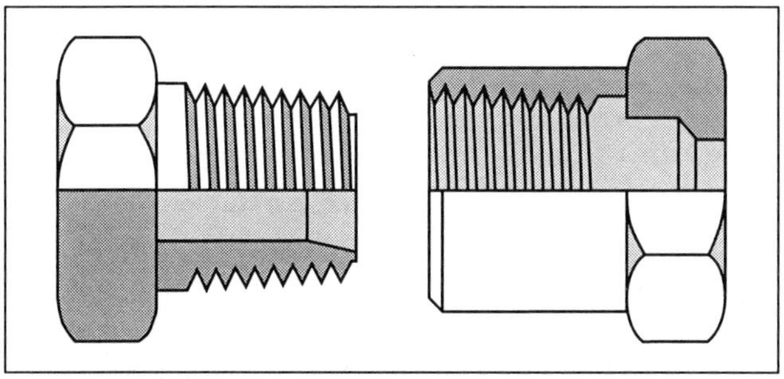

Figure 9-26: Pipe threads seal by metal-to metal interference.

JIC 37 Degrees

The fittings (see Fig. 9-27) seal by conical, metal-to-metal contact between mating male and female seats. They are the most widely used connector in the U.S. Simple and inexpensive, JIC couplings perform well at low-to-moderate pressures.

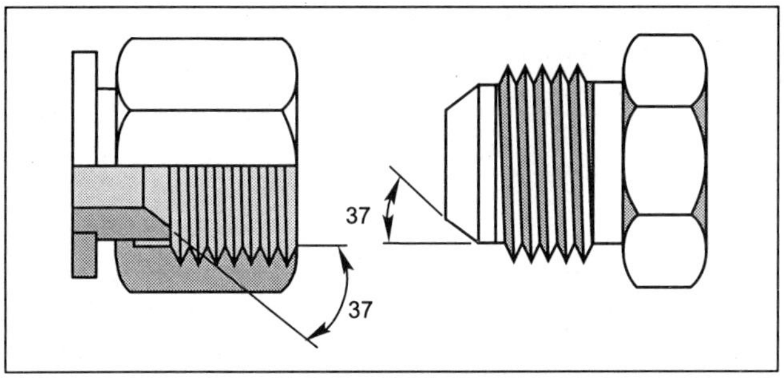

Figure 9-27: JIC couplings seal by metal-to-metal contact between mating seats.

Line Sizing

Pipe and tubing must be large enough to carry the required flow, and they must be strong enough to withstand internal pressures. Line size can be determined from the amount of fluid that must be carried and the maximum velocities at which the fluid may travel. Normally, good design practice dictates that fluid velocities should not exceed 10 to 15 fps in pressure lines and 2 to 5 fps in inlet or suction lines.

Average pipeline velocity can be approximated by the following equation:
$Q = 2.45d^2V$

Where
Q = flow rate in gpm
D = the tube internal diameter, inches
V = velocity in fps

To determine the required wall thickness of pipe and tubing, apply this basic equation:
Tt = pdoM/2s

Where
P = pressure, psi
Do = outside diameter, inches
M = safety factor
S = material tensile strength, psi

Flat-Face O-ring

These connectors have a recessed O-ring seal held in a circular groove on the male connector (see Fig. 9-28). It mates with a flat, finished surface on the female end when assembled. These are suitable for 6,000 psi service.

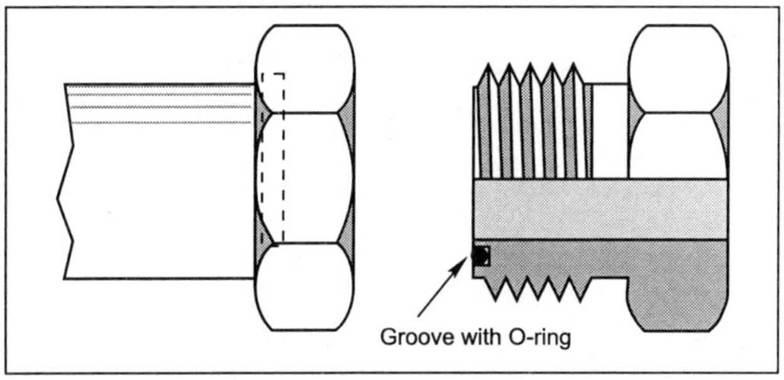

Figure 9-28: Flat-face O-ring fittings seal by pressing the O-ring against a finished metal surface.

SAE Flange

The couplings are available in two domestic styles: Code 61 for 3,000 to 5,000 psi service (depending on size) and Code 62 for 6,000 psi service. In both, an elastomeric O-ring set in a groove on the flange mates with the flat machined surface of a pump, valve, or housing (see Fig. 9-29). Four small mounting bolts attach the flange to the port.

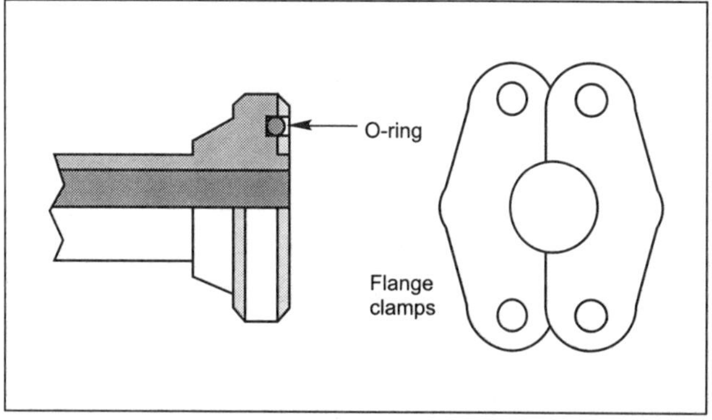

Figure 9-29: An elastomeric O-ring in a flange fitting mates with the flat machined surface of a pump, valve, or housing.

SAE Straight-Thread

This port (see Fig. 9-30) seals with an O-ring between threads and flange. It is used on all sizes and is suitable for 6,000 psi service.

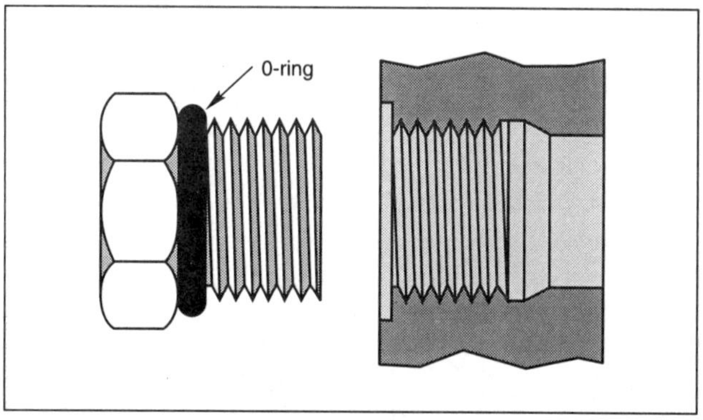

Figure 9-30: The straight-thread port seals with an O-ring between threads and wrench flats.

Hydraulic and Pneumatic Filters

Filters are a must in air and hydraulic systems because dirty fluids have a short life, accelerate component wear, and lead to premature failure.

Typically, a hydraulic system is cleaned by a single filter. The most common placement of the filter is in the return line. This is a low cost unit that catches any particles that the system ingests or generates and prevents them from re-entering the system through the pump. Ten micron filtration is the most common level of filtration. In proportional or servo systems, 3 to 5 micron filtration is required on the pressure line to prevent any silting on the moving parts of these valves. Many systems also have a suction filter. This filter is on the inlet of the pump and typically has a very coarse filtration rating. Its purpose is to remove large debris that might get into the reservoir. Many pump manufacturers object to this filter location because it creates a pressure drop on the pump inlet that can cause cavitation and failure of the pump.

In addition, care must be taken in long pipe or tubing that is run in hydraulic systems. In a 50-foot run, the total conductor length will be two times the distance because the fluid must travel out and back. Typical pressure drop at 15 feet per second velocity of the oil is about 1.5 psi per foot. If we multiply this times the total distance, we see that in a 50-foot distance from the pump to the actuator we have 100 feet of conductor length. If we multiply this times 1.5 psi per foot, we get a pressure drop of 150 psi. At higher oil velocities, the pressure drop per foot is more profound. At 30 feet per second, the pressure drop will be 5 psi per foot or even more. Now we have a 500 psi problem that requires more horsepower to transmit the fluid.

Nominal Rating

Determined by the filter manufacturer, nominal filtration rating is an arbitrary value. The rating system refers more to the types and sizes of holes in the filter medium than actual filter performance. Nominal filter ratings have many limitations. First, they do not present a clear indication of the largest-size particle that can pass through a filter. Second, it is a nonstandard system that lacks consistency from one manufacturer to another.

Filter Performance

The Beta ratio, a system for rating filter performance, is expressed by:

$\beta x = Nu / Nd$

Where
Bx = Beta filtration rating for particles
Larger than x microns,

Nu = number of particles larger than x microns upstream of the filter, and

Nd = number of particles larger than x microns downstream of the filter.

The Beta filtration rating can also be converted into an efficiency rating, Ex. Efficiency rating is expressed as a percentage of the filter medium's ability to remove particles larger than x microns:

$Ex = ((\beta - 1)/\beta) \times 100$ percent.

A filter with a reported rating of $\beta 3 > 100$ removes a minimum of 99 percent of all particles larger than 3 µm. Similarly, a filter rated at $\beta 10 > 20$ captures a minimum of 95 percent of all particles larger than 10 µm. Clearly, the element rated $\beta 3 > 100$ removes many more small particles than the element rated at $\beta 10 > 20$, especially in the 3 to 10-µm range.

Filtration for pneumatic systems is handled quite differently. In most industrial settings, compressed air is supplied from a single compressor to a large number of operating systems. Individual filters are used on the separate systems. Often the filters are combined with regulators and sometimes lubricators, forming a filter-regulator-lubricator (FRL) for the system. Typical conditions for an industrial compressed air supply is a filter regulator, an oil removal filter, and a vapor removal filter. Many pneumatic components are pre-lubed and require no additional lubrication.

Filters are rated by the size of contaminant particle that they can retain.

When specific filtration requirements are stated for a system, contact your filtration manufacturer for correct filtration system and components.

Fluid-Power Standards

Today, U.S. standards still dominate the American fluid-power industry, covering virtually every component and aspect, from cylinder dimensions to valve configurations. The central clearinghouse for these standards is the National Fluid Power Association.

International Standards Organization (ISO) standards, widespread in Europe, are gaining importance here with the influx of off-shore machinery to the U.S. Also, many domestic manufacturers now offer fluid-power components made to ISO standards.

Hose and connectors are most often made to SAE standards. However, a note of caution is in order here. With an increase in foreign-made machinery being imported to the U.S., a problem that has surfaced in recent years is the proliferation of hydraulic connectors from suppliers around the world. This dramatically increases the possibility of mismatching threads and seats, guaranteeing leakage. Even though couplings made to different standards are close enough to be assembled, under no circumstances should they be coupled as they may cause a failure or personal injury. When dealing with offshore couplings, check styles and dimensions. Most manufacturers and distributors of connectors have kits to measure thread pitch and seat angles, and these are a good investment.

Additional Information

Addresses and phone numbers for associations are provided in the Association Appendix.

- The FPDA Motion & Control Network
  www.fpda.org
- National Fluid Power Association www.nfpa.com
- SAE International
  www.sae.org
- The Association for Hose and Accessories Distribution (NAHAD)
  www.nahad.org

Chapter **10**

Linear Motion

Chapter Objectives

When you have finished Chapter 10, "Linear Motion," you should be able to do the following:

1. Understand new terminology introduced in the "Key Terms" section.
2. Remember one or more of the application acronyms and how they relate to sizing and selecting a system.
3. Remember the key questions to ask when selecting a motion system.
4. Understand the difference between plain, recirculating elements, and cam roller bearings, and their advantages and disadvantages in specific applications.
5. Understand the features and benefits of each bearing type and the application that best suits it.
6. List and describe the purpose of the major components of a linear motion system.
7. Describe the types of motors most commonly used with linear motion systems.

Introduction

Historically, engineers considered power transmission to be primarily rotary systems consisting of shafts, gears, belts and pulleys, chains and sprockets, and motors. Linear motion systems were thought of as ancillary parts of rotating machines, such as a lead screw that moves a cutting tool on a lathe, or as stand-alone machines that move heavy objects, such as a screw jack. Modern applications call for increasingly complex motion and means for controlling that motion.

For example, a machine tool cutting tip often must follow a prescribed three-dimensional path. In such cases, the motion of the tip may consist of simultaneous linear motion in two or more axes created by multiple stages of independently controlled linear actuators. Furthermore, many modern processes call for unattended operation, very precise movement, high throughput, flexibility for short runs, and integrated manufacturing. Such cases require modern sensors and controls coupled with precise linear and rotary motion devices. As a result, designers must consider linear motion as a separate but integral part of industrial motion and control. Equipment manufacturers have responded by offering a wide range of equipment dedicated to linear motion.

The major components of linear motion systems fall into one of three categories: actuators, linear bearings, and control systems. Figure 10-1 shows a typical linear motion system consisting of a motor-driven ball screw that drives a table along a pair of profile linear guides. (The control system is not depicted in the figure.)

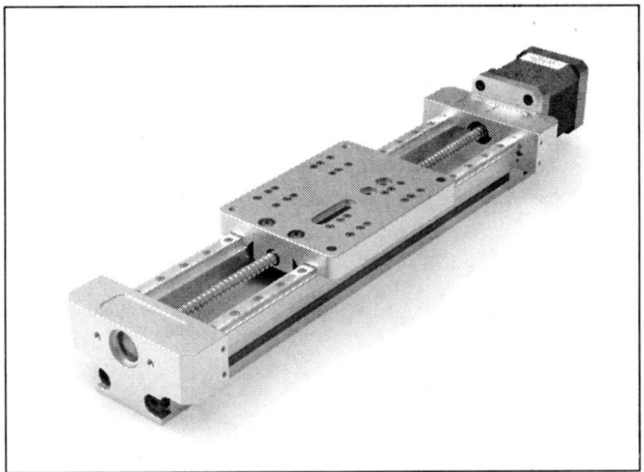

Figure 10-1: Typical linear motion system

Linear bearings covered in this chapter include plain linear bearings, linear bushings, linear ball bushings, and linear guides. Controls and sensors are covered in detail in Chapter 13. Much of the control technology applicable to rotary motion systems applies to linear systems. This chapter will focus on some control aspects specific to linear motion systems.

Most manufacturers employ application engineers to assist customers in selecting the appropriate components for their application. To avoid misunderstandings and unnecessary back-and-forth correspondence, the customer should have at least a fundamental understanding of linear motion components prior to selecting a system. It is the purpose of this chapter to provide that information, as well as an overview of the most common linear motion products available today.

General Selection Criteria

Some readers may be familiar with the acronyms LOSTPED and LOPSTED, which contain letters representing **L**oad, **O**rientation, **S**peed, **T**ravel, **P**recision, **E**nvironment, and **D**uty Cycle. These are seven of the most basic key elements to consider when selecting and sizing a robotic or motion system. Each should be considered independently and jointly, as the combination of requirements will sometimes lead to the qualification or disqualification of a potential solution. These acronyms have been in use for some time now, although there are two omissions: a "U" for unknown and an "S" for safety.

To remedy these omissions, we can add a "U" and an "S" to the end of the two existing acronyms to form LOSTPEDUS and LOPSTEDUS, or we can create a new acronym: POSTLUDES. Many are familiar with the word prelude, meaning an opening performance, action, or event. Postlude is simply the opposite, meaning a closing performance, action, or event. Any of these three helper acronyms will do, so choose the one that's easiest to remember.

Precision (a.k.a.: accuracy, repeatability)

Precision is an often misunderstood and misapplied element. Instead of the word precision, let's use the words accuracy and repeatability, although people often confuse the two. In general, repeatability is more important than accuracy. That said, customers often request an accurate system. Accuracy can be defined as the difference in position between where a system actually is and where the controller thinks it is. Repeatability is defined as the difference in position when a system returns to a location under the same circumstances (i.e., same direction and motion profile). Because of backlash and "slop" within the mechanics, there is often a big difference between single-directional repeatability and bi-directional repeatability. Most systems have a bi-directional repeatability rating that is much worse than the single-directional repeatability rating.

When combined with the speed element, things can get complicated. Most systems with high dynamic performance levels (high velocity and acceleration) tend to have difficulties with high accuracy requirements. These systems tend to overshoot their target location then reverse direction to "hunt" for the intended position. This can be problematic for some applications. Note: Linear motor-based systems do not typically have this problem; however, they are much more expensive than other comparable systems.

Key questions to ask yourself or the customer:

- What is more important: accuracy or repeatability?
- What is the accuracy/repeatability requirement?
- Are these values realistic, based upon the desired motion profile?

Orientation

The orientation of a system is often taken for granted—the assumption being that the normal (horizontal) orientation will be utilized. This is a faulty assumption because a system that works in the horizontal orientation may not work if it is inverted and likely will not work as intended if the system is vertical. When orientation is combined with load, there can be a dramatic impact on the performance of a system. For example, not all systems can support the same payload in the normal (horizontal) orientation as they can when the payload is inverted or mounted on its side. A system that works fine in a horizontal orientation may not work properly if moved to a vertical orientation, as the motor must now directly overcome the force of gravity.

Key questions to ask yourself or the customer:

- How will the system be mounted? (e.g., "normal," on its side, inverted, vertical, at an angle, etc.)
- How will this affect the load requirements?

Speed (a.k.a.: velocity, acceleration, motion profile)

The speed of a system includes much more than the maximum velocity. If you were to ask a customer to define maximum velocity, maximum acceleration, or desired motion profile, they might find it difficult, but they could easily explain how they need their equipment to produce X parts per hour or move from points A to B in so many seconds. Once this requirement is established, simple mathematical calculations will reveal the maximum velocity, acceleration, and motion profile. If a customer needs to move a distance of two meters, there's a significant difference between doing it in, say, 10 seconds versus a single second.

An often overlooked factor is "jerk," which is the rate of change of acceleration. For example, slowly depressing a vehicle's accelerator results in a low jerk value; quickly depressing the accelerator yields a medium jerk value; and smashing into a brick wall would produce a very high jerk value.

Motion profile is a term used by control engineers to describe the motion of a system by defining position, velocity, and acceleration (sometimes jerk) versus time. The two most common motion profiles are triangular and trapezoidal.

Key questions to ask yourself or the customer:

- What is the maximum speed and acceleration required?
- What is the maximum jerk allowable?
- What motion profile (shape) is desired?

Travel (a.k.a.: stroke, overtravel, envelope)

Travel is one of the easiest elements to define; however, there are elements to this variable that are often overlooked. The three elements that need to be defined are stroke, overtravel, and overall envelope. Stroke is the distance the system needs to travel. Overtravel is additional travel used to compensate for errors during installation (e.g., misalignment) and extra travel during an emergency stop situation. Overall envelope is the total space available to the motion system. Some systems have very little additional space around the motion system, which can make component selection difficult.

Key questions to ask yourself or the customer:

- What is the required travel (stroke)?
- What is the overall envelope allowed?
- How much overtravel (safety zone) is required?

Load

Load refers to the forces acting upon a system. The load starts with the End of Arm Tooling (EOAT) and the payload. The EOAT is often permanently attached to the robot or motion system and performs some type of work. Typical examples of EOAT include grinders, welders, suction cups, grippers, dispensers, vacuums, and spindles. The EOAT usually has some kind of additional load applied to it. This load can come in the form of a weight being transported by the system or forces acting upon the system if the EOAT is a cutting tool or pusher.

In multi-axis systems, the weight of additional linear motion components (e.g., actuator, slide, bearings) is often forgotten and not included in the load numbers.

There are actually three components to "load": static, dynamic, and impact. People often forget about the dynamic condition, and almost everyone forgets about the impact condition. Dynamic and impact load conditions are often forgotten because they cannot be seen on a drawing.

A static load condition occurs when the system is fully loaded and at rest. This is the easiest condition to describe. Additional forces, such as acceleration and deceleration, act upon the system while it's in motion. There's also the impact load condition to consider. What will happen when the system crashes? Is it expected to survive? How many crashes is the system expected to survive? Should a "weak link" component be designed as a failure point?

Key questions to ask yourself or the customer:

- What is End of Arm Tooling (EOAT) and where is it located?
- What additional forces are seen by the system during use (e.g., cutting or pushing forces)?
- What do the static and dynamic free body diagrams look like? Have all loads been accounted for?
- What is expected of the system after an impact?

Unknown

No matter how much planning goes into a system, there will always be unexpected variables that can negatively affect performance. The most common factors are actually seen by most systems. Two of the most common are extra load caused by misaligning parallel linear guides and extra drag from cables and cable carriers. Both of these can be overcome by allowing for an adequate safety factor when sizing system components. These factors are often referred to as the "known unknowns" because you know they exist and will affect the system, but you don't know to what extent.

In addition, designers need to consider how their system might be misused. Is it likely the end user will make the system go faster or carry a heavier load? Will a maintenance technician stand on an important piece of equipment to gain access to something else? These things do happen, and it's important to plan for them.

Key questions to ask yourself or the customer:

- What are the known unknowns? What is a reasonable value for these unknowns?
- How will someone misuse this system?
- What could possibly go wrong?
- What else could go wrong? (Repeat this question until you can't think of anything else.)

Duty Cycle/Life Cycle

Duty cycle is often a miscalculated value. For example, a factory runs a robot for eight hours a day. During those eight hours, the robot is in almost constant motion. The most common mistake people make is to incorrectly assume the duty cycle is 33 percent (8/24 hours). In reality, the duty cycle is 100 percent because when the system is in use, it's in constant motion. The easiest way to calculate duty cycle is to look at a single move. How long will it take to make the single move, and how long will the system rest between each move? Many motion control component manufacturers publish a defined maximum allowable duty cycle, and some will even reduce the rated load capacity when their component is used in a high duty cycle application.

The second aspect of duty cycle to consider is how long a customer expects the system to last. The customer will typically state a certain amount of time, and it will be necessary to work backward to figure out the total distance traveled in that time and the total hours of operation. Motion components have a lifetime in meters or inches, and electrical components have a lifetime listed in hours.

Key questions to ask yourself or the customer:

- What is the actual duty cycle for the system?
- What is the expected lifetime?

Environment

There are two ways to consider the environmental effects for a motion system. First, you have to consider how the environment will affect the system; and second, you have to consider how the system will affect the environment. Is the environment at a temperature extreme (i.e., hot or cold)? Are there contaminants in the area such as dirt, chips, or liquids? Is a vacuum applied? Are there vibrations or other shock loads that could be applied directly to the system or to the area around the system? Will the system be installed in a clean environment where particulate generation could be an issue?

Finally, where and how will the system be installed? What surrounds the system? Can a technician easily gain access to perform preventative maintenance? Does the system require lubrication storage because there will be no maintenance?

Key questions to ask yourself or the customer:

- In what environment will the system be installed?
- Are there any hazards in the environment?
- What is the maintenance schedule, and is the system accessible for maintenance/lubrication?
- Are there contaminants in the environment that can damage the motion system?
- Will the system disburse contaminants into the environment that could damage other equipment or products?

Safety

In today's "litigation happy" world, addressing safety issues is more important than ever. Nothing can sink a company faster than a lawsuit, and many insurance companies refuse to cover an accident if a company has intentionally neglected to install required safety equipment. Tougher standards have been imposed upon the industry by governmental regulation and protection agencies (e.g., OSHA, CCOHS, EU-OSHA) and industry standards (e.g., ANSI Z##, ISO13849, IEC61508, EN61508, EN954, UL, CE). To determine which safeguards to include, it is very important to ask a customer if there are any special requirements to which their system needs to conform, especially if human beings will interact on any level.

Also consider what might happen during unusual and/or unexpected events. What will happen to this system and the payload during a power outage or natural disaster? Will the payload be safe? Will the people around the system be safe? What needs to be done to make the system safe, and how much will it cost?

Key questions to ask yourself or the customer:

- Are there any safety standards to which this system needs to conform?
- What could happen if the system fails? Are there safeguards that need to be installed for a system failure?
- Could people be injured by this system? If so, how will the people be protected?

Linear Guidance

Linear motion systems are supported and guided by linear bearings. Linear bearings come in many shapes and sizes, but all perform the same basic function, which is to bear loads while ensuring stable and accurate linear motion. When selecting a suitable linear motion solution, the chosen solution must meet both technology and shape requirements. In this instance, "technology" refers to the type of bearing system. There are three basic categories of bearing technology that will be discussed in this chapter: recirculating element, plain, and cam rollers. The "shape" of the bearing system refers to the form factor in which the technology is packaged. The four basic shapes discussed in this chapter are round shaft (a.k.a. round rail), profile rail (a.k.a. square rail), cam rollers, and drawer slides (a.k.a. telescopic slides). A user must first determine the technology type and shape of the solution for his or her application.

Finally, a user must determine if he or she wants to fully design and assemble the components or buy a preassembled system. Depending on the manufacturer, preassembled systems are called actuators, stages, slides, tables, or positioners. Most users find it much more cost effective to buy a preassembled solution for low-volume applications. The make versus buy decision becomes more complicated as the potential volume increases.

Before we discuss the specifics regarding different technologies and shapes available, we will discuss some of the basic selection criteria. Bearings and linear guides come in a wide variety of shapes and sizes. Selection will depend on many factors, including, but not limited to: loads, speeds, duty cycle, required life, rigidity, accuracy, machine configuration, and installation limitations.

Selection Criteria

Size

Linear guides vary in size from less than 3 mm to more than 100 mm. A larger linear guide will typically have a higher load capacity than an otherwise similarly configured smaller linear guide. Other factors determining load capacity include the material type (typically aluminum, steel, or stainless steel), slider length, number of balls, number of rows of balls, ball size and raceway design for a ball-based system or surface area, and material properties for a plain bearing.

Stroke

The maximum linear guide stroke refers to the distance the carriage can travel from one end of the linear guide to the other. This distance is equal to the rail length minus the carriage length. The maximum linear guide stroke should be selected to meet or exceed the required stroke for the application. If the length just barely meets the requirement, the rail must be installed perfectly for the application to function properly. Thus, it's often better to have a few extra millimeters of rail length to allow more flexibility during installation.

Accuracy Grade

The accuracy of a linear guide or ball screw refers to both dimensional and running accuracy. Dimensional accuracy refers to the deviation and variation of the mounting height and mounting width. Running accuracy refers to the horizontal and vertical running parallelism of the ball slide along the rail. **Accuracy grades are not consistent between different manufacturers**. Accuracy grades should be selected based on the need for positioning accuracy in the application. For example, machine tools require a relatively high degree of positioning accuracy, while equipment in steel mills requires significantly less.

Preload

Linear guides may be preloaded in order to eliminate radial play and increase rigidity. Preload is established by inserting balls that are slightly larger than the space provided by the ball grooves. The requirement for rigidity should determine the required preload amount. Although rigidity improves with increased preload, it is important not to use too much preload. An increase in preload will result in an increase in drag torque and heat generation and a decrease in service life.

Life

Life calculations for linear rolling bushings and linear (ball or roller) guides are very similar to those for radial bearings and based on the same principles. Fatigue life in distance is a function of the basic dynamic load rating, applied load, and a load factor. The basic life equation is as follows:

$$Life_{Ball}(meters) = \left(\frac{C}{P}\right)^3 \times 50 = \left(\frac{Max\ Allowable\ Load\ (N)}{Applied\ Load\ (N)}\right)^3 \times 50$$

$$\textbf{or}\quad Life_{Roller} = \left(\frac{C}{P}\right)^{\frac{10}{3}} \times 50$$

Here, "C" is the basic dynamic load capacity, and "P" is the applied load. Life in total distance can be converted to life in total hours if speed information is known. It is important to note that the main variable here is the applied load, and even a small increase or decrease in the load can have a big effect because it is being cubed.

Static Load Limitations

Linear bearing selection must also take into account static load limitations. The static load limit of a linear guide is based on avoiding permanent deformation of the ball contact points. Limits based on permanent deformation of ball contact points can be determined using the basic static load ratings provided by the linear rolling guide manufacturer. The static load limit of a plain bearing is the maximum static load the system can bear prior to plastically deforming (damaging) the bearing material.

Technology

One of the first choices to be made is to determine which technology is best suited for an application. The three choices discussed within this chapter are recirculating element, cam rollers, and plain bearings. Each technology choice has advantages and disadvantages. It's important to remember that, for a majority of applications, any of the three options will work. The decision regarding which type of technology to use typically comes down to total installed cost and product availability. In the following sections, we will discuss each technology type in greater detail.

Recirculating Element

Ball Bearings

Ball bearing systems have been around for many years, and they are still in use today. The name "recirculating element" comes from the method in which the balls travel through the carriage. The balls enter and leave the load-carrying area of the carriage, known as the "load zone." Ball-based recirculating element bearings are popular because of their low friction, long life rigidity (preload ability), and accuracy. The small friction value allows for the use of smaller motors and amplifiers, which reduces cost. In addition, these products are readily available from a number of manufacturers. Ball-based systems also provide the additional benefit of a predictable life expectancy. An approximate lifetime can be calculated based on formulas found in most manufacturers' catalogs. Recirculating ball systems are typically limited to a maximum velocity of 3 m/s–5 m/s, depending on the manufacturer and type of product. Recirculating element bearings require lubrication and need to be protected from environmental contamination. Lack of lubrication and damage from contamination are the two leading causes of ball bearing failure.

Roller Bearings

Roller bearings are very similar to ball bearings, except that, instead of balls, they incorporate cylindrical rollers or needles inside the bearings. For linear motion applications, roller bearings are more rigid, carry more load, and are more tolerant of shock loads; however, they typically do not tolerate as much misalignment and cost more than a comparably sized ball bearing product. For more information on ball and roller bearings, see Chapter 2: Bearings.

Cam Rollers

Cam rollers are based on ball or roller bearing technology, so they offer many of the same benefits as recirculating element bearings. However, cam roller bearings do not have recirculating balls or rollers. The bearings inside the cam roller simply rotate around a central stud. There are several benefits to this type of bearing arrangement for linear motion applications. The cam roller bearings are always in contact with both the inner and outer race of the bearing, and there is no major shock to the system when the balls enter and exit the load zone. This allows the bearing system to travel faster than a recirculating element system. Most recirculating element bearings are limited to between 3 m/s (10 ft/s) or 5 m/s (16 ft/s), depending on the product, whereas a cam roller product can travel much faster, up to 10 m/s (32 ft/s).

Plain Bearings

Plain bearings are the oldest, simplest, and least expensive linear bearing and still have a wide range of applications. With plain bearings, a thin lubricant film is formed and maintained between two surfaces in sliding relative motion to each other. Because sliding contact bearings have a greater friction coefficient than rolling bearings, plain bearings will require larger, more expensive motors and controls.

The simplest plain linear bearing is the flat way. It is perhaps the oldest device that allows one machine element, such as the bed of a planer, to move easily on another. However, it is difficult to manufacture flat ways to be highly accurate over a long distance. Early machine ways were hand-scraped to remove high spots. The flat way must also incorporate a means to resist transverse loads. A common method is to provide two or more flat surfaces butted at an angle to each other. Examples are v-shaped, box, and dovetail ways.

Plain linear bearings are being steadily replaced by rolling-element linear bearings in applications requiring high-speed, high-precision, and long-term maintenance-free operation. The disadvantages of plain linear bearings include high friction, poor positioning accuracy, difficulty in determining accuracy, rigidity, and predicted life.

In general, there are three classes of plain bearings:

- Class I An outside source of lubrication is required (oil, grease, etc.).

- Class II Lubrication is impregnated within the walls of the bearing (bronze, graphite, etc.). Typically, these bearings require an additional lubricant.
- Class III Self-lubricating bearings that do not require additional lubrication.

Shapes

Each type of bearing technology can be packaged into several different shapes. Each shape has unique advantages or disadvantages for a specific application. Table 10-1 shows some of the commonly compared characteristics and how four different types of products compare. An important point to remember is, for most applications, a variety of products will work. Often, the choice comes down to price and availability.

Round Shaft (a.k.a.: round rail, linear bearings, linear ball bushings)

The simplest type of rolling-element linear bearing is the linear ball bushing (see Fig. 10-2), often referred to as linear ball bearings or round-rail linear guides. A linear ball bushing has three or more circuits of balls in rolling contact with the shaft and the bearing raceways. A retainer prevents the balls from falling out if the bushing is removed from the shaft. This makes linear ball bushings as easy to handle and install as plain linear bushings. Most linear ball bushings are solid cylinders, though some are split axially to allow for preloading. The latter is preloaded on a shaft when the bearing is forced into an interference-fit housing. In the modern bearing world, this type of bearing is the oldest and most accepted form of linear motion bearing.

Figure 10-2: Linear ball bushing

Ball Spline—Ball splines are a subcategory of round shaft products. Typical round shaft bearings have two degrees of freedom; that is, they are free to move back and forth along the shaft and can also rotate around the shaft. Ball splines have grooves down the length of the shaft, which prevent rotation and limit the product to only one degree of freedom. Ball splines are more expensive than traditional round shaft bearings because additional manufacturing is required for the shafts; however, there are also a few key benefits.

Benefits of ball splines over traditional round shaft bearings:

- Ball splines can transmit torque.
- Ball splines will not slip in the rotational direction.
- Due to the increased contact angle, ball splines can carry more load than a comparably sized traditional bearing, or they will have a longer life for the same applied load.

Profile Rail (a.k.a.: square rail, linear guides)

Profile-rail linear guides consist of carriages (a.k.a. sliders, trucks, runners, blocks, etc.) containing rolling elements that ride on continuously supported rails. The basic operating principle of a profile-rail linear guide is the same as that for a round-rail ball bushing and ball spline. Rolling elements roll along raceways formed by grooves along the rail and carriages. The rolling elements are turned around by return guides in the end caps and redirected back to the other end of the raceway, providing continuous circulation (see Fig 10-3).

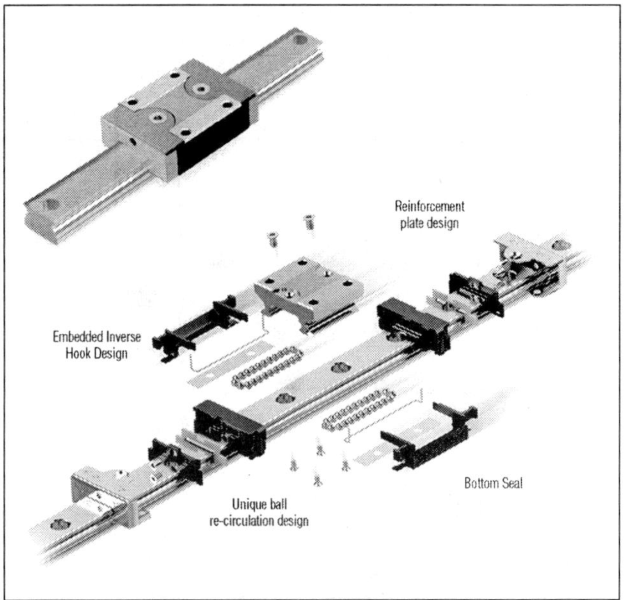

Figure 10-3: Linear guide exploded

Profile-rail linear guides are referred to by size, which is the width across the base of the rail in millimeters. Carriages come in either a clearance condition or in any of several preload levels. Preload should be specified based on the application requirements for rigidity. Different accuracy classes are available and should be specified based on the application requirements for accuracy.

The primary difference between profile-rail linear guides and round rail linear guides is the shape of the inner race. In place of the shaft used in round-rail linear guides, a profiled-shape rail with precision ground ball conforming grooves is used. The ball conforming grooves increase the contact area of the rolling elements on the inner race, greatly increasing the load capacity compared to a linear guide with a round shaft. This increase in load capacity comes with a tradeoff. Because the balls must travel in the ball conforming grooves in the rail, the rotational degree of freedom found in the round-rail ball bearing is lost. This means the allowable installation tolerances for a profile-rail linear ball guide are tighter than those of round-rail linear guides, although it has the added benefit of anti-rotation. This allows a user to employ a single-profile rail where two parallel round shaft bearings would otherwise be required.

Linear roller guides (see Fig. 10-4) offer approximately twice the load capacity and rigidity compared to the same size profile-rail linear ball guides. This is due to the increased contact area of a roller compared to a ball.

Cam Roller

Cam rollers are popular choices for many designers because they offer more design freedom and can often be installed at a lower total cost than other products discussed in this chapter.

Guide wheel technology for linear applications offers many performance characteristics sought after by today's equipment designers. Vee guide wheels are capable of exceeding the maximum velocity offered by more common recircu-

TABLE 10-1 — Comparing characteristics of four different bearing types

	Recirculating Element			
Characteristic	Round Shaft	Profile Rail	Plain Bearing	Cam Roller
Accuracy	Good	Excellent	Fair	Good
Catastrophic Failure	Yes	Yes	No	Yes
Clean Environment	Good	Good	Good	Good
Compliance	Good	Poor	Fair	Excellent
Cost	Good	Poor	Excellent	Excellent
Dirty Environment	Poor	Poor	Excellent	Excellent
Ease of Assembly	Good	Good	Good	Excellent
Friction	Very Low	Very Low	Moderate	Low
High Load	Good	Excellent	Good	Good
High Speed	Good	Good	Excellent	Excellent
High Temperature	Good	Poor	Excellent	Excellent
Long Lengths	Good	Poor	Good	Excellent
Low Profile	Poor	Excellent	Good	Excellent
Noise	Good	Good	Excellent	Excellent
Preloadable	Yes	Yes	No	Yes
Rigidity	Good	Excellent	Fair	Good
Required Maintenance	Yes	Yes	No	Yes
Rotary Capability	No	Yes[1]	Yes	Yes

[1] Profile Rail products DO have rotary capabilities; however, they are very expensive.

Figure 10-4: Linear roller guide

lating ball guides. Vee guide wheels, combined with linear motor actuators, offer a very good system solution for high-speed applications. Depending on the diameter of the wheel, speeds up to 39.4 ft/s (3m/s) may be attained. The vee guide wheel design consists of retained balls moving in a circular path. This design helps reduce vibration and improve the smoothness of the linear system. A linear guide design that offers smooth movement usually generates less noise. Vee guide wheels can run on a variety of track types. Standard tracks and common round shafting can utilize the inner vee of the wheel. Some linear system designs may favor using the outer vee of the wheel—running in a 90 deg. groove machined into a base. Even steel angle iron can serve as track. Vee guide systems also work well when combined with aluminum machine frame systems.

Guide wheel technology enables today's equipment designers to create guide systems with high-performance characteristics. Guide wheels are rolling element bearings with profiled circumferences that roll on track with complementary profile-running surfaces. Running-surface profile types include flat or crowned roller wheels with flat track, round profile wheels with round track, and vee profile wheels with vee track.

The design of guide wheels presents unique characteristics. The complementary running-surface profiles enable all guide wheels to have continuous line contact with their track, permitting higher load capacities and smoother rolling action than linear-bearing or profile-rail guide systems, which have smaller contact regions between their rolling elements and running surfaces. Guide wheels are ideal choices for operation in dirty environments because their rolling elements travel through circular, internal raceways that are permanently sealed off from the operating environment. Linear-bearing and profile-rail guides are more susceptible to environmental contamination because their rolling elements run on surfaces that are exposed to the environment most of the time. Guide wheels can be quieter and attain higher speeds and accelerations because their rolling elements recirculate at relatively slower speeds and follow more uniformly curved paths than those in linear bearings and profile-rail linear guides.

Cam rollers can be constructed from either ball bearings or roller bearings. When ball bearings are used, there can be either one or two rows of balls (one or two tracks) typically called one- or two-row ball bearings. Going from a single row to a double row will significantly increase the radial and axial load the roller is able to handle. Typically, when loads are applied only in the radial direction, a single-row bearing will be sufficient. However, when there's combined loading (when loads are applied both radially and axially) or just axially, a double-row bearing is often employed, as it has a significantly higher load capacity than a similarly sized single-row bearing.

A basic guide wheel system consists of a guide wheel carriage running on a double-edged or paired track assembly. Each guide wheel carriage consists of a carriage plate (which serves as a mounting frame for guide wheels and customer payload) and a set of three or more guide wheels. Guide wheel carriage plates can be simpler in design and manufacturability than linear-bearing and profile-rail guides, which require small but complex tight tolerance features for circulating and supporting their rolling elements against their running surfaces. The preload between the rolling elements and running surfaces in linear bearing and profile rail systems is fixed and determined by the components' manufactured dimensions, whereas the preload between the wheels and track in a guide wheel system are user adjustable. Preload adjustability allows guide wheel carriage plates to be made with lower precision, complexity, and cost while permitting users to customize system preload to meet their individual needs, such as decreased preload for reduced drag resistance or increased preload for enhanced system rigidity. The adjustable preload property also provides unrivaled component interchangeability by eliminating the need for the components in each guide system to be selectively matched and assembled together, even when the application requires a rigid structure and precise motion.

Guide wheels are unique among linear motion technologies because they can also be used in rotary and curvilinear applications. Guide wheel carriage plates can be custom made with wheel mounting positions that match the curve of a specific track, but it is possible for guide wheel carriages designed for straight track to also be used on curved track with little to no component modification or adjustment. Operation on straight track, track with nearly any curvature, or a curvilinear system with a combination of both is possible for a three- or four-wheel carriage, provided the widths of the entire track match the carriage wheel spacing and there are no problems with carriage plate interference. These capabilities are not possible in linear-bearing and profile-rail systems. Because linear-bearing and profile-rail carriages must keep long rows of numerous, closely spaced rolling elements in constant, uniform contact with their running surfaces, the rolling element circuits of each carriage must closely match the profile and curvature of the track. Such closely matched features prevent these carriages from operating (or even fitting) on track with alternate profiles or curvatures, so curvilinear systems cannot be made with these types of guide systems.

The dynamic load capacity for a guide wheel is generally specified in terms of the allowable load for an operating life of one million revolutions.

Flat and crowned roller-guide wheels can use any flat-surfaced object of adequate strength as track. Flat roller-guide wheels have cylindrical running surfaces. Crowned roller-guide wheels have slightly convex curvatures on their running surfaces in the axial direction, which reduce stress concentrations in the wheels and track. Each wheel can provide only radial constraint, thus flat and crowned roller-guide wheels are typically used in applications where axial movement and rotation in any orientation is desirable. However, flat or crowned roller-guide wheel-based carriages can be constrained against movement in some or all of those orientations through the addition of properly positioned and oriented sets of wheels and track.

Round profile-guide wheels feature concave, semicircular running surfaces and are used on rods or convex curvature rails with matching radii. The round contact profile permits the wheels to support axial loads, radial loads, and yaw moment loads. Self-aligning forces in these orientations automatically develop when the wheels are pressed against the track. Round profile-guide wheel-based carriages are further constrained against roll moment loads when two oppositely loaded sets of wheels and track are utilized.

Vee profile guide wheels feature one or two vee profiles on their circumferences and roll on complementary-angle vee profile track that has two offset angle surfaces. The vee-shaped contact profile permits the wheels to support axial, radial, and yaw moment loads. Self-aligning forces in these orientations automatically develop when the wheels are pressed against the track. The most common angles in vee profile guide wheels are 70 and 90 degrees. Wheels with a 90-deg. vee can run on nearly anything with a right angled straight edge, which opens the possibilities for using simple angle iron or even existing components as track.

Crossed Roller

Crossed roller slides or bearings are the most accurate form of mechanical linear motion. Crossed rollers work similarly to linear ball bearings, except the bearings are cylinders instead of balls—they crisscross each other at a 90-deg. angle, and they move between two separate and parallel guide ways. Crossed rollers are almost always used in pairs. In general, crossed roller slides can offer twice the load capacity of similarly sized ball units, up to five times the life expectancy, less noise, and increased vibration resistance and rigidity. The downside to crossed roller systems is that they only offer limited travel (short strokes) and are typically much more expensive and harder to assemble than similarly sized ball units.

Telescopic Slides (a.k.a.: drawer slides)

One of the niche categories within linear motion is telescopic slides, also known as drawer slides. There are generally two types of drawer-slide products: commercial and industrial. Commercial drawer slides are typically made from formed sheet metal and are designed for lighter loads and lower duty cycles. These are the drawer slides you would find in furniture, tool boxes, server racks, etc. Industrial drawer slides are typically made from thicker steel or aluminum and are designed to carry much heavier loads. Industrial drawer slides are typically used to support tooling and fixtures, large batteries, etc.

Typically, most drawer slides do not fully extend, so if you desire, say, 12 in. of extension, the drawer slide must be more than 12 in. long. There are some drawer slides that do offer a full stroke, and these are typically referred to as fully extending or full stroke drawer slides. Most telescopic systems extend in only one direction. Special systems allow for extension in both directions, and these are known as double-stroke or double-acting drawer slides.

Others

Fluid Bearings (a.k.a.: air bearings, hydrostatic bearings)

Fluid bearings rely on a thin layer of liquid or gas to lift the system off the mating surface, which allows for nearly frictionless motion. Fluid bearings are typically broken down into two categories: hydrostatic bearings and fluid dynamic bearings. Hydrostatic bearings are externally pressurized fluid bearings, where the fluid is usually air, oil, or water; and the pressurization is done by a pump. Hydrodynamic bearings rely on the high speed of the journal self-pressurizing the fluid, usually oil or water, in a wedge between the faces. Fluid bearings are typically more expensive and thus are used as a last resort as opposed to being the most desirable choice for any given application. Common applications for air bearings include Coordinate Measuring Machines (CMMs), optical spectrometry, semiconductor wafer handling/processing, optical grinding, and extremely high-end digital printing.

The greatest advantages of fluid bearings include:

- No solid-to-solid contact, which means less noise, zero wear, and no particles or lubricants (for air bearings).
- If properly used, the bearings have a nearly infinite life and don't require regularly scheduled maintenance.
- Negligible static and dynamic friction allow for very high acceleration and velocity.
- Negligible stiction (stick-slip), which is very beneficial for applications with quick, short-distance moves.
- Very high stiffness and vibration dampening capabilities.

Some additional issues with fluid bearings include:

- Overall power consumption is typically higher compared to ball bearings because these bearings require a pump to be in constant operation.
- Power consumption and stiffness or damping vary greatly with temperature, which complicates the design and operation of a fluid bearing in wide temperature range situations.
- Fluid bearings can catastrophically seize under shock situations. Ball bearings deteriorate more gradually and provide acoustic symptoms.
- Fluid leakage, as keeping fluid in the bearing can be a challenge. Oil-fluid bearings are impractical in environments where oil leakage can be destructive or where maintenance is not economical. Environmental standards (e.g., ISO 14001) frown upon oil usage where leakage can escape into the environment.
- Fluid bearing "pads" often have to be used in pairs or triples to prevent the bearing from tilting and losing fluid from one side, thus increasing installation and maintenance expense.
- The bearings may not require maintenance, but the "support systems" (e.g., pumps, filters, etc.) require regular maintenance. If a support item fails, the bearing can also fail.

Round Tracks

Circular motion is not discussed in any other chapter, so it will briefly be discussed here. Several of the products discussed can also be formed into circular-motion products. These products are priced at a premium compared to the pure linear version and are most commonly used on assembly lines that are oval in shape because there can be a mix of linear and curved motion. In general, if an application is purely circular, a large diameter bearing (like those discussed in Chapter 2: Bearings) would be much more cost effective and appropriate.

Linear Propulsion

Another major component in a motion system is the linear propulsion device. The most common forms of linear propulsion are screws (ball or acme/lead), belts, chains or cables, rack and pinions, and linear motors. Belt and chain drives are discussed at length in Chapters 3 and 4, respectively, so they will not be covered in great detail in this chapter. For additional information on these two products, see their dedicated chapters. Ball screws are more common in demanding applications, and they will be covered first.

Screws

Ball Screws

Ball screws serve as mechanical actuators in linear-motion systems. They are precision mechanical components that convert rotational motion to linear motion with highly accurate positioning. This is usually accomplished by rotating the shaft (fixed in the axial direction) to drive the ball nut. In some cases, the ball nut is rotated, and the shaft is driven. Steel balls are inserted between the grooves in the screw shaft and nut to establish rolling contact with very low friction compared to acme feed screws subject to sliding friction. During operation, the balls are redirected within the ball nut in order to prevent them from rolling out the back of the nut. This is accomplished through the use of return tubes around the exterior, deflectors within the ball nut body, or end caps at both ends. This allows for the continuous circulation of the balls within the ball nut during operation. Balls may circulate in one or more circuits. Each circuit may consist of one or more rows of balls. Each additional circuit will require additional recirculation components. In general, the more rows of balls, the higher the load capacity of the ball screw.

Types and Features

Ball screws are often referred to as either precision ground or rolled. The manufacturing process refers only to the ball screw shaft and whether it is precision machined and ground, or roll formed. In general, precision ground ball screws have better accuracy and are more expensive than rolled ball screws. Ball screws can be further classified by ball recirculation system and the preloading method.

Ball Recirculation System

Ball recirculation methods include the return tube type, the deflector type, and the end cap type.

Return tube-type ball screws use tubes exterior to the ball nut to redirect balls from one end of the ball circuit to the other end of the ball circuit (see Fig. 10-5). The return tube type is the most popular method of ball recirculation and is available in a wide range of shaft diameters and leads. Ball nuts with return tubes can have more than one ball circuit and more than one row of balls per circuit.

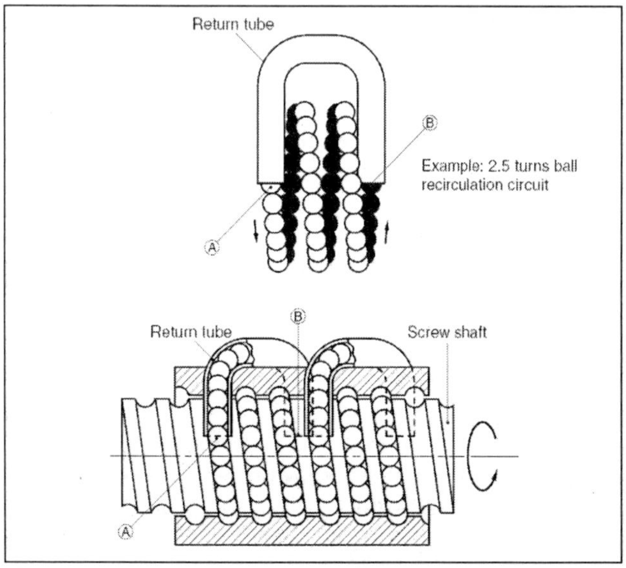

Figure 10-5: Return tube type ball recirculation system

Deflector-type ball screws use deflectors interior to the ball nut to redirect balls from one end of the ball circuit directly to the other end (see Fig. 10-6). The deflector type is more compact in nut diameter than the return tube type due to the lack of exterior recirculation tubes, but it is limited to finer pitch leads. Ball nuts with deflectors can have one or more ball circuits but are limited to one row of balls per circuit.

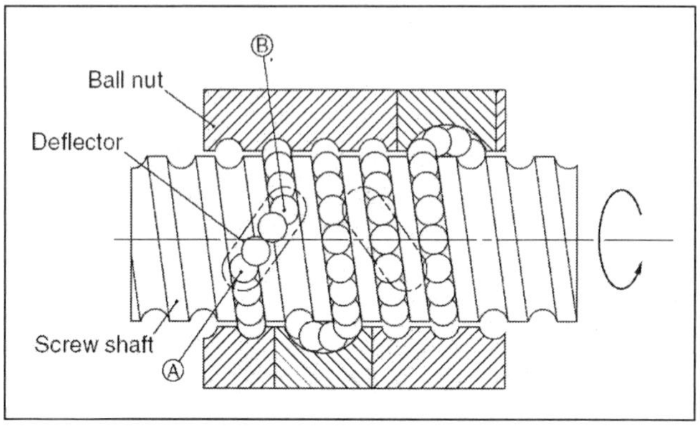

Figure 10-6: Deflector type ball recirculation system

End cap-type ball screws use the end caps to redirect balls from one end of the ball circuit through the length of the ball nut to the other end (see Fig. 10-7). Because each ball circuit runs the length of the ball nut, the number of circuits will be the same as the number of thread starts. High-lead ball screws will sometimes incorporate multi-start threads to ensure an adequate number of ball rows, as nut lengths are limited by manufacturing constraints.

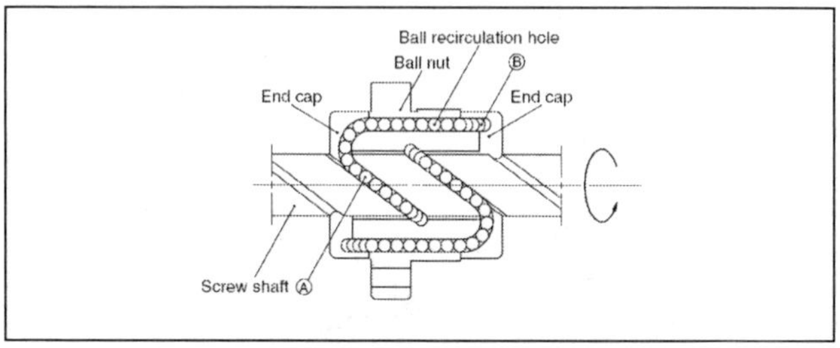

Figure 10-7: End cap type ball recirculation system

Preloading Method

Ball screws may be preloaded to a predetermined axial load in order to eliminate axial play (backlash) and increase rigidity. This may be accomplished with one of four preloading methods: the double nut preload, double nut spring preload, offset lead preload, and oversize ball preload.

With the double nut system, a tensile preload is established by inserting an oversized spacer between two nuts (see Fig. 10-8). The spacer forces the two nuts outward, resulting in the balls being squeezed between the ball nut and shaft. The amount of preload is determined by the width of the spacer. A wider spacer will produce a higher preload. Sometimes an undersized spacer will be used to establish a compressive preload. The double nut system is well-suited for higher preloads.

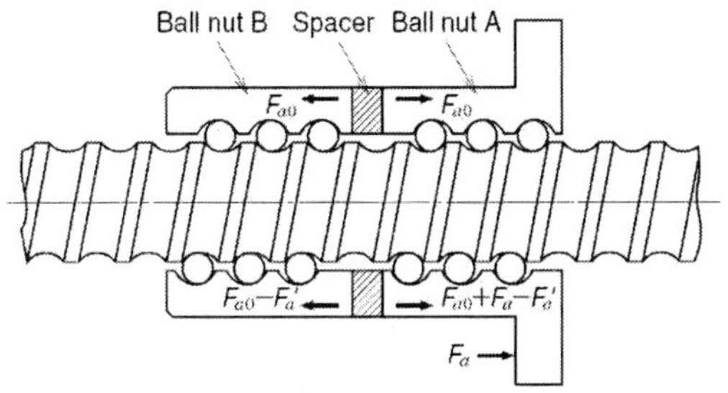

Figure 10-8: Double nut preload

The double nut spring method is similar to a double nut method except that the spacer between the two nuts is replaced by a spring (see Fig. 10-9). The amount of preload is determined by the spring force. The double nut spring method provides the most consistent torque characteristics with minimal variation over an extended period of time.

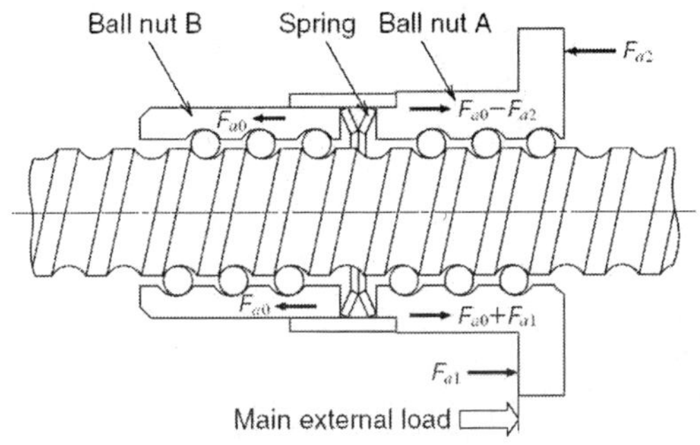

Figure 10-9: Double nut spring preload

The offset lead method employs a single nut with an increase in the thread lead at the middle of the nut in between the two ball circuits (see Fig. 10-10). The oversize lead produces the same effect as the spacer or spring in the double nut method. The amount of preload is determined by the increase in lead. This method results in a more compact ball nut. Manufacturing limitations for the length of a single nut limit the number of ball circuits and rows.

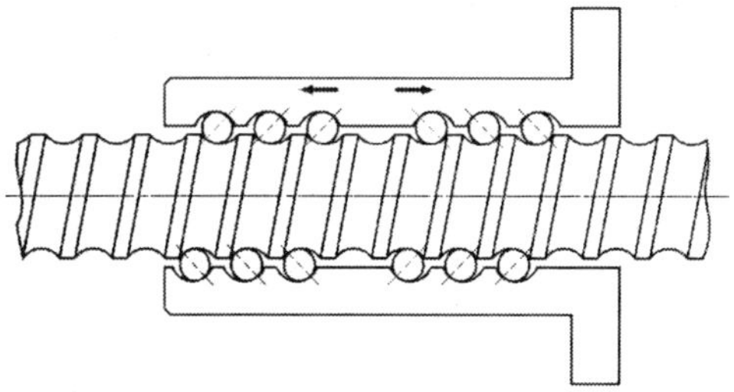

Figure 10-10: Offset lead preload

With the oversize ball method, preload is established by inserting balls that are larger than the space between the ball grooves (see Fig. 10-11). The preload is determined by the amount by which the balls are oversized. This system results in the most compact nut length but is limited to lower preloads.

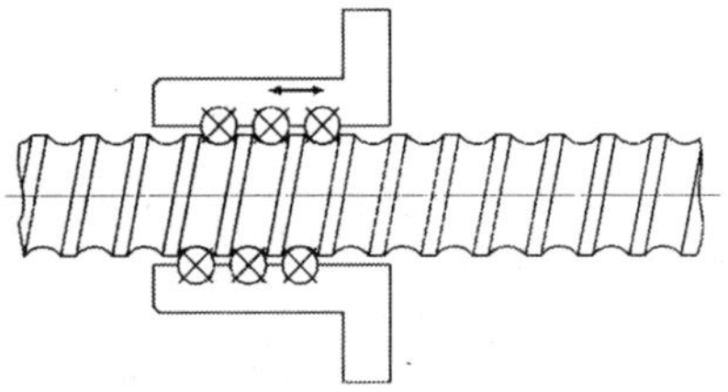

Figure 10-11: Oversize ball preload

Selection Procedures

While the exact process for selecting a ball screw will vary depending on the application, the main considerations and selection criteria will be the same for most situations. Ball screw selection can be described as an iterative procedure in that tentative selections are made based on initial considerations and later modified as required based on additional considerations. For example, the initial decision to go with a rolled ball screw because of lower cost may have to be revisited based on the need for better accuracy or load capacity (rolled ball screws are limited in size due to manufacturing reasons). The decision between precision ground or rolled will typically be based on cost, accuracy and preload requirements, and availability. The decision between the various ball recirculation methods and preloading systems previously described will typically be based on cost, nut size, required preload, and availability.

Once these initial decisions have been made, the ball screw can be specified based on diameter, lead, stroke, accuracy grade, preload amount, and shaft end configuration.

Diameter

The diameter of a ball screw refers to the nominal diameter of the ball screw shaft. Ball screw diameters can range from about 4 mm to about 250 mm. A larger diameter ball screw will have higher load capacity than an otherwise similarly configured, smaller diameter ball screw.

Lead

The lead of a ball screw refers to the distance traveled by the nut with each rotation of the shaft. Leads vary from less than 1 mm for smaller diameter ball screws to 50 mm or more for larger diameter ball screws. Leads can be categorized in comparison to the shaft diameter. Ball screws with leads that exceed the shaft diameter are considered high-lead ball screws, while ball screws with leads that are less than the shaft diameter are considered medium- or fine-lead ball screws. The lead of a ball screw determines the relationship between the linear speed of the nut and the rotational speed of the shaft. The relationship is as follows:

Linear speed of nut (mm/min) = rotational speed of shaft (rpm) x lead (mm)

Lead should be selected based on required linear speed and be consistent with the specifications of the motor that will be used to drive the ball screw. In general, high-lead ball screws are suited for high-speed operations, while medium- or fine-lead ball screws are suitable for medium- or slow-speed operations. In general, a ball screw with a helix angle of 6 deg. or more will back drive.

Stroke

Ball screw stroke refers to the maximum distance the nut can travel from one end of the ball screw to the other. This distance is equal to the threaded length of the shaft minus the nut length. The ball screw stroke should be selected to meet or exceed the required stroke for the application.

Accuracy Grade

The accuracy of a ball screw depends on the lead accuracy of the grooves along the length of the shaft. For example, the lead of a ball screw with a nominal lead of 5 mm may actually vary from 4.997 mm to 5.002 mm along the length of the shaft. The actual travel of the ball nut along the length of the shaft will deviate from the specified travel (based on the nominal 5 mm lead) due to these lead errors. Ball screw accuracy grades specify the amount by which the actual travel may deviate (vary) from the specified travel along the length of the shaft. Ball screws are primarily made through two methods: rolling and grinding. Rolled screws are less expensive than ground screws, but they have less accuracy. Ground screws are more accurate and also more expensive.

Accuracy grade should be selected based on the need for positioning accuracy in the application. For example, machine tools require a relatively high degree of positioning accuracy, while equipment in steel mills requires significantly less. You should take care to select the lowest [least accurate] accuracy grade required as the price can increase significantly with an increase in accuracy level.

Preload Level

As mentioned previously, ball screws may be preloaded in order to eliminate axial play (backlash) and increase rigidity. The precision of a ball screw actuator depends largely on the amount of backlash between the nut and screw. The requirement for rigidity should determine the required preload amount. Although rigidity improves with increased preload, it is important not to use too much preload. An increase in preload will result in an increase in drag torque and heat generation and a decrease in service life.

Shaft End Configuration

The configuration of the shaft ends will depend on how the ball screw will be mounted. Shaft ends may be supported by fixed or simple supports or be left free (unsupported). Fixed supports usually consist of a pair or more of angular-contact ball bearings. The outer rings will be fixed in their housings to prevent axial movement of the ball screw shaft. Simple supports usually consist of a single deep-groove ball bearing allowed to float in its housing to allow axial movement of the ball screw shaft (in response to thermal expansion of the shaft). Unsupported shaft ends are sometimes used in lower speed and/or vertical applications and for short stroke applications. The most popular ball screw mounting arrangements are fixed-simple (fixed support on one end of shaft, simple support on the other end) and fixed-fixed.

The shaft ends should be configured to accommodate the selected mounting arrangement and be consistent with appropriate design guidelines for bearing journals. Shaft ends may need to accommodate shaft nuts, snap ring grooves, wrench flats, seals, hexagon holes, key ways, couplings, and pulleys.

Maximum Rotational Speed

Maximum rotational speed is the lower of the critical speed and the limiting speed. The critical speed is the rotational speed at which the shaft will begin to resonate and cause damage to the ball screw. It is dependent on the shaft diameter, (unsupported) length, and mounting conditions. The limiting speed is the speed beyond which the ball recirculation components may become damaged. It is based on the maximum DxN (shaft diameter x rotational speed), which can vary by ball screw type and grade.

Life

Life calculations for ball screws are very similar to those for bearings and based on the same principles. Ball screw fatigue life in total revolutions is a function of the basic dynamic load rating, applied load, and a load factor. The basic life equation is as follows:

Life = [Ca / (Fa x fw)] **3

Here, "Ca" is the basic dynamic load capacity, "Fa" is the applied axial load, and "fw" is a load factor based on operating conditions. Life in total revolutions can be converted to life in total hours or running distance if speed information is known.

Static Load Limitations

Ball screw selection must also take into account static load limitations. The static load limit of a ball screw is the lower of the maximum loads calculated for shaft buckling, shaft yielding, and permanent deformation of the balls and raceways. Shaft buckling limits are a function of shaft diameter and (unsupported) length and the mounting conditions. Shaft yielding limits are a function of shaft diameter (cross-sectional area) and material strength. Limits based on permanent deformation of balls and raceways can be determined using the basic static load ratings provided by the ball screw manufacturer. Shaft buckling and yielding limitations for a particular ball screw can also be obtained from the manufacturer.

Drive Torque

Drive torque, or the torque required to drive a ball screw under load, can be determined from the following equation:

Drive Torque = Operating Torque + Drag Torque

Drive Torque = [(Fa x l) / (2 x π x E)] + Drag Torque

Here "Fa" is the applied axial load, "l" is the lead, and "E" is the ball screw efficiency (generally assumed to be 0.90 for a ball screw). Drag torque is due to friction, and preload and can be provided by the manufacturer.

When specifying a drive motor, the drag torque of the support bearings and the gear or pulley ratios, if applicable, should be considered.

Roller Screw

A roller screw, also known as a planetary roller screw or satellite roller screw, is a low-friction precision mechanical device for converting rotational motion to linear motion, or vice versa. Planetary roller screws are most commonly used as the actuator mechanism in electro-mechanical linear actuators. Due to its complexity, the roller screw is a relatively expensive propulsion mechanism, but it may be suitable for high-precision, high-speed, heavy-load, long-life and heavy-use applications. A roller screw is a mechanical actuator that is similar to a ball screw that uses rollers instead of balls as the load transfer elements between nut and screw. The rollers are typically threaded, but they may also be grooved, depending on the roller screw type. Roller screws provide more bearing points than ball screws within a given volume, thus can be more compact for a given load capacity. They provide similar efficiency (75 percent to 90 percent) at low to moderate speeds and relatively high efficiency at high speeds. Roller screws can surpass ball screws in regard to positioning precision, load rating, rigidity, speed, acceleration, and lifetime. Standard roller screw actuators can achieve dynamic load ratings above 130 tons of force (exceeded in single-unit actuator capacity only by hydraulic cylinders).

The three main elements of a typical planetary roller screw are the screw shaft, nut, and planetary roller. The screw, a shaft with a multi-start v-shaped thread, provides a helical raceway for multiple rollers radially arrayed around the screw and encapsulated by a threaded nut. The thread of the screw is typically identical to the internal thread of the nut. The rollers spin in contact with and serve as low-friction transmission elements between screw and nut. The rollers typically have a single-start thread with convex flanks that limit friction at the rollers' contacts with screw and nut. The rollers typically orbit the screw as they spin (in the manner of planet gears to sun gear) and are thus known as planetary or satellite rollers. As with a lead screw or ball screw, rotation of the nut results in screw travel, and rotation of the screw results in nut travel.

For a given screw diameter and quantity of thread starts, more rollers correspond to higher static load capacity, but not necessarily to a higher dynamic load capacity. Preloaded split nuts and double nuts are available to eliminate backlash.

Lead/Acme Screw

Lead screws provide a simpler, quieter, less expensive alternative to ball screws, although less accurate and efficient with a reduced load capacity. It is important to remember that ball screws can be manufactured through a rolling or grinding process. The rolling process for ball screws and lead screws is very similar, and therefore can produce the same accuracy grades. Lead/Acme screws are typically not made through a grinding process.

Lead Screw Selection: Thread Types

Lead screws are typically classified by thread type. Vee threads generate more friction than Acme threads, and thus are often unsuitable for industrial use, although the self-locking characteristics of the vee thread design helps prevent back driving. Acme screw threads, machined at a 29-deg. angle, are easier to manufacture than square threads, though they are not as efficient. Square threads, though difficult to machine and expensive, create comparatively little friction and are quite efficient. They are often used for screws with high power requirements.

Lead Screw Selection: Other Considerations

Because the lead screw and nut rub against each other, they produce comparatively more friction and stiction than mechanical mechanisms that mate with bearings and other rolling surfaces. Typical lead screw efficiency is around 25 to 70 percent, with higher pitch screws displaying increased efficiency. The ball screw provides a higher performance, although it is a more expensive alternative to the lead screw. Lead screws, though not the best choice for continuous high-speed operation (as they tend to overheat), are often desirable in applications where back driving is unacceptable, such as vertical loads. A second nut or tensioning spring can be added to reduce backlash and create preload.

Lead Screw Advantages

- Simple
- Precise
- Easy to machine
- Quiet
- Often self-locking
- Minimal maintenance
- Abundant load capability

Lead Screw Disadvantages

- Produce friction
- Comparatively inefficient
- Not recommended for continuous power applications

Lead Screw Alternatives

- Ball screws
- Roller screws
- Piezoelectric and electromagnetic actuators
- Worm gears
- Rack and pinion

Timing Belt (a.k.a. toothed belt)

Timing belt systems are popular choices for linear motion systems because of their relatively low cost and ability to support applications requiring high acceleration and velocity over long lengths. Designers sometimes choose belts over screws in order to increase system length or reduce total installed costs. Belt systems typically require less alignment during installation, which speeds assembly and reduces cost.

For more specific details about timing belts, refer to Chapter 3: Belt Drives.

Chains and Cables

Chains and cables are infrequently used for linear motion applications; however, there are a small percentage of applications where this type of product is the best choice. Chains and cables are an efficient propulsion mechanism for applications with long length requirements, or for use in dirty environments, as they are not as susceptible as screws or belts to contamination-related failures.

For more specific details about chains, refer to Chapter 4: Chain Drives.

Rack and Pinion

A rack and pinion is a type of linear actuator that comprises a pair of gears that convert rotational motion into linear motion. The circular pinion engages teeth on a linear rack. Rotational motion applied to the pinion will cause the rack to move to the side, up to the limit of its travel. For example, in a rack railway, the rotation of a pinion mounted on a locomotive engages a rack between the rails and pulls a train along a steep slope. Racks are typically chosen for linear motion systems because of their ability to offer a nearly unlimited range of motion without the stretching issues sometimes encountered with belts or cables. These systems handle contamination well and are often used in industrial settings. Alignment is critical during installation, so costs can be higher than with belt systems.

For more specific details about rack and pinion, refer to Chapter 8: Gears.

Linear Motor

An electric linear motor is essentially an unrolled electric rotary motor. It produces linear motion directly, making the mechanical conversion of rotary to linear motion unnecessary. This eliminates the use of mechanical devices such as screws, belts, and gears; which in turn reduces space, energy, and cost. Linear motors also incorporate their own support systems, eliminating the need for separate linear bearings. Linear motors consist of two elements: a forcer (primary) and a platen (secondary), which move relative to each other. The forcer and platen are separated by an air gap and are not in direct contact with each other. Either element can be fixed in place, although a fixed platen is more common. With the exception of brush universal motors, the forcer is powered, and the platen is not. The forcer must be supported and guided along the platen by a guidance system (i.e., linear bearings). Linear motors are less vulnerable to wear because there is no rotary to linear conversion.

Motion in a linear motor comes from the interaction of two or more magnetic fields that are out of phase (offset) from each other. These magnetic fields are generated by electromagnets (coils or windings) in the forcer that are attracted or repelled by an induced magnetic field or permanent magnets arranged along the platen. The physical structures (electromagnets and cores) that generate and focus the magnetic fields are called poles. The force generated by a motor is proportional to the surface area of the poles. Commutation is the act of turning on and off the various electromagnets to maintain the magnetic fields in and out of phase condition in order to create continuous motion. Brushed motors are commutated mechanically, while brushless motors are commutated electronically.

Types and Features

Several types of linear motors in common use are listed in Table 10-2. All types rely on the interaction of magnetic fields (flux) to produce a force and resulting linear motion. For the voice-coil, DC force, and step-motor types, this flux comes from a permanent magnet. For the induction motor, AC current excites a coil that produces flux. In turn, this flux interacts with flux produced by induction (like a transformer) and generates a force proportional to the combined strength of the interacting fluxes.

TABLE 10-2 — Typical linear motor performance specifications

Type	Max stroke	Max force, lb	HP	Acceleration, g	Resolution, in.
Voice coil	0.5 in.	100	Less than 1	Several	Less than 0.001
AC linear induction	Several ft	1-3,000	1-10	Less than 1	Less than 0.1
AC linear induction servo	100 ft	25-1,800	1-10	10	Less than 0.0002
DC linear force	0.5-2 ft	500-1,000	1-5	1	Less than 0.001
Step	0.5-2 ft	10-25	Less than 1	1	0.025
Microstep	Several ft	5	Less than 1	1	0.00004

Voice-Coil Motors

Constructed like solenoids, these motors come in moving-coil and moving-magnet configurations and provide more precise motion than solenoids. Voice-coil motors maintain high linearity between applied current and developed force and operate efficiently. Those built with high-energy magnet materials develop high forces and acceleration rates; however, as with a solenoid, the mechanical structure precludes long strokes.

Linear Induction Motors

A linear induction motor resembles a rotary induction motor that has been split axially and rolled out flat. It has two major components: a wound structure (much like a conventional motor stator) called the primary that typically operates on three-phase power, and a metallic structure (much like a rotor) called the secondary. The secondary generally moves, but in principle, either structure can be the moving part of a linear induction motor.

Linear induction motors can operate at line voltage with a fixed speed-thrust characteristic. They can also operate in an open loop from an adjustable-frequency source for adjustable-speed applications, as do servo systems with a closed-position loop. Figure 10-12 shows such a servomotor system in which the linear measuring system could be an encoder that feeds position data back to an AC vector drive (see Chapter 12).

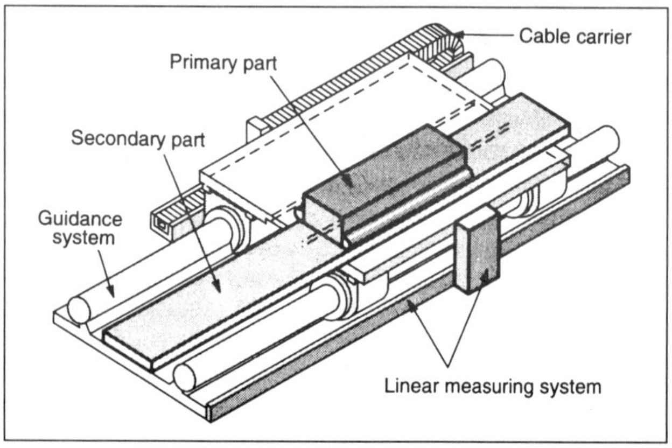

Figure 10-12: Typical linear motor

The linear induction motor (see Fig. 10-13), sometimes called an asynchronous linear motor, is a single-coil device with the primary part holding the windings (similar to a stator in a rotary motor). The secondary (like a rotor) consists of iron and short-circuit rods of copper or aluminum. In most cases, the secondary coil is stationary, and the primary moves with the moving part of the system.

Figure 10-13: Linear induction motor

Linear Force Motors

As with the linear induction motor, a linear force motor is like a conventional DC motor that has been split axially and rolled out flat, (see Fig. 10-14 and Fig. 10-15). Both brush and brushless types are used. In some cases, the magnet moves, and in other cases the coil moves. The precise thrust-to-current characteristic of a linear force motor, especially when controlled by a microprocessor, suits it for precise positioning applications.

Figure 10-14: Linear force motor

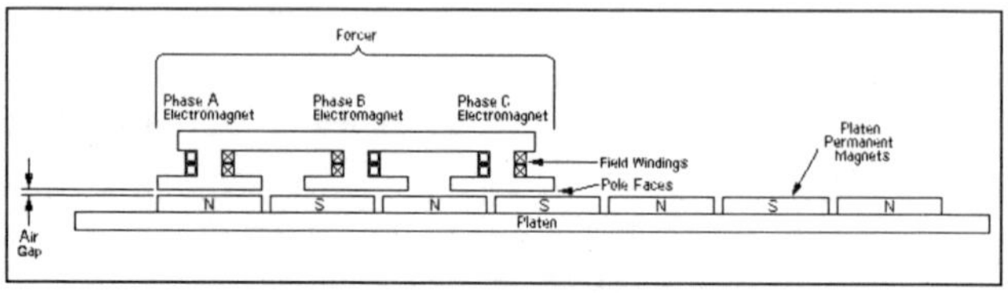

Figure 10-15: Linear force motor configuration

Linear Step Motors

Linear step motors (see Fig. 10-16 and Fig. 10-17) provide the same incremental point-to-point precision as their rotating counterpart. A linear step motor has a toothed, magnetic pole structure on the stator (platen) and on the slider (forcer). Platen and forcer tooth structures are slightly mismatched. For example, the platen may have 11 teeth in the same length in which the forcer has 10. By sequentially energizing two coils that operate in conjunction with a permanent magnet (oriented parallel to the axis of motion), the step motor can be made to move in 1/4 tooth-pitch increments. And, with micro-step controls, the motor can achieve extremely fine resolution (up to 25,000 steps/in. or 1,000 steps/mm).

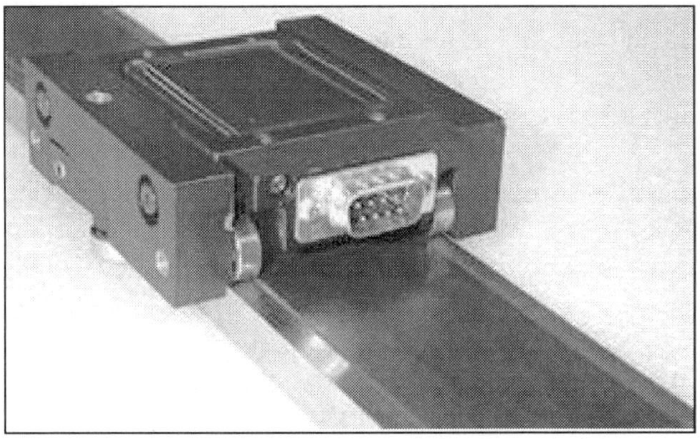

Figure 10-16: Linear step motor

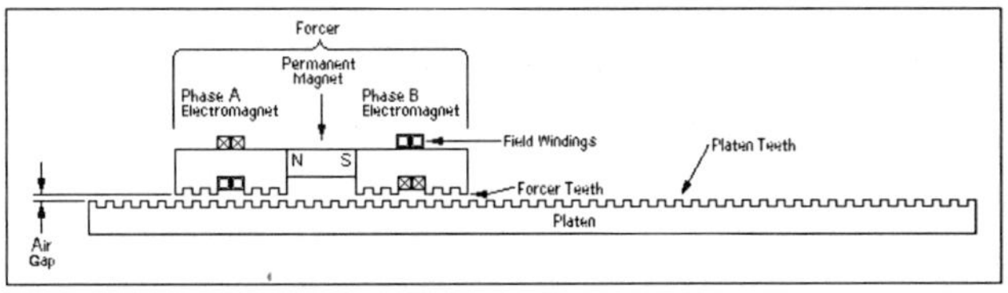

Figure 10-17: Linear step motor configuration

Selection Procedures

Compared to other means of generating linear motion, linear motors are faster, more responsive, and have greater precision. Because there are no mechanical linkages, they are stiffer and have zero backlash and require little or no maintenance. Disadvantages include cost, low force relative to their size, and little to no friction, which can reduce stability. However, linear motors are the only solution for many of the most demanding applications.

Voice-coil motors are compact and can develop high forces and acceleration rates, but their mechanical structure precludes long strokes. Linear induction motors have a wide speed range with unlimited travel and are common in many types of conveyance systems. Linear force motors have a low profile and small cross section, are available in a wide range of lengths and widths, and are capable of high speed and high acceleration of lighter loads. Linear step motors are common in positioning applications requiring high speed and acceleration. They can provide precise open-loop operation with repeatability to one micron.

Choosing a linear motor for an application involves defining the application requirements, calculating the required forces (of which the force of acceleration is usually the largest), the maximum speed, and the distance of travel. A further consideration is the environment and how the motor technology may interact with that environment. How the linear motor will be controlled must also be taken into account. In the end, several different linear motor technologies may work, and the final selection may simply be related to cost.

Actuators (Electromechanical, Stage, Slide, Table)

Linear actuators are available in a variety of configurations, but all convert some kind of power into linear motion. Many are suitable for industrial, scientific, and commercial applications requiring precise control of thrust, speed, and position.

Electromechanical Actuators

Electromechanical actuators provide automated linear positioning, transfer, and guidance for a variety of diverse applications, including food and beverage, factory automation, packaging, agricultural, construction, aviation, automotive, semiconductor, pharmaceutical, medical equipment, and lab testing.

Although multiple design configurations and profile sizes are available, a typical electromechanical actuator would consist of an aluminum housing equipped with a load-bearing carriage propelled by a motor-driven ball screw, lead screw, or toothed belt. A magnetically attached stainless steel strip is often provided to help seal out contamination. A variety of positioning sensors and multi-axis mounting options are also offered.

Ball Screw Advantages

- Accurate
- Efficient
- High load capacity

Lead Screw Advantages

- Simple, quiet
- Less expensive
- Self-locking
- Low maintenance

Belt Drive Advantages

- High acceleration, speed, rigidity
- Low friction, noise, vibration
- Long travel length

Miniature Linear Actuators

For precise linear travel in a small footprint, miniature lead screw-driven actuators are now available for applications such as medical and biotech testing, packaging, military/ defense, and small-scale automation. Compact, lightweight, and remarkably powerful despite their small size, these mini-actuators are designed for both precision and performance in single and multi-axis assemblies.

Actuator Motors

Depending on the application requirements, many types of motors are suitable for a linear actuator system, including stepper, DC brush or brushless, and AC induction. Electromechanical linear actuators used in laboratories or other facilities requiring fine resolution and high accuracy may require a low horsepower stepper with a fine-pitch lead screw, while applications requiring abundant force and speed may do well with an AC induction motor. It is essential to know specific design and application requirements to determine suitable components. Proper gearing allows a motor to spin the screw under a heavier load than the motor would otherwise be capable of driving directly. Motors are covered in greater detail later in this chapter.

Static Load Capacity

Linear screw actuators have a static loading capacity, which means when the motor stops, the actuator locks in position. This braking force varies with the angular pitch and design of the screw threads and cannot be dynamically altered. Acme threads have a high static load capacity, while ball screws with a low load capacity can be nearly free-floating.

Linear Slides

All linear slides provide motion based on bearings. Movement for motorized linear slides such as machine slides and roller tables is typically provided by some kind of drive mechanism. That said, not all linear slides are motorized. Non-motorized roller and ball bearing slides can provide linear movement for manually operated equipment of various types.

Rolling-Element Bearings

A rolling-element bearing has an outer sleeve ring and several rows of balls retained in cages and runs on hardened or stainless steel raceways. Ball bearing slides and roller slides are both examples of rolling-element bearings.

Ball Bearing Slides

Ball bearing slides offer precise motion and are self-lubricating, which enhances reliability. Applications include instrumentation, robotic assembly, and cabinetry. A ball bearing drawer slide is a common example. Typically, two linear rows of ball bearings are contained by rods on the sides of the base to facilitate smooth carriage travel. Power is supplied by hand or a drive mechanism. Ball bearing slides have a relatively low load capacity for their size compared to other linear slides and are less resistant to wear.

Roller Slides

Roller slides, often powered by hand, provide linear movement for non-motorized mechanisms. A typical configuration consists of steel-crossed rollers situated between bearing races on the linear base and carriage. Roller slides function similarly to ball bearing slides, except the bearings are cylindrical and offer improved repeatability and the ability to handle heavier loads, though at a higher cost. Roller slides are versatile and can provide linear motion on more than one axis. They can be modified to accommodate numerous applications, including material handling and automation machinery.

Advantages

- Economical
- Easy maintenance
- Smooth motion, low friction
- Long life

Disadvantages

- Require seals and lubrication
- Sensitive to contamination

Plain Bearings

Plain bearings slide without balls or rollers and may or may not require lubrication. They are less rigid and can handle a wider temperature range than rolling-element bearings. They are also corrosion-resistant and less sensitive to contamination.

Linear Stage

A linear stage consists of a platform and a base joined by some form of linear bearing, which restricts the platform to a single axis of linear motion with respect to the base. A variety of different bearings can be utilized, each with benefits and drawbacks, depending on the application. Examples include ball bearings, recirculating ball bearings, and crossed-roller bearings.

Position Control

Positioning of the platform relative to the fixed base is typically controlled by a linear actuator equipped with a lead screw or toothed belt. The most common method is a lead screw passing through a ball nut with matching helical threads. Rotation can be controlled either manually or with some type of motor.

Motors

In some applications, a stepper motor may be used in place of a manual knob. A stepper motor moves in fixed increments or steps, so it is actually much like a manual index knob. Another alternative would be a DC motor with encoder.

Multiple Axis

Linear stages may be used together to allow position control in more than one direction. A dual-axis stage can be assembled from two linear stages, with one mounted perpendicular to the first.

X-Y Tables

X-Y tables provide precision-controlled horizontal motion for automated machinery in manufacturing facilities worldwide. Applications include material handling, machinery building, and industrial automation.

X-Y tables are typically mounted on ball bearing or roller slides and are available with a variety of linear bases. X-Y tables are lightweight, easy to use, and generally require little maintenance, although ball bearings may need to be replaced occasionally.

Other Linear Actuators

Other common linear actuators include, but are not limited to, rodless cylinders, rodless mechanical actuators, magnetostrictive actuators, and short-stroke actuators.

Rodless Cylinders

Unlike conventional cylinders, rodless cylinders have no external piston rod. Rather, they have an internal rod that is connected to an external carriage. Compressed air or hydraulic fluid is used to drive the piston, thereby moving the carriage. These cylinders require much less mounting space than conventional cylinders—in some cases, almost 50 percent less. And, they are well suited for long-stroke applications where there is insufficient space for cylinder rod extension. Rodless cylinders have been available for years, but have become widely used only recently as suppliers began offering more variations and options. These cylinders now compete effectively with mechanical and electrical actuators in many applications, including those requiring high load or high speed. For information on conventional rod-type cylinders, see Chapter 9.

A cable cylinder, the earliest type of rodless cylinder, has a piston with taut cables attached at each end. They pass through seals at the cylinder ends, run over pulleys, and join at a payload carrier. Air or hydraulic pressure drives the piston to move the carrier.

In today's rodless cylinder (see Fig. 10-18), the cylinder wall has a slot running along its length that lets the piston connect to an external carriage. Dynamic strip-type seals cover the slot's outer and inner surfaces to keep out contaminants. Carriage travel opens the seal section beneath the carriage, but as the carriage passes, the seals reseat on the slot and continue to seal.

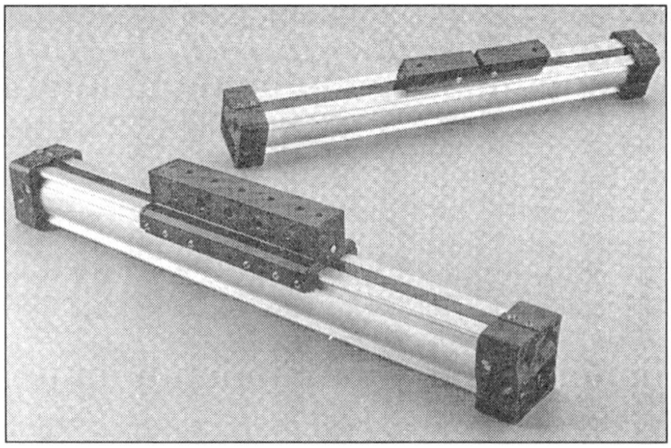

Figure 10-18: Rodless cylinder

A dual-chamber band cylinder uses a steel band in place of the cable to reduce stretch. The cylinder is enclosed so there is no slot, and therefore, less leakage.

Many improvements have been made in rodless cylinders, reducing complexity and overcoming early obstacles such as the inability to handle transverse loads. The most recent type is a magnetically coupled rodless cylinder (see Fig. 10-19). Here, the piston couples magnetically to the external carriage, making slots and dynamic seals unnecessary.

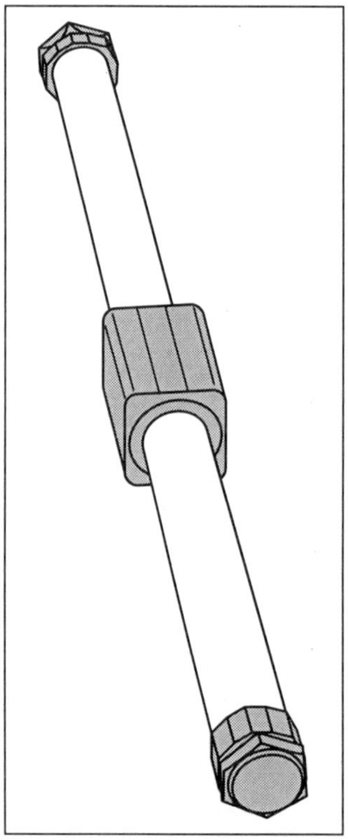

Figure 10-19: Magnetically coupled rodless cylinder

Rodless Mechanical Actuators

Like fluid-powered rodless cylinders, motor-powered rodless mechanical actuators (see Fig. 10-20) offer substantial space advantages over conventional rod-type cylinders.

The external follower (saddle) of a mechanical rodless cylinder travels beside a fixed housing instead of protruding from the housing end. Thus, overall length is the same whether the actuator is extended or retracted. The saddle connects through a slot in the housing to the drive, typically a ball screw nut or timing-belt attachment. Conversely, the saddle can be stationary and the housing movable.

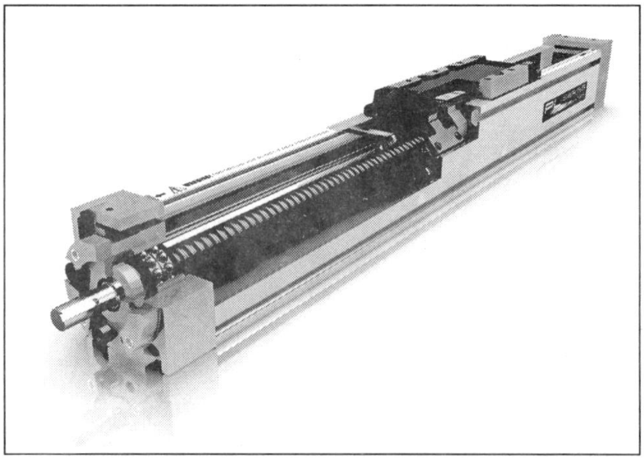

Figure 10-20: Rodless mechanical actuator with ball screw drive

A magnetic strip covers the slot, sealing out contaminants. Inside the saddle, a wedge lifts the strip off the slot, letting the saddle and the ball nut or belt follower pass by, then the strip reseats on the slot while still inside the saddle.

In general, AC, DC, servo, step, or hydraulic motors can drive rodless mechanical linear actuators, but conventional electric motors are the most widely used. Several actuators can be synchronized to share load or move simultaneously, offering opportunities for clamping or framing. In many applications, a device mounts on the actuator to do a special job, such as hoisting or holding.

Rodless mechanical actuators function well in parts-feeder applications, such as moving bars into a welding process. The easily controllable feed rate available with an adjustable-speed motor in a closed-loop system is advantageous. Slicing of bulk or sheet products is another application. An adjustable-speed drive and programmable controller provide precise slicing control and process flexibility.

Short-Stroke Actuators

The short-stroke linear actuator is commonly used in applications ranging from doorbells to jet aircrafts.

Short-stroke linear actuators, or solenoids as they are commonly called, are the most widely used devices for applications that require electromechanically based linear motion. These actuators are simple, cost-effective, come in many configurations, and often have long life—up to 100 million or more actuations in some cases.

A typical linear actuator (see Fig. 10-21) consists of a coil that generates a magnetic field, an iron shell that provides a magnetic circuit, and an iron plunger that moves linearly in response to a magnetic field. There are four main types of short-stroke actuators: low profile, tubular, open frame, and magnetically latched (see Fig. 10-22).

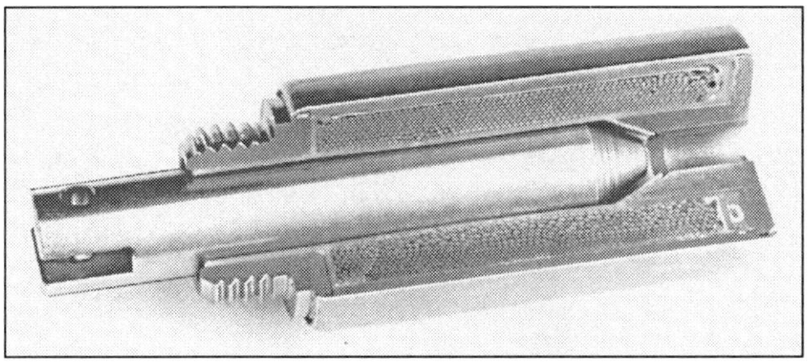

Figure 10-21: Tubular short stroke linear actuator

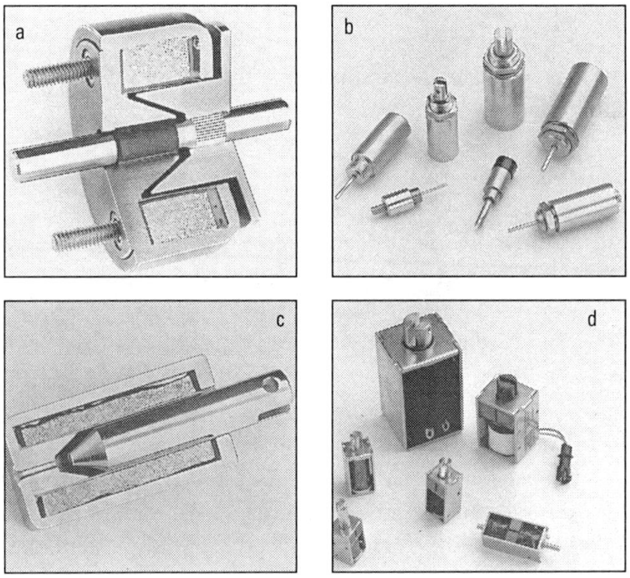

Figure 10-22: Short-stroke actuator types: low profile (a), tubular (b), open frame (c), and magnetically latched (d).

Low Profile

Offering the highest force of all four types, low-profile actuators are also the most expensive. Available models generate forces ranging from 5 lb (2.27 Kg) for a 3/4 in. (19.05 mm) OD unit to 350 lb (158.76 Kg) for a 3 3/8 in. (85.725 mm) OD unit. Life expectancy of this type of solenoid can range up to 25 million actuations. Maximum strokes range from 0.140 to 0.700 in. (3.556 mm to 17.78 mm).

Tubular

These are typically available in 1/2 (12.7 mm), 3/4 (19.05 mm), and 1 in. (25.4 mm) OD, although much larger sizes have been produced for special applications. Forces for standard sizes range from 20 oz to 8 lb (0.57 Kg to 3.63 Kg). Life expectancy can reach 25 million actuations. Strokes to 0.7 in. (17.78 mm) are feasible.

Open Frame

Open-frame actuators are the least expensive short-stroke linear actuators, with a life expectancy of 50,000 cycles. Sizes range from 1/4 to 4 in. (6.35 mm to 101.6 mm) with forces from 4 to 7 lb (1.8 Kg to 3.2 Kg). Strokes range from 0.1 to 1 in. (2.54 mm to 25.4 mm). More than 30 sizes are widely available, each with different plunger configurations.

Magnetically Latched

These actuators incorporate an integral permanent magnet. When a momentary pulse is applied to the solenoid, the plunger is pulled to the energized position, where the magnet holds it. A secondary pulse of opposite polarity releases the plunger. This suits applications where the actuator will be used only intermittently and where the energized position must be maintained.

Linear actuators use the magnetic attraction between poles of opposite polarity to perform either push or pull functions. In a pull-type actuator, one pole (the stator) is held stationary, so the force pulls the plunger toward the stator pole. A push-type actuator uses a modified construction that enables it to perform a push function using the same magnetic attraction.

Factors involved in choosing the right linear actuator for a job include force, life, stroke, duty cycle, maximum on time, operating speed, temperature, other environmental conditions, and power available (AC or DC).

Short-stroke linear actuators are used in many applications, such as home sterilizers, packaging machines, fluid-analyzing equipment, vending machines, and copy machines. But they are not suitable where force must be applied slowly or in steps. Also, harsh environments (abrasive chemicals) should be avoided because these actuators are usually not sealed. They are generally best suited to short strokes and are almost never used with strokes beyond 3 in.

Magnetostrictive Actuators

Employing materials that stretch when subjected to a magnetic field, magnetostrictive actuators (see Fig. 10-23) are used in applications that call for very short linear movements, typically 50 to 100 min.

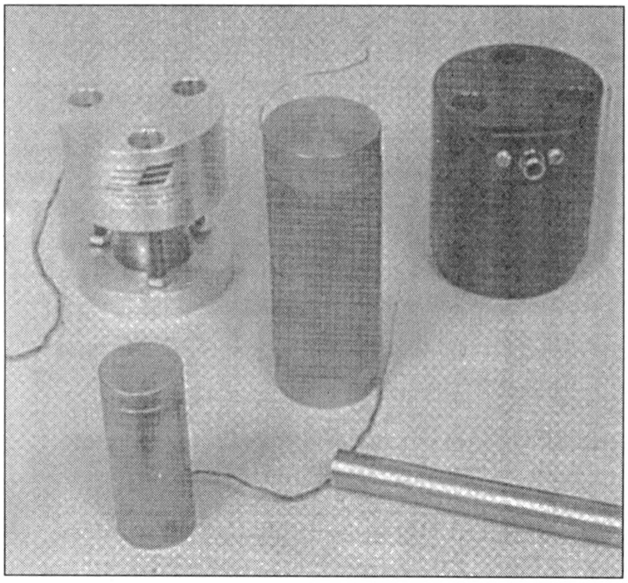

Figure 10-23: Magnetostrictive actuators

They are similar to piezoelectric actuators, which use rods that stretch when an electric voltage is applied to the ends. But piezoelectric actuators are effective only within a narrow temperature range, and they exhibit adverse aging and flowing characteristics. Moreover, though early magnetostrictive materials had limited stretching ability, new versions can stretch up to five times more than commonly used piezoelectric materials.

A typical actuator consists of a magnetostrictive rod surrounded by a solenoid coil (see Fig. 10-24). Externally powered, the coil generates a magnetic field that causes the rod to become longer and thinner. The actuators generally require a magnetic return path (usually a permanent magnet). But a return path through air often suffices for those having a long thin rod.

These actuators, having rods from 0.2 to 2 in. (5.08 mm to 50.8 mm) diameter and 1 to 8 in. length (25.4 mm to 203.2 mm) compete effectively with piezoceramic and voice-coil actuators, particularly when operated at 150 to 2,000 Hz frequencies. And, a unit with a 1/2 in. (12.7 mm) diameter rod generates up to 400 lb (181.4 Kg). Response time can be as little as 10 msec. The actuators typically operate with a magnetic bias (provided by a permanent magnet) that positions it at the midpoint of the travel range. This enables both positive and negative displacement (extension and retraction).

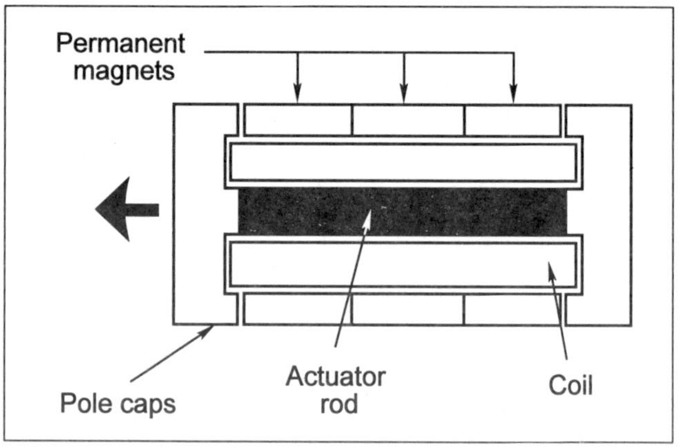

Figure 10-24: Magnetostrictive actuator rod extends linearly in response to a magnetic field.

Maximum performance is achieved by applying a mechanical preload to the rod, usually with springs. This preload increases the material displacement and produces a more linear response (displacement proportional to coil current).

The new magnetostrictive materials are suitable for use in linear motors, servo valves, and vibration dampers.

- **Linear motors** – Unlike conventional motors, magnetostrictive linear motors require few components and produce linear motion directly. And, they provide more motion than actuators that require alternating feed and clamping strokes. They also combine high speed with relatively long strokes and achieve high positional accuracy.

- **Servovalves** – Hydraulic valves have been built with magnetostrictive actuators that drive the valve spool, thus eliminating the electromechanical linkage found in conventional servovalves. The magnetostrictive material generates enough force to make a self-cleaning valve, crushing debris that might otherwise impede spool action.

- **Vibration dampers** – Magnetostrictive materials can eliminate vibration in machine tool tables and chatter at the cutting tool.

Screw Jacks

Screw jacks are one of the oldest actuation devices and consist of an input rotating shaft, an output linear motion shaft (screw), and support bearings, all enclosed in a housing.

Types and Features

Three types of screw jacks commonly used in linear motion systems are machine screw jacks, ball screw jacks, and roller screw jacks.

Machine Screw Jacks

In the common machine screw jack (see Fig. 10-25), rotation of the input (worm) shaft turns the worm gear and drive nut, which connects rigidly to the worm gear. The lead screw (also called a lifting screw or stem) has an Acme or modified square-thread form. It is threaded through the drive nut and converts rotary motion of the nut to linear motion. Rolling-element bearings support the input shaft and worm gear to minimize frictional loss. Thrust bearings support the load. A stem cover stores lubricant and protects the stem from damage and contamination.

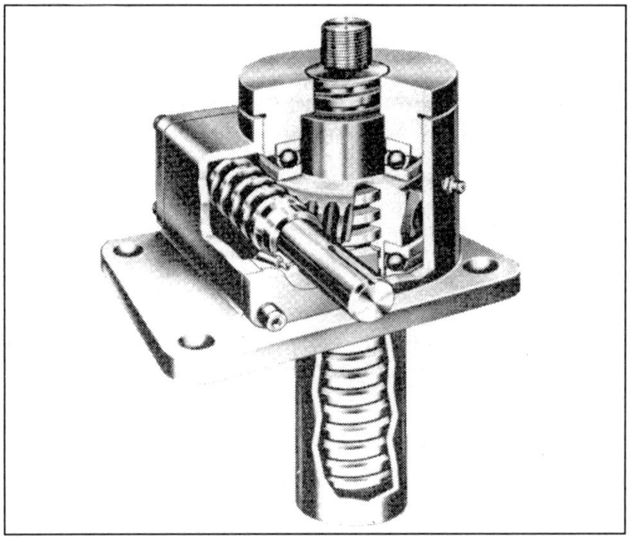

Figure 10-25: Machine-screw jack converts rotary to linear motion.

Machine screw jacks come in various sizes with load ratings from less than 1 to more than 300 tons (300 ton = 272,000 Kg). A jack can be mounted stationary so the stem moves linearly. In an alternate design, an external nut is used so the stem rotates, and the nut (attached to the load) moves linearly (see Fig. 10-26). Machine screw jacks are generally self-locking so they restrain the load in the event of a power failure. A limitation of machine screw jacks is low efficiency, generally about 25 percent. A sliding action between the drive nut and screw generates heat, which restricts machine screw duty cycles to 5 to 7 1/2 min/hr at full load.

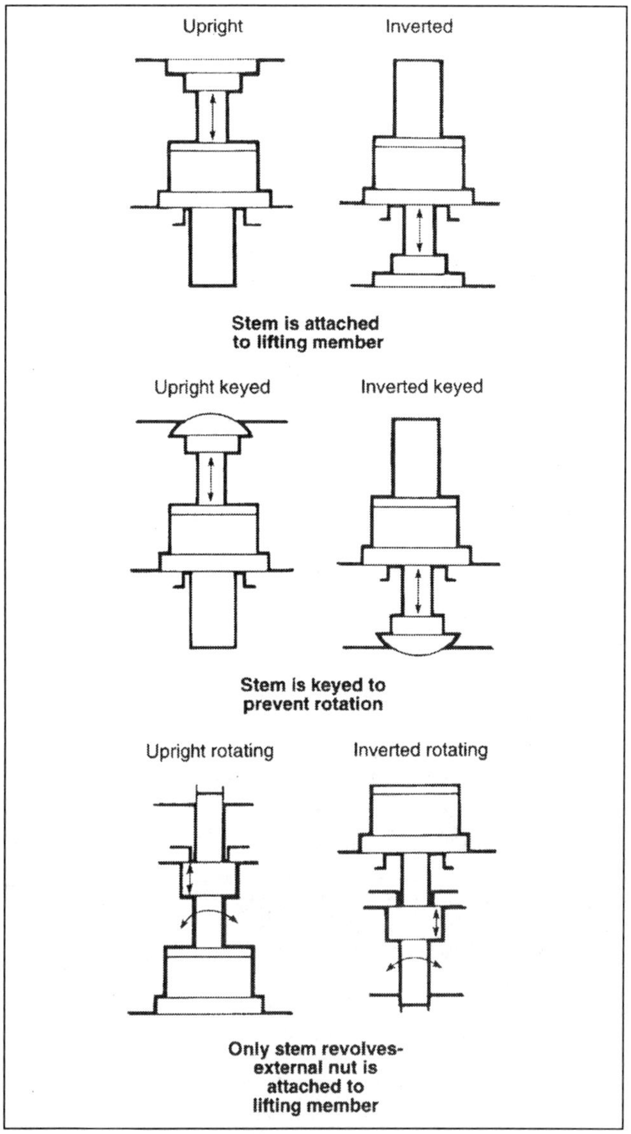

Figure 10-26: Common screw jack mounting configurations

Ball Screw Jacks

Where an application calls for a longer duty cycle or higher efficiency than available with a machine screw jack, a better choice may be a ball screw jack (see Fig. 10-27), where the screw and nut are separated by a series of recirculating bearing balls.

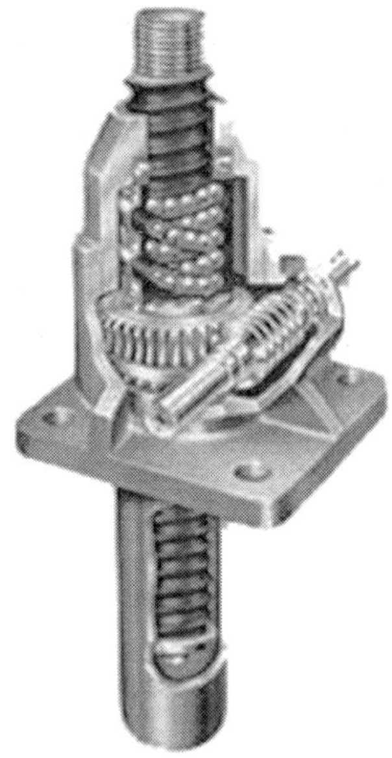

Figure 10-27: Typical ball-screw jack

Unlike a machine screw jack, the rolling action of the bearings in a ball screw exhibits low friction and generates little heat. This permits higher lifting speeds and higher efficiency, typically above 90 percent. This higher efficiency reduces the input power requirements to about two-thirds of what is required for a machine screw jack. Ball screw jacks also have a lower ratio of starting to running torque. And, because of low friction, many ball screw jacks can be back-driven.

Roller Screw Jacks

A variation of the ball screw concept, roller screw jacks also exhibit higher efficiency than machine screw jacks. Roller screws use threaded rollers, rather than balls, to serve as the intermediate connecting member between the lead screw and nut in a roller screw. Figure 10-28 shows a planetary roller screw in which the screws remain in constant contact with the threaded shaft. In another version, roller screws circulate into and out of the load zone.

Though roller screws first appeared in the 1950s, their popularity increased significantly in the last decade or so. Roller screws are more expensive than ball screws and are used mostly in applications where load capacity, axial stiffness, speed, or acceleration and deceleration rates are especially stringent. They are generally similar to ball screws in preload configuration, backlash, lost motion, left-hand and right-hand thread, back driving, efficiency, torque, and power requirements.

Figure 10-28: Typical planetary roller screw. In this design, roller screws maintain contact with the threaded shaft. In a recirculating type, roller screws circulate into and out of the load zone.

Screw Jack Selection and Application

To select a screw jack, use a system design manual that has illustrations, column strength charts, power and torque data, and sample calculations. Before selecting a jack, determine:

- Number of lifting points.
- Total load per jack.
- Load direction (tension or compression).
- Speed at which load must move.
- Distance load must move.

Also, consider the duty cycle and environmental conditions such as temperature and contaminants. Next, calculate the input torque, speed, and power requirements. Then, check column strength and select the jack. Selection of the best arrangement for an application is usually based on space availability and motor accessibility. Figure 10-29 depicts a typical lifting arrangement, sometimes called a T-system, in which one motor powers several synchronized lifting points by using couplings and right-angle gearboxes.

Here are some screw jack application guidelines:

- Keep the load direction parallel to the screw axis as much as possible.
- Keep the span between drive components as short as practical. That keeps connecting shafts short and limits the chance of a critical speed problem.
- When necessary, use pillow-block supports, dynamically balanced shafting, or both, to help avoid critical speed vibrations.
- Select shaft couplings with high strength-to-bore ratios (e.g., gear couplings) to minimize system inertia.

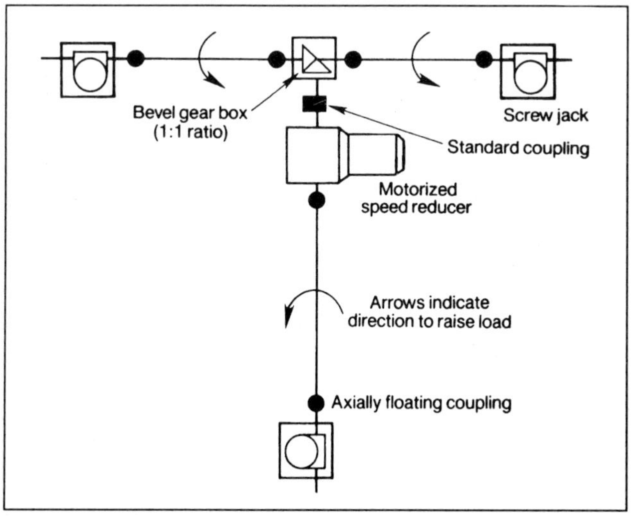

Figure 10-29: Screw jack lifting arrangement

- Use limit switches to restrict extremes of stem travel.
- Use a slip coupling between motor and jack input shaft to protect against overload.
- Add torque-sensing devices at a jack or motor to protect the system from overload.

Linear Motion Control Systems

Most screw-driven linear motion systems accept NEMA 23, 34, or metric 60 mm frame motors. The motor can be a full-step, half-step, or micro-step motor; or a DC-motor-based analog or digital servo system, depending on resolution, speed, and torque requirements. Positioning applications may require constant monitoring of motor and system position. Controllers generally provide circuitry needed for a position sensor, such as an optical encoder, to detect position. Servo systems usually include an integral brushless resolver for feedback.

Much of the control technology applicable to rotary motion systems also applies to linear systems. For example, programmable controls and computers simplify motion control and permit simultaneously controlled motion along several axes. A case in point is a multi-axis positioning system that includes both rotary and linear motion. Some sensing devices, such as linear variable differential transformers, optical linear encoders, and force transducers are especially suited for linear motion applications. Others, such as proximity switches, may work equally well for rotary or linear systems.

In a closed-loop system, an encoder generates pulses as it translates or rotates, indicating true position. Hence, encoder position rather than motor pulse count determines table position. The position controller determines any error between the desired and actual positions. The controller then commands the motor to adjust speed to eliminate the error. See Chapters 11, 12, and 13 for more on rotary motors, drives, controls, and sensors.

Motors

Motors are an essential part of a linear motion system. The motor provides the power as rotary motion, which is then converted to linear motion by a belt, screw, or other component discussed earlier in this chapter. Motors will be discussed at length in Chapter 11: Motors, so we will offer only a brief overview here, as follows:

DC and Can Stack

DC motors and can stack are generic terms for smaller sized, motors, typically DC. These motors are most often used in high-volume OEM applications, as they offer the best price/performance point. The downside to using these types of motors is the drives are typically not available as prepackaged units ready to integrate with PLCs and motion controllers. Most OEMs that use this type of motor build the control logic directly into printed circuit boards (PCBs) within the control of the machine. The two most common control methods are either to have limit switches at each end of travel or to mount an encoder on the back side of the motor. When sensors are used, the motor is controlled by turning it on, and it runs until it reaches the stop sensor.

DC Stepper (a.k.a. stepper, step motor)

Steppers or step motors are one of the most popular motor choices for linear applications. These motors are especially popular because of their low cost and prepackaged drive systems. These motors integrate easily with almost any PLC or motion controller. Most step motor drives are fed a pulse and direction signal from the control unit. Most drives are able to count the steps (fraction of a revolution) the motor makes, which can serve as an approximate positioning system. Step motors are available in both open-loop and closed-loop (servo) configurations. Most step motors are limited to 3,000 rpm. One of the biggest drawbacks to step motors is the significant decrease in the torque output of the motor as the rotations per minute (rpm) increase.

AC Brushless Servo (a.k.a. servo)

The term "servo" is often misunderstood. Servo essentially means "closed loop"; however, it is often misapplied to mean an AC brushless motor with an encoder. Servo motors are significantly more expensive than step motors, yet they remain popular for more demanding applications because they yield no significant drop in torque output as rotations per minute increases. In addition to more torque, servo motors run quieter and use less energy. Some companies are now spending more money up front for a servo motor because it will save electricity costs over the life of the machine.

Rotary Encoders

Rotary encoders are installed on the back side of a motor to provide feedback to the control unit and/or drive. This allows for a "closed-loop" system. Rotary encoders are the most popular type of encoder because they are readily available as preinstalled options by most major motor manufacturers. Encoders will be covered in greater detail in other chapters.

Linear Encoders

The best way to monitor and control position with a high degree of accuracy is to mount a linear encoder to a linear table. This configuration allows the encoder to directly read true table position, not shaft angular position. The latest innovations in optical encoders are interferential systems. For applications demanding very high accuracy and resolution, laser interferometer measuring systems offer resolution of 0.01 μm. Common applications are inspection machines, extreme precision machine tools, and wafer-slicing machines for semiconductor manufacturing. These devices require special consideration of the laser's environment, such as air temperature, pressure, and cleanliness.

Controls

Some step motors come with a simple control system prepackaged within the motor. This allows some simple machines to be controlled without a more expensive PLC or motion controller. When additional I/O functionality or complex movements are needed, users can upgrade to a PLC or dedicated motion controller. From a motion system standpoint, PLCs and motion controllers are similar devices, as both will control the motion of a system. Chapter 13: Controls & Sensors contains more detail about these types of systems; however, we will impart some quick advice first. Controlling motion is very demanding on the processor within the control unit; therefore, it is not a good idea to attempt to control a large number of axes, complex motion paths, or continuous motion from a general duty PLC. Though costlier, it is much more efficient to install a dedicated motion controller to handle the motion within a machine then to have a PLC act as the upper control system to manage I/O and send higher-level commands to the motion controller.

Key Terms

Accuracy: The difference in position between where a system actually is and where the controller thinks it is.

Acme Screw: A type of lead screw featuring a 29 deg. thread angle.

Ball Screw: A precision mechanical component that converts rotational motion to linear motion either by rotating a shaft to drive a nut or by rotating a nut to drive a shaft. Ball screws utilize tiny balls to reduce friction.

Backlash: Axial play between two mating components (e.g., screw and nut, teeth in a gear train, linear guide and carriage, etc.). Also known as "play" or "slop."

Bi-Directional Repeatability: The difference in position when a system returns to a location from either direction. Typically, this value is larger (worse) than single-direction repeatability.

Duty Cycle: Percentage of time a motion system is performing work. This is defined per cycle of motion, not per hour, work shift, day, etc.

EOAT: End of Arm Tooling.

Fatigue Life: The number of cycles of stress that a metal component can sustain prior to failure under a stated test condition.

Lead Screw: Also known as a power screw, the lead screw translates turning motion into linear motion.

Linear Bearing: Mechanical component that bears loads while ensuring stable and accurate linear motion.

Linear Motor: An electric motor that produces linear motion directly, making the mechanical conversion of rotary to linear motion unnecessary.

MTBF: Mean Time Between Failures is a prediction of the reliability of a product.
Payload: The load carried by the system.

Precision: See Repeatability.

Preload: A load used to eliminate backlash and increase rigidity by creating elastic compressive deformations in the balls and ball grooves.

Repeatability: The difference in position when a system returns to a location under the same circumstances (i.e., same direction and motion profile).

Resolution: Smallest possible movement of a system. Also called a "step" or "step size."

Rodless Mechanical Actuators: Linear actuators whose load is attached to a fully supported carriage. Rodless mechanical actuators provide linear motion via a motor driven ball screw, acme screw, or belt drive assembly.

Stroke: Maximum amount of travel from one end to the other.

Summary

Linear motion systems exist in a wide range of applications, from aerospace and automotive, to machine tools and industrial automation. The major components of linear motion systems fall into one of three categories: actuators, linear bearings, and control systems.

Actuators are used to initiate linear motion along an axis. The actuators discussed in this chapter include ball screws, lead/Acme screws, screw jacks, linear slides/stages/tables, electromechanical acutators, rodless cylinders, rodless mechanical actuators, short-stroke actuators, and magnetostrictive actuators. Ball and lead screws are precision-mechanical components that convert rotational motion to linear motion with highly accurate positioning. Screw jacks are one of the oldest actuation devices and consist of an input rotating shaft, an output linear motion shaft (screw), support bearings, and lubricant, all enclosed in a housing. Electromechanical actuators equipped with either a screw or belt drive provide precise linear positioning, transfer, and guidance for a variety of diverse applications. Actuator selection will depend on loads, speeds, accuracy, duty cycle, required life, efficiency, machine configuration, and installation limitations.

Linear bearings provide the support for linear motion systems. Plain linear bearings, linear rolling bushings, and linear rolling guides are common types of linear bearings. Plain linear bearings and bushings are sliding-contact bearings, while linear ball bearings and linear (ball or roller) guides are rolling-element bearings. Linear bearings come in a variety of shapes and sizes. Selection will depend on the loads, speeds, accuracy, duty cycle, required life, rigidity, machine configuration, and installation limitations.

Linear motors produce linear motion directly, making the mechanical conversion of rotary to linear motion unnecessary. This eliminates the use of mechanical devices such as screws, belts, and gears, which in turn reduces space, energy, and cost. They also incorporate their own support systems, eliminating the need for separate linear bearings.

Control systems provide constant monitoring of motor and system position. Controllers generally provide circuitry needed for a position sensor (such as an optical encoder) to detect position. Much of the control technology applicable to rotary motion systems applies to linear systems. For example, programmable controls and computers simplify motion control and permit simultaneously controlled motion along several axes.

A growing tendency among suppliers is to offer complete linear-motion systems as well as linear-motion components. Some systems include the actuator, its bearings, framework, drive and control, and sensors for control and safety limits.

Chapter 11

Motors

Chapter Objectives

When you have finished Chapter 11, "Motors," you should be able to do the following:

1. Describe important factors that are common to all motor types.
2. Identify the parts and features of AC motors.
3. Describe the features of polyphase motors.
4. Describe the features of single-phase induction motors.
5. Describe the features of brush-type and brushless DC motors and servomotors.
6. Describe the features of and applications for step motors.
7. List important criteria for motor selection.
8. Identify associations that specify motor standards.

Introduction

All electric motors operate on the principle that a current-carrying conductor produces an electromagnetic field that may be attracted or repelled by another electromagnetic field. Although a wide variety of motor types are available, they differ only in the methods used to create the magnetic field and control the current. While electric motors all depend on this basic principle, there are a number of performance and physical variables that need to be understood when choosing the right motor for an application. The following will discuss some of these variables, including NEMA performance standards. Replacement motors and new motor applications will first need to dimensionally fit into the required envelope of the equipment. Next, the requirements of starting toque, speed, power supply, and duty cycle will need to be considered. Selection of the best fit motor is much less difficult if the intended use is known and if these fundamentals are applied to your motor selection decision.

Common Factors

The electric motor is the workhorse of the power transmission industry. It transforms power received from electric companies into rotary motion for use in homes and industry. All motors share certain characteristics, and these must be identified when specifying them.

There are two basic classes of electric motors: those that are operated on alternating current—referred to as AC—and those that operate on direct current, or DC. AC types include single-phase and three-phase, which are also called polyphase. DC types include permanent magnet, shunt wound, series wound, and compound wound. Other designs include servo, step motors, switched reluctance, and synchronous reluctance.

Unit Systems

Ratings for motors used in North America are based on the foot, pound, and second, often called British Engineering Units. Motors built or sold for use in North America will most often use NEMA (National Electrical Manufacturers Association) standards. These standards will cover both mechanical as well as electrical performance and safety standards. Virtually all the rest of the world uses a version of the metric system called SI, or International Standard, known as IEC standards. British Engineering units use inches and horsepower. The International Standard uses the millimeter and kilowatt.

Input and Output Ratings

AC motors typically operate directly on AC power lines; therefore, they are rated in voltage, such as 115, 208, 230, 460, and 575 VAC, as well as in frequency. Frequency is expressed in hertz (Hz), which is the number of times in one second an alternating current voltage switches from positive to negative.

Motors for North and Central America, Mexico, Caribbean countries, the Philippines, and South Korea generally operate on 60 Hz power systems. South America, Japan, and Saudi Arabia use both 50 and 60-Hz systems. When specifying motors for these countries, be sure to check the frequency and voltage. Many countries use 50 Hz.

The output of general-purpose North American motors is rated in horsepower (HP) and in revolutions per minute (rpm). The rated output of motors using the International Standards is in kilowatts, watts, or Newton Meters.

Frames

Many of the physical size and operating characteristics of motors manufactured in the United States are established by NEMA. These standards enable motor users to order and apply motors with a high degree of understanding about a motor's performance and physical size. This enables interchangeability among motors from different manufacturers.

Frame sizes are often defined by numbers and letters, for example, 145TC. Two-digit frame sizes (48, 56) indicate the shaft center height in 16ths of an inch. For three-digit frame sizes, the first two digits indicate the shaft center height in fourths of an inch. The third digit refers to the mounting holes on the motor base. The letters indicate special features (see Table 11-1).

For special applications—such as washers, dryers, and other appliances—non-NEMA frame motors are often custom designs produced in large quantities and do not have to be interchangeable with motors produced by other manufacturers. Subfractional horsepower gear motors also do not meet NEMA frame sizes.

Table 11-1: Common NEMA frame suffix letters for AC motors

Suffix	Explanation
T............	"Long" shaft, suitable for belted and direct connected loads when fitted with a ball bearing at the drive end. Suitable for belted service only when fitted with a roller bearing at the drive end.
TS.........	"Short" shaft, suitable for direct connection
C...........	C-face motor
TC.........	C-face motor with T, or long shaft
JM, JP...	Close-coupled pump motors
U...........	Frame size-horsepower relationships predating 1963 and existing North American standards

Motor Heating

During the process of changing electrical energy into mechanical energy, heat is produced. A motor that is properly sized for its application will be capable of dissipating this heat. The user can verify that the motor will not overheat from being overworked by measuring the current draw under full load, which should be equal to or less than the current rating shown on the motor's nameplate.

Other factors that can cause a motor to overheat are high or low supply voltage, slow or repeated starts, high ambient temperatures, blocked ventilation holes, and voltage imbalances on polyphase AC motors.

Types of Loads

Loads can be classified as constant-torque or variable-torque. A constant-torque load produces a demand for horsepower that is in direct proportion to the speed of the load. For example, a conveyor that is moving at 100 fpm and requiring 5 HP will require 10 HP if the speed is increased to 200 fpm. Other constant-torque loads are hoists and positive displacement pumps.

A variable-torque load requires an increase in horsepower in greater proportion than the increase in speed. For example, if a centrifugal fan speed is doubled from 600 rpm to 1,200 rpm, the horsepower required might triple from 5 HP to 15 HP. Other examples are centrifugal pumps, blowers, and compressors.

Constant-horsepower loads are found on machines whose torque requirements decrease as their speed increases. Winding applications and metal working applications are good examples.

Motor Insulation

The windings of electric motors are coated with insulation to prevent shorting to adjacent conductors or the motor frame. The determining factor of how much heat a motor can withstand is based on the temperature limits of this insulation. Motor nameplates carry a letter designation that identifies the maximum allowable rise above 40° C (104° F) that a motor can withstand. As shown in Table 11-2, a motor wound with Class B insulation and operating at 1.0 service factor may safely run with a surface temperature of 120° C (240° F).

TABLE 11-2 — Allowable Insulation temperature rise above 40 deg C

Ambient temperature

Insulation class ⁰ °C	Allowable Rise ⁰ C 1.0 S.F. Motor	Allowable Rise ⁰ C 1.15 S.F. Motor
A 105	60	70
** E 120	75	—
B 130	80	90
F 155	105	115
H 180	125	—

Notes

* IEEE Standard 117
** Used in European equipment

1. "Hot spot" allowance is 10 ⁰ C more than above listed temperatures at 1.0 S.F. for class B&F insulation systems
2. These temperature-rise values are based on the resistance method for measuring coil temperatures.

Service Factor

Frequently abbreviated as "SF" on the motor nameplate, service factor defines the amount of overload a motor can continuously deliver without overheating. For example, a motor with a 1.0 SF is designed to deliver its rated load continuously. A motor with a 1.15 SF is able to deliver 11 percent of rated load without overheating. This extra capacity enables the motor to deliver full load, yet be protected against damaging overloads. This is not so with a 1.0 SF motor, which is more difficult to protect while delivering near the rated output power. Service factors above 1.0 can only be used if other conditions are met. These conditions include operation of the motor within the voltage, frequency, and ambient temperature shown on the motor's nameplate. Also, the service factor should not be used in altitudes greater than 3,300 feet above sea level.

Speed-Torque Curves

One of the most important application tools used to match a motor's performance to its load is the speed-torque (S-T) curve (see Fig. 11-1). An S-T curve defines the maximum available torque a motor can deliver at any given speed. It indicates a motor's short-term torque capability, including the torque available for starting and accelerating the load and the continuous torque capability at rated voltage and frequency.

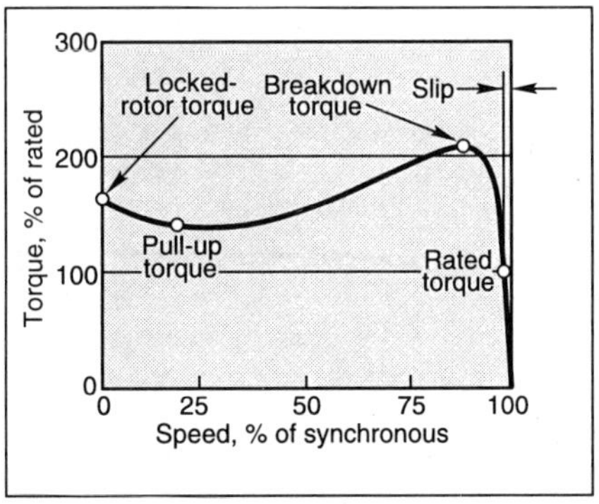

Figure 11-1: Speed-torque curve of a typical induction AC motor.

Enclosures

Induction motors are available in various types of enclosures, each providing a different level of mechanical protection and cooling. Most types are guarded to prevent personnel from touching a moving or electrically live part.

DP enclosures

DP, or open, drip proof enclosures, which are most frequently referred to as an ODP motor (see Fig. 11-2), are suitable for clean, dry, indoor industrial environments. Ventilation holes are constructed so that water droplets or solid particles falling at 15 degrees or less from vertical will not enter the motor.

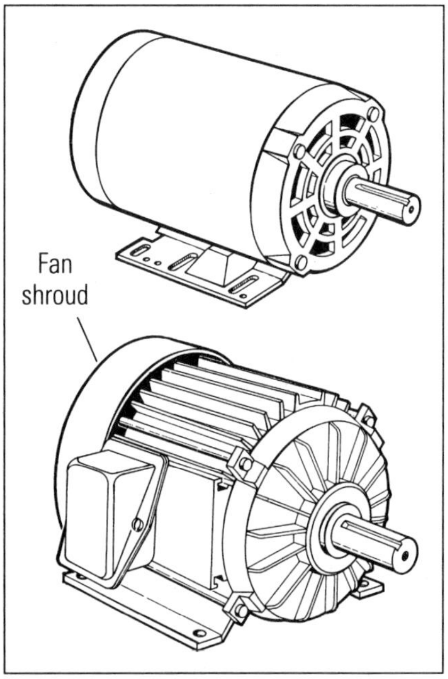

Figure 11-2: Typical dripproof (DP), top, and totally enclosed, fan-cooled (TEFC), bottom, enclosures. The shroud of the TEFC enclosure houses an external fan, which is mounted on the motor shaft. This fan blows cooling air over the fins on the outside of the motor. See Fig. 11-4.

Weather-Protected Machines

Type WP I is an enhanced version of the DP enclosure. Its ventilating passages are constructed to minimize the entrance of rain, snow, and air-borne particles into the electric parts. Its ventilated openings are constructed to prevent the passage of a cylindrical rod 0.75 inch in diameter from entering the interior.

Type WP II has, in addition to the enclosure defined for a weather-protected type I machine, its ventilation passages at both intake and discharge arranged so that high-velocity air and air-borne particles blown into the machine by storms or high winds can be discharged without entering the internal ventilating passages that lead to the electric parts of the machine. The flow-path of ventilating air passing through to the electric parts of the machine is manufactured to provide at least three abrupt changes in direction, none of which to be less than 90 degrees. Velocity of air flow is also restricted to a maximum of 600 feet per minute. This minimizes the possibility of moisture or dirt being carried into the electric parts of the machine.

DPBV enclosures

Motors subject to overheating due to frequent starts, overloads, or slow speeds often need a separate motor-driven blower. These motors are classified as DPBV (drip proof, blower ventilated). This is typically used for DC motors but is also frequently required for AC motor adjustable-speed applications.

TENV enclosures

Wet, dirty, and outdoor applications call for totally enclosed, non-ventilated (TENV) motors. A TENV motor has no external fan. It dissipates all its heat through the frame and mounting surfaces.

TEAO enclosures

These are similar to TENV, except that airflow from an external source blows air across the motor to cool it. A typical application for this type of motor would be a fan mounted on the motor's output shaft.

TEFC enclosures

A totally enclosed motor has an external shaft-driven fan that blows cooling air over the outside of the motor frame. This type is called a TEFC (totally enclosed, fan cooled) enclosure and is the most frequently used frame construction in use today (see Fig. 11-2).

TEFV enclosures

Designed for motors operated over an extended speed range, the TEFV enclosure utilizes a constant-speed fan to force-ventilate the motor, regardless of motor speed.

Hazardous-location motors

Induction motors are available for operation in the presence of explosive liquids, gasses, and dusts. North American standards require special operating characteristics, enclosures, and test regimens for these motors, which are specially designed for the purpose. They are not simply modified standard motors in special enclosures.

It should not be assumed that a motor tested, listed, or recognized as meeting North American standards for hazardous locations will meet IEC standards or vice versa.

Optional Mounting

To conveniently mount motors to some types of driven loads (gear reducers, pumps, and the like), motors are offered with well-defined mounting configurations. The following are two of the most common.

C-face motors

C-face motors (Fig. 11-3) have a finished face built into one endbell (sometimes referred to as an endshield or bracket) for attaching the motor to reducers and pumps. Threaded bolt holes are provided to allow for attachment.

Figure 11-3: C-face, top, and D-flange, bottom, for mounting motor to driven loads. These often include speed reducers, pumps, and special machinery.

D-flange motors

The D-flange motor incorporates a mounting flange, which allows the user to bolt the motor directly onto the driven machinery. The bolt holes in this flange are clearance holes (not threaded) and have a larger bolt hole circle than on the C-face motor.

Modifications

Electric motors are available with a broad array of pre-engineered motor modifications. Mechanical options include special bearings, special shaft extensions, breathers (to enable small amounts of air to enter and leave the enclosure as it heats up and cools down), drains (so condensed water can drain out), seals, slingers, high-temperature insulation, winding-temperature detectors, vibration detectors, and numerous others.

Environmental modifications include tropicalization and a variety of special enclosures. Electrical modifications include special voltages or frequencies, special duty-cycle capabilities, higher-than-normal torques, and many others.

Motor Protection

There are a variety of ways to protect a motor from overheating. In fractional horsepower motors, a temperature-sensing device may be internal to the motor that disconnects power to the windings if the motor overheats. These devices are manual thermal overloads (MTO) or automatic thermal overloads (ATO). MTOs require a reset button on the thermal be pushed in order to start the motor. ATOs will automatically start the motor once the motor cools and the thermal resets itself. Care must be taken in applying the correct motor/thermal combination to the application. Motors with ATOs should not be used in applications where the operator using the equipment would be injured if the motor starts automatically once the thermal resets.

On large horsepower internal motors, overload devices operate as a function of heat as a result of the motor load. These overloads sense the motor heat and relay that information to a control device to turn off the motor. Common overload devices of this type are thermostats, thermistors, or resistance temperature detectors. These are mounted on or in the motors windings. When using thermostats, normally closed or normally open contact thermostats should be specified depending on the requirements of the control being used.

Mountings

The life of a motor may depend partly on the type of base selected. Keeping a motor in proper alignment and free of vibration can help reduce bearing wear. A motor may need a mounting base to provide a means for anchoring the motor in position, provide a means for tightening a belt drive, provide alignment of the motor and load, and reduce vibration on the motor or reduce motor vibration noise.

The most common mounting is an integral base that is attached to the motor. This can be a rigid base that is welded or bolted to the motor frame. Another type of base used in air handling is a resilient base, sometimes referred to as a cushion base or cradle mounting.

NEMA frame sizes were reduced in 1952 and again in 1964 due to better insulating materials used in motors. Because of this, separate bases have been developed to properly align the newer NEMA "T" frame motors when they replace the older motor frames. These conversion bases consist of a right-hand and left-hand piece that are each mounted to the NEMA "U" frame mounting holes. The base of the "T" frame motor then mounts to the conversion base. Using these bases compensate for the mounting differences between the two different frames and provides proper alignment for the "T" frame motor shaft.

Another type of separate base is an adjustable motor base. This type of base is used to simplify mounting and belt tension adjustment for applications having belt driven loads. The rigid base of the motor is bolted to the adjustable base at right angles to the load. The belt is then tensioned by turning an adjusting screw in the base.

Both the conversion base and adjustable base are purchased as separate items from the motor and ordered by the motor frame size needed.

Brakes

A brake is a mechanical or electrical device that causes the shaft of the motor to stop turning when power to the motor is turned off.

The mechanical brake is mounted to the motor and consists of discs between pressure plates. When the power is turned off, friction stops the shaft. Brake leads may be brought out separately (as in explosion-proof motors) or connected internally to the motor. This brake stops and holds the load. Typical applications for this type of motor are hoists or inclined conveyors.

The electrical brake, also known as a dynamic brake, activates the field winding of the motor with DC to stop the motor shaft. This type of brake stops the load but does not hold the load. The brake is normally not on the motor itself. Typical applications for this type of brake are for stopping large circular saw blades or grinding wheels.

AC Motors

The flow of electrons is electric current. Alternating current (AC) is what power companies transmit through the electric wires to our homes and factories.

As the name implies, AC travels in both directions through the wires. AC is used so that transformers may raise and lower the voltage (pressure). Electric generators produce low-voltage electricity. Transformers are used to raise the voltage for long-distance transmission. This prevents power losses due to resistance in the wires. The voltage is then lowered by transformers to make it usable by the consumer.

The power supplied at electrical outlets in our homes is 115 V or 230 V, single-phase. Single-phase means that only one voltage waveform is applied to the motor.

Three-phase 230 V, 460 V, 575 V, or higher voltages are an alternative service provided. Three wires supply voltage waveforms instead of one, and each supplies electricity at different times. Three-phase electricity is provided to industrial sites for heavy-duty equipment with three-phase motors. This system is more efficient and therefore more economical than single-phase power.

The important points are that single-phase and three-phase motors cannot be interchanged, and three-phase should be used if it is available.

The AC motor consists of two main parts: the rotor and the stator. As the name implies, the stator is the outer shell of the motor that remains stationary. It contains the windings, which transform the incoming electricity into a magnetic field. This magnetic field causes the rotor to become magnetized in the opposite polarity and repel and rotate. The stator may be wound with two or more sets of windings called poles. The number of poles dictates the rpm of the motor. Standard synchronous speeds available are 900; 1,200; 1,800; and 3,600 rpm. Under load, the induction motor would have speeds lower than the synchronous speed.

The rotor consists of aluminum or copper bars that run lengthwise to the motor and are connected at their ends. As the magnetic field generated by the stator crosses these bars, electric currents are induced into the bars, which create a magnetic field in the rotor. This inducing of current gives the induction motor its name. The increasing and decreasing of the fields in the stator caused by the application of AC power gives the effect that the outer field is continuously rotating. The magnetic fields in the rotor rotate in an attempt to keep up with this effect.

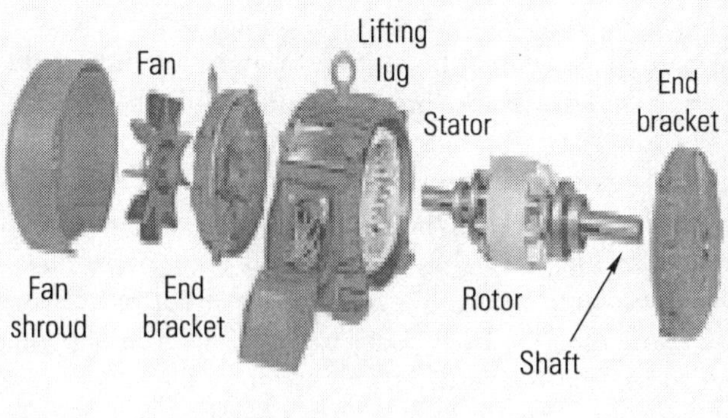

Figure 11-4: Typical 3-phase, AC induction motor. The protrusions on each end of the rotor act as fans for circulating internal cooling air.

Synchronous speed

The rotational speed of the stator's magnetic field is called the synchronous speed of the motor, a value determined by:

$$Ns = 120f/p$$

where:
Ns = Synchronous speed, rpm
f = Frequency of the applied power, Hz
p = Number of poles per phase

Table 11-3 lists the synchronous speeds for popular 60 Hz and 50 Hz motors.

The load determines where the motor operates on the motor's speed-torque curve (Fig. 11-1). As this curve shows, the higher the load, the slower the speed.

Table 11-3 — Synchronous speeds for motors operating on 50 and 60-Hz plant power supply

Number of poles	Synchronous speed, rpm	
	50 Hz	60 Hz
2	3000	3600
4	1500	1800
6	1000	1200
8	750	900

Slip

The difference between the synchronous speed and the actual rotor speed is termed slip. There must be some slip for an induction motor to operate. This slip is needed to induce current in the rotor. Most AC induction motors have 3 to 5 percent slip at full load. Motor data plates and manufacturer's catalogs will give rpm ratings of motors with the slip taken into account. The standard motor speeds given are 850; 1,140; 1,750; and 3,450 rpm. Slight variations occur among motor designs.

Critical torque levels

The speed-torque curve (Fig. 11-1) displays the four torque values that are critical for motor selection and application: locked-rotor torque, pull-up torque, breakdown torque, and rated or full-load torque.

Locked-rotor torque is that torque available at zero speed for accelerating the load. Pull-up torque is the minimum available during acceleration from rest to the speed at which breakdown torque occurs.

Breakdown torque is the torque produced by a motor just before it stops turning due to the application of a heavy or sudden load.

Breakdown torque is three to four times the full rated torque and occurs at speeds from 50 to 80 percent of synchronous speed.

Polyphase Motors

Because of its high efficiency and low cost, the three-phase AC induction motor is most commonly found in industrial applications.

These motors are available in ratings from fractional horsepower to thousands of horsepower. The following explains how manufacturers apply different motor designs to work best with various load parameters.

The following are the major factors encountered in selecting and applying these workhorses of industry.

Design types

By changing the design of the rotor, the shape of the speed-torque curve can be altered to produce different characteristics needed for different applications.

North American standards recognize four common designs of induction motors: Design A, Design B, Design C, and Design D. These are defined similarly by both NEMA (National Electrical Manufacturers Association) and CEEMA, the Canadian equivalent. Fig. 11-5 displays speed-torque curves for all four, and Table 11-4 summarizes their characteristics for a 10 HP, 1,800 rpm, four-pole motor.

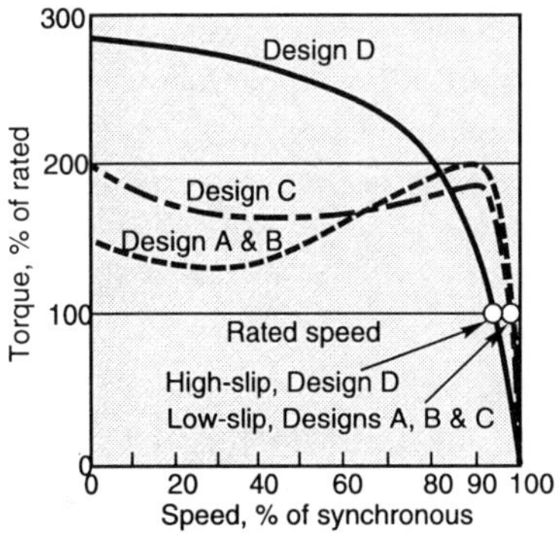

Figure 11-5: Speed-torque curves of typical Design A, B, C, and D, 3-phase AC motors.

Table 11-4 — Three phase induction motor design characteristics for a 10 HP 1800 rpm 4 pole motor

Design	Starting torque	Breakdown torque	Slip percent
	As a percentages of full load torque		
A	165	200	3 to 5
B	165	200	3 to 5
C	250	190	3 to 5
D	275	275	8 to 13

The starting torques and breakdown torques vary in value depending on horsepower and speed (rpm). For example, a NEMA Design A or Design B motor at 1 HP at 1,800 rpm has a starting torque of 275 percent and a breakdown torque of 300 percent of full-load torque, whereas a 200 HP motor at 1,800 rpm has a starting torque of 100 percent and a breakdown torque of 200 percent of full-load torque.

Designs A, B, and C tend to have similar horsepower-frame relationships. Design D motors are larger and more costly.

Design A motors have a speed-torque curve similar to the Design B motors, and the main difference is that the Design A motors carry a higher locked-rotor current with a higher breakdown torque than that of the Design B motor.

Design B motors, the most widely used, are typically applied to fans, blowers, centrifugal pumps and compressors, and other easy-to-start loads. Design C motors are used with loaded conveyors, crushers, mixers, agitators, reciprocating compressors, positive displacement pumps, and similar hard-to-start loads. Some manufacturers also build Design B motors with Design C starting torques. Design D motors are installed on punch presses, shears, hoists, oil well pumps, and other high-peak-load machines. Design D motors have the most slip of any motor design at full load.

Multispeed motors

Three-phase AC induction motors are also available for operation at two or more speeds. Motors of this type are single voltage only. These motors contain stator windings that can be interconnected to provide a differing number of poles. So-called single-winding, multispeed motors provide two rated speeds, where the lower speed is generally one-half the higher speed.

Two-winding multispeed motors, on the other hand, can operate at speed ratios other than 2:1. Such motors are also available with three and four rated speeds.

Two-winding multispeed motors should not be confused with reconnectable motors. The latter can be reconnected for operation from, for example, either 230 or 460 VAC. Multispeed motors, on the other hand, are reconnectable (by using multispeed motor starters) to provide two different speeds when operating from the same voltage and frequency.

Ratings

Design A, B, and C motors are available in ratings from less than 1/4 HP to over 5,000 HP. Design D motors are generally available in the lower ratings, but costs increase rapidly for ratings over 150 HP.

Induction motors built to North American standards are typically rated 230, 460, or 575 VAC. three-phase 60 hertz for use on power systems with nominal voltages of 240, 480, and 600 VAC, respectively. Normal voltage drop accounts for the difference between motor and power system voltages.

Most manufacturers offer reconnectable motors that can be arranged for operation on either 230 or 460 VAC. (They are marked 230/460 V to indicate dual voltage.) Motors rated up to 500 or 750 HP typically operate on these low-voltage systems. Those rated above this level typically are powered by medium-voltage systems, such as 2,300 or 4,000 Vac.

Adjustable–Speed AC Drives

To meet the need for adjustable-speed operation of AC motors, the motor controller, commonly called an inverter, was developed.

The inverter adjusts the input frequency to a three-phase AC motor to control its speed. It also reduces the applied voltage by a proportional amount to prevent the motor from overheating. Inverter-duty motors are now being manufactured to provide increased reliability under the demands of reduced-speed operation. Care should be taken when using these controls with a brake motor that have internally connected brakes. Because the control adjusts the voltage as well as frequency to the motor, there may not be enough voltage to energize the brake coil to release the brake when the motor is turned on. Brake motors used with inverters should have the brake leads connected separately to the power source.

Single-Phase Induction Motors

The single-phase motor operates under the same basic principles as the polyphase motor. However, the rotating magnetic field effect generated by the stator does not exist until running rpm has been reached. Because there is no starting torque available, manufacturers have to include design mechanisms to cause the motor to start. There are various designations given to motors that use these different techniques, including shaded-pole, split-phase, capacitor-start, and permanent split-capacitor.

Shaded-pole motors

A copper loop called a shading ring surrounds a portion of each stator pole in a shaded-pole motor. This configuration causes a shift in position of the applied magnetic field in relationship to the rotor, creating starting torque. These low-cost motors are used in fans, small appliances, and other applications that accept low starting and breakdown torque.

Split-phase motors

In light-duty applications, the split-phase motor is a commonly used single-phase type. The split-phase motor has two sets of stator windings. During start up, a set of windings called the "start" windings are energized in addition to the "run" windings. The start windings are positioned 90 degrees to the run winding and shift the magnetic field of the stator by about 25 degrees, which induces starting torque to the rotor. At full rpm, the start windings are a detriment to normal operation; therefore, a switch is built into the motor to disconnect them at about 3/4 of synchronous speed. Split-phase motors are less expensive than equivalent capacitor start motors, but draw more current and are less efficient. They are suitable for small-horsepower, easy to start applications, such as fan loads.

Capacitor-start motors

The most common single-phase motor used in industrial applications is the capacitor-start motor. Like the split-motor, it has both start and run windings. However the capacitor start motor includes a capacitor in series with the start windings. The addition of the capacitor causes a shift of the stator's magnetic field to about 82 degrees. This larger magnetic field shift during start-up increases start-up torque to 234 percent of split-phase motors. A common variation of this motor is the capacitor start capacitor run motor.

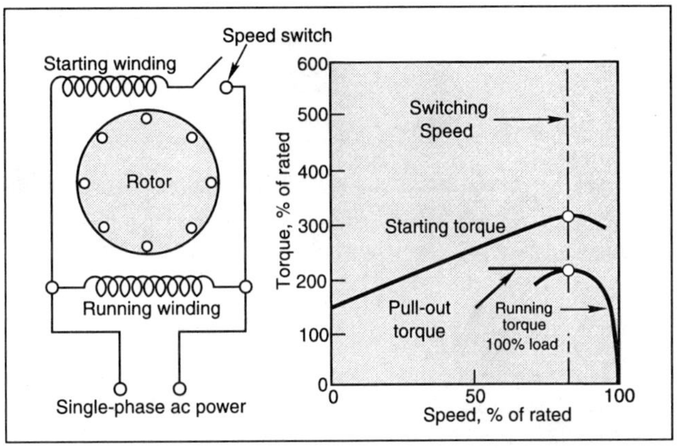

Figure 11-6: Schematic and speed-torque curve for a split-phase motor.

This motor uses two capacitors to provide magnetic field shift (phase shift) in both the start and run mode to produce added torque. A speed-sensitive switch, either mechanical or solid state, is set in both motors to disconnect the start windings after motor acceleration.

Permanent split-capacitor motors

The permanent split-capacitor motor uses identical main and auxiliary windings in combination with a capacitor to provide starting torque. Because both windings are used continuously, no start switch is required. These motors have little starting torque and are used in variable-torque applications such as fans and pumps in the HVAC (heating, ventilation, and air conditioning) and refrigeration industries. They are manufactured in sizes less than 1 HP.

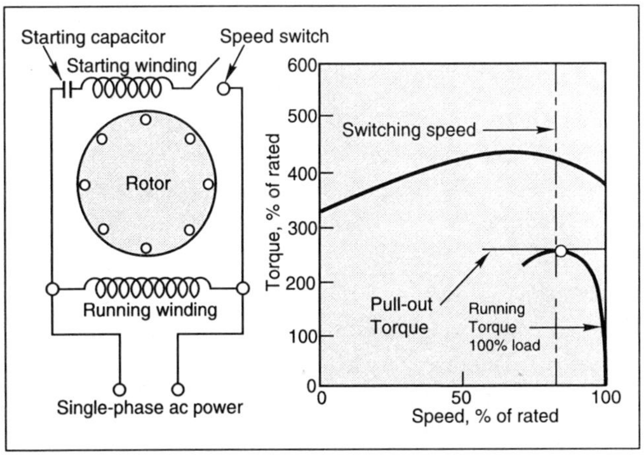

Figure 11-7: Schematic and speed-torque curve for a capacitor-start motor. Note the higher starting torque of this motor over the split-phase motor. The starting capacitor is often mounted on top of the motor in a long, cylindrical "can."

DC Motors

DC motors, as the name implies, use direct current instead of AC current. Direct current does not change direction and always moves in the same direction. The direction of the flow is determined by the polarity of the source. A battery, for example, consists of two plates, an anode (+) and a cathode (-), separated by an electrolyte. The electrolyte produces a chemical reaction, causing electrons to move from the anode to the cathode inside the battery. If a DC motor were connected to the battery terminals, electrons (current) would then flow through and operate the motor to balance inequality within the battery.

There are many sources of DC power; however, in industrial applications, DC is provided by changing plentiful and inexpensive AC power into DC power through controllers that use electronic circuits called silicon-controlled full-wave rectifiers (SCRs).

Although more complex and costly and requiring more maintenance than corresponding AC induction motors, DC motors can operate at adjustable speed when connected to relatively simple controllers. There are two basic types of DC motors—brush-type and brushless.

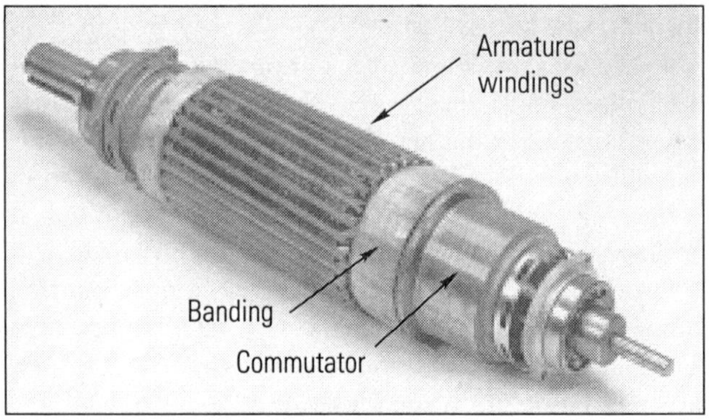

Figure 11-8: Wound armature of dc motor. Carbon brushes ride on commutator to conduct current to armature windings. Bandings hold windings in place during high-speed operation.

Brush-Type DC Motors

There are two types of DC brush-type motors: permanent magnet (PM) and wound field. Brush-type DC motors should be applied to run at the rated horsepower/torque of the motor. Running the motors either underloaded or overloaded shortens the normal expected brush life.

Permanent magnet

DC permanent magnet motors use permanent magnets as the stator. The magnetic field that they create is fixed and constant. The rotating portion of a DC motor is called the armature. The armature is made up of steel laminations with lengthwise slots that hold copper windings. These windings are attached to a segmented copper cylinder called the commutator. Electricity is fed into the armature through carbon brushes that ride on the commutator. As the armature rotates, the commutator acts as a switching device to route the current to the winding that will create a magnetic field, which will cause the armature to rotate within the fixed field of the permanent magnets.

Permanent magnet motors are typically available from 1/4 to 5 HP. Controllers will vary the speed of these motors by varying the voltage supplied to the armature from either 0 to 90 V or 0 to 180 V, depending on the motor design.

Figure 11-9: Stator of DC motor. Field poles contain both shunt and series windings. The interpoles are series windings that reduce sparking between brushes and commutator for longer brush and commutator life.

Wound field

The magnetic field for brush-type wound-field DC motors is produced by wire-wound poles (called field windings) on the stator (see Fig. 11-8). Passing current through these windings produces the magnetic field. The armature is similar to a PM-motor armature.

There are three distinct types of wound-field DC motors: series-wound, shunt-wound, and compound or stabilized shunt-wound. These terms in each case define the technique employed for producing the magnetic field.

Series-wound

The field winding connects in series with the armature winding (see Fig. 11-10). The arrangement provides high starting torque, but the motor speed varies widely with a change of load. This causes series DC motors to overspeed when unloaded. This type motor is ideal for metal rolling and traction drives (streetcars and electrified trains).

Shunt-wound

The motor field winding is generally connected in shunt, or parallel, with the armature (see Fig. 11-10). In some cases, however, the two operate from different power sourc-

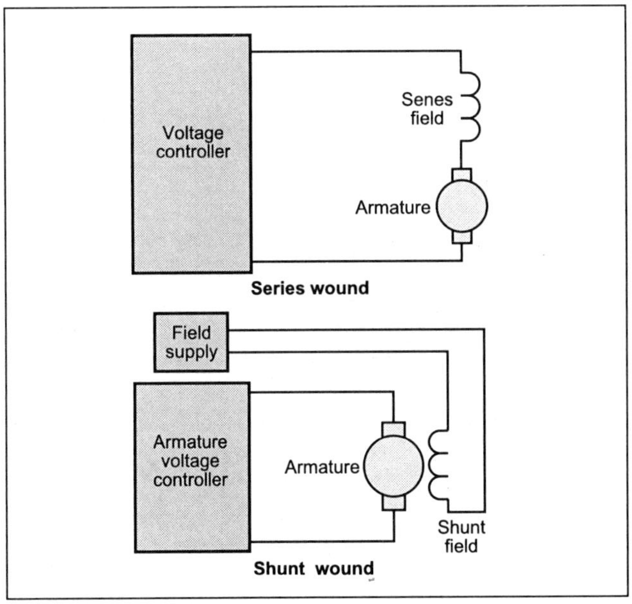

Figure 11-10: DC motors include series windings, shunt windings, or both. Motors with both are compound wound motors.

es. In either case, motor speed remains relatively constant throughout its normal load range. Shunt motors do not overspeed when lightly loaded, but they may run away (overspeed) upon loss of shunt field power.

Speed is usually adjusted by independently varying voltage applied to the armature. Sophisticated control systems vary both armature and field voltages.

Compound and stabilized shunt-wound

Motors having both shunt- and series-wound windings typically exhibit a speed characteristic that varies little with load, yet still have the ability to increase torque output when heavily loaded. These motors are in applications such as elevators and cranes because they can start heavily loaded but not run away when unloaded.

Rating factors

Brush-type DC motors manufactured to North American standards are available in the same horsepower ratings as AC motors. Base (rated) speeds are similar to rated speeds for 60 Hz induction motors. Armature and field voltage ratings coincide with voltages produced by half-wave and full-wave rectification of standard 60-Hz voltages in the drive controller (see Chapter 12).

DC motors are classified as subfractional, fractional, and integral horsepower ratings. As power ratings increase, field and armature voltage ratings increase as well. Table 11-5 lists the typical ratings available from a broad range of suppliers.

Table 11-5 — Typical power and voltage ratings for brush-type DC motors

		Vdc	
Categeory	**Hp range**	**Armature**	**Field**
Subfractional HP	1/50 to 1/8	90 & 180	PM
Fractional HP	1/4 to 1	90 & 180	PM, 50, 90, 100
Integral HP	1 to 700+	180, 240, 500	PM, 150, 240, 300

Note: PM = Permanent magnetic fields.
Others are wound-field motors.

Brushless DC Motors

Like brush-type, permanent magnet motors, brushless DC motors employ permanent magnets. But unlike brush-type motors, the magnets are mounted on the rotor, and the windings are in the stator.

Moreover, this type motor has neither commutator nor brushes. A solid-state controller electronically provides the commutation function in response to rotor-position signals. Brushless DC-motors have the capability to deliver rated torque at higher speeds than corresponding conventional permanent-magnet motors.

Ratings

General-purpose brushless DC motors are widely available with ratings up to about 20 HP at 1,800 rpm.

Enclosures

DC motors are supplied in different enclosure types, each providing a different level of cooling and mechanical protection. These enclosures include DPG (drip proof, guarded), DPFG (drip proof fully guarded), DPBV (drip proof, blower ventilated), TENV (totally enclosed, nonventilated), and TEFC (totally enclosed, fan cooled).

Options

Performance options include tachometers for speed feedback, blowers, additional thermal capacity, special brush grades, and heavy-duty bearings. Protection options include winding and bearing thermostats or RTDs (resistance temperature detectors), space heaters, breathers, and drains.

Servomotors

A special class of motor, a servomotor has low inertias and high response, which produces the capability to accelerate from zero to full speed in milliseconds. They must be used with a dedicated servo control. They typically exhibit higher maximum speeds and faster response than conventional motors. Servomotors are widely used for rapid-reversing and precision-positioning applications, such as machine tools, packaging machines, inspection equipment, and pick-and-place applications.

Servomotors and controllers are usually offered as a package. Two types of servomotors are popular—DC brush-type and brushless. (In addition, AC induction motors—with low-inertia rotors—powered by flux-vector controllers function as servo drives. These systems are described in Chapter 12.)

Brush-Type Servomotors

Most brush-type servomotors have permanent-magnet fields. Except for having higher slot fill, lower armature inertia, and faster response, it has the same basic construction as the general-purpose DC motors previously discussed.

Brushless Servomotors

The brushless motors in some cases are called DC and other times AC, but the two are identical in construction. Details of the basic brushless motor construction are given in the previous section on DC motors.

As previously discussed, brushless servomotors (frequently called BLDC for brushless DC) contain permanent magnets on the armature (or rotor) with windings mounted on the stator. To meet the precise needs of servo operation, an encoder, resolver, or other feedback device is mounted on the motor shaft. This feedback device sends a signal that indicates precise motor speed and position (for electronic commutation) back to the servo controller. Fig. 11-11 shows a typical speed-torque curve.

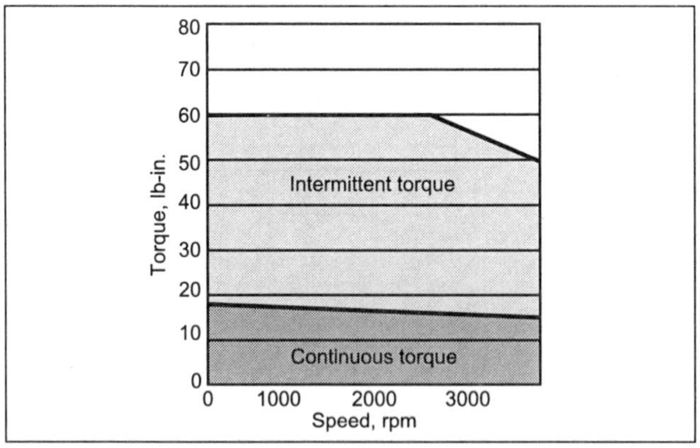

Figure 11-11: Speed torque curve of typical brushless (BLDC) servomotor.

Rating Factors

North American servomotors are typically rated in output torque (lb-in. or oz-in.). In both cases, speed ratings may be in either radians per second (Rad/s) or revolutions per second (rps).

The input voltage to the motor must be compatible with the output of the controller.

Frames and Enclosures

Servomotors are typically supplied in TENV (totally enclosed, nonventilated) enclosures. The DC permanent magnet servos are typically round, and the brushless servos are typically square. Diameters of North American DC servo motors are based on inches and have mountings such as 23, 34, and 42, whereas diameters of brushless servo motors are based on either metric units or inches. As is typically required for precision applications, the motors include shaft-mounted feedback devices, typically an encoder or resolver. These are discussed in Chapter 13, Controls and Sensors.

Step Motors

Step motors are widely used industrially and commercially in typewriters, printers, disk drives, and a variety of other applications that do not require the high speed and high torque offered by servomotors. Unlike conventional motors, step motors convert electrical pulses into discrete angular motion. (A discussion of linear step motors is included in Chapter 10, Linear Motion, although the same basics apply to both rotary and linear motors.)

The angular rotation produced by an electrical pulse is called the step angle. The most widely used step motors rotate 1.8 deg. per pulse or 200 steps per revolution. Step angles available from many manufacturers range from 7.2 to 0.18 deg. There are three basic types of step motors: permanent magnet, variable reluctance, and hybrid.

Permanent Magnet Step Motors

The stator of a permanent magnet step motor contains two or more windings, and the rotor supports multiple permanent magnets. When pulses are applied sequentially to the windings, the magnets rotate step-by-step, matching up for each step with the polarity of the pulsed winding. Permanent magnet step motors are simple and inexpensive, but they exhibit resonance characteristics at certain speeds and excessive temperature rise at high speed. These limitations can be minimized with suitable closed-loop control techniques.

Variable-Reluctance Step Motors

The stator of a variable-reluctance step motor resembles that of a permanent-magnet step motor, but the rotor consists of salient steel poles rather than magnets. Here, when the windings are pulsed, the rotor rotates step-by-step, seeking a position of minimum reluctance. The torque-to-inertia ratio of a variable reluctance motor is high, but the torque produced is less than that for comparable permanent magnet and hybrid types.

Hybrid Step Motors

Combining the best characteristics of both permanent-magnet and variable-reluctance step motors, hybrid step motors are popular for high-torque applications. To show the basic concepts of how these motors operate, Fig. 11-12 displays a schematic diagram for a two-phase hybrid step motor. Here, the rotor contains two sets of three salient poles—one set offset from the other by 60 deg. A permanent magnet in the shaft magnetizes one set as north poles and the other as south poles. Pulsing the phases alternately rotates the shaft at 1.8 deg. per pulse.

Half-Step and Microstep Control

With suitable control techniques, a motor can be made to rotate a half step or an even smaller portion of a step for each pulse. For example, by energizing one phase on the motor in Fig. 11-12, then both phases, then the first phase again but with reversed polarity, and so on, the motor rotates 0.9 deg, or a half step, with each pulse. Even smaller steps result when so-called microstep techniques are employed. Here, for each step, one phase is weakened only slightly and the other phase strengthened correspondingly. With a typical motor, one with 200 full steps per revolution, half-step and microstep control techniques can increase the number of steps from 400 to 100,000 steps per revolution.

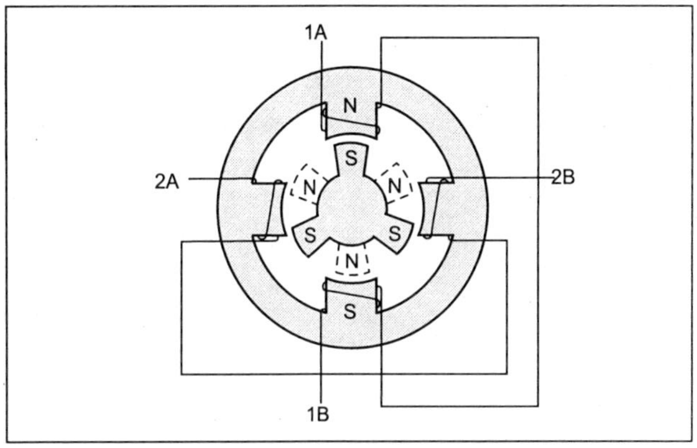

Figure 11-12: Schematic diagram of two-phase hybrid step motor.

Speed-Torque Characteristics

Step motors exhibit different types of speed-torque characteristics than other types of motors. Fig. 11-13 is an example of one for a typical hybrid step motor. The curves here define values for holding torque and for two speed ranges—the start-stop range and the slew range.

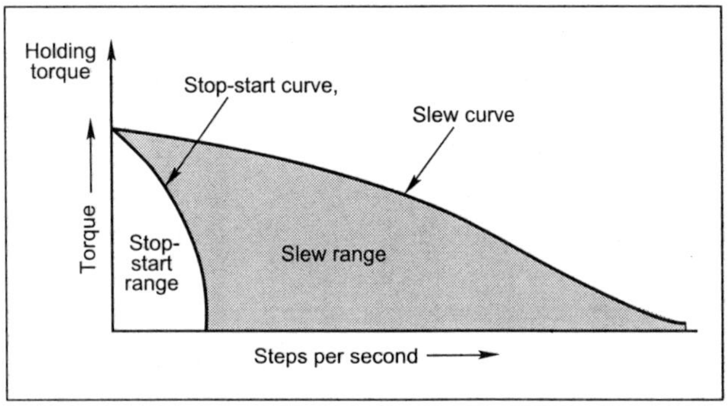

Figure 11-13: Typical hybrid step-motor, speed-torque curve.

Holding torque is the maximum torque a motor can deliver with rated excitation at zero speed without shaft rotation. The start-stop range is a torque vs. pulse-rate area where the motor can be stopped within a single step. The slew range is an area where the motor precisely responds to a pulse rate. A motor loses steps if pulse rates exceed those indicated by the slew curve. A motor must be slowed from the slew area to the start-stop area before it can be stopped without losing steps.

Detent torque, not depicted on a speed-torque curve, is the maximum torque a motor delivers when de-energized. Only permanent-magnet and hybrid motors exhibit detent torque.

Bipolar Coils

Most stepping motors have two sets of windings per pole, a so-called bipolar arrangement. The two sets can be connected either in series or parallel. Connecting the windings in series provides higher output torque at lower pulse rates than the parallel connection but lower pull-out torque values at high pulse rates. Pull-out torque values are those defined by the slew curve in Fig. 11-13. Those values are the maximum torque the motor can produce at the corresponding pulse rates without pulling out of synchronism and losing steps.

Resonance

A step motor is a spring-mass system that oscillates when pulsed at its natural frequency. At this pulse rate, a step motor bounces or rings with every pulse. When ringing is severe, the motor may pull out of synchronism and stall. Half-step and microstep control techniques avoid or minimize resonance problems.

Ratings, Frames, and Enclosures

North American step motors are typically rated in output torque (lb-in. or oz-in.). Speed ratings may be in either radians per second (Rad/s) or revolutions per second (rps).

Industrial step motors are available with torque ratings from 14 to more than 3,000 oz-in. Some torque ratings approach those for integral-horsepower motors. North American step motors use frame sizes 17, 23, 34, and 42, generally in TENV enclosures. Step motors compete favorably with servomotors in many low- to moderate-torque applications.

Application and Selection

Selecting a motor for a specific application calls for consideration of numerous parameters. Standard specifications such as horsepower, rpm, voltage, enclosure, and frame are easily determined. However, factors such as starting torque, load characteristics (continuous or intermittent), duty cycle (number of starts per hour), ambient temperature, altitude (motor cooling is less efficient in thin air), and environment (clean, dry, wet, explosive gasses, etc.) must be considered to assure proper motor performance and motor life.

Servomotor and step motor selection calls for a detailed analysis of the driven load and application objectives.

Induction Motors

Locked-rotor, pull-up, and breakdown torque values must be sufficient to start, accelerate, and run the expected load.

If any form of reduced-voltage or reduced-current starter is used, the effect of the starting technique on the torque ratings must be taken into account. For example, reducing the starting current to one-half normal cuts the starting torque to one-fourth normal.

In many industrial applications, operating cost, and thus motor efficiency, are also important factors. All major motor manufacturers offer high-efficiency Design A, B, or C motors that meet or exceed regulated efficiency levels. The U.S. electric motor regulation is based on the Energy Independence and Security Act (EISA). The Act was implemented December 19, 2010 and covers two distinct types of polyphase motors. It includes motors built to NEMA as well as IEC standards regardless of country of origin. Motors imported as part of a piece of equipment are also required to meet the EISA efficiency requirements. The two type of motors covered are general purpose, referred to as type one and seven additional categories referred to as type two motors. Motor type one must meet minimum efficiencies per NEMA table 12-12 (referred to as NEMA Premium) while type two motors are required to meet minimum efficiencies of NEMA table 12-11 (referred to as Epact92 levels). Motors sold in the United States are required to include the NEMA nominal efficiency (NNE) on each nameplate. They are also required to have a manufacturer's compliance number on the name plate. The compliance number (often referred to as the CC number) is a unique identifier issued to the motor manufacturer by the U.S. Department of Energy as recognition of meeting the requirements of U.S. federal energy code (CFR10 part 431). In addition, NEMA has trademarked "NEMA Premium," and licensees may also include this mark on the name plates of products that comply fully with all regulatory requirements.

Canadian and Mexican motor efficiency requirements follow the basic requirements of the U.S. regulations.

Motor starting equipment protects a motor from running overloads and limits the damage caused by short circuits in the windings. In addition, the high cost of large motors often justifies use of winding and bearing-damage detection devices, which further decreases the likelihood of motor damage due to motor or power-supply malfunction.

After a motor is selected, the locked-rotor current of the motor—typically six or more times full-load current—must be determined before suitable starting equipment can be specified.

Adjustable-Frequency Drives

Selecting induction motors for use with adjustable-frequency drives calls for additional considerations. Here, the motors are powered by inverters that convert power from a constant-frequency system to one where frequency is controllable (see Chapter 12). Although this technique permits adjustable-speed operation of induction motors, the non-sinusoidal current waveform generates more motor heating than do constant-frequency applications, which operate from pure sine-wave current.

Design B induction motors meeting EPACT should have a smaller speed range when powered by adjustable frequency controls. Many motor suppliers publish special application rules for their motors when so operated. The motor manufacturer should also be informed where an induction motor may be run over or underrated speed (operated at a frequency higher or less than 60 Hz).

Motors intended for inverter duty often include separately powered blower-ventilation systems to maintain constant cooling-air flow, regardless of motor speed. The arrangement minimizes a need to derate the motor. In addition, AC motors intended for flux-vector adjustable-frequency drives generally call for an auxiliary shaft-position or speed-feedback device.

Design C motors using double squirrel cage rotor construction are not suitable for inverter applications.

However, many motors achieve Design C torque levels without using double squirrel cage rotors. Single-cage Design C motors can be used with inverters.

International Standards

North American motor standards (compiled and published by NEMA and CEEMA) are consensus standards and are generally accepted within the United States, Canada, and other countries using 60 Hz power.

However, for companies exporting machines and equipment outside the United States and Canada, motor specifiers must be aware of standards, such as IEC.

IEC standards are accepted by treaty and have the force of law within the Europe Economic Community (typically Western Europe). IEC standards are also accepted by consensus (not binding by law) in areas outside the EEC where 50 Hz power is used. This includes sections of Eastern Europe, Africa, and portions of Asia.

IEC motor standards differ from North American standards in numerous ways. This includes:

- IEC standards use the SI system of units. (See Unit systems.)
- IEC standards cover motors for both 50 and 60 Hz.
- Standard IEC motors exhibit speed-torque characteristics similar but not identical to NEMA Design A and B motors; there are no IEC equivalents to Design C and D motors.
- Standard IEC motors often exhibit higher values of locked-rotor current than North American equivalent Design B, but are similar to Design A.
- IEC standards define nine types of duty cycles. The IEC definition of continuous duty (Type S1) essentially matches that for North American motors.
- Shaft extension, C-face, and D-flange dimensions for IEC motors differ from North American motors.
- IEC frames carry designations from 56 to 355L. The numbers represent the approximate distance from the floor to the shaft centerline in millimeters.
- For flameproof and other hazardous-location motors, the EEC countries have developed stringent standards, EN 50 014 and EN50 018. It should not be assumed that a motor tested, listed, or recognized as meeting North American standards for hazardous locations will meet IEC standards or vice versa.

Chapter 12

Adjustable-Speed Drives

Chapter Objectives

When you have finished Chapter 12, "Adjustable-Speed Drives," you should be able to do the following:

1. Define an adjustable-speed drive.
2. List three reasons adjustable-speed drives are used.
3. List eight specific questions to ask and answer when evaluating the possible benefits of adjustable-speed drives.
4. Describe four adjustable-speed drive applications.
5. Differentiate between open-loop and closed-loop systems.
6. Explain the features and applications of belt and chain drives.
7. Differentiate thyristor from transistor characteristics.
8. Define 16 terms and functions unique to adjustable-speed drives.
9. Describe the features and applications of traction drives.
10. Describe the features, rating, and application factors for eddy-current, wound-rotor, and DC adjustable-speed drives.
11. Describe the features, types, ratings, and application factors for AC adjustable-speed drives.
12. Explain four categories of application criteria, regardless of drive technology.
13. Differentiate between a servo drive, its application, and a standard drive system.
14. Define and describe five terms that explain how a drive performs.
15. Summarize some of the major performance capabilities of common drive types.

Introduction

Drives are widely employed in a variety of industrial operations and carry ratings from a fraction of one horsepower to thousands of horsepower. Some are based on mechanical techniques, and others are based on electrical. The mechanical drives covered in this chapter include both belted and traction types. The electrical drives include eddy-current, wound-rotor induction motor, DC, AC, servo, variable-speed, and engineered drives. Hydraulic drives are covered in Chapter 9.

Why Use Adjustable-Speed Drives?

Adjustable-speed drives are used for one of three reasons. The user needs or wants to vary the speed of a motor-driven shaft. They can also be used to save money through energy efficiency and reduced maintenance costs. Finally, a user or OEM may be able to gain a competitive advantage by using an adjustable-speed drive.

Generally, adjustable-speed drives are users' top choice because of their simplicity, cost, high efficiency, and low maintenance. Adjustable-speed drives are used when:

1. It is more convenient than changing belts.
2. A slow initial startup speed may be required.
3. Several finite-operating speeds may be required.
4. Shaft speed may be required to stay constant despite a changing load.
5. It may be necessary to overcome high inertia on startup.
6. Torque must be held constant even though there is a change in shaft speed, such as a wire-winding operation.
7. Two shafts must be synchronized to run at the same speed.
8. Costs are decreased by reducing operating and maintenance costs.
9. Product quality will be improved with its use.

Typical Applications

The following four applications typify the range of adjustable- and variable-speed uses encountered in general industrial and commercial installations. They also introduce many of the issues users and power transmission specialists encounter when selecting a drive technology and defining drive specifications and requirements.

Application A is a simple belted conveyor (see Fig. 12-1). Typically, a conveyor is a constant-torque load driven through a speed-reduction device, which minimizes the effects of load inertia. Both mechanical and electrical adjustable-speed drives are widely used in conveyor applications.

Application B typifies a mixer or blender in a process that requires high torque levels at low speeds (see Fig. 12-2). As the blending process nears completion, the torque is often reduced so the speed can be increased.

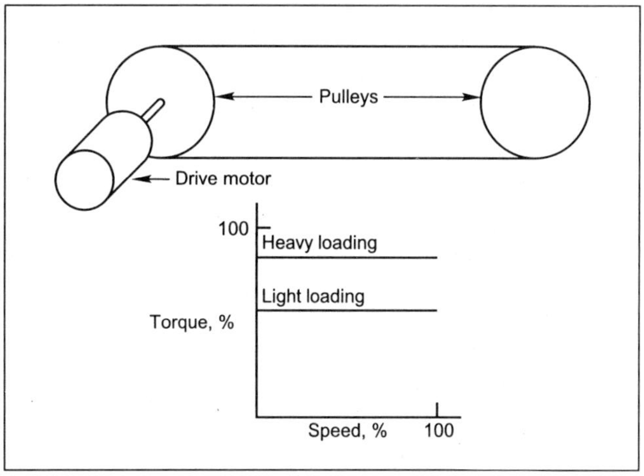

Figure 12-1: Belted conveyor.

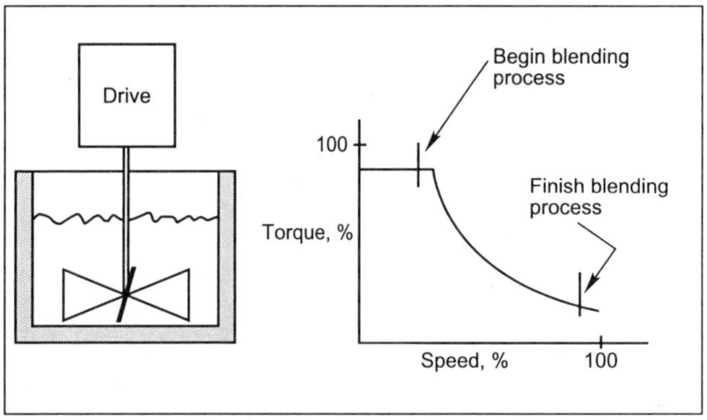

Figure 12-2: Mixer or blender.

In other processes, reactions increase torque requirements as the blending process continues. Either way, blenders and mixers are either constant-torque or constant-horsepower loads. For some, the control scheme is very sophisticated. The control may store recipes that specify and control the speeds and times for various compounds. When food industry standards apply, smooth motor enclosures and hose-down capability become part of the application requirements. NEMA-rated enclosures are available on most drives to meet environmental requirements. This application uses both mechanical and electrical drives.

Application C covers a raw sewage lift pump (see Fig. 12-3). These no-clog pumps are variable-torque loads. That is, as the speed increases, so does the required torque. Horsepower requirements for the pumps range from less than 100 HP to more than 1,000 HP. At ratings greater than approximately 500 to 750 HP, users generally prefer medium-voltage (2,300 or 4,000 VAC) drive motors. Usually operated as a closed-loop control, such drive systems include eddy-current, adjustable-frequency, and wound-rotor drives—with and without slip-loss recovery control.

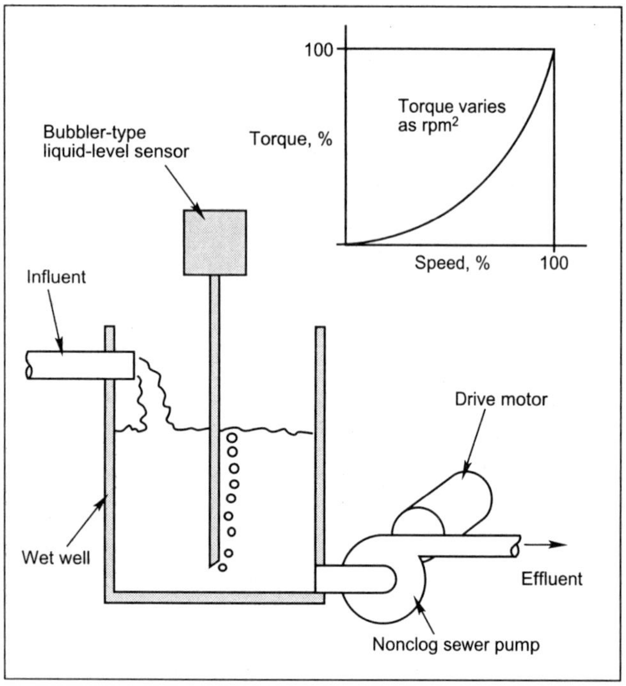

Figure 12-3: Waste water lift pump.

Application D shows a high-speed cutoff machine (see Fig. 12-4). In this high-performance application, a moving sheet or pipe is cut to precise length by a "flying" cutoff knife. The drive accelerates the knife to the speed of the product and positions the knife precisely. The external cutoff then performs a clamp-cut-release operation. Then, the drive brakes and reverses the knife back to the starting point, and the process is repeated. Process line speeds are often measured in thousands of feet per minute, and close-cutting tolerances are essential. This is a difficult and demanding drive application that requires servo drive response. In some applications, hydraulic drives compete with electrical servo drives to provide the best drive solution.

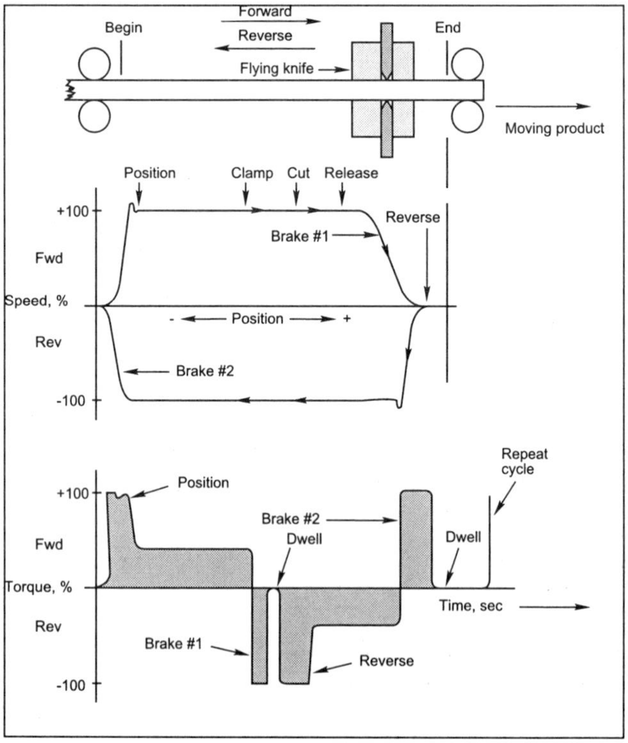

Figure 12-4: High-speed cutoff machine.

Mechanical Adjustable-Speed Drives

Two types of mechanical adjustable-speed drives are widely used: the belt drive and the traction drive. Both drives employ mechanical techniques to adjust the speed ratios between a drive shaft and a driven shaft.

Belt Drives

Open-belted mechanical drives

Fig. 12-5 shows two configurations of open-belted mechanical adjustable-speed drives. The unit on the left is a motor that is mounted on an adjustable motor bed and is turning a spring-loaded variable-ratio pulley. As the center distance between the driver and the driven pulley is increased, the belt is forced to the bottom of the driver pulley, increasing the drive ratio and slowing down the driven pulley. If the motor is moved closer to the driven pulley, the driver pulley forces the belt to the outer portion of the driver pulley. This decreases the ratio and speeds up the driven pulley.

The unit on the right employs a fixed motor with an adjustable pulley mounted on it. The width of the pulley is adjusted by turning a control knob on the pulley. As the pulley width is changed, the drive belt rides on the smaller or the larger diameter of the pulley. This changes the drive ratio, as with the previous unit. In this unit, however, the driven pulley is also spring-loaded to uphold the slack caused by adjusting the driven pulley.

Open-belted mechanical variable-speed drives range from 1/4 through about 15 HP. Compound drives are generally available through 5 HP. Maximum speed ratios of fixed diameter pulley units are about 3:1, whereas compound drives have maximum speed ratios of about 9:1. Either standard—single or three-phase induction motors—can be used.

Application factors

Data required for application analysis include a complete description of output load torque and speed requirements (Application Factors, p. 12–233). Open-belted drives are ordered as a compatible group of components, and the power transmission specialist is responsible for selecting the components to perform as an integrated drive system. Component selection for each drive type involves at least five steps:

1. Define the installation configuration,
2. Determine horsepower and rated speed of drive motor,
3. Determine driven shaft speed, and select output pulley,
4. Determine center distance(s) between drive and driven shafts, and
5. Determine the proper base or countershaft assembly.

Drive manufacturers offer a number of drive modifications, including remote speed adjust and extended chain-driven base hand wheels.

Enclosed belted mechanical drives

A cutaway view of an enclosed mechanical adjustable-speed drive (see Fig. 12-6) displays the three basic elements—the drive motor, a variable speed transmission, and the output section. The motor is a single- or three-phase squirrel cage, typically having a C-face to facilitate mounting and replacement.

The transmission contains a controllable top pulley. Here, rotating the leadscrew in the speed control mechanism moves the top pulley faces either closer together or farther apart. When the top pulley faces move closer together, a spring forces the lower pulley faces farther apart (the belt length remains constant), and the output shaft increases speed. The converse occurs when the speed control moves the top pulley faces apart.

The output section generally includes a number of reduction gear combinations. Both parallel and right-angle reducers are available. The belt displayed in Fig. 12-7 is re-

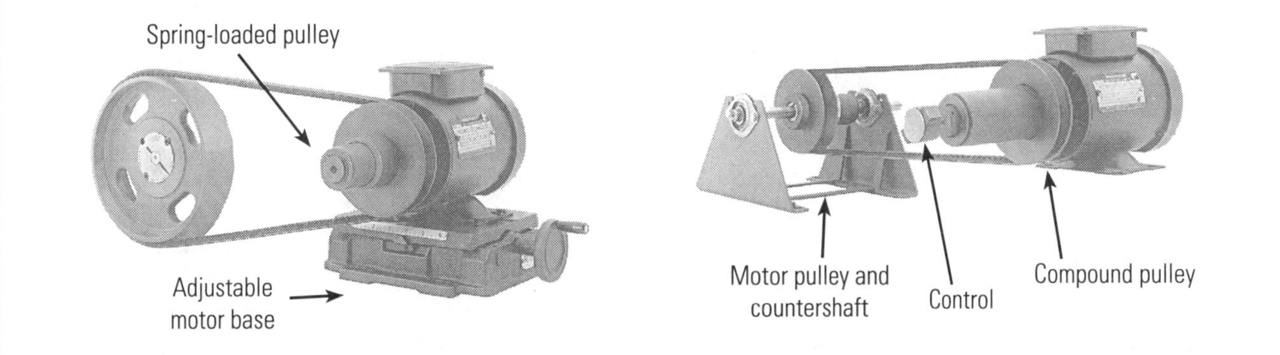

Figure 12-5: Open belted mechanical drives; adjustable- base unit (left), compound drive unit (right).

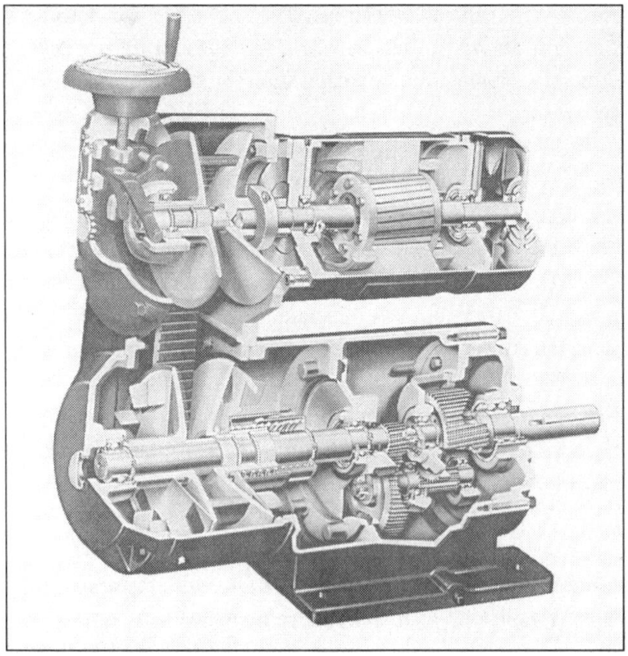

Figure 12-6: Cutaway — enclosed belted mechanical adjustable-speed drive.

inforced rubber, but wooden blocks on a composition belt, metal chains, and steel belts are also employed. Each belt type performs better than the others in certain applications. Enclosed drives are designed for the quick replacement of belts and other components subject to wear.

Drive ratings: Belted adjustable-speed drives are widely available in ratings from 1/4 through about 50 HP, with motor voltages of 200 through 690 VAC, three-phase, 60 Hz. The drives are available with speed ratios of up to 10:1 and higher. Rated output speed depends on the output section gearing. Typical reduction ratios with parallel gearing run from 1.4:1 to 38:1.

Options: Some belted adjustable-speed drives are available with remote electrical speed adjustment and speed indicators. Other versions incorporate tachometer feedback and electronic interface modules that respond to zero to 5 VDC, 1 to 5 mA, or 4 to 20 mA signals from PLCs and process computers. These systems control speed within ± 1 percent of rated values. Other options include pneumatic speed control for hazardous locations, wash down capability, and open or fully enclosed motors.

Application factors: Data required for application analysis include the description in Application Factors. Special attention is needed to define the peak loading magnitude and frequency, duty (continuous, intermittent, frequency of stops and starts), Wk2 of load, and reversal frequency. Belted adjustable-speed drives can withstand moderate shock loading, while traction drives are sensitive to shock loading. Mechanical arrangements and the overhung loading on built-in speed reducers are also part of the drive application requirements. Defining the environmental operating conditions, necessary protection, and selection of drive modifications completes the application analysis.

Traction Drives

Fig. 12-7 shows cutaway views for two types of traction drives. Both types are driven by standard AC induction motors. The bottom drive, a flat-disk type, employs a C-face motor. The top drive, a ball-and-cone type, calls for a footed motor and coupling, but versions using a C-face motor are available. Both drives depend on friction between a speed-adjusting mechanism and specially shaped input and output plates to achieve adjustable-speed.

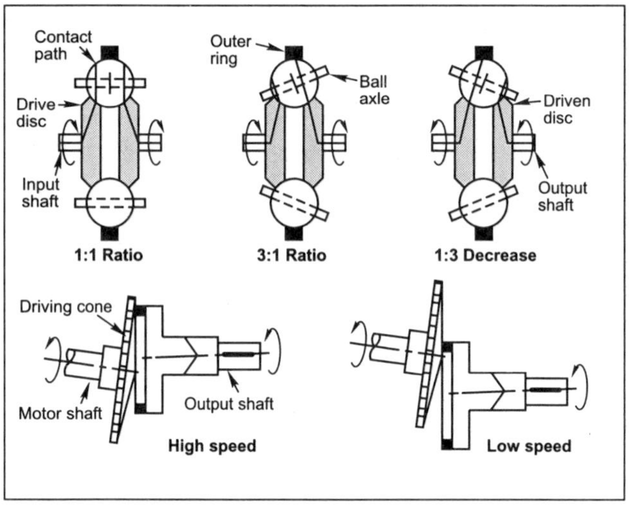

Figure 12-7: Flat disk (bottom) and ball-and-cone (top) traction drives.

Traction drives exhibit relatively high efficiency. They are best suited for applications with steady loading. Peak loading must be within a drive's continuous rating. All the wearing components can be replaced by competent mechanics.

Drive ratings

Flat-disk traction drives are available with ratings from 1/4 to 5 HP and higher with speed ranges of up to 10:1. Ball-and-cone traction drives are available with ratings from 1/3 to 15 HP with speed ranges up to 8:1. Both drives employ standard four-pole (1,800 rpm) induction motors. Speed reducers and remote control options are a few of the many modifications available with traction drives.

Application factors

Traction drives should not be subjected to shock loading. Constant-torque loads call for higher capacity drive transmissions than constant-horsepower loads do. A dough mixer is a typical constant-horsepower application, and a bottling line conveyor is a typical constant-torque application.

Data required for application analysis includes load torque and speed characteristics, peak loading magnitude, stop and start frequency, and load inertia. When the drives are integrated with speed reducers, mechanical configuration and overhung loading of the reducers must also be defined.

Power Semiconductors

Both AC and DC drives employ power semiconductors that fall into two families—the thyristor family and the transistor family.

Thyristors

Also called silicon-controlled rectifiers (SCRs), thyristors exhibit characteristics shown in Fig. 12-8. They begin to conduct at a point in the positive half cycle of an AC voltage wave when a gate pulse is applied. Conduction ceases when the voltage goes negative. Available in ratings of more than 1,000 volts and more than 1,000 amperes, thyristors are widely used in DC drives and were once the only practical power semiconductor for AC drives. Thyristors exhibit a characteristic whereby conduction, once started, continues as long as the applied voltage remains positive. This characteristic is undesirable in adjustable-frequency drives because it requires a dual set of power semiconductors—one to initiate conduction and another of equal rating to stop it. Therefore, AC drives with thyristors are often large, complex, and costly. Despite these drawbacks, thyristors are still widely used in a variety of AC and DC drives due to their ruggedness and high power handling capability.

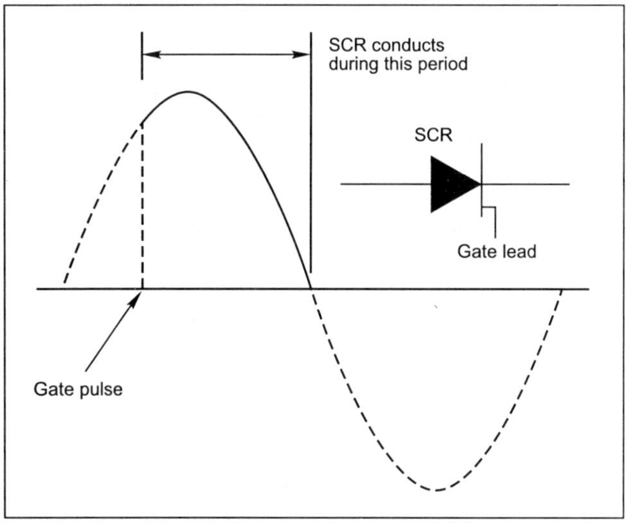

Figure 12-8: Characteristics of Thyristors.

The thyristor family includes gate turn-off devices (GTOs). With one exception, GTOs behave like thyristors, but they have an ability to stop forward conduction upon command. However, a high current pulse is required to stop this conduction. In some cases, the "stop" pulse is 30 percent of the GTO's forward current rating. Nevertheless, GTOs are used in high-power, low- and medium-voltage adjustable-frequency drives. Integrated Gate Commutated Thyristors (IGCTs) combine some of the desirable properties of thyristors and transistors and are also popular in higher-power and voltage-rated drives.

Transistors

Using a totally different approach, power transistors can start and stop forward conduction upon command and at a faster rate than comparable thyristors. However, transistors typically have more losses then thyristors, and as such, were once limited to drives rated up to 460 Vac and under 150 HP. But, the more recently developed Insulated Gate Bipolar Transistor (IGBT) is capable of meeting higher power demands and is currently employed in drives rated up to 2,000 HP and higher—and at motor voltage ratings of up to 690 Vac.

A disadvantage of power transistors and power thyristors is the maximum frequency at which they can be switched on and off—called chopping or carrier frequency. This frequency can vary from approximately 50 Hz to over 20 kHz. Much of this is in the audible range. This is not a problem in DC drives, but it may be in adjustable-frequency AC drives that use pulse-width modulation (PWM) tech-

Open or Closed-Loop

Adjustable-speed drives operate as either open-loop or closed-loop systems. In general, mechanical adjustable-speed drives are open-loop, but closed-loop control is available as an option with some mechanical drives.

Open-loop

Drive output speed is adjustable in an open-loop system. But, at any speed setting, changes in drive loading, input line voltage or frequency, ambient temperature, and time affect drive speed.

Closed-loop

A closed-loop system compares a speed reference input signal with a feedback signal representing actual drive output speed. When the two signals differ, the speed control equipment causes the drive motor to accelerate or decelerate to a speed that eliminates that difference. Other feedback signals, such as load compensation, are usually included to stabilize the drive. As a result, closed-loop drive output speeds are only slightly affected by changes in loading, voltage, frequency, temperature, and time.

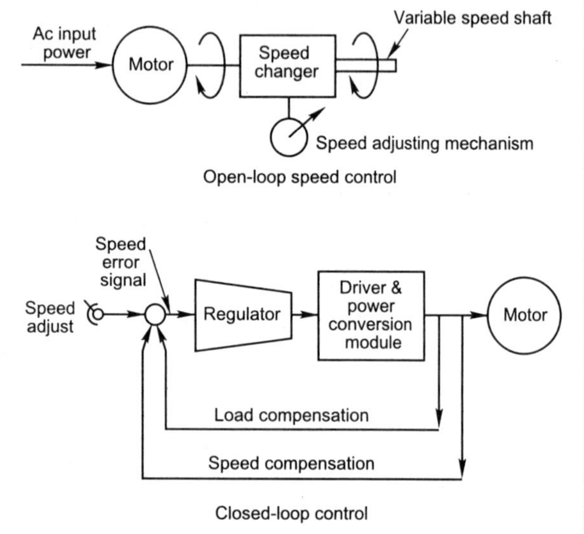

Open-loop speed control

Closed-loop control

Therefore, closed-loop systems have better speed regulation than open-loop systems due to load changes. Speed regulation due to load changes is defined by the following equation:

$$R = \frac{N\,(nl) - N\,(fl) \times 100}{N(fl)}$$

Where:
R = Speed regulation, percent
N (nl) = No load speed, rpm
N (fl) = Full load speed, rpm

Drive manufacturers generally specify speed regulation as a percent of the drive motor's base or rated speed. For example, a specific drive has a base speed of 1,750 rpm, a controlled speed range of 20:1 (87.5 rpm to 1,750 rpm), and a specific speed regulation of +0.5 percent of 1,750 (+8.75 rpm). Speed regulation at the minimum set speed of 87.5 rpm is +8.75 rpm, not +0.5 percent of 87.5 rpm, which is + 0.4 rpm.

niques. Consequently, PWM adjustable-frequency drives using any of these devices may tend to be noisy at certain carrier frequency settings.

Functions and Options

As with any technical product, adjustable-speed drives have unique terms and functions. Some of the more common functions include:

Automatic restart

Adjustable-speed drives generally remain stopped following a momentary power interruption or overload trip. Normally, restarting the drive calls for the operator to manually depress the Start or Run pushbutton (this safety feature prevents unexpected drive operation). However, drives can be arranged to restart automatically after a power interruption or overload trip.

Automatic tuning

Starting up a complex drive system for the first time can be a technologically challenging task. Many drive suppliers provide circuitry in their products that measure drive parameters and adjust stabilizing circuits accordingly. This functional arrangement simplifies the startup task. Even so, startup may call for a combination of experience, skill, and a thorough understanding of the user's driven machinery and processes.

Braking

Adjustable-speed drives often provide control for mechanical brakes added to the drive motor to hold or stop a load. In addition, both AC and DC drives often provide for either dynamic or regenerative braking as drive control options. However, dynamic braking, for example, aids in stopping a motor but provides no holding torque at zero speed.

A DC drive's dynamic braking uses a DC motor to act as a generator when driven by its load. When a stop is initiated, the dynamic braking circuit disconnects the motor's armature from the drive's output and reconnects it to a power resistor. Energy stored in the load inertia is dissipated in the resistor as heat. Braking torque diminishes with speed, dropping to zero at zero speed.

An AC induction motor's resistor-type dynamic braking uses the motor's ability to operate as an induction generator. When a stop command is issued, the motor remains connected to the power source and is driven by the load inertia. Load energy is dissipated as heat in power resistors connected to the DC bus, an action similar to dynamic braking a DC motor.

DC injection braking with AC drives requires that DC power be applied to one of the drive motor's phase windings after the motor is disconnected from its AC source. Here, braking torque is low at high speed, peaks at about 20 percent of rated speed, and drops to zero at standstill.

Regenerative braking (see four-quadrant operation) also employs a motor (either AC or DC) to act as a generator. Here, the system is reconfigured to pump load energy back into the input AC power line. This braking system provides braking or negative torques at any speed.

Controlled speed range

Typically, drive/motor combinations have a controlled speed range listed as a ratio such as 3:1, 10:1, 20:1, 1,000:1 or 2,000:1. This ratio defines the relationship between minimum and maximum controlled speeds that is usually based on the drive motor's rated or base speed. Within the controlled speed range, the drive is expected to deliver both rated (or a specified value of) torque and specified speed regulation.

The drive system must obviously have the capability to accelerate from zero speed to the minimum controlled speed. However, in a closed-loop drive, speed regulation outside the controlled speed range is not guaranteed.

Other drive functions listed in this box are often included or available as options, drive technology permitting.

Follower inputs

This function enables the drive to use an external speed reference input. Drive output speed is proportional to either a controlled voltage (e.g., 1 to 10 VDC) or a process controller's controlled current output signal (e.g., 1 to 5 mA, 4 to 20 mA, or 10 to 50 mA of direct current).

Four-quadrant operation

Adjustable-speed drives generally operate in Quadrant I of the four quadrants. Here, the drive motor runs in a forward direction, producing positive torque.

When a motor operating in Quadrant I is dynamically or regeneratively braked, operation transfers to Quadrant IV. The motor produces negative torque while still running in the forward, or positive, direction. Similarly, when a motor operating in the reverse direction (Quadrant III) is dynamically or regeneratively braked, operation moves to Quadrant II, where it generates positive torque.

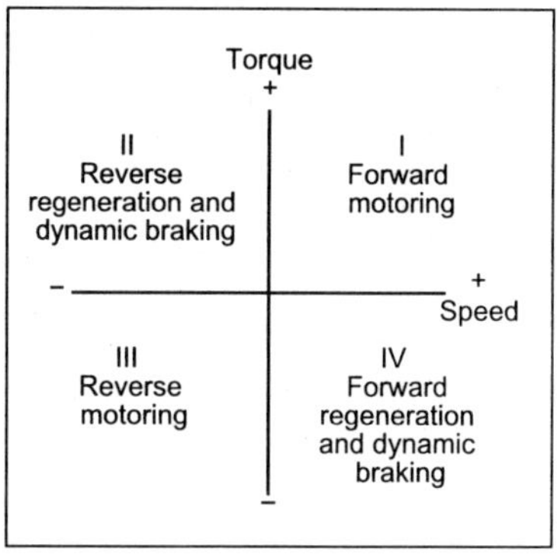

Figure 12-9: Four-quadrant operation.

Continuous, contactorless, four-quadrant operation calls for sophisticated power conversion equipment and control strategies. Typical AC drives operate in the regenerative quadrants by storing the energy as voltage on the DC bus capacitors. When the bus voltage gets too high on AC drives with a diode or thyristor bridge front end (not an active converter), the energy must be switched into dynamic braking resistors or the drive must be shut down. Thyristor-based DC drives typically require a second anti-parallel power bridge to operate in the regenerative quadrants and put the energy directly back onto the power grid.

Inching or threading

When provided, the inching or threading function permits an operator to run at slower-than-normal speed, as with the jogging function, but without the need to hold the Start pushbutton down.

Jogging

The jogging function permits a drive to operate at slower-than-normal speed for set-up operations. The operator may select the function with a Run/Jog selector switch. When the Start pushbottom is depressed, the drive runs at jog speed, and when the pushbottom is released, it stops.

When configured for jog at set speed, the drive jogs at the speed set by the primary speed-adjusting device. When configured for jog at preset speed, the drive jogs at a specific preset jogging speed that is different than the normal running speed. Some drives provide multiple operator-selectable preset jog speeds.

Network communications

Adjustable-speed drives are capable of network communications through a Local Area Network (LAN) or a Wide Area Network (WAN). A network connection may come standard in the drive or may require an optional module. Adjustable-speed drives can communicate with remote devices such as a drive-controller, operator interface, programmable logic controller (PLCs), or an industrial computer. A drive accepts a digital speed or torque command through the network communications. Other critical information may also be communicated, such as drive status, current demand, internal or external temperatures, or the status of protective circuits, to name a few. There are many open and closed industrial networks currently available.

Multi-motor drives

Some complex machines require two or more motors to operate. Their speeds must be coordinated with one other. For example, the speed of one or more motors is a function of the speed of a lead motor. Such drives are frequently used in machine tools that have coordinated spindle and feed drives. These coordinated drive systems involve complex stability and performance evaluation and are not covered in this section.

On/off/speed adjust

All adjustable-speed drives provide a means for starting, stopping, and adjusting the drive's output speed. Drives generally operate in the set-speed mode. Output speed varies as the speed adjustment device is turned.

With some mechanical drives, output speed cannot be adjusted unless the drive is running.

Preset speed

When a drive operates at preset speed, its output speed responds to a speed command defined prior to energizing the drive system. In some cases, a drive has multiple preset speeds—either selected by an operator or in response to some external control device.

Protection

Drive systems may include overload relays for motor protection and current-limiting fuses or circuit breakers for branch-circuit (short-circuit) protection. Because overload relays operate too slowly to protect solid-state equipment, current-limiting circuits protect internal components from excessive current during acceleration and deceleration.

In addition, thermostats provide some protection to the control equipment from excessive ambient temperatures or loss of cooling air. Some drive motors are equipped with over-temperature devices, which must be coordinated with the drive control circuitry.

Drives using wound field DC motors must include field loss protection to prevent destructive over speeds in the event of a field power supply malfunction. Adjustable-frequency drives often include skip frequency circuits to avoid operation at or near resonant speeds of the drive motor and connected load.

Other protective functions include elaborate status displays that use indicator lights, light-emitting diodes (LEDs), and liquid crystal displays (LCDs). Internal microprocessors monitor drive operations and provide data for the display of both normal and abnormal drive functions.

Reversing

A number of techniques are available for reversing a drive. In the simplest case, the operator selects "forward" or "reverse," and electrical power contactors configure drive motor power connections accordingly. Generally, a drive with contactor reversing must be brought to a stop before reversing its direction of rotation.

A plug reverse function permits the drive to be reversed while running in the opposite direction. When plug reversing is provided, contactors may or may not be involved. Circuitry may limit the rate of acceleration or deceleration, especially when the drive powers a high inertia load.

Timed acceleration

A timed-acceleration function permits an operator to preset the length of time a drive requires to accelerate or decelerate from one speed to another (see Fig. 12-10). With linear-timed acceleration, a drive operates as described by the upper curves. The lower curves show how a drive operates with S-curve control.

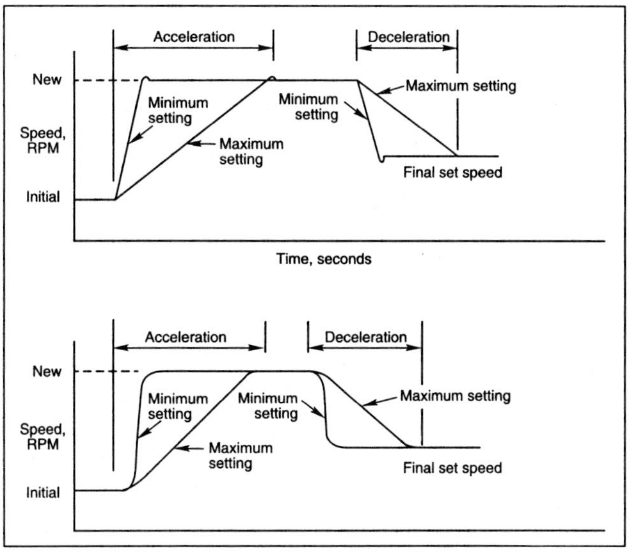

Figure 12-10: Time acceleration.

Neither of these functions provide for decelerating an overhauling load where load inertia forces a speed higher than set speed unless the drive is capable of operating in a regenerative quadrant.

Torque control

A drive can be converted from speed control to directly controlled motor torque through a torque control option.

Eddy-Current Drives

A simplified cross-section view and a schematic diagram (Fig. 12-11), displays the primary elements of an eddy-current drive. Here, an input cylinder surrounds, but does not touch, the output cylinder. A coil, powered by a DC exciter, generates a magnetic field that envelops both cylinders. When the input cylinder rotates faster than the output cylinder, eddy currents induced in one cylinder interact with the magnetic field, producing a driving torque in the output cylinder.

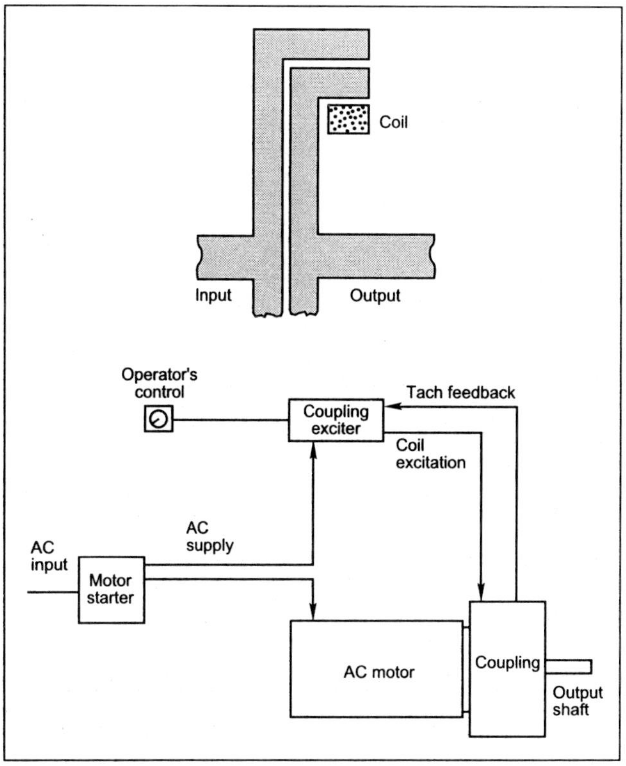

Figure 12-11: Eddy-current coupling.

The drives are closed-loop systems. Once a speed is set, the drive control equipment receives shaft speed data from a tachometer on the output shaft, and any difference between set and actual speeds produces an error signal, which in turn, increases or decreases the output of the exciter. Increasing the exciter's output increases the drive's output speed. Eddy-current drives are offered with excellent speed regulation, and they are electrically simpler than both DC and AC electrical adjustable-speed drives.

Drive Ratings

Eddy-current drives, usually integrated with appropriate AC squirrel cage induction motors, are available in a wide range of horsepower ratings from less than 1 HP to well over 1,000 HP. The drives typically use low-voltage (230 to 575 Vac, three-phase, 60 Hz) drive motors. But, some employ medium-voltage (2,300 and 4,000 volt) motors, including vertical motors, which can be a major application advantage. Speed ratios for variable-torque loads range up to 10:1. Unlike some mechanical drives, the top speed of an eddy-current drive cannot exceed the speed of the drive motor.

Application Factors

Eddy-current drives always exhibit slip. Thus, maximum rated output speeds typically run from about 95 to 97 percent of the input speed. Eddy-current drives dissipate their losses through air cooling up to about 100 HP and through water cooling at higher ratings. Eddy-current drives are best suited for variable-torque loads such as centrifugal pumps and fans. When powering constant torque loads, slip losses occur, and therefore, heating increases rapidly with decreasing speed.

When applied to pump and fan loads, the controlled variable is often flow, pressure, or liquid-level rather than speed. Consequently, output speed signals for these drives are frequently derived from the user's process control equipment, including process computers.

These drives can withstand high shock and brief overloading. Application analysis requires torque and speed specifications, duty cycle analysis when loading is intermittent, frequency of stops and starts, load inertia, environmental conditions, protection, and modifications.

Wound-Rotor Drives

The speed of a wound-rotor induction motor is controlled by varying the current through the motor's rotating armature. Transistors in series in the rotor circuit control the current to provide smooth acceleration and deceleration. Some new systems allow for energy losses due to slip to be recovered and fed back to the AC input lines, providing high efficiency of the system. The diagram shows this power recovery (Fig 12-12).

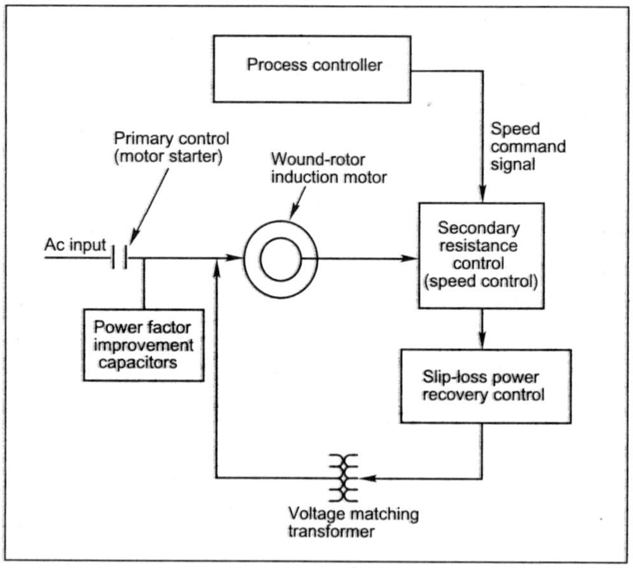

Figure 12-12: Block diagram — slip-loss power recovery drive.

Drive Ratings

Wound-rotor drives are available in ratings from 50 HP and higher. Wound-rotor motor drives are generally used in the U.S. and at ratings above approximately 500 HP. Maximum top speeds of wound-rotor induction motor drives are about 95 percent of synchronous speed, and minimum speeds are about 50 percent (roughly a 2:1 controlled speed range) of synchronous speed. They are most suitable for variable-torque loads. Generally, they are selected as the result of a comprehensive engineering analysis of user requirements, including operating costs. When both high horsepower ratings and medium voltages are preferred in a pump or fan application, wound-rotor motor drives with or without slip power recovery represents an attractive technology.

Application Factors

Wound-rotor motor drives require considerable application analysis. While it is possible to provide controlled speed ranges wider than 2:1, pump and fan applications seldom require more speed range. The controlled variable is seldom speed, so the speed commands and closure of the drive's control loop are generally integrated through the user's process controller.

DC Adjustable-Speed Drives

Early high-performance DC drives employed the so-called Ward-Leonard drive system, where an AC motor drives a DC generator. The generator is an adjustable-voltage power supply for a DC drive motor. This type of drive is electrically simple, allows continuous operation in all four quadrants, and covers a wide range of power ratings. Many Ward-Leonard drives are still used to power elevators, hoists, and cranes.

The majority of general-purpose and special-industry DC drives uses solid-state power conversion modules for both armature and field power supplies. Thyristor-type power modules are compact and economical. They are used for drive ratings from less than 1 to more than 2,000 HP.

A DC motor with a constant field voltage increases its speed when the voltage across its armature circuit increases, and vice versa. Above the motor's base speed (maximum rated speed with rated load, armature, and field voltage), speed can be increased by reducing field voltage (field weakening control). Although speed increases and power remains constant, torque degrades in this realm. Because motor speed increases as the field current decreases, safeguards must be put into place to prevent motor "run-away" in the event that all power is lost in the field circuit.

Fig. 12-13 shows output power versus speed characteristics of a typical stabilized shunt wound DC motor. Note that at higher speeds, output horsepower must be reduced because of motor commutation limits.

Under some operating conditions, a wound field DC motor could "run away" to damaging over speeds if the field power supply is interrupted. Many general-purpose lower horsepower drives now use motors with permanent magnets that supply field power. Drives with wound field motors include protective circuits to automatically shut the motor down in the event field power is lost. The motor is easily reversed by simply changing the polarity of the armature voltage power source.

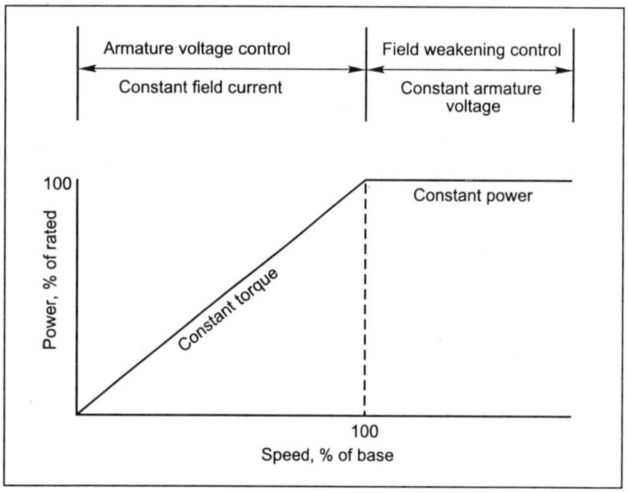

Figure 12-13: Output power versus speed of DC drive with armature voltage and field control.

The two diagrams show two common ways that DC motor speed may be precisely regulated. Fig. 12-14 shows that current and voltage sensors can be used to monitor a DC controller's output to provide feedback for speed regulation, and Fig. 12-15 shows a current sensor and a motor tachometer used for the same purpose. Fig. 12-16 shows the full-wave rectifier circuits of a DC controller. During forward direction, only the thyristors on the left are turned on by a signal sent to the thyristor gates. For reverse, only the thyristors on the right are allowed to fire and provide current to the motor.

Most general-purpose drives use digital regulators (see box, Analog Versus Digital Regulators). However, some DC drives that need to provide servo-like performance retain analog regulators because of their superior dynamic capabilities.

Brush-type DC motors call for more maintenance than AC squirrel cage induction motors. But, DC motor maintenance has been consistent for decades. There are no surprises regarding DC drive systems for an experienced user.

Industry traditions, and the technical interactions between the drive motor and controls, are such that a DC drive is considered a packaged drive system. A DC drive system consists of both a DC motor and a dedicated speed control assembly even at the fractional-horsepower, general-purpose level.

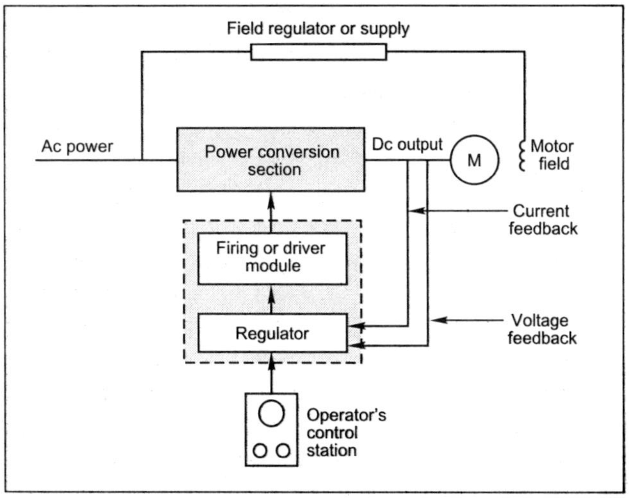

Figure 12-14: DC drive with voltage and current feedback.

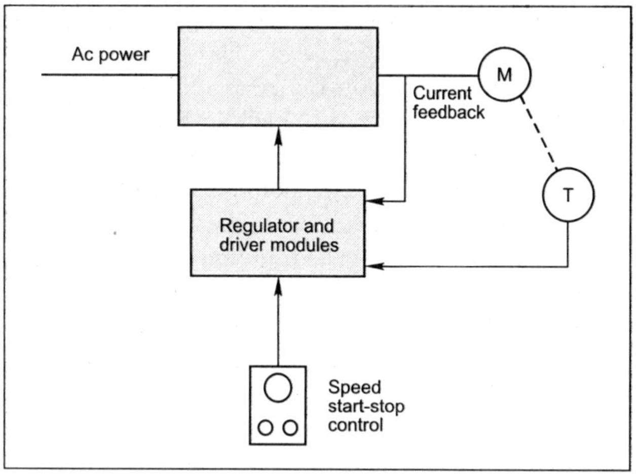

Figure 12-15: DC drive with current and tachometer feedback.

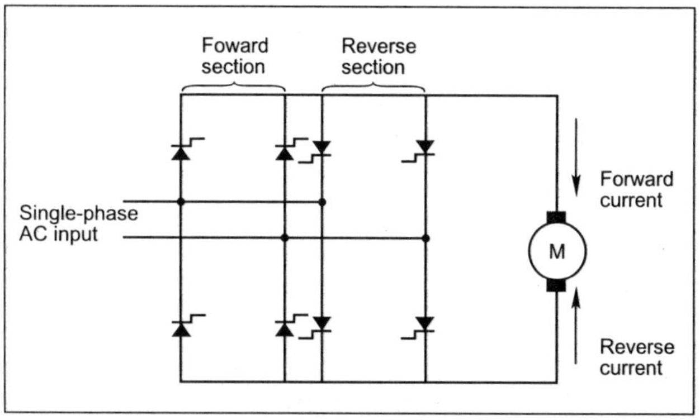

Figure 12-16: DC drive with regenerative power (forward and reverse) modules. Single-phase input. Field power supply not shown.

Drive Ratings

DC drives are available from about 1/4 HP to 10 HP for operation from single-phase 115 Vac and 230 Vac power. Drives from about 3 HP through about 2,000 HP operate from 230 Vac, 460, 575, and 690 Vac, three-phase inputs. Both 60 Hz and 50 Hz input powers are generally accommodated. Motor base speeds vary from 3,450 down to 300 rpm, but not all horsepower ratings are available at all base speeds. A controlled speed range of 20:1 is commonly available, but 100:1 is obtainable with the addition of tachometer feedback. These speed ranges can be extended by field weakening, which increases the maximum speed by factors of up to 5 times.

Application Factors

Selecting a DC drive can involve considerable application analysis depending on the user's requirements. High-performance drives must define the entire load, input, physical arrangement, and environmental criteria listed under Application Criteria. High-performance drives allow extensive modification and include many performance and display control equipment options. General-purpose DC drives include fewer motor and control options and are intended to support simple constant torque, 10:1 to 20:1 speed-range applications—with and without reversing. Consequently, the application analysis required for general-purpose drives is structured to be fairly simple.

When a drive system is closely integrated into the user's manufacturing system, including communications with external intelligent devices, the user and the power transmission specialist are encouraged to consider engaging the services of a system integration professional.

AC Adjustable-Speed Drives

The speed at which an AC motor rotates is based on the number of poles built into it by the manufacturer and the frequency of the AC voltage applied to it. Power companies supply a very closely regulated 60 (or 50) cycles per second (hertz) alternating voltage. Therefore, in order to change the speed of AC motors, AC controllers are used to change the frequency of the applied signal by first changing the 60 hertz signal to DC and then changing the signal back to AC of varying and controllable frequency and voltage. These controllers are known as inverters. Most inverters use pulse-width modulated (PWM) technology, but current-source inverters (CSI) are also available, mainly at higher HP ratings. Flux-vector drives incorporate a more accurate control strategy than strictly controlling the output volts and frequency for PWM inverters.

Fig. 12-17 on the previous page shows a functional block diagram of an adjustable-frequency drive, applicable to all inverter technologies. The converter section rectifies incoming AC power, producing a DC bus. The voltage of the DC bus may or may not be a controlled value, depending on the inverter technology. The filter section smoothes the DC waveform and, in the case of a CSI inverter, contains two large power chokes. The inverter section converts the DC back into AC, where both frequency and voltage are controlled. Output current and voltage waveforms differ depending on the inverter technology employed. The driver and regulator modules process control and feedback signals to adjust and maintain motor speeds.

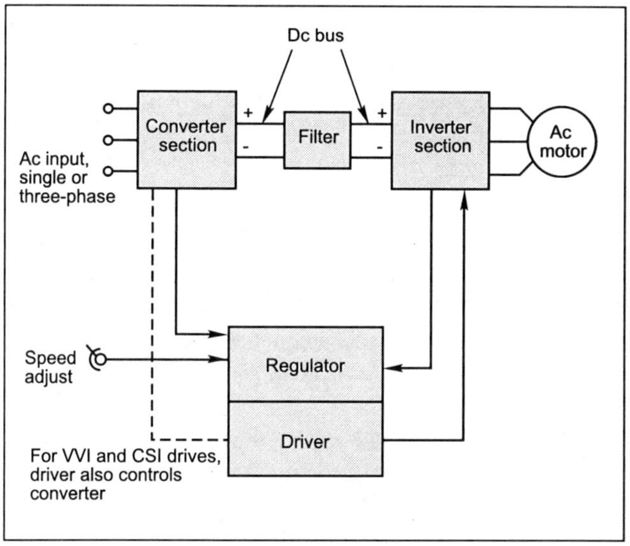

Figure 12-17: Adjustable frequency drive.

In all adjustable-frequency drive systems, the voltage at the drive motor varies directly with output frequency, up to the rated frequency of the drive motor. There are two exceptions to this general rule. During acceleration, some drives boost the input voltage above the constant volts-per-hertz ratio to produce faster acceleration. And, when drives op-

Analog Versus Digital Regulators

A drive regulator is the circuitry responsible for processing input data—from both command and feedback signals—and generates both speed error (change) signals and drive status information.

In DC drive systems, the regulator sends commands to the thyristor firing circuit module (or driver) to increase, decrease, or maintain the output DC voltage.

In adjustable-frequency drive systems, both the output voltage and frequency are controlled by the regulator.

Regulators are generally classified as either analog or digital.

Analog regulators

Analog regulators process current and voltage signals in analog form (a continuous signal that varies in magnitude). This technique has two main advantages: it exhibits no appreciable update time lags in converting a speed error signal into a speed change command, and there is no resolution loss as occurs with any digital signal. A disadvantage of analog regulators is the dependence on potentiometers, operational amplifiers, and other components subject to value tolerances and drift with changes in time and temperature. Thus, drives with analog regulators may call for fairly sophisticated initial startup routines.

Digital regulators

Digital regulators convert and process all incoming signals in digital form (a binary number, as used in computers). These regulators do not depend on potentiometers and analog circuit components to stabilize a drive. As a result, digital regulators are less affected by input voltage and ambient temperature changes than analog types. They can be programmed to search for all drive system stability criteria and essentially self-tune the drive during start up.

Digital regulators have a data update time lag not found in high-performance analog regulators. When drive system variables undergo major changes during this update time, digital regulators may be unstable. Thus, high-performance servo drives often use analog regulators.

erate above rated motor frequency, they generally maintain constant motor voltage. Fig 12-18 shows output power versus speed plot of a typical adjustable frequency drive.

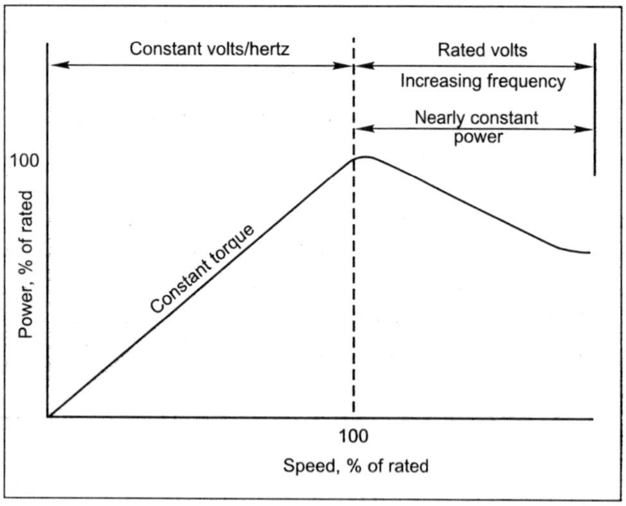

Figure 12-18: Output power versus speed of adjustable-frequency drive.

CSI Drives

CSI drives regulate current rather than voltage. As the frequency of the AC signal that is applied to the motor increases or decreases to control motor speed, the inverter regulates the amount of current that will flow through the motor to prevent motor damage. The inverter section produces three quasi-square-wave output current waveforms of the desired frequency, as shown in Fig. 12-19. Voltage waveforms resemble a "notched" sine wave.

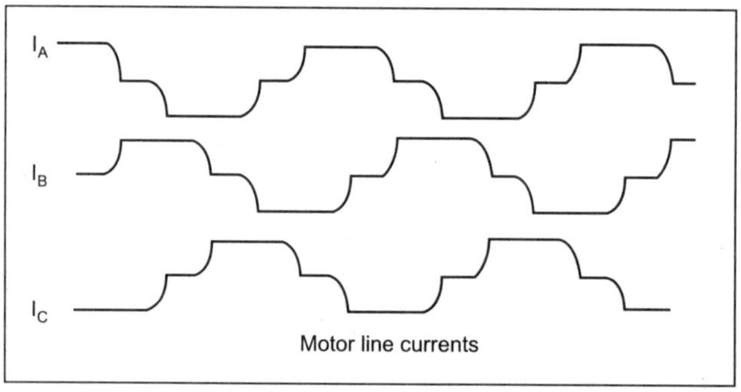

Figure 12-19: Typical current waveforms for a CSI drive.

CSI drives inherently can regenerate power back to the input AC line so that braking an overhauling load does not depend on the thermal capabilities of dynamic braking resistors. The economics of CSI drives made them particularly attractive in ratings more than 200 HP.

PWM Drives

Output voltage and current waveforms of a PWM drive are shown in Fig. 12-20. In the PWM drive, motor speed is again controlled by the frequency of the inverter's output. However, motor current is controlled by rapidly switching the inverter on and off many times during each cycle. Thus, effective rms voltage is a function of the total "on" time. The current waveform depicted is controlled to prevent motor overheating and is void of damaging harmonics.

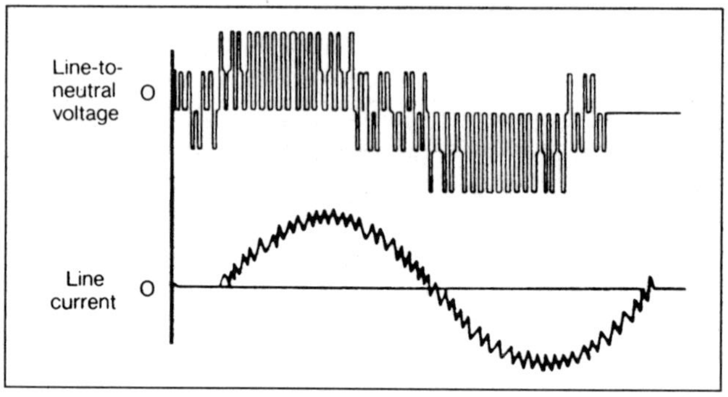

Figure 12-20: Typical voltage and current waveforms for a PWM drive.

A typical converter section consists of a full-wave rectifier section, which is unregulated but reasonably constant. The inverter section is controlled to pulse the power semiconductors on and off at carrier frequencies from about 2 to 20 kHz, where IGBTs and IGCTs are used.

When a load may cause a motor to over speed, such as in the "down" mode of an electric crane application, the PWM drive may allow the motor to act as a generator that puts power back onto its DC bus. The drive may route excess power through dynamic braking resistors that are switched across the DC bus. Drives with an active rectifier front end (e.g., six IGBTs configured like the inverter output section) can put regenerative energy directly back on the power grid. These drives can also eliminate most line current harmonics that are typical with diode or thyristor front end drives.

Flux Vector Control Drives

Flux vector control drives represent an advanced control strategy for PWM drives, not a different drive technology. The AC voltage impressed on the windings of a squirrel cage induction motor consists of two vectors, a torque vector, and a magnetic flux vector, which are mutually perpendicular. Neither vector can be directly measured. However,

System Response and Stability

Adjustable-speed drives can often be adjusted to provide a wide range of responses—the speed with which a motor shaft responds to changes in commands, usually to speed commands.

On the surface, this sounds simple. However, considering that most mechanical systems are composed of three mechanical characteristics—inertia, drag (usually caused by friction), and a spring—the operation of this complex system can make it difficult to set up if the system will operate in a high-performance mode. That is, if the system is to accelerate from, say, standstill to 2,000 rpm in 0.2 sec, then the adjustments must be just right. If the gain is set too low, the acceleration time may take 3 sec. If it is set too high, the drive may be unstable and never stay at 2,000 rpm. Instead, it keeps hunting for the right speed by going faster than slower then speeding up, and it keeps repeating this hunting.

Ringing may not be as easy to detect as hunting, which can often be heard.

Stabilizing adjustments

An analog drive regulator typically uses potentiometers to adjust for the motor and system characteristics. These adjustments affect the damping and gain characteristics. The higher the gain, the "stiffer" the system and the faster its response. Increasing the amount of damping reduces the tendency to ring or hunt.

Drives with digital regulators either tune themselves (using a self-tuning feature) during start up, or they use a keyboard or computer to insert numerical values for system constants.

In some complex situations, stabilizing the drive requires assistance from the drive supplier.

The following are terms that explain how a drive is performing.

Stable responses

Fig. A shows a stable response to a rapid change in the command signal. This can be speed, position, or any other system variable sensed by the drive.

Ringing

Fig. B illustrates a slight instability, often called "ringing." Here, the motor hunts during a short period for the commanded signal and finally settles at the desired speed.

Hunting

Fig. C depicts unstable operation, commonly known as "hunting." In this most undesirable situation, the motor continually hunts around the desired value, but never settles there.

Fig. A

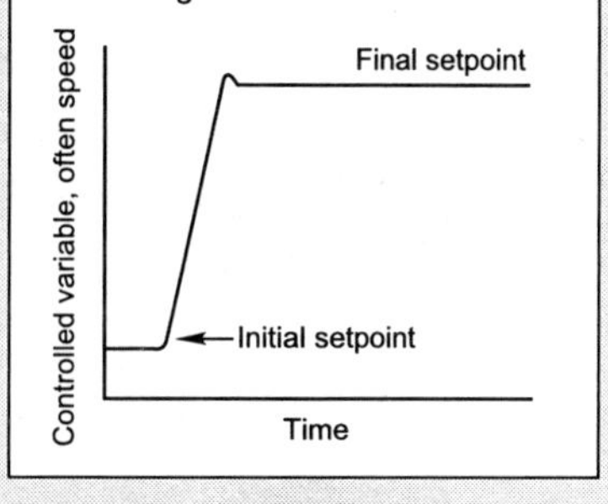

Fig. B

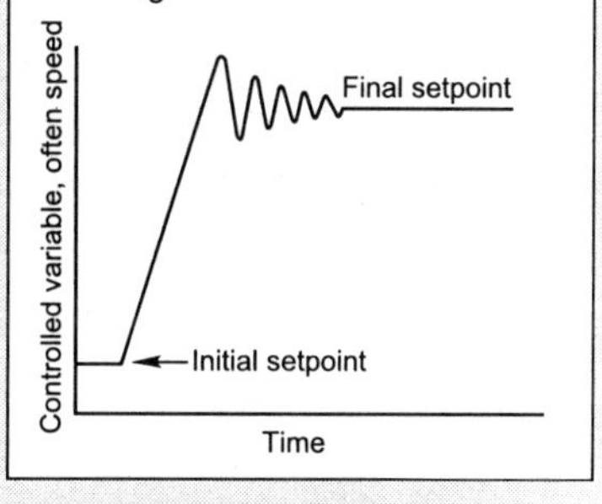

Fig. C

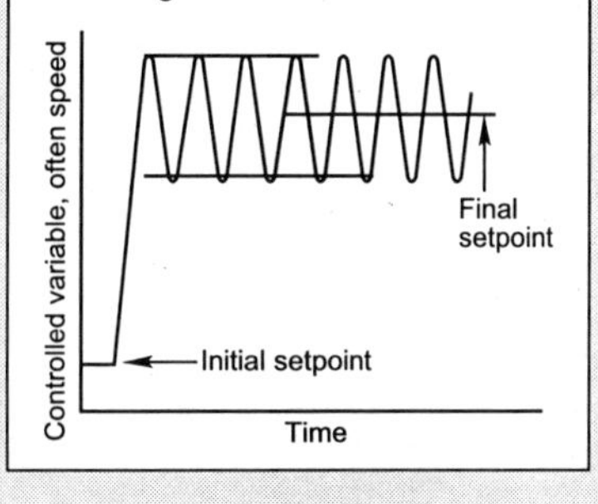

Fig. D

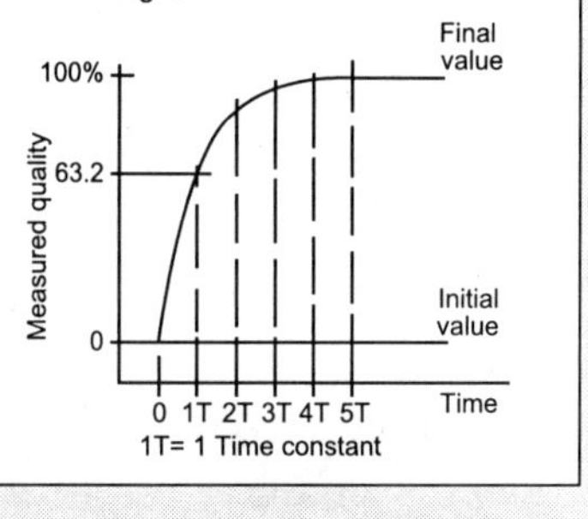

their values can be calculated when drive frequency, shaft position or instantaneous speed, and instantaneous rms voltage are known.

Initially, most flux vector drives had a speed sensor such as a precision shaft encoder or resolver mounted on the drive motor's shaft. A digital regulator performs the necessary calculations to determine an optimum instantaneous output voltage for the motor's instantaneous speed. The object is to keep the magnitude of the motor's flux vector constant to prevent torque ripples. This strategy is implemented by a microprocessor, which performs thousands of calculations per second. With the more recent use of sensorless control algorithms, shaft feedback is not mandatory for this control strategy. Superior drive performance, however, can still generally only be achieved with precision shaft feedback.

System Response and Stability

Response terminology

The most general terms to describe the response are either in hertz (Hz) or radians per second (rps) (1 Hz = 2Pi (6.283) radians/sec). This gives the maximum frequency to which the drive can respond and have less than a specified amount of lag between the output and input, approximately 45 deg.

To describe the operation in more specific terms, Fig D. shows the derivation of the term time constant for a control device or a drive system.

Time constants are widely used to describe the speed of response for drives, control systems, sensors, and user processes. Time constant is the length of time (usually in seconds or fractions of a second) required to reach 63.2 percent of the final value. In a so-called first-order control system, 86.5 percent of the final value is achieved in two time constants, 95 percent in three time constants, 98.2 percent in five time constants, and 99.9 percent in five time constants. A typical first order system is the coast down curve of speed versus time for a de-energized drive motor.

Energized drive systems are so-called second-order systems containing inertia, a spring-like response characteristic, and damping. Inertia exists as the sum of the inertias of the drive motor's rotor and the driven loads as modified by speed changers. The spring components exist electrically in the motor and mechanically in such things as the twisting of shafts in high-performance drives. Damping is a combined function of drive control characteristics, the drive motor, and the driven system. The response to a rapid change in operating conditions (speed reference input or load) resembles that shown in Fig. E. Five specific terms are widely used to define the desired response of any closed-loop control system:

1. Delay time: The time required to reach 50 percent of the final value.

2. Rise time: The time required to go from 10 percent to 90 percent of the final value.

3. Maximum overshoot: The maximum peak value above the final desired value, expressed in percent of the final value.

4. Peak time: The time required to reach the first peak overshoot.

5. Settling time: The time required to reach and stay within the allowable tolerance band around the final value. Allowable tolerances can be any specified value, such as 2 percent, 1 percent, etc.

When all of these times and values are specified, the dynamic response of a stable drive or control system is clearly defined.

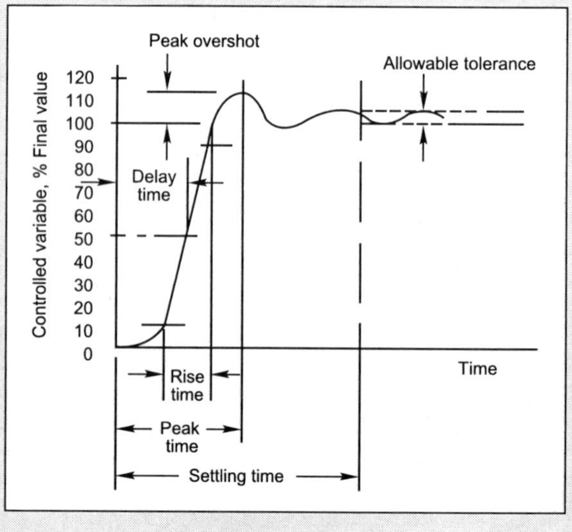

A fundamental advantage of flux vector control is the ability to deliver a wide controlled speed range, say 2,000:1. Standard control techniques produce 10:1 speed ranges for CSI drives, and up to 2,000:1 for PWM drives. Drives with sensorless flux vector control can provide good torque control down to approximately 1 Hz, while good control all the way to 0 Hz is achievable using speed feedback.

Drive Ratings

Adjustable-frequency drives are available in all integral horsepower ratings from 1 to more than 2,000 HP. Generally, the rated output voltage is either 230 Vac, 460 Vac, 575 Vac, or 690 Vac three-phase at 60 Hz. Large drives using GTO (Gate Turnoff Thysistor) or IGCT (Integrated Gate Commutated Thysistor) power semiconductors and CSI drive technology are available in medium voltages

Energy Saving Opportunity with Variable Speed Drive

Adjustable-speed drives can save energy in a few of the more prevalent industrial applications. These applications are: fans, centrifugal and axial pumps, and centrifugal compressors. These applications have one factor in common: The faster the motor turns, the more torque it takes to rotate the load. This is called a variable torque load. Variable torque loads offer the greatest potential for energy savings due to inherent operating characteristics. Many applications requiring these drive mechanisms are not required to operate at full speed 100 percent of the time. Traditionally, the motor speed remains constant, and the input or output of the systems is dampered or throttled to control the flow. These applications adhere to a set of rules called the Affinity Laws.

- Volume flow is proportional to **speed.**
- Pressure is proportional to the **speed squared.**
- Power used is proportional to the **speed cubed.**
- Audible noise is proportional to speed to the **fifth power.**

So, how do you save energy? If you decrease the speed of the motor using a variable speed drive by just 20 percent, you can decrease the power used to drive the load by about 50 percent. If you apply this concept to a ventilation system, the system is traditionally sized for the hottest/worst condition. On a mild day, the fan does not need to operate at full speed. It is easy to decrease the speed by just 20 percent, which decreases the flow by just 20 percent and saves around 50 percent on energy. Decreasing the speed more results in more savings. A side benefit is that the fan noise is also decreased substantially, speed to the fifth power. Figure 12-20 graphically shows the results of decreasing the motor speed electronically. These energy saving characteristics apply to the other noted applications.

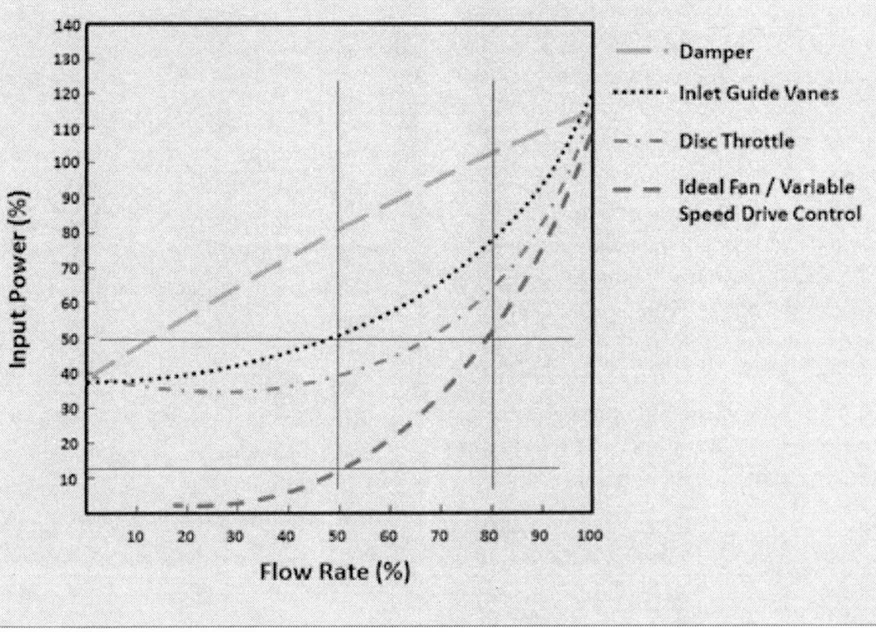

(2,300 Vac). Output frequency adjustment runs from 0 to 120 Hz, and in some cases, up to 4,000 Hz.

Application Factors

When specifying an AC inverter, there are several factors that must be known. These include the complete continuous and intermittent torque and speed profiles of the driven load and the load inertia. Also, the braking requirements must be defined. When more than 60 Hz, the overspeed capabilities of the drive motor must be known, not assumed. Drive manufacturers offer a variety of control options, some designed as function modules to support specific industry needs. Drives with digital regulators often include self-tuning, elaborate diagnostics and network communications (available in DC drives as well).

All adjustable-frequency drives produce higher levels of motor heating than an identical motor that runs on a pure sine-wave power source. Because of this, it is important to specify an "inverter duty" motor or down-rate a standard motor when used with an AC inverter. Additionally, applications that require high torque at very low speeds may cause motor overheating and therefore shorten its life.

The drive regulator is the circuitry that maintains a constant speed of the AC or DC motor under changing load conditions. It processes signals from the command controls, such as the operator's control station or the computer controller, and from a feedback loop if used. Using these signals, the regulator sends commands to the drive output circuits to increase, decrease, or maintain the output voltage, current, or operating frequency applied to the motor.

In DC drive systems, the regulator sends commands to the thyristor firing circuit module to increase, decrease, or maintain the output DC voltage. In AC drive systems, the frequency is controlled by the regulator. The regulator in an AC drive will also control either the current or voltage (depending on the drive) to protect the motor from overheating.

Analog regulators measure current and voltage as a continuous signal that varies in amount, similar to the display of a speedometer in a car. This system's advantages are that there is no time lag in converting a speed error signal into a speed change command.

A disadvantage of analog regulators is that they depend on potentiometers and other analog circuit components to establish and modify drive performance and stability characteristics. Thus, drives with analog regulators may call for fairly sophisticated initial startup routines.

Digital regulators convert and process all incoming signals in digital form (a binary number, as used in computers). These regulators do not depend on potentiometers and analog circuit components to stabilize a drive. As a result, digital regulators are less affected by input voltage and ambient temperature changes than analog types. They can be programmed to search for all drive system stability criteria and essentially self-tune the drive during start up.

Digital regulators have a data update time lag not found in high-performance analog regulators. When drive system variables undergo major changes during this update time, digital regulators may be unstable. Thus, high-performance servo drives often use analog regulators.

Regardless of the type of AC or DC drive used, the list of application criteria contains factors that apply to all. They fall into several categories:

- Output performance
- Incoming power available
- Mechanical arrangement and restraints
- Environmental restraints

Output performance

Requirements for the output of an adjustable-speed drive system are defined when the user establishes numerical values for all of the drive's output characteristics. These generally include:

- Operating and controlling speed range.
- Speed-torque characteristics of the load (constant or variable torque and horsepower).
- Speed range over which the drive must deliver its stated speed regulation at rated torque.
- Required dynamic performance and steady-state load regulation.
- Speed regulation limitations that must be maintained over specific input voltage and frequency variations. This may be the single issue that mandates use of higher-performance drive technologies.
- Directions of operation: forward only or forward and reverse. If reversing is required, can contactors be used to switch from forward to reverse motoring, or is regeneration required?

- Possibility of overhauling load—when the load tries to turn the motor.
- Duty-cycle of the driven load.
- Probability of severe impact loading.
- Number of motors that must be coordinated in a system.
- Incoming power.

What are the nominal values and variations of the incoming voltage and frequency? Also, how many phases are available at the drive site? (NEMA standards for drives indicate a usual voltage tolerance is +10 percent/-5 percent, which is narrower than for AC motors, ±10 percent.)

Mechanical arrangement

The physical arrangement of the drive and its mechanical interface with the driven load is a critical part of all drive specifications. The following are some of the most pertinent questions.

- Will the load be directly driven?
- Will the drive motor be integrated with a gear reducer, pulley, or sheave?
- Can the drive motor be physically separated from the drive's controller?
- What motor frame size limitations and shaft extensions are needed for the application?

Environmental

The physical environment of both the drive motor and its speed regulating control must be clearly specified. They are frequently different. The following are the major areas.

Minimum and maximum ambient temperature of the surrounding air directly adjacent to both the drive motor and the control equipment. This is not outside air temperature. NEMA "usual" ambient temperatures have a maximum value of 40° C (104° F). While both AC and DC drive motors have the ability to withstand moderate overheating at the expense of reduced insulation life, solid-state speed-control equipment must be designed for worst-case, short-time, highest ambient temperatures. Values of 50° C and 55° C (122° F and 131° F) are not uncommon. Low ambient temperatures pose fewer problems. However, if sustained ambient temperatures below 5° C to 10° C (41° F

TABLE 12-1 — Relative Response of High-Performance Adjustable-Speed Drives

Drive type		Relative performance		Typical values	
Motor	Control	Response band width	Maximum accel. rate range	Cont. speed range, rpm	Max. cont. torque, lb-in.
DC-WF	3P-HW	Low	Low	0 - 1750	6500
DC-PM	3P-HW	Medium	Medium	0 - 2000	500
AC-IN	PWM-FV	High	Medium	0 - 3000	1250
DC-MC	PWM	Very high	High	0 - 3000	200
DC-PM FER/RE	PWM	High	Medium	0 - 3000	500/750
BLDC FER	PWM-FV	High	High	0 - 9000	1800
BLDC RE	PWM-FV	Very high	High	0 - 9000	800

Abbreviations and notes:

DC-WF = Brush-type, dc, wound-field
DC-PM = Brush-type, dc, permanent magnet field
AC-IN = Ac squirrel-cage induction
DC-MC = Brush-type, dc, moving coil
FER = Brush-type or brushless, dc, with ferrite-type permanent magnet field
RE = Brush-type or brushless, dc, with rare earth-type permanent magnet field
BLDC = Brushless-type dc motor

Control:

3P-HW = 3-phase, half-wave, thyristor dc
PWM = Pulse-width-modulated
PWM-FV = Pulse-width-modulated, flux-vector control

Relative performance:

Value	Response band width, Hz	Max. acceleration rates, rad/sec²
Low	15	500
Medium	25	4,000 to 12,000
High	100 to 150	90,000 to 175,000
Very high	200	—

to 50° F) are involved, review the environmental conditions with the drive supplier.

The user should specify the NEMA or IEC-type control enclosure that provides acceptable protection from dust, dirt, and noncorrosive falling or splashing liquids. Specify hose down or food industry applications as special situations.

Installing solid-state power-conversion equipment outdoors, particularly where it may be exposed to solar heating, is a significant application challenge that should be discussed with the drive supplier.

Drive motors built to NEMA standards have enclosures designed to easily provide specific levels of ventilation and protection (DP, DPFG, DPBV, TEBC, TEFC, and TENV.)

Servo Drives

Servo drives can be considered variations of basic DC and AC drives. The differences between a servo drive and a standard drive system are that a servo drive accelerates, decelerates, and reverses a load faster and arrives at a defined position with greater precision. DC drives and flux vector controlled AC drives often provide servo-like performance (see box, System Response and Stability). Servomotors are covered in Chapter 11—Motors. Step motors and their controllers compete with servo drives in some applications, and they are also covered in the chapter on motors.

The majority of servo drive controllers deliver a pulsed DC or PWM AC output to either a two-phase or three-phase stator winding. A form of flux vector control is popular with brushless servomotors.

The choice of digital versus analog regulators (see box on this subject) is important in the evaluation of a servo drive application. Where load disturbances occur during the update time of a digital regulator, analog speed and positioning regulators are used, even in the most modern servo drive systems.

Servo Drive Ratings

Servo drives are generally rated in terms of continuous and intermittent torque values over the controlled speed range and maximum acceleration rate rather than horsepower. Servo drive ratings may be expressed in either British Engineering Units or SI units. Standard North American and international voltages are accommodated. Many servo drive controllers have dual voltage—50 Hz or 60 Hz inputs.

The choice of motor and control technology establishes limits on the controlled speed range, the maximum acceleration rate, and the maximum output torque. Servo drives generally are reversible, and some operate in all four quadrants. Top speeds as high as 9,000 rpm are possible with brushless servo drives, as high as 3,000 rpm for moving coil DC drives, and up to 2,000 rpm for thyristor controlled brush-type DC drives.

Application Factors

When specifying servo drives, the application criteria used for DC and AC variable speed drives applies. However, because of the increased precision of acceleration, deceleration, and precision of starting and stopping location, other factors must be considered. These include:

- Speed and torque requirements for all possible positions.
- Maximum acceleration and deceleration rates.
- Load inertia capabilities.
- Electrical disturbances from surrounding equipment.
- Input control signal source.
- Type of feedback needed for the controller.
- Accommodation for multi-axis motion, if needed.

Key Terms

SCR	Silicon Controlled Rectifier
PWM	Pulse Width Modulated
GTO	Gate Turn Off Thyristor
IGBT	Insulated Gate Bi-polar Transistor
IGCT	Integrated Gate Commutated Thyristor
LAN	Local Area Network
PLC	Programmable Logic Controller
AC	Alternating Current
DC	Direct Current
CSI	Current Source Inverter
RMS	Root Means Squared
Wk2	Inertia

Drive Performance Summary

To select the best high-performance adjustable-speed drive for a specific application, Table 12-1 lists some of the major performance capabilities of the common drive types. Included are several types of DC drives, including both brush and brushless and flux-vector controlled AC drives. For brevity, this table excludes the more general-purpose drives because speed of response and fast acceleration rates are seldom an issue in general applications.

Engineered Drive Systems

Controlling metal rolling mills, float glass lines, web-fed printing presses, multi-axis robots, and other complex machines requires coordinating the speed and torque of several drives. Without this coordination, strips may break, and parts may be damaged. Typically, such drive systems are controlled by a computer. They also often continuously feed information on process status and quality to a general management computer.

To achieve this complexity, firms invest in engineered drive systems designed for a specific machine. These drives, which can include AC, DC, and all types of servo drives, involve extensive system and application engineering during all phases of pre- and post-contract system design, installation, start up, and operation.

Chapter 13

Controls & Sensors

Chapter Objectives

When you have finished Chapter 13, "Controls & Sensors," you should be able to do the following:

1. Describe the two functions of AC motor starters.
2. Explain what NEMA standards define and require.
3. Compare and contrast motor starter types.
4. Describe the functions and features of circuit breakers.
5. Describe the functions of various control devices such as relays, pushbuttons, and switches.
6. Compare and contrast the differences between PLCs and computers.
7. Explain the features of four key display technologies.
8. Describe the use of sensors encountered by power transmission specialists.
9. Describe the purpose of codes and standards.

Introduction

Driven equipment is generally powered by electric motors. All but the smallest require a controller of some sort. Where a drive operates at constant speed, the controller may be a simple motor starter. Such controllers often also include auxiliary devices—fused switches or circuit breakers, auxiliary contactors, or relays—and remote components such as pushbuttons, selector switches, indicating lights, and a variety of displays and special-purpose sensors.

Adjustable-speed drives generally power systems where motor speed must be changed to meet differing operating conditions. The drives, which sometimes include relatively complex controllers, are covered in Chapter 12.

Both motor starters and adjustable-speed drives sometimes operate in coordination with one or more other starters or drives. And, in some cases, they operate in conjunction with programmable logic controllers (PLCs) and computers.

Motor Starters

AC Motors starters, available for both single- and three-phase operation, consist of two main components and perform two functions. The contactor portion provides the connection of the motor to the incoming power. The overload relay protects the motor from overheating due to overloading.

The overload portion of the starter causes the contactor portion to electrically disconnect (trip) when it senses a higher than normal current flow to the motor. Manufacturers use several types of overload devices, which will be discussed later in this chapter.

AC starters are available for both single- and three-phase operation in all electric motor voltages.

Full-Voltage Starters

Often called an across-the-line starter, a full-voltage, nonreversing starter (FVNR) connects the motor directly to a power line (see Fig. 13-1). Fig. 13-2 shows a manual motor starter rated for 5 HP, 460 VAC, three-phase, 60 Hz. Fig. 13-3 shows a typical NEMA magnetic starter, rated for 10 HP, at 460 VAC, three-phase, 60 Hz.

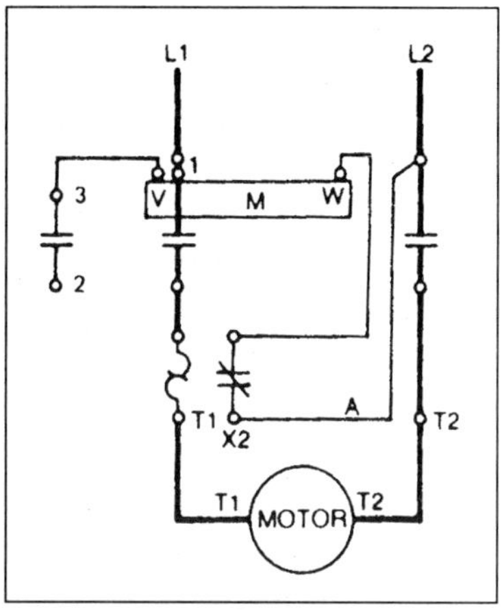

Figure 13-1: Across-the line starter

Manual starters are limited to single-phase motors through about 5 HP at 230 VAC and three-phase motors through about 10 HP at 460 and 575 V.

Starters generally are designed to meet either NEMA or IEC standards (see Codes and Standards in this chapter). Applications in the U.S. typically call for NEMA starters, and those in Europe generally employ IEC standards, but both types are used in both areas and throughout the world. The two

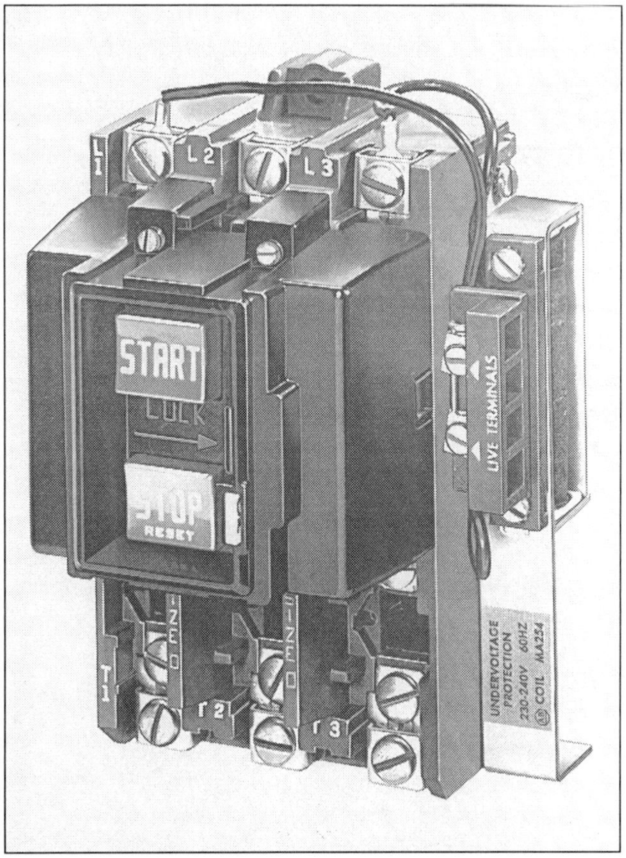

Figure 13-2: Manual motor starter.

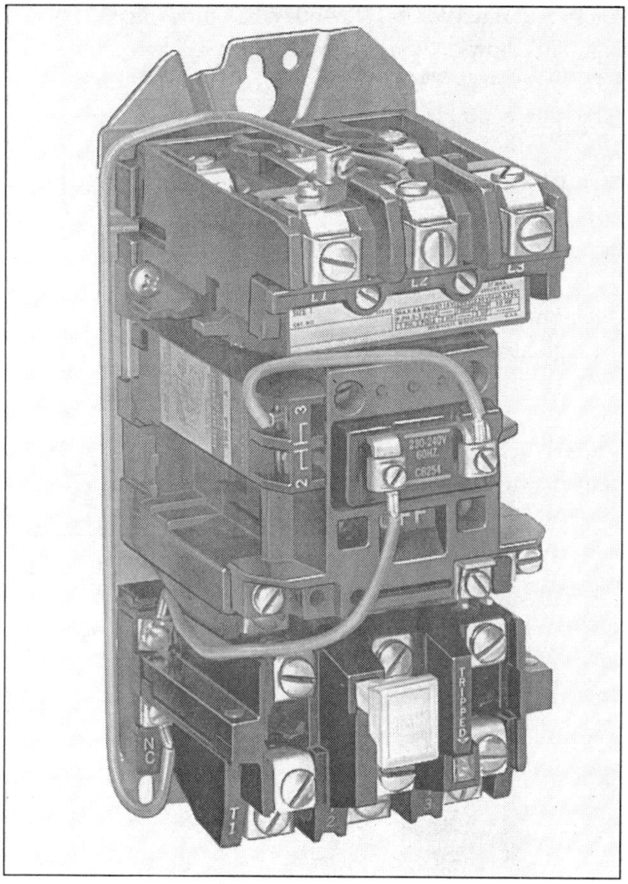

Figure 13-3: Magnetic Size 1, full-voltage starter.

starter types differ in three important characteristics—ratings, life, and overload types.

Frame Ratings

NEMA standards define 11 magnetic starter sizes (00 through 9) for low-voltage starters and specify horsepower ratings for each size as a function of various line voltages, but IEC standards permit the user to apply a starter based upon a technical evaluation of the application.

International ratings for IEC starters provide 15 sizes and require for the consideration of rated current and operations per hour.

Therefore, the physical size of the IEC contacts may be smaller based on the fact that the application may be of a light-duty, low-use application.

Contactor Life

NEMA standards require manufacturers to design all contactors for heavy service. (However, plugging and jogging are considered severe service.) As a result, NEMA contactors are larger than corresponding IEC ones and have more massive silver-alloy contacts.

IEC standards define various levels of service called utilization categories for starters. The most common categories are AC3, for normal starting and stopping service, and AC4, for plugging and jogging service. Generally, manufacturers provide a specific electrical-life curve for each contactor, plotted in millions of operations, for each utilization category. Users select and apply starters based on this data.

NEMA starters generally exhibit much longer mechanical and contact life than IEC contactors, when subject to identical service. This is not to imply that IEC starters are inferior. The differences in physical size between NEMA and IEC starters are grounded on proper engineering principles and are not an indication that one device has too much material or the other too little.

Overload Relays

Manufacturers use several types of overload devices.
The motor control industry has essentially abandoned the heater-element overloads in favor of electronic solid-state overloads. This type of overload provides a greater degree of protection than the old-fashioned heater design. The electronic overload monitors actual motor current and takes the motor offline in three seconds or less when it exceeds the preset current rating of the motor, thereby dramatically reducing the possibility of motor damage. In addition to sensing current, solid state overloads also provide protection against phase-loss, phase imbalance, and short circuit. They are also insensitive to ambient temperature, so compensation is not required for applications in high-temperature areas.

The earliest overload designs placed a piece of resistance wire called a heater in the phases of the input circuit.
During starting and normal running, the heaters heat up. If too much current flows, producing heat, it would either melt a lead alloy material or bend a bi-metallic strip, causing the motor contactor to trip. Later designs utilized a transformer circuit to measure current flow and trip the contactor.

NEMA standards call for overload relays to have interchangeable heaters or electronic overloads to exhibit Class

20 trip characteristics at 600 percent of full load current. Class 20 overload heaters trip in 12 to 17 seconds for use with motors that have a 1.15 factor. IEC overload designs have built-in adjustable heaters that operate with Class 10 trip characteristics. At 600 percent of full-load current, Class 10 overloads trip in six to eight seconds, suitable for 1.0 service factor motors. Most electronic overloads have field-selectable trip classes from 5 through 30.

Reversing Starters

Three-phase motors are reversed by interchanging any two of three power leads to the motor. Full-voltage reversing starters (FVR) employ two contactors, one providing run-forward, and the other providing run-reverse (see Fig. 13-4). Here, pressing the "For" button energizes the "F" contactor, and pressing the "Rev" button energizes the "R" contactor. (The two contactors are mechanically and electrically interlocked to prevent both being activated simultaneously.)

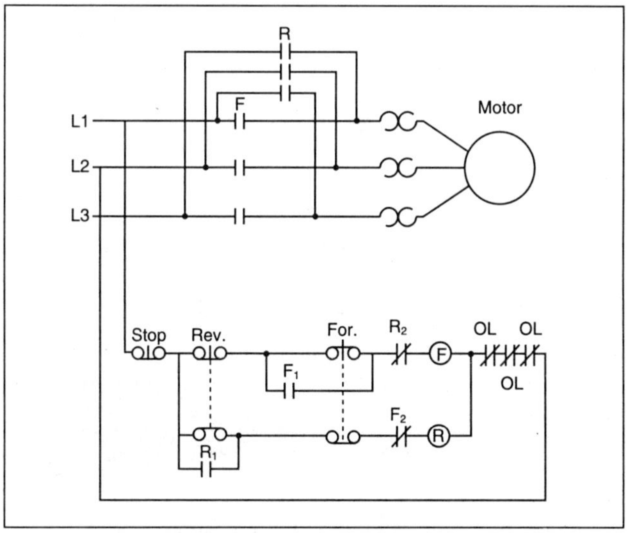

Figure 13-4: Full-voltage reversing starter.

When a motor is running in one direction and the contactor for the opposite direction is energized, the action is called plugging. Here, the motor decelerates rapidly and accelerates in the opposite direction. Where an application calls for rapid deceleration but not the subsequent reverse rotation, a motor can be equipped with a plug-stop switch. The plug-stop switch is a centrifugal switch that applies opposite rotational power to the motor for rapid deceleration but drops out completely when the motor speed nears zero.

Reduced-Voltage Starters

When a motor is started across the line, the large initial current inrush can cause the voltage to drop in the circuit providing the power, affecting other electrical apparatus on that circuit. This large inrush current is caused by the very low internal electrical resistance of a motor at rest. Once a motor is rotating at full speed, its internal resistance (impedance) increases and limits the current. During motor starting, circuit breakers may open, lights may flicker, or control devices may drop out. To prevent this from occurring, a reduced voltage starter may be used.

Primary-Resistor Starters

The simplest reduced-voltage starter inserts resistors in series with a motor during the start phase (see Fig. 13-5). Here, when the start contactor (S) closes, the motor is connected to the line through a set of resistors, reducing the applied voltage and thus the inrush current. After a brief period, the run contactor (R) closes, shorting out the resistors as the motor approaches rated speed. Thereafter, the motor runs at rated voltage.

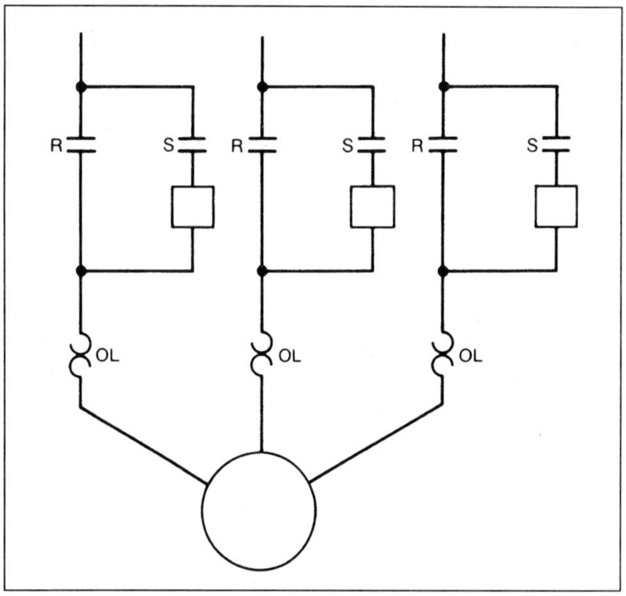

Figure 13-5: Reduced-voltage starter, primary-resistor type.

With this type of starter, the system dissipates power in the form of heat during the starting phase. Thus, applications where the losses would be unacceptable often employ reactors rather than resistors. However, reactor starters cost more than resistor types and exhibit poorer power factor during startup.

Autotransformer Starters

During motor acceleration, power is supplied to the motor through an autotransformer that supplies a reduced input voltage. The lower applied voltage limits the current, preventing the overtaxing of the motor circuit. Once the motor running speed has been reached, a second contactor, controlled by a timer, energizes to bypass the transformer and supplies full voltage to the motor. Typically, autotransformers will have several taps to provide various starting voltages, depending on the starting torque required and power available. During the transition of power source from starting contactor to running contactor, a third contactor is often employed to fill the time interval during the switch. This is done to prevent current surges. This type of starter is called a "closed transition starter." If the third contactor is not used, it is an "open transition starter."

Reduced-Inrush Starters

Wye-Delta Starters

A three-phase motor has three sets of windings in its stator, or outer frame. During start-up, the wye-delta starter connects them in series (wye connection) to increase the electrical resistance of the motor and limit the inrush current. As the motor reaches running speed, a timer connects them in parallel (delta connection), and all three sets of windings receive the same across-the-line voltage. The starting torque and current is about 30 percent of normal; therefore, wye-delta starters are used on low starting torque applications such as blowers or centrifugal pumps. Open- and closed-transition starters are available.

Part-Winding Starters

Part-winding starters require the use of motors that have been specially wired to allow the starter to make connections to only a portion of the motor windings during start-up. As the motor accelerates, a timer causes a second contactor to close, energizing the remainder of the motors windings. This type of starter is the least expensive to purchase; however, the starting current is higher than the other types of reduced voltage starters, and it must be confirmed that the motor is wired for this type of service.

Solid-State Starters

A schematic diagram of a solid-state starter, which uses SCRs as variable-voltage "valves," is shown in Fig. 13-6. These starters incorporate adjustable-voltage acceleration and deceleration ramps to slowly increase motor voltage and speed to avoid shock loading and limit inrush current. Solid-state starters can use either current-limit ramping or tachometer feedback to produce different acceleration and deceleration characteristics. Solid-state soft-start devices are available as stand-alone units, to be used when an across-the-line starter is already in use.

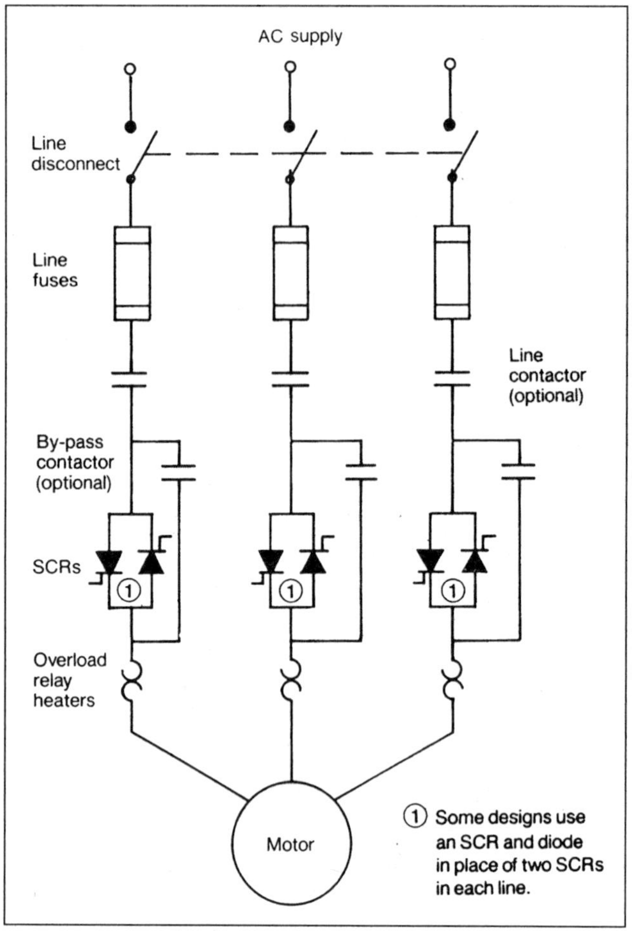

Figure 13-6: Solid-state starter.

In addition, an initial voltage quick ramp-up—or pedestal voltage feature provides a fast voltage rise to a value that will produce rotation. To overcome static friction, (as on some conveyor loads), a short duration kick-start function produces sufficient torque to break away the motor load and produce easier acceleration (see Fig. 13-7).

Some solid-state starters can produce low-frequency outputs (7 percent or 14 percent of input frequency) for short-duration (two minutes or less) jogging operations. Many include built-in DC braking for quick stopping of the driv-

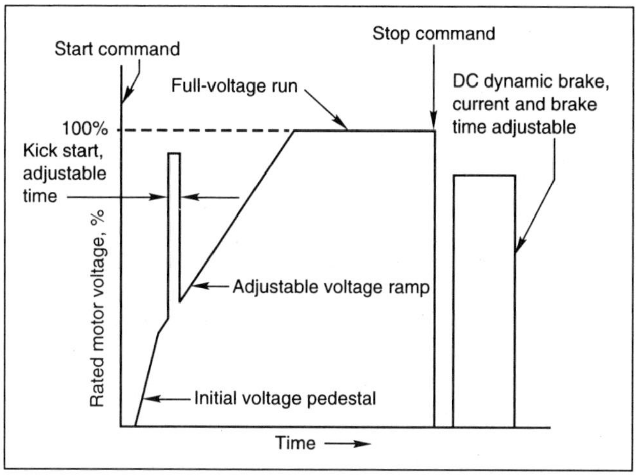

Figure 13-7: Typical plot of voltage vs. time solid-state starter.

en load, which reduces terminal voltage on lightly loaded motors to increase both their power factor and efficiency.

Solid-state starters incorporate a wide range of adjustability, both in the magnitude of their current limit settings (from 150 percent to 500 percent of rating) and in tailoring the acceleration and deceleration ramp characteristics (from 0.5 to 100 seconds). Some duty-cycle analysis is usually necessary, such as coordinating the starting ramp settings made with motor-overload and motor-heating characteristics. This avoids nuisance tripping or thermal damage.

Solid-state starters are popular in pumping applications—particularly those with plastic pipe—as well as variable-torque, constant-torque, and even hard-to-start loads. They are available in ratings from 5 to 600 HP at voltages of 200, 208, 230, 460, and 575 Vac, 60 Hz, three-phase. Comparable ratings are also available for IEC motors with voltages of 220, 380, and 415 Vac, 50 Hz.

TABLE 13-1 – Induction Motor-Starter Comparison

Type	Values, percent			Advantages	Limitations
	Volts	Amperes	Torque		
FVNR	100	600	100	Lowest cost. Highest starting torque.	Highest inrush. Max.voltage dip and shock loading.
AXFMR	80	380	64	High torque. Best torque per ampere	Expensive. Heavy. Complex.
Primary resistor or primary reactor	80	380	42	Moderate cost. Closed-transition.	Low starting torque. Resistors add heat. Reactors reduce power factor.
Wye-delta	33	200	33	Moderate to low cost. High torque per ampere.	Special motor. Fixed, low value starting torque. Not for constant torque loads.
Part winding	100	390	45	Low cost. Closed-transition.	Special motor. Performance varies with motor design.Unsuitable for high inertia loads.
Solid-state	Adj.	Adj.	Adj.	Ramp starts & stops. Low inrush. No shock loading. Standard motors.	Moderate cost. Low-voltage motors only, at present.
Adjustable frequency drive	Adj.	Ad.	Adj.	Same advantages as solid-state starters, plus adjustable-speed.	Higher cost than other controllers. If adjustable-speed, special motor maybe required.

Abbreviations:

FVNR = Full-voltage, non-reversing type AXFMR = Autotransformer, closed-transition type Wye-delta = Wye-delta, closed-transition type

Starter Performance Overview

Table 13-1 provides a comparison in the types of motor starters previously discussed. It provides comparative information based on cost, starting torque available, motor requirements, motor ramping, and speed control. Adjustable frequency drives are included because they are used as a combination soft-starter and speed control. Notice that all numbers represent a percentage of the full load rating.

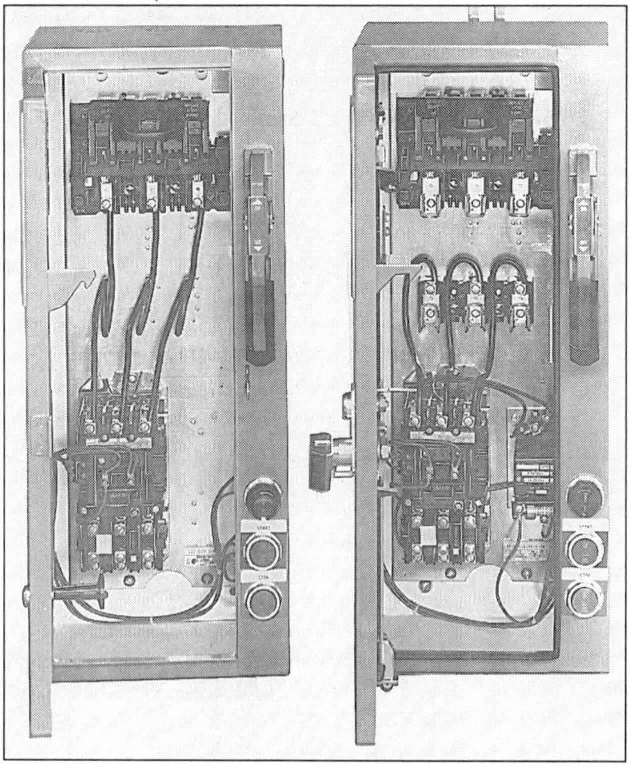

Figure 13-8: Combination starter with disconnect (left) and with fusible disconnect (right)

Combination Starters

North American electrical codes require that a branch circuit containing a motor have both a short-circuit protective device and a disconnecting device in addition to the motor starter. The overload section of a starter is a slow reaction time device designed to protect a motor from overheating due to an overload condition. If a short- circuit condition occurs, such as when there is an insulation failure on a power line, additional protection is required in the form of either a fuse or a circuit breaker. The fuse or circuit breaker protects the starter, motor, and all associated wiring from damage due to extremely high current flow. The disconnecting device required by code is either a manual disconnect switch used with fuses, or it is a circuit breaker which has a built in disconnect switch. When the disconnecting device, short-circuit protective device, and motor starter are furnished as an integrated equipment assembly, that assembly is called a "combination starter."

Fused Disconnects

For mostly cost reasons, fused disconnect switches (see Fig. 13-8) are preferred over circuit breakers in many geographical areas. Their first cost is often lower than that of equivalent circuit breakers. Fuses can be found to match a wide range of time versus current characteristics. Time-delay fuses permit carrying heavy loads for a short period of time (e.g., motor starting) and provide long-term overload protection. Fuses with current-limiting capabilities—the ability to clear high values of short-circuit current—also contribute to the popularity of fused disconnects.

Circuit Breakers

Offering more convenience with higher price, circuit breakers provide a means for disconnecting a motor and starter from the power system and protect the branch circuit from excessive current in one unit. They are rated by voltage, current, interrupting current, and construction type. Breaker standards are set by UL, CSA, IEC, VDE, and JIC. But many molded-case circuit breaker models meet all applicable domestic and international standards.

Figure 13-9: Typical molded case circuit-breaker rated 600 amp @ 600 vac.

There are three voltage classes—low (under 600 V), medium (600 through 15,000 V), and high (over 15,000 V). Domestic low-voltage circuit breakers are offered in three types of construction—molded case, insulated case, and low-voltage power. Each type of construction has specific advantages, functions, and range of current and interrupting ratings.

Classes and Characteristics of Fuses

Industrial fuses used in combination starters, or dedicated fuse boxes, are generally referred to as cartridge fuses. There are two different types of cartridge fuses: the ferrule and the knife-blade. Unlike circuit breakers, fuses must be replaced if an over-current causes them to open. A fuse that has opened is commonly referred to as a blown fuse.

To determine the fuse size required for short-circuit protection of motors, multiply the full load amperage of the motor by 1.75. Fuses that are very slightly overloaded will open or "blow" based on time-current curves supplied by fuse manufacturers. Short-circuited fuses blow almost instantaneously. There are a large variety of special purpose fuses. Consult manufacturers' catalogs for more information.

The following are the more common types of fuses and their characteristics:

Class H

Class H fuses (also frequently called NEC fuses) have renewable fuse links, and short-circuit interrupting ratings of 10,000 amperes, rms symmetrical (the proper method for specifying transient current values). However, they act too slowly to be classed as current-limiting.

Class K

Class K fuses have similar dimensions to Class H fuses and usually fit into Class H fuse clips. Because of the possible substitution with Class H fuses, UL does not allow them to be listed as current-limiting fuses, even though they do limit peak current and have high interrupting ratings.

Class R

Class R fuses are current-limiting fuses with high interrupting ratings. Switches equipped with Class R fuse clips have rejection features that prevent Class H or Class K fuses from being installed. However, Class R fuses can usually be installed in a switch equipped with Class H fuse clips.

Class J

Class J fuses are current-limiting fuses that are shorter in length than Classes H, K, or R. Class J fuse clips are shorter in length and use clips that are closer together, which prevent insertion of fuses other than Class J in the disconnect switch.

Class L

Class L fuse is the most commonly encountered current-limiting fuse above 600 A.

A typical molded-case breaker, rated 600 A, 600 V, 60 Hz, is shown in Fig. 13-9. The rated amperage is the amount of current that a circuit breaker can carry on a continuous basis. The breakers trip, or disconnect, when the current exceeds the breaker rating following a time delay that is inversely proportional to the current overload. Circuit breakers also have an interrupt rating, which indicates the amount of current that the breaker can safely carry during the time that it takes to trip. Many other current values (amperages) are available from 15 amps to 2,000 amps.

Control Devices

Because the heavy wiring that carries the current to run a motor must pass through a starter, the starter is generally located in close proximity to the motor. However, the circuit of the starter that turns the starter on or off uses a small amount of current, usually less than 10 amps, and can be located some distance from the motor and use smaller wires. They may operate at 120 or 240 volts, but often these circuits will be low-voltage circuits for operator safety, operating at 12 or 24 volts. The control components can be pushbuttons, relays, pressure switches, or manually operated switches.

Pushbuttons

A heavy-duty, industrial, oil-tight pushbutton is shown in Fig. 13-10. Pushbuttons also are available with reed or mercury-wetted contacts for low-power applications.

Figure 13-10: Heavy-duty, oil-tight pushbutton switch.

Pushbutton lines generally include rotating selector switches and indicating lights. All items in the line are generally available in a wide range of colors and a variety of different open and closed contact variations. (The "normal" state is when a pushbutton has not been pushed.) Thus, a normally closed (NC) contact will be passing power in its "normal" stage. When the pushbutton is pressed, the NC contact opens, removing power from that control circuit and shutting off whatever device it was operating. A normally open (NO) contact works just the opposite; membrane, capacitive, and hall-effect type pushbuttons are also available for electronic circuitry.

Relays

Relays are switches that contain electrical contacts that are closed or opened by energizing the coil of an electromagnet within the relay. They are used to provide remote contactor control, interlock electrically isolated circuits, or provide a sequencing function.

Relays may contain one or more sets of contacts. If relay contacts are open when no voltage is applied to the coil, they are considered normally open. If relay contacts are closed when no voltage is applied to the coil, they are normally closed. A relay that is either mechanically or magnetically latched will stay in the last activated position until it receives another signal to return to its opposite position.

Industrial relay contacts are typically rated at 10 amps or lower. If more than 10 amp capacity is required, magnetic contactors are used to act as a relay.

Solid-State Switches

Solid-state relays (switches) provide the same function as electromechanical relays. Switches employing various solid-state devices are often used in power circuits. The switches typically fall into one of two categories—transistor based or SCR based. Transistors are current amplifiers where a small current switches a DC circuit on or off. Transistors can also switch AC circuits but are rarely used for this.

SCRs, or a combination of SCRs called the "triac," are widely used only for AC switching (see Fig. 13-11). Here, a small control current turns the device on anywhere during the positive half cycle of an AC voltage, and turn off automatically when the voltage goes negative. However, the device is costly to use in DC circuits because a second SCR and additional circuitry is needed to force it to switch off.

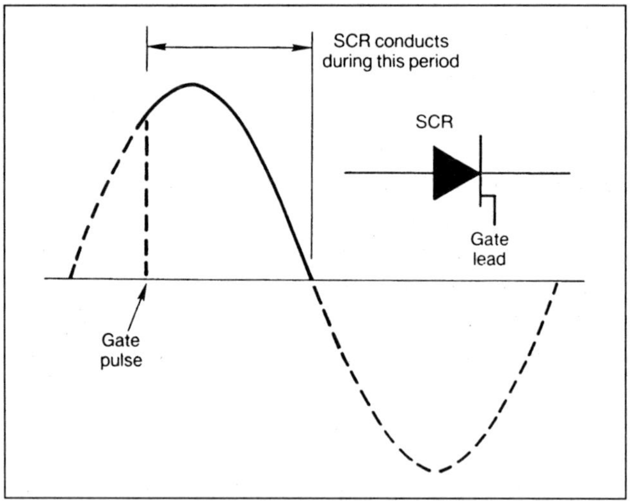

Figure 13-11: SCR characteristics.

Unlike manually and magnetically operated switches when turned off, solid-state switches leak a small amount of current. The characteristic calls for caution to avoid shock because seemingly open circuits may be well above ground potential.

Programmable Controllers and Computers

Power transmission specialists divide programmable logic controllers (PLCs) and computers into roughly four categories: programmable logic controllers, board-level industrial computers, personal computers (PCs), and industrial computers.

There are fundamentals that bind PLCs and computer control devices together, and it is helpful to examine the similarities before discussing differences.

- Each are programmable so that the operator can tell the controller when devices should be turned on, off, sequenced, or adjusted to fulfill the needs of the process being controlled. The controllers then activate switches to cause events to happen at the right time. Users program the system using simple language programs that are provided by the PLC manufacturers.

- They have four common features—recallable memory, inputs and outputs, user programmability, and internal bus systems.

- They store data and access data when necessary. Data can be anything from instructions or commands to status information from external devices.

- Via inputs and outputs, receive and transmit data to external devices. These can be displays, transducers, sensors, other computers, or a variety of other devices.

- They respond to a set of stored instructions, or programs, that define what information they will accept, how that information will be processed, and under what conditions various outputs will be energized or de-energized.

- They contain communication buses to transmit data to various sections of the systems. External communication buses, using twisted pair wires, coaxial cable, or fiber optics, enable communications between external I/O devices. Data networks, such as LANs (local area networks) and WANs (wide area networks) offer communications between various intelligent devices such as other computers, PLCs, and workstations. Standard communication routines and protocols have been developed to permit data exchanges between products of different suppliers.

PLCs

Developed to be a long-life replacement for electro-mechanical relays in logic circuits, PLCs are a special type of industrial computer specifically designed for operation in a factory-floor environment. Fig. 13-12 illustrates the schematic of a representative PLC.

Programmable logic controllers:

- Interface directly with external input devices (limit switches, pushbuttons, thermocouples, and encoders) without additional signal conditioning. Output commands from a PLC go directly to output control devices, such as relays, starters, solenoids, and adjustable-speed drives.

- Use a variety of programming languages. So-called ladder logic programming resembles the electrical schematic diagrams of conventional relay-based control circuits, consisting of normally open and normally closed contacts and relay coils. The programs are based upon rapid scans of all input signals, comparison of the input signals with the program instructions retained in memory, subsequently sending the proper output commands at the conclusion of the updating process. The scanning and updating process is fast—often 1,000 "words" of input data are scanned in less than one millisecond.

- Contain counter and timer circuits, which expand the capabilities of relay logic control circuits; intelligent I/O modules, such as thermocouple modules, drive interface modules, and high-speed counters; and PID (proportional, integral, and differential) loop control modules.

- Are capable of using higher language programming than ladder logic. Not only does this include BASIC or C, but also symbolic function block programming developed for PLCs to help simplify repetitive control routines.

PLC users classify the broad offerings from dozens of vendors by a number of different objective yardsticks. They include the following:

Memory

The size of the PLC memory is usually defined in so many meg millions or "K," or thousands of eight- or sixteen-bit "words." For example, with 32K of RAM, there can be different levels of memory for the applicable program, register, or special data storage, and for active data. Memory can be classified by various types of "RAM" and "EEPROM."

RAM needs battery backup to maintain data in the event of a power loss, whereas EEPROM does not.

Input / Output Capability

Input/output (I/O) capability is generally expressed in terms of a quantity—8, 16, 64, 128, all the way up to 10,000 or more. An input can be a discrete AC or DC on/off signal from a pushbutton or limit switch. Also, inputs can be analog—so many volts or milliamperes—or compatible with a thermocouple. Inputs can come from intelligent devices, such as high-speed counters or drive positioning modules, where the input device contains its own microprocessor.

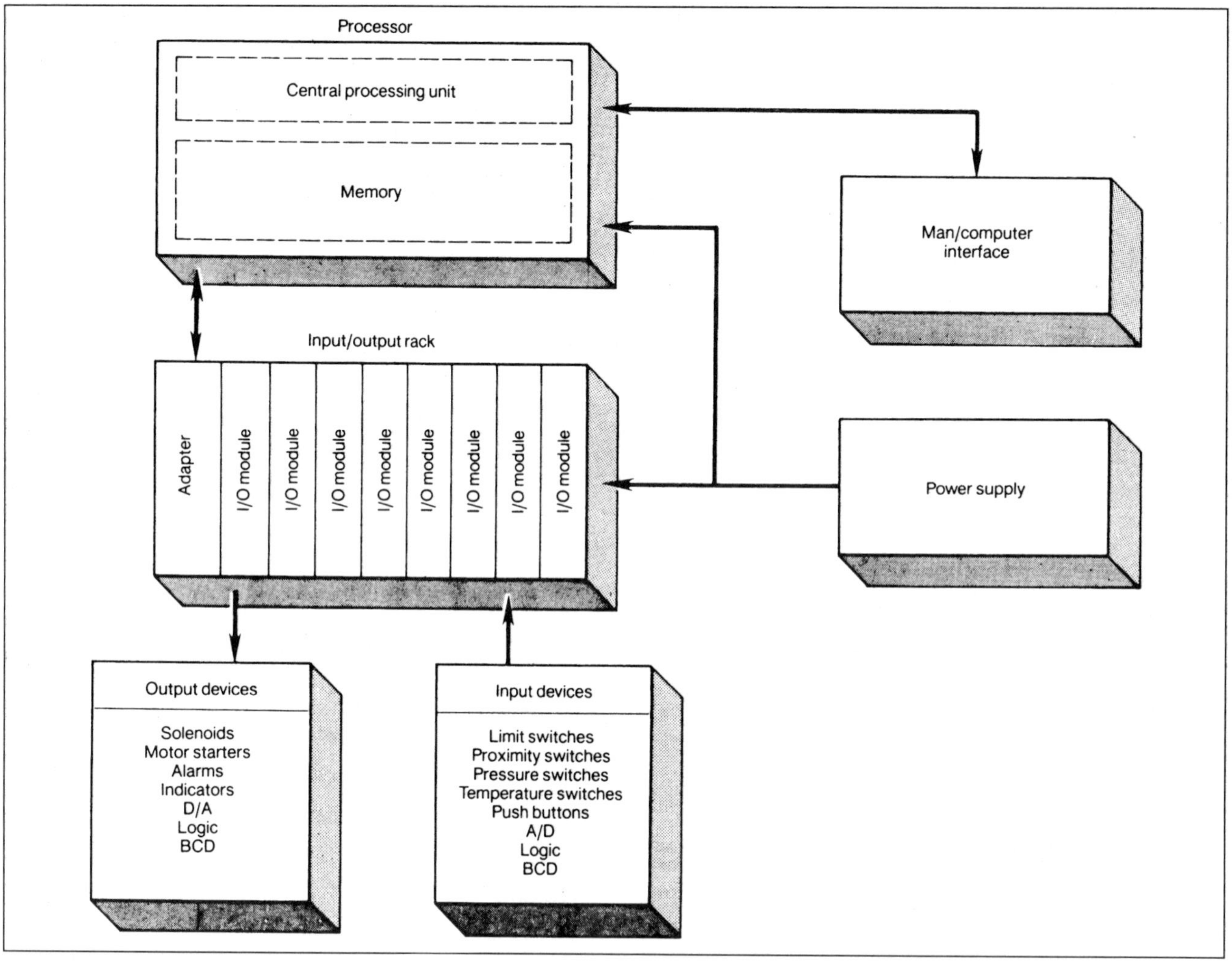

Figure 13-12: Schematic block diagram of a PLC.

PLC inputs (and outputs) generally require no additional signal conditioning. PLC outputs can be discrete DC or AC and rated for appropriate power devices, such as relays, contactors, or motor starters. Most PLC suppliers offer inputs to receive analog and intelligent outputs, such as speed reference commands from a drive positioning module. In some cases, discrete outputs (on or off) must be isolated through the use of relays because solid-state outputs are often not electrically isolated—a common user requirement.

Users often classify PLCs as small, mid-sized, and large, either by their I/O count or memory. However, the flexibility of even the smallest PLCs defies simple classifications.

Programming Languages

Users need to know the programming languages used by the PLC. The simplest units use either ladder logic or Boolean statement (AND, OR, NOR, etc.) programs. However, newer forms of functional programming and standards are available, such as IEC 61131-3, along with the use of high-level languages, which greatly improve code writing efficiency.

There are many types of programming devices used with PLCs. The simplest PLCs may require the use of a dedicated programming unit and entering the programs directly to the PLC. However, most PLCs now use some form of personal computer to develop programs off-site and enter them into PLC memory by simply transferring the data from a computer to PLC.

Networking

Virtually all PLCs work with some form of local area network (LAN) communications. The issue becomes which LAN protocol is used and whether it is better to use an open (multivendor) LAN or a proprietary LAN that is offered by only one vendor.

Other Features and Functions

PLCs are also classified by their ability to handle math functions (add, subtract, divide, multiply, compare values, double precision number handling), diagnostics, program documentation, and fault tolerance capabilities.

Users easily encounter an information overload as they sort through competing claims from various PLC suppliers. Frequently, qualitative judgments, such as past personal experience, user support, and reputation, become decisive in making a vendor selection.

Personal Computers

Many industrial users have found personal computers (PC) to be a useful industrial automation tool, and low-cost, high-performance industrial application programs have been developed that ease the task of programming the PC.

PCs have several disadvantages. First, personal computers do not have an I/O structure. A board-level I/O must either be interfaced to process or an industrial I/O system must be interfaced to the computer. In addition, control language is not integral to the PC; therefore, the user must choose an appropriate software package.

Commercial PCs are not designed for operation in a factory floor environment. Industrial PCs are available that are "hardened" to accept the harsh environment of the factory floor—temperature, shock, and voltage variations.

Board-Level Computers

A number of different board-level computer systems are widely employed industrially. Some compete with PLCs, and some PLCs are designed to be compatible with them. All have some degree of multivendor standardization so that with the same product family, the products of different suppliers can interface with each other.

IEC 61131-3

IEC 61131-3 is the third part of the open international standard IEC 61131 for programmable logic controllers and was first published in December 1993 by the IEC. Part 3 of IEC 61131 deals with programming languages and defines two graphical and two textual PLC programming language standards:

- Ladder diagram (LD), graphical.
- Function block diagram (FBD), graphical.
- Structured text (ST), textual.
- Instruction list (IL), textual.
- Sequential function chart (SFC), has elements to organize programs for sequential and parallel control processing.

Board-level computer hardware consists of multiple printed circuits for CPUs—central processing units—and various I/O modules mounted in standardized electronic racks. Standards permit products of various manufacturers to be compatible.

Some of the important "backplane" communication standards include: VME/VXIbus, Multibus II, STDbus, STD 32 NuBus, Futurebus, and EISA (Extended Industry Standard Architecture). Board-level computers have the capability to perform multitasking and multiprocessing tasks.

Industrial Computers

Industrial computers are integrated computer systems designed specifically for factory automation applications. Their operating systems, such as UNIX, support multitasking and multiprocessing routines. Some industrial computers have fault-tolerant designs (designed to protect the controlled process in the event of specific hardware failures, e.g., a power supply or a disk drive).

Generally, manufacturers of industrial computers offer application software support designed for their equipment rather than depend on third-party software.

Operator Displays

Operator displays provide an information interface between a controlled process and human operators. Four key display technologies are outlined: indicators, message displays, operator interface units, and workstations.

Indicators

Indicators use light to illuminate a device that is actuated by some control event. The control event can be the result of normal operation, such as the status of a motor starter, or the normal operation of a machine, as detected by one or more sensors. The control event can also be a warning to the operator of a malfunction. Pilot lights are one of the most common indicators, but light-emitting diodes (LEDs) are also used. Fig. 13-13 shows the construction of an LED lamp and a bar graph display where 10 to 15 LEDs are used to produce a word or numerical legend. When a small amount of current flows through the LED, it produces a specific color (yellow, green, red, or orange), depending on the materials used in its construction.

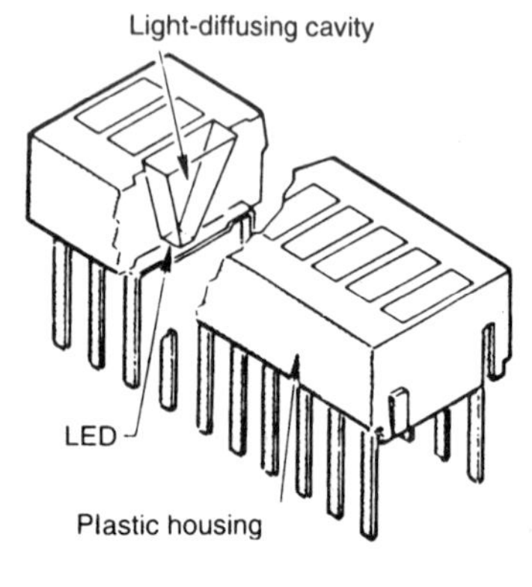

Figure 13-13: LED light and display.

Character displays use other technologies such as liquid crystal displays (LCDs) and electroluminescent displays (ELDs).

Message Displays

As the name implies, message displays provide operators with plain English (or the language of choice) messages in response to various control events.

Large characters—up to 4 inches high—are common in so-called marquee displays. Smaller, multiple-line displays are also popular on individual control devices. Fig. 13-14 shows a typical, multiple-line message display using multiple LEDs as point light sources. LCD and ELD technologies are also used.

Figure 13-14: LED message display unit.

The sophistication used to generate messages depends upon the creativity of the equipment designer. Some messages are fixed and energized in response to a single control event. Other message displays are stored in the memory of an external device, such as a PLC, or in the programmable message display unit itself. Complex messages can be displayed or scrolled in response to a wide variety of control events.

Operator Interface Units

An operator interface unit is an intelligent portable control terminal that enables an operator to access and change the control parameters of an item. Fig. 13-15 shows a small hand-held operator interface unit using a multiple-line LCD display and data entry keys.

Operator interface devices communicate directly with an external intelligent device (PLC, PC—personal or industrial computer) through a data network.

Figure 13-15: Operator interface unit.

Workstations

Carrying operator interface devices at least one step further, workstations offer increased display capabilities—such as color graphics and touch screen capability—and full access to system data. Workstations are generally considered as stationary devices mounted at an operator's normal place of business. Workstations can be proprietary devices, part of a manufacturer's total automation system, or they can be manufactured by third parties to be compatible with the products of several vendors.

Workstations frequently contain on-board computers. They can extend the capabilities of operator interface devices (data access, data display, and set-point adjustment) to centralized control of a total manufacturing process.

Fig. 13-16 shows a full-capabilities workstation that contains an on-board computer with an 80-MB hard drive for data storage.

Sensors

Sensors are devices that respond to a physical stimulus (heat, light, pressure, and flow, for example). They react by producing a signal that is used for measurement or control. There is virtually no physical property that cannot be detected and measured by the right sensor. This chapter covers only sensors frequently encountered by the power transmission specialist. These include sensors that detect speed, position, physical presence, liquid level, pressure, and force. This chapter also covers signal conditioning, often an adjunct to sensor operation.

Tachometer Generators

Fig. 13-17 shows two types of tachometer generators (frequently called "tachs") that are widely used for measuring the speed of a rotating shaft. Illustration A is a DC tachometer generator, and illustration B is a pancake tachometer generator. In some cases, output from such a tach powers a display that indicates speed of a machine or process or provides a speed feedback signal for an adjustable-speed drive. DC tachometer generators are generally rated 50, 100, or 200 VDC per 1,000 rpm. They are available with ratings for either 40° C or 65° C ambient temperatures, ripple voltage from 8 percent to 0.5 percent rms, and with various temperature coefficients (the amount of voltage change per

Figure 13-16: Workstation.

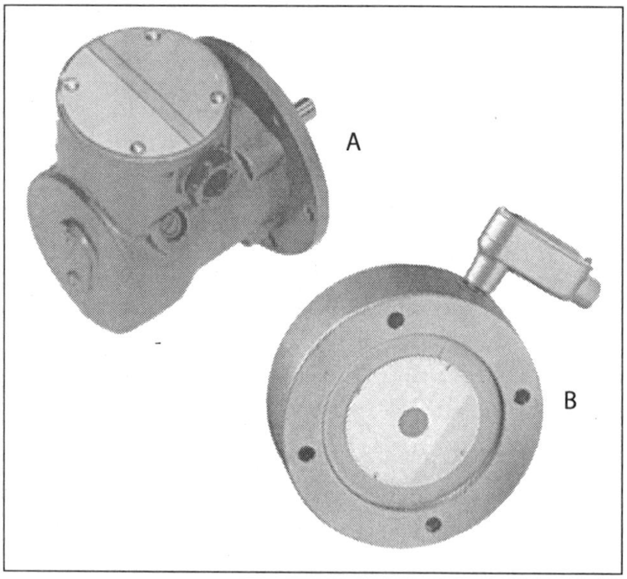

Figure 13-17: Tachometer-generators.

degree C). These factors affect tachometer cost and weight, which range from about 1.7 to 35 pounds. AC brushless tachometer generators produce a variable voltage which, when rectified, generally provide 22.5, 45 or 90 VDC per 1,000 rpm.

Encoders

Used for both position and speed feedback, encoders produce a pulse each time a shaft rotates through a specified angle. Both optical and magnetic-type encoders are available. Magnetic encoders are less expensive and mechanically simpler, but their resolution is limited to about 1 degree (360 pulses per revolution). By contrast, optical encoders produce up to 5,000 pulses per revolution.

Incremental Encoders

There are two types of optical encoders—incremental and absolute. Fig. 13-18 is a schematic diagram of an incremental encoder. They are relatively inexpensive, resistant to shock and vibration, and provide position and speed data down to zero rpm.

Incremental encoders employ a rotatable disk which has etched rectangular gratings and a stationary mask with etched slots. A light, either from light-emitting diodes (LEDs) or incandescent lamps, shines through both the mask and disk slots. The resulting pulses of light are detected by phototransistors as the encoder's shaft rotates. The device typically generates pulses for channels A and B, which are in quadrature to indicate the direction of rotation. This arrangement produces four sets of edge pulses, which provides a resolution angle that is one-fourth that of the etched disk.

The two-channel technique indicates both direction and speed. A third channel, an index or Z-channel, produces one pulse per revolution. This is a zero-position or reference signal.

Square-wave encoders offer resolutions up to 5,000 lines per revolution; however, operating at more than a unit's maximum rate, called slew speed, accuracy decreases rapidly, and the possibility of mechanical damage increases.

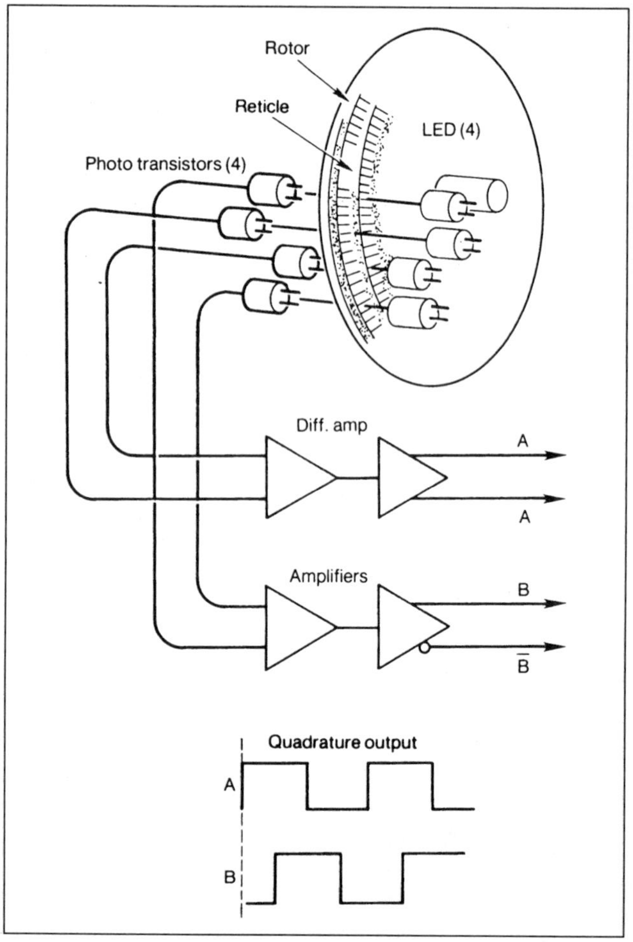

Figure 13-18: Basic components in an incremental optical encoder designed to sense bidirectional motion.

Incremental encoders have one limitation. The circuits with which they are used count pulses to determine position. Hence, a power interruption causes a loss of position information because the counter loses count.

Absolute Encoders

To avoid the limitation of incremental encoders, absolute encoders employ a radically different disk pattern. These encoders provide multiple tracks, each with an independent light source. The alignment of light pulses from the multiple tracks, when viewed through a masking slot, is unique for each increment of shaft position, and the flow of binary data (pulses) for one directional rotation differs from the other. Thus, position data is not lost on power failures.

Gearing additional disks to the primary disk allows an absolute encoder to count turns. Adding a second disk provides data on eight turns, a third disk on 64 turns, etc. According to one manufacturer, data up to 512 turns is practical.

Absolute encoders also provide more precise position data when operating in electrically noisy environments. They do not rely on counting circuits, and thus can be used in higher-speed applications than incremental encoders. Also, they do not require a referenced starting point, as do incremental encoders—a major advantage in some motion control applications, but absolute encoders cost more than incremental encoders.

Resolvers

The resolver is constructed like an electric motor in that it has a rotor and a stator. However, the rotor contains a primary winding that has an AC signal applied to it that may range from 400 Hz to several thousand Hz. The stator contains two secondary windings that are displaced from each other by 90 deg. Fig. 13-19 shows the schematic of a typical resolver. As the rotor rotates, the output voltages produced in the stator create signals, which can be counted by associated circuitry to provide precise shaft position data with resolutions of up to 65,000 counts per revolution.

Resolvers are often enclosed in rugged housings and frequently installed in machine tools.

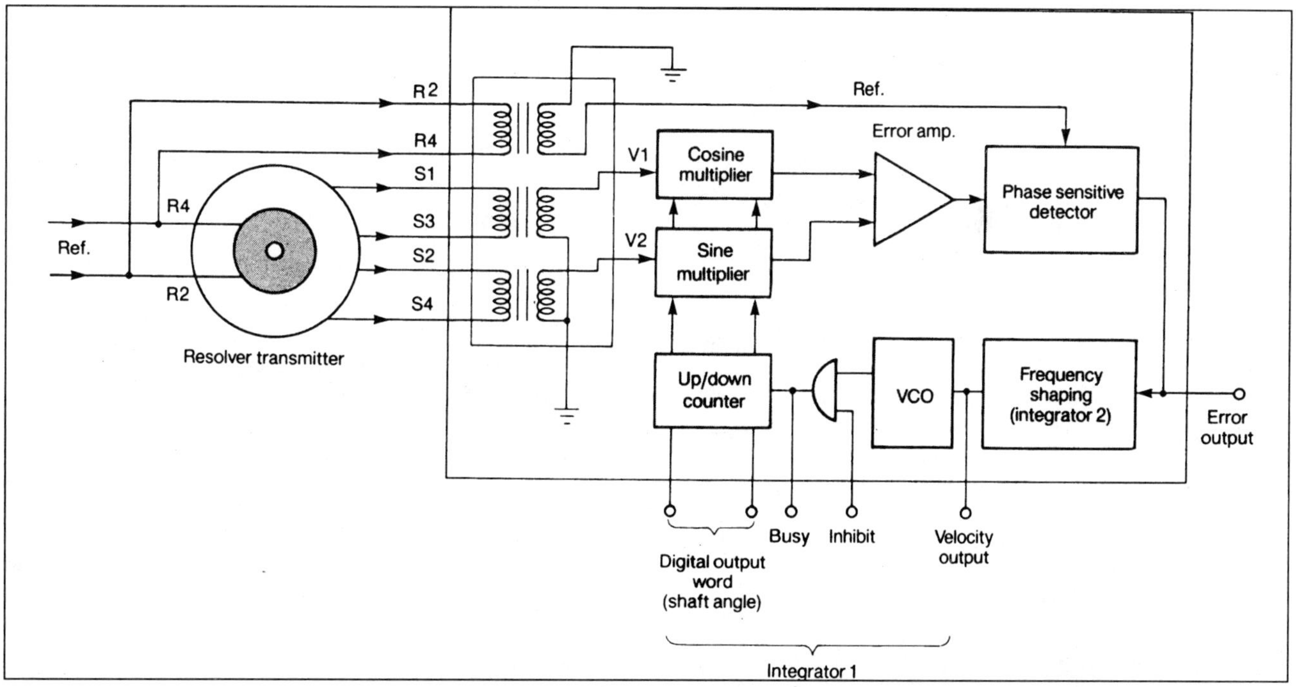

Figure 13-19: Resolver with resolver-to-digital converter circuitry.

Proximity Sensors

A wide variety of proximity sensors are available that detect the presence of an object without physically touching it. Sensing distances can range from fractions of an inch to several feet. The sensors buried in the street pavement to detect cars for controlling traffic signals are usually proximity sensors.

According to one survey, eleven different sensing technologies are used by different proximity sensors (capacitive, inductive, eddy-current, hall effect, Weigand effect, photoelectric, ultrasonic, radar, field effect, radio absorptions and magnetic). Fig. 13-20 shows one type of inductive proximity sensor.

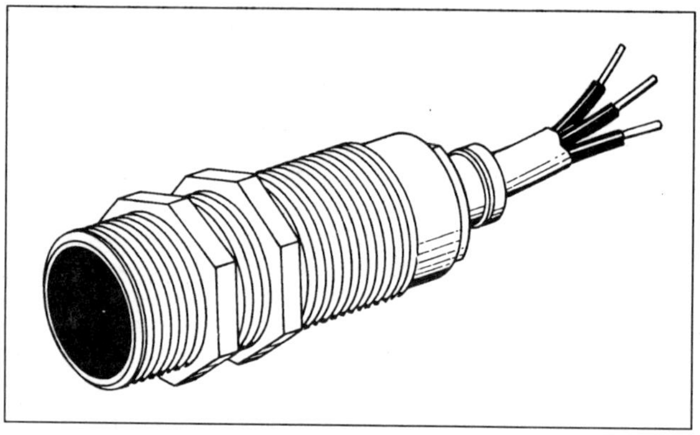

Figure 13-20: Typical inductive proximity sensor.

Proximity sensors exhibit higher operating speeds than mechanical limit switches, can sense objects approaching from any direction, and may not be dependent on the mass or size of the object to produce a signal output. Because proximity sensors are a non-contact sensing technology, they are not generally subject to mechanical damage or wear.

A number of different issues need to be considered when selecting a proximity sensor. These include the sensing speed required, the level of contamination to which the sensor will be exposed, the type of enclosure required, the maximum distance from the target object to the sensor's probe, target materials (some sensors—e.g., inductive—work only with metallic objects), exposure to chips (here, some produce false signals), exposure to RF interference, welding fields, and exposure to mechanical impact from target objects. Proximity sensor manufacturers provide menus or checklists to assist in proximity sensor selection.

Limit Switches

Limit switches translate motion into switch actuation. The motion can be vertical, horizontal, axial, or angular (cam operated). Fig. 13-21 shows typical push-roller and rotary-limit switches.

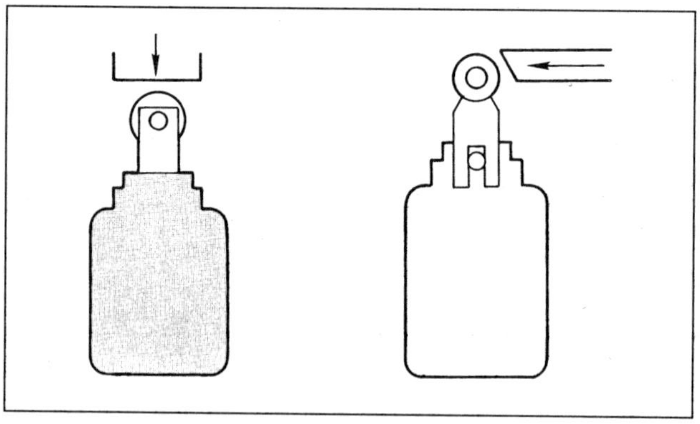

Figure 13-21: Typical push-roller and rotary limit switches.

Switches are used to detect that a specific control event has occurred—a limit switch, for example, controlling an indicating lamp to show when a machine part reaches a certain point. Others are used to initiate control operation of a machine process, for activating reversal of direction, or when an object has reached a specified position. Limit switches are also widely used to trigger emergency stop circuits.

Liquid Level Sensors

Widely installed in continuous and batch control process systems, liquid level sensors provide two different types of information. Some types indicate when liquid levels have reached a specific point. For example, the water and waste-treatment industries use liquid level sensors to control the output speed of adjustable-speed pump drives.

One type, the float switch, is used to automatically control pump motors which regulate liquid levels. These switches may be applied so that when the liquid rises contacts close (for sump pump applications) or open (for open-tank pump applications).

Another type, the bubbler sensor (see Fig. 13-22) is popular for water and waste treatment applications. The sensors are simple. They continuously purge the probe of contamination by suspended solids, provide continuous level measurement, and are compatible with automatic controllers, such as adjustable-speed drives. The principle of a bubbler sensor is this: if a constant flow of compressed air flows through a pipe inserted in a liquid, the pressure in the air line is proportional to the depth of immersion.

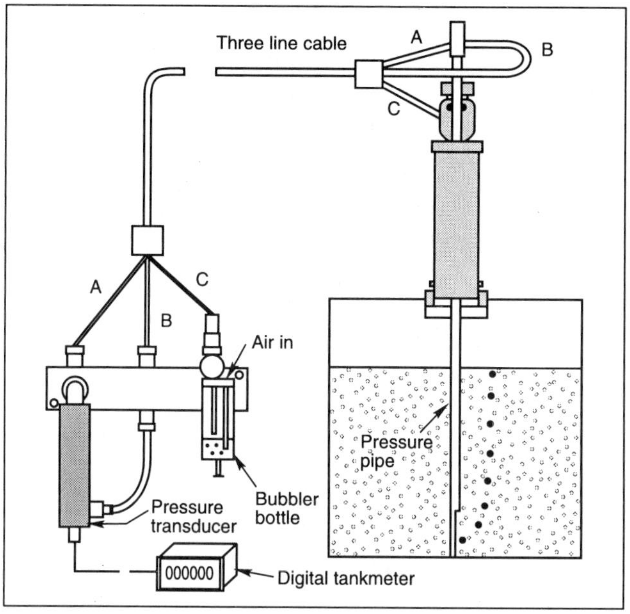

Figure 13-22: Typical bubbler-type liquid-level sensor.

More sophisticated level sensing devices employ capacitance/admittance, differential pressure, radar, electromechanical devices (e.g., magnetic floats, resistive tapes, etc.), and ultrasonic and optical technologies. Many sensor manufacturers include extensive on-board signal processing. Each sensing technology has a set of performance capabilities, advantages, limitations, and costs that become part of the evaluation criteria.

One factor determining the choice of sensor technology is a clear definition of the liquid being sensed. For example, from bottom to top, the properties of a liquid can change from a sludge, to a heavy liquid, to an emulsion interface, to a lighter liquid, to a surface sludge, to a surface foam, to a vapor, and then to air. The issue becomes which liquid surface is the one to be sensed. Various sensor technologies work best with one or two, but not all liquid interfaces.

In addition to defining the interface of the liquid level to be sensed, the user needs to define the temperature and pressure of the liquid medium. The corrosive nature of the measured liquids must also be defined.

Pressure Sensors

There are more than a dozen different pressure sensor technologies ranging from simple bellows units to more sophisticated electrical/electronic units. But only the three popular high-performance technologies (see Fig. 13-23) are discussed here. They are the strain gauge type, the piezoresistive type, and the capacitive type sensors. Each type provides good repeatability, high accuracy, and reasonable costs. Both strain gauge and piezoresistive sensors use the principle that electrical resistance varies with pressure. Both types employ a sensor that is bonded to a movable diaphragm. An unbalanced pressure on the diaphragm alters sensor resistance directly proportional to the pressure differences between the two sides of the diaphragm. Both types employ temperature compensation, because temperature as well as strain affects the resistance of the sensing element. Strain gauges are rated by the amount of pressure that they are able to withstand and should be applied accordingly.

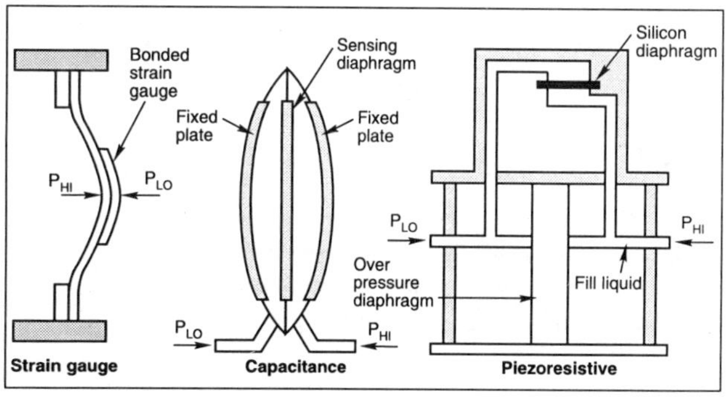

Figure 13-23: Strain gauge, silicon crystal piezoresistive and capacitance-type pressure transducers.

An important variation of the strain gauge sensor is the so-called single-crystal diaphragm. This sensor is more accurate and avoids the strain gauge bond cracking sometimes exhibited by other strain gauges.

Capacitive sensors are widely employed for measuring low values of pressure. Here, two plates are electrically charged, and movement of the sensing diaphragm changes the capacitance values of both plates. The difference in the capacitance of each plate, divided by the sum of their capacitance values, is proportional to the differential pressure.

Force Sensors

Two devices are widely used in the measurement of force, weight, and tension. They are linear variable differential transformers (LVDTs) and strain gauges. Strain gauges are also employed for measuring torque, a task for which LVDTs are unsuited.

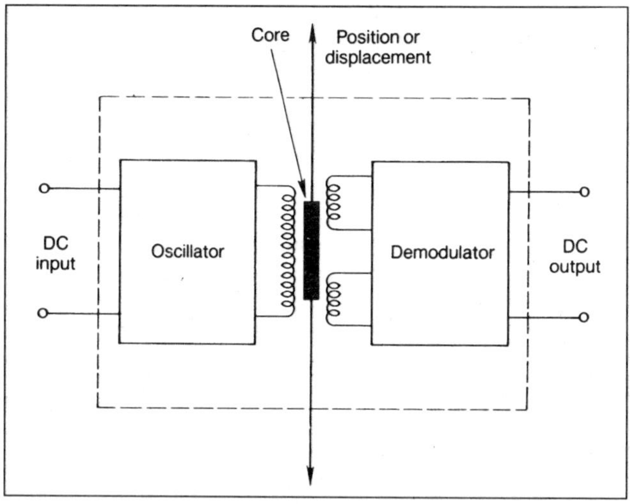

Figure 13-24: Principle of a DC-to-DC LVDT.

Fig. 13-24 is a schematic diagram of an LVDT. The primary, or exciting winding, is powered by an oscillator. When the movable axial core is precisely centered, the two secondary windings at the right, which are wound in opposition to each other, produce opposing voltages that cancel each other out. When the core is displaced by an external force in either direction, the voltage of one winding increases, and the other decreases proportionately. The summation of the voltages are demodulated (rectified), producing a DC voltage whose magnitude and polarity are functions of the axial core displacement and direction. The sensors are small and can precisely measure small values of displacement. Fig. 13-25 shows a load cell type of force transducer using an LVDT to measure force or weight. Fig. 13-26 shows two types of torque transducers using strain gauges.

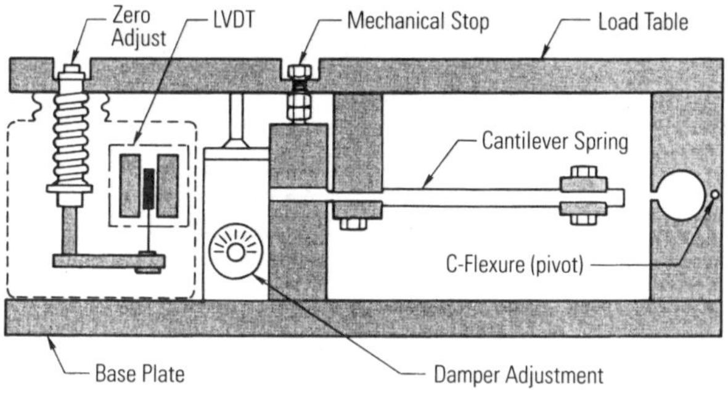

Figure 13-25: Typical load cell or force transducer.

Fig. 13-27 shows two types of strain-gauges, the bonded filament and thin film foil types. Other types of strain-gauges exist, which provide specific application advantages. However, their operational theory is the same. Distortion of a gauge changes its value of electrical resistance.

The Wheatstone bridge circuit, used in many strain gauge measuring systems, is shown in Fig. 13-28. It consists of three precision resistors, a source of DC voltage, and a galvanometer. When the strain gauge is not distorted, the current divides evenly in the four resistor bridge circuits, and no current flows through the galvanometer. When the strain gauge is distorted, its resistance changes, causing a current imbalancè between the upper and lower paths of the bridge circuit. The current flowing through the galvanometer is proportional to the distortion of the strain gauge.

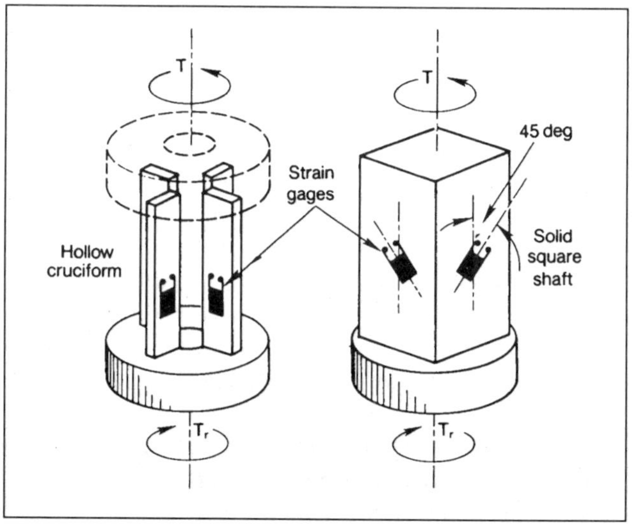

Figure 13-26: Typical torque transducers using strain gauges.

Evaluation of force, weight, tension, and torque transducers requires complete definitions of the values to be measured, the dynamic response required, needed accuracy and resolution, environmental conditions, and, in the case of torque transducers, whether slip rings can be tolerated.

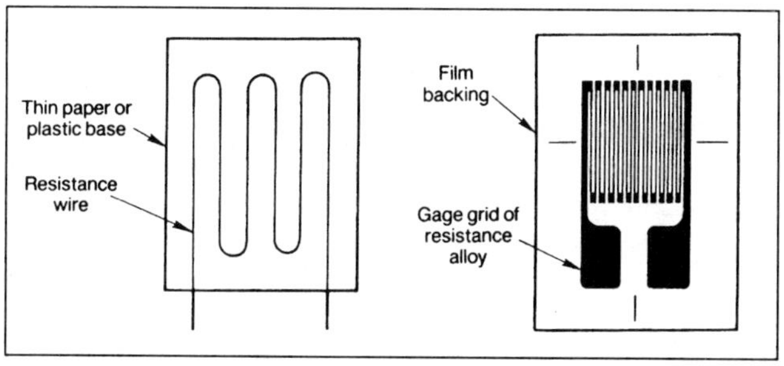

Figure 13-27: Strain gauges.

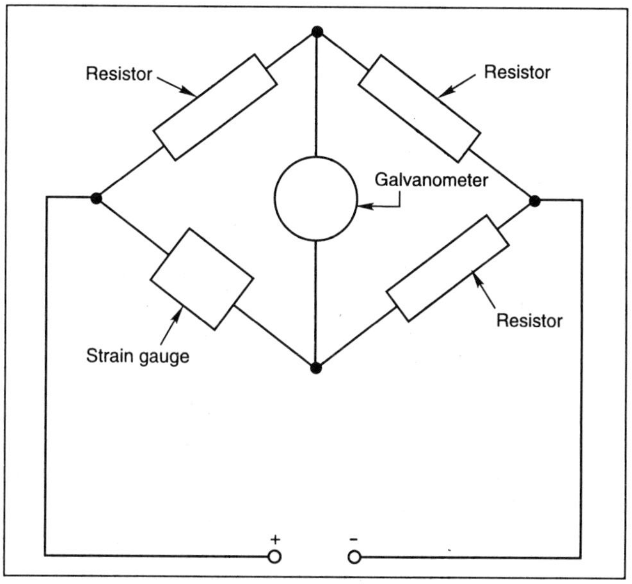

Figure 13-28: Wheatstone bridge.

Photoelectric Sensors

Photoelectric sensors are a type of proximity sensor, but because of their popularity, they are treated separately. Photoelectric sensors contain special materials that produce a voltage when exposed to light. The sensors are generally used to detect the presence of a moving target, employing one of two techniques. With one technique, the target interrupts light going from a source to a sensor. In the other method, the object reflects or absorbs light via a reflective path (see Fig. 13-29).

Four variations of a reflective path are shown in Fig. 13-30. Specular reflection occurs when a beam of incident light strikes a mirrored or shiny surface. Here, the angle of incidence for both incident and reflected light are equal. Retroreflection uses a prismatic or special reflector that reflects light directly back to the source. With diffuse reflection, a dull surface, such as cardboard, reflects light in all directions. Absorption occurs when all light is absorbed by the object (e.g., by a dull black object).

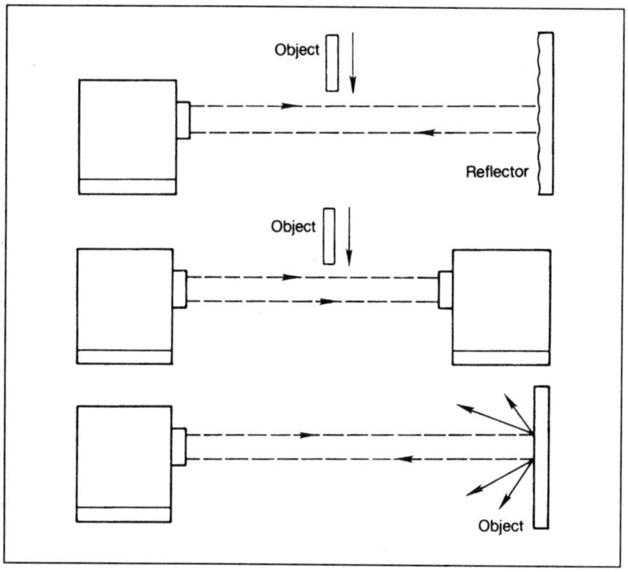

Figure 13-29: Photoelectric sensor scanning techniques.

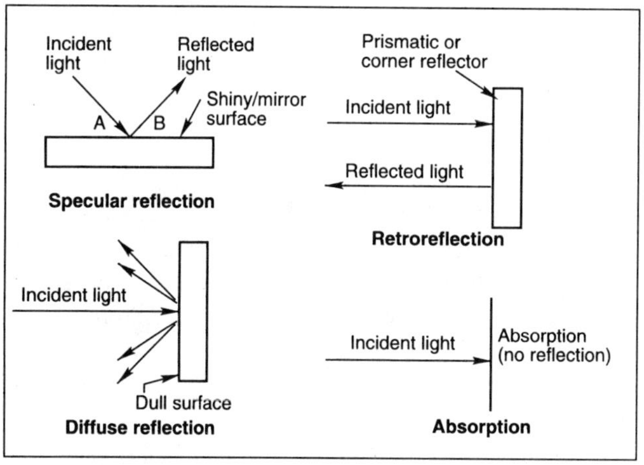

Figure 13-30: Specular reflection, retroreflection, diffuse reflection, and absorption.

Photoelectric systems require considerable application analysis using manufacturers' data on limits for misalignment, distances, and target surfaces. Most systems are insensitive to light from extraneous sources, such as sunlight and ambient room lighting. It is important to consider environmental operating conditions, such as temperature, shock, and dirt.

Common Types of Sensors Used in Automation Systems

Inductive

Detects various metal targets using an inductive field. Range is 0–40 mm. Advantages: very rugged design, able to operate in harsh environments. Disadvantages: short range operation, ability to detect metal objects only. Applications include general purpose welding, metal cutting, and rpm pickup.

Capacitive

Detects object based on a change in density. Range is 0-40 mm. Advantages: best for detecting dense materials and liquids, simple to use, and adjustable. Disadvantages: short ranged, affected by humidity changes. Applications include water level, glue, and label detection.

Magnetic

Detects a magnet. Range based on power of the magnet. Advantages: best for pneumatic cylinders, detects magnet through most metals except steel. Disadvantages: short range, requires a magnet target. Applications include cylinder position detection, rpm pickup.

Mechanical

Detects object through physical contact. Advantages: Most repeatable switching point, very rugged. Disadvantages: subject to wear, sensor is usually large. Applications include machining centers, gantries, and robotics.

Photoelectric – Diffuse

Detects any object capable of reflecting light. Range is 5-2,000 mm. Advantages: handles many applications, detects small objects. Disadvantages: short range, adversely certain colors, and dirt. Advantages include general purpose use, color mark detection, and label detection.

Photoelectric – Retroreflective

Detects any target capable of blocking light. Range is 0–12 m. Advantages: best for conveyor operations and transparent targets. Disadvantages: requires a reflector, adversely affected by dirt build up. Applications include conveyors, large openings, and bottle detection.

Photoelectric – Thru Beam

Detects target capable of blocking light. Range is 0 - 50 m. Advantages: best for long range or to see through a translucent object, tolerant of dirty environment. Disadvantages: requires both an emitter and receiver. Applications include height and configuration verification, fill levels, and loop control.

Vision Systems

Especially helpful for improving product quality and productivity, vision systems integrate industrial computers with video cameras that use either charge-coupled devices (CCDs) or vidicon tubes. A tube contains thousands of pixels, and each pixel provides data about the light it receives.

Individual pixel outputs produce analog signals that, when processed by a converter, indicate various shades of gray or one of the three primary colors. The computer thus enables the vision system to recognize specific items regardless of physical orientation.

Signal Conditioning

Signal conditioners process raw data from various field sensors and transmit that data to external intelligent devices, such as computers, for further processing. Designed for factory environments, signal conditioners feature the following:

- Isolation—They incorporate transformer or optical isolation to protect circuits from damaging voltage transients (typically providing 1.5 to 3 kV protection), eliminate shock hazards, and avoid problems caused by ground loops between field sensors and the data processing equipment.
- Stable output—They deliver a stable output, capable of being transmitted over reasonable distances. For example, using RS-422 communication channels, output signals can be transmitted up to 4,000 feet.
- On-board intelligence—Conditioners may include on-board data processing, analog-to-digital (A/D), and digital-to-analog (D/A) conversions from eight to 20 bits of data, analog displays, alarm outputs, and multiplex multiple channels of data over a communication bus.

Fig. 13-31 illustrates three criteria of the single term, "accuracy," which is one of the more important performance criteria of a signal conditioner.

Linearity quantifies the ability to produce a perfectly linear relationship between input and output signals. The maximum difference, measured as a percent of the full-scale input, is the linearity error. The gain error is the difference in slope, expressed as a percentage value, between a perfectly linear and the actual input-output relationship. The offset error is based on the amount of input signal necessary to begin producing an output from the signal conditioners. Gain and offset errors can be compensated for in many signal conditioners.

Users are increasingly interested in real-time control, and they consume vast amounts of data at a high speed. When data must be communicated at high speed, or where multiple channels are communicated by multiplexing on a single bus, the user can encounter limits on data throughput. Higher gain signal conditioners have slower dynamics than their lower gain cousins and take longer to reach a stable output.

Selection of a signal conditioning strategy involves evaluation of all of the factors described here, as well as others dependent on the type of sensor information to be transmitted and processed by the external computer or PLC.

Control Enclosures

Two sets of widely different standards for control enclosures impact all control devices and equipment used in North America. This handbook concentrates on similarities and differences between North American and IEC control enclosures.

NEMA, CSA, and UL Control Enclosures

Table 13-2 lists the types of enclosures available to meet North American standards for nonhazardous locations. Tables 13-3 and 13-4 list intended levels of protection provided by indoor- and outdoor-type enclosures, respectively.

Table 13-5 lists the North American classifications for hazardous locations. The procedure to certify that a product and its enclosure are suitable for operation in a hazardous location requires that the combination be tested for many factors, such as surface temperature.

NEMA and UL Type 9 enclosures not listed in tables are intended for Class II, Groups E and G hazardous locations.

There is also a Class III hazardous location, where ignitable fibers are present but are not likely to be in suspension in sufficient concentration to produce ignitable mixtures.

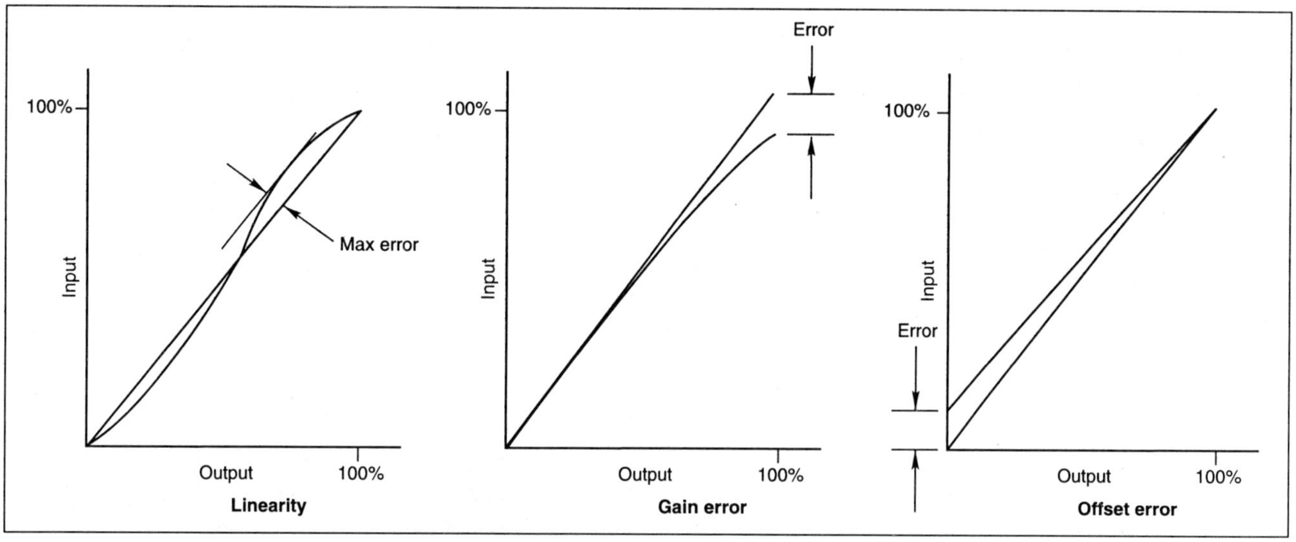

Figure 13-31: Methods of measuring signal errors.

TABLE 13-2 – Cross-reference of enclosure type availability			
Enclosure type	NEMA	UL	CSA
1	XX	XX	XX
2	XX	XX	XX
3	XX	XX	XX
3R	XX	XX	OO
4	XX	XX	XX
4X	XX	XX	OO
5	OO	OO	XX
6	XX	XX	OO
12	XX	XX	OO
13	XX	XX	OO

XX = Available; OO = Not available

TABLE 13-4 -- Levels of protection for outdoor control enclosures
(Excluding 6 and 6P submersible enclosures)

Protection against these conditions	NEMA, UL or CSA Type				
	3	3R	3S	4	4X
Incidental contact with enclosed equip.	X	X	X	X	X
Rain, snow & sleet (1)	X	X	X	X	X
Sleet (2)	O	O	X	O	O
Windblown dust	X	O	X	X	X
Hosedown	O	O	O	X	X
Corrosive agents	O	O	O	O	X

Notes: 1 - External ice-covered controls may not be operable.
2 - External ice-covered controls are operable.

TABLE 13-3 – Levels of protection for indoor control enclosures

Protection against these conditions	NEMA, UL, or CSA Type									
	1	2	3	4	4X	5	11	12	12K	13
Incidental contact with enclosed equipment	X	X	X	X	X	X	X	X	X	X
Falling dirt.	X	X	X	X	X	X	X	X	X	X
Falling liquids & light splashing.	O	X	X	X	X	O	X	X	X	X
Dust, lint & flyings.	O	O	O	X	X	X	O	X	X	X
Hosedown & splashing water.	O	O	O	X	X	O	O	O	O	O
Oil & coolant seepage.	O	O	O	O	O	O	O	X	X	X
Oil/coolant splashing & spraying.	O	O	O	O	O	O	O	O	O	X
Corrosive agents.	O	O	O	O	X	O	X	O	O	O

X = Suitable, O = Unsuitable

TABLE 13-5 – North American classification of hazardous locations

Class I Combustible gases, liquids and vapors	Class II Combustible dusts
Group A – Acetylene Group B – Hydrogen & hydrogen-based gases Group C – Ethyl ether, ethylene, etc. Group D – Natural gas, petroleum products, acetone, etc.	Group E – Metal dusts, e.g., magnesium Group F – Carbon-based dusts,e.g., coal Group G – Other dusts, e.g.,flour & wood

IEC Control Enclosures

Enclosures designed to meet IEC standards have a nomenclature bearing the prefix "IP" followed by two numbers. The first (see Table 13-6) describes the degree of protection from mechanical intrusion (for example, a finger or a wire). To quantify this, IEC tests for mechanical intrusion with four sizes of probes, as indicated in the table. The second number (see Table 13-7) indicates protection from water.

Tables 13-6 and 13-7 list all IEC control enclosures except those for shipboard applications.

Table 13-8 lists an approximate correlation between IEC and NEMA, CSA, and UL enclosures. This listing ignores the differences in the effects of oils and coolants versus water or the use of corrosion-resistant materials in some NEMA/CSA/UL enclosures.

This table shows, for example, that a NEMA Type 1 enclosure should meet the general requirements for IP21, IP22, and IP23 enclosures, and vice versa. This table does not suggest that an enclosure manufacturer is relieved of testing responsibilities to demonstrate compliance to specific enclosure standards.

TABLE 13-6 -- IEC enclosure mechanical protection
(Data courtesy Hoffman Engineering Company)

First numeral	Description
0	Unprotected
1	Protected against objects larger than 50 mm dia. (fist)
2	Protected against objects larger than 12 mm dia. (finger)
3	Protected against objects larger than 2.5 mm dia. (#10 wire or rod)
4	Protected against objects larger than 1 mm dia. (#18 wire)
5	Dust protected (ingress of dust won't affect operation)
6	Dust-tight

TABLE 13-7 -- IEC enclosure protection from water
(Data courtesy Hoffman Engineering Company)

Second numeral	Description
0	Unprotected
1	Dripproof (water falling vertically)
2	Dripproof (water falling 15 deg from vertical)
3	Protected from water spray (falling at any angle up to 60 deg from vertical)
4	Splashproof (water splashing from any angle)
5	Hosedown (water jets from any angle)

TABLE 13-8 – Approximate NEMA (including UL and CSA) versus IEC enclosure cross-reference
(Data courtesy Hoffman Engineering Company)

NEMA / UL / CSA enclosure	IEC enclosure classifications, "IP - -"			
	21, 22 & 23	31 & 32	33, 41, 42, 43, 51, 52, 53, 54, 55, 56, 61, 63 & 64	65
1	X	O	O	O
2	X	O	O	O
3R	X	X	O	O
3S	X	X	X	O
3	X	X	X	O
4	X	X	X	X
4X	X	X	X	X
12	X	X	X	O
13	X	X	X	O

X = Degree of protection afforded, O = Unsuitable for this condition

PID Control

Industries that use equipment to hold products on spools, such as paper, foil, or plastic film, require devices that maintain constant tension on the product. The spools are supported by winders and unwinders that rotate at ever-changing speeds as the spools get larger or smaller. The devices that maintain tension and hold the very close tolerance required are called PID controls. PID is the abbreviation of the words proportional, integral, and differential (also known as derivative). Without the use of PID controls, the final product will stretch or break with excessive tension or, it will have waves because the tension was too low.

Proportional Control

When the tension is controlled by a proportional device, a correcting signal is fed to the tension control that is proportional to the amount of tension change. If tension should increase or decrease, the output-correcting signal is applied within a few milliseconds. Gain, the variation in tension control correction signals away from direct proportionality, can be adjusted up or down. However, excessive gain can cause the tension control to be unstable and vary in its outputs.

Proportional Plus Integral Control

Proportional plus integral control improves over the simple proportional control. As tension changes, the integral control provides a correction signal that increases in size over a period of time to catch up with the tension error. This is denoted by the curve in the correction signal in the diagram. Again, gain can be adjusted by the user.

Differential Plus Integral Plus Proportional Control

A tension device that contains differential control produces a correction signal that changes in size in proportion to the speed of change of the tension error. If the amount of the tension error is increasing, the differentiator circuit will increase the amount of error correction signal. If the tension error is getting larger at a decelerated rate, the differential circuit will not affect the correction signal; however, the integrator will increase the correction signal over a period of time to catch up to the error.

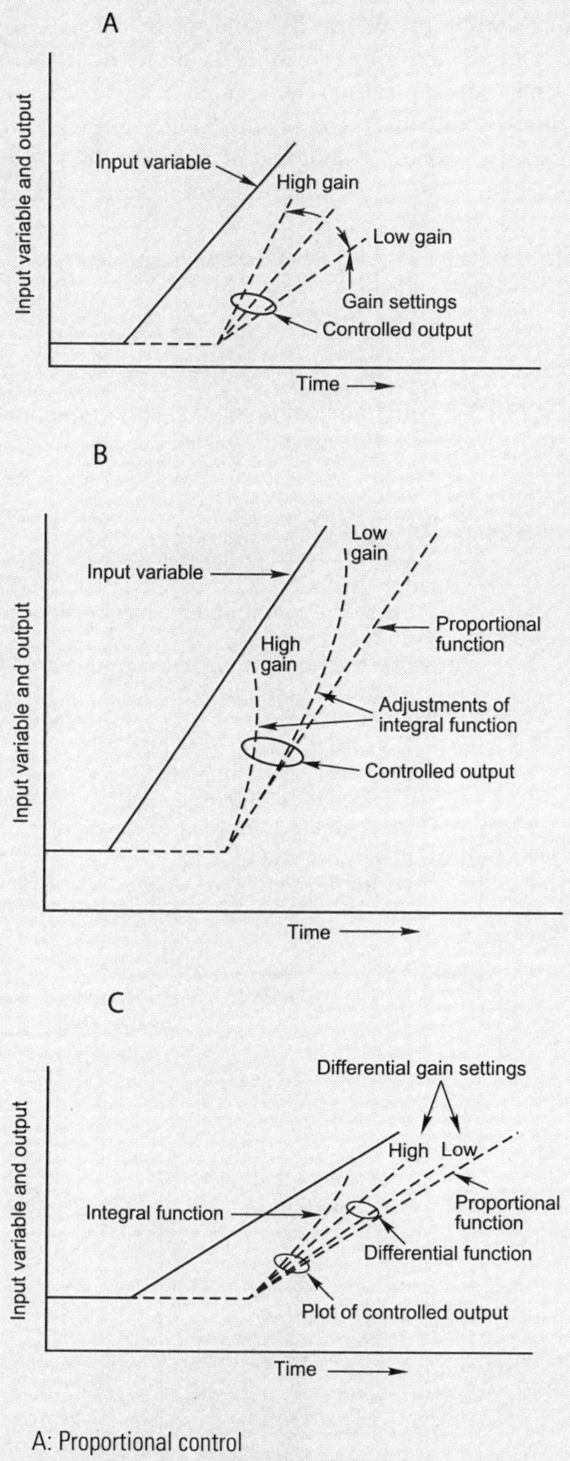

A: Proportional control
B: Proportional plus integral control
C: Differential plus integral plus proportional control

Codes and Standards

In North America, a code is a set of rules adopted by a lawmaking body to regulate some element of human behavior. A standard is a set of rules adopted by consensus of a special group to simplify or regulate use of the products manufactured or used by that group. Such standards are not necessarily enforceable by law.

In the U.S., an example of a code is the National Electrical Code (NEC) published every three years by the National Fire Protection Association (NFPA). The NEC has been adopted by U.S., local, state, and federal authorities into law, wherever it applies. The objective of the NEC is to promote fire safety and minimize the possibilities of electrical shock. But the NEC, for instance, does not apply to trains, automobiles, ships, utilities, mines, or airplanes. Individual cities and states are free to adopt more stringent codes than the NEC, which Chicago and California, for example, have done.

Standards developed by various groups specify product characteristics to comply with the NEC requirements and they often define tests to demonstrate compliance. Underwriters Laboratories, Inc. (UL) is one organization that develops standards for products covered by the NEC. UL serves as a third-party certifier to establish that specific products meet the applicable NEC standards. Products manufactured in North America are often required to meet international standards when they are to be installed offshore. Products manufactured offshore and imported into North America may also have to meet North American standards. In North America, some European standards, covering the DIN rail for example, have been adopted as practical solutions, outside the requirements of any lawmaking body.

North American Standards

Two somewhat parallel organizations exist in the U.S. and Canada. They are National Electrical Manufacturers Association (NEMA) in the U.S., and Canadian Electrical and Electronic Manufacturers Association (CEEMA) in Canada. Both organizations develop standards for electrical machinery, such as motors, drives, and controls.

In Canada, the Canadian Standards Association (CSA) functions as both a code and standards-setting organization (combining the functions of both the NFPA and UL). CSA also certifies products to U.S. and international standards. Generally, a control product used in Canada must have CSA certification.

Two other consensus standards organizations are important to North American control products, manufacturing, and usage. They are the Institute of Electrical and Electronics Engineers (IEEE) and the American National Standards Institute (ANSI).

International Standards

The European Economic Community (EEC) has been moving toward full integration of various national standards. Germany has two important standards-setting bodies, Verband Deutcher Electrotechniker (VDE), an association of German electrical engineers, and Deutshe Industrie Normen (DIN), German Industrial Standards. Other countries with important national standards-setting bodies include the United Kingdom (BSI or ISO), France (NF), Sweden (SEMKO), Switzerland (SEV), and the Netherlands (KEMA).

Two bodies, the International Electrotechnical Commission (IEC), headquartered in Geneva, and European Committee for Electrotechnical Standardization (CENELEC), headquartered in Brussels, are working in harmony to resolve differences between IEC standards and those of the member countries of the EEC.

There are many control products, motor starters, for example, built to IEC standards. By treaties within the EEC, certain IEC standards have the force of law, and products certified as meeting specific IEC standards can be sold and used anywhere within the EEC. Once CENELEC has resolved differences between IEC standards and applicable national standards, Euro-Norm (EN) standards are issued. For example, EN standards 50-014 and 50-016 cover flame-proof motors for hazardous locations.

Countries outside the EEC are free to adopt North American, IEC, EN, VDE, BS, etc., standards—or develop their own—as they see fit. For example, Eastern Europe tends to follow IEC and CENELEC standards, and Hong Kong follows BS standards.

Japan has its own Japanese Industry Standards (JIS). Products used in Japan must meet JIS requirements. While there are many common denominators between UL, IEC, and JIS standards, special attention to specific Japanese requirements is essential.

Another standard—ISO-9000—is receiving wide attention. The International Organization for Standardization (ISO) is headquartered in Geneva. A set of quality—not product—standards has been published under the lead standard ISO-9000. Considerable user pressure is exerted on manufacturers to register their facilities as meeting ISO-9000, regardless of their particular geographical spheres of interest. The ISO-9000 series of standards extend through every phase of design, manufacturing, quality assurance, and testing.

Differences

In the area of motor control, as just one example, there are major differences in the hardware designed to meet NEMA/EEMA and IEC standards. IEC starters are smaller, employ a different type overload, and are applied with a more detailed evaluation of an application and the user's desired life expectancy.

Additional Information

Addresses and phone numbers for the organizations mentioned are provided in the Association Appendix.

Chapter 14

Sealants & Adhesives

Chapter Objectives

When you have finished chapter 14, "Sealants & Adhesives," you should be able to do the following:

1. Describe the common applications for sealants and adhesives.
2. Identify the key selection factors for threadlocking adhesives.
3. List the types of sealants used for sealing pipe threads.
4. Understand the difference between anaerobic and RTV sealants.
5. Understand the difference between compression and form in place gaskets.
6. Identify the two distinct types of bonded cylindrical assembly.

Introduction

Machinery designers and technicians are constantly challenged to ensure equipment will deliver reliable performance. This chapter describes adhesives and sealants engineered to increase the performance and reliability of typical machinery components—fasteners, fittings, flanges, and bearings. These innovative products are effective because they address the "root cause" of mechanical failure: the inner space between components. Inner space includes the clearance that exists between threaded assemblies, in the unfilled space in press fits, and in the surface imperfections between flanges. Machinery adhesives solve the problem of mechanical failures caused by the clearance between close-fitting parts in assemblies by filling this space with a thermoset plastic engineered to provide predictable properties of adhesion, elongation, and environmental resistance.

Types of Sealants and Adhesives

The following provides a summary of sealant and adhesive types by application.

Threadlocking

Threadlockers prevent the most common forms of fastener failure: loosening, leaking, and seizure. A threadlocker is a one-component anaerobic adhesive. It is applied to the threads of the fastener as a liquid where it fills the gap between the mating threads. The liquid then anaerobically cures to form a tough thermoset plastic adhesive. This adhesive fills the gap between the mating threads and also creates an interfacial connection that bonds to the surface roughness. This combination of gap filler and adhesive prevents any relative movement (loosening) of the threads. A significant addition, the threadlocker also fills and seals the helical gap between the threads, which eliminates leakage and thread corrosion. A bolt secured with an anaerobic threadlocking agent will last the life of the assembly but will not vibrate loose or seize.

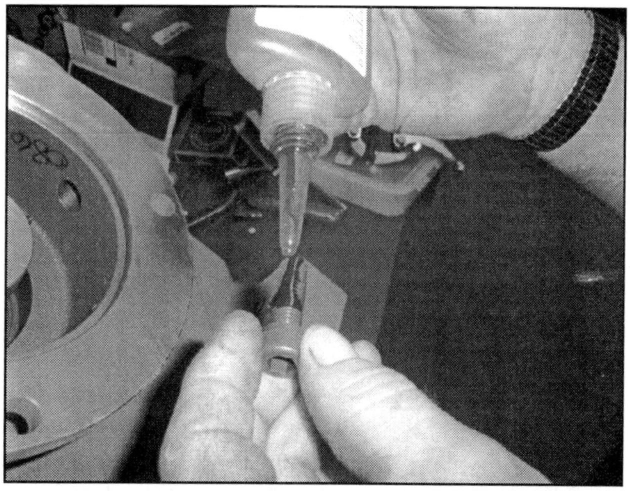

Figure 14-1: Threadlocking adhesive is being applied to the threads of a bolt before assembly

Key factors in selecting the most appropriate threadlocking adhesive:

- Strength: Blue, medium strength is usually the best choice. It reliably prevents loosening, leakage, or seizure, and fasteners can be dissembled/serviced with standard tools. Red, high strength should be reserved for permanent assemblies because localized heat of 500° F (260° C) is often needed to aid disassembly.
- Application method: Liquid threadlocker is most commonly applied to the fastener when it is assembled. Newer no-drip paste, gel, and tape forms are alternatives to liquid. Wicking grades are designed to be post-applied to assembled fasteners. A variety of manual, semi-automatic, and fully automatic applicators are available to ensure consistent product application. Pre-applied threadlockers are an alternative to applicators. Pre-applied threadlockers are a dry microencapsulated film that is pre-coated onto the fastener or nut by the supplier. The fastener arrives on the production line ready for assembly.
- Unique requirements: Grades are available for specialized requirements including high temperature, agency-certified (Mil Spec, UL, NSF, food, low halogen, etc), and plastic fasteners.

Threadlockers are a reliable and highly cost-effective alternative to mechanical threadlocking devices, which include a wide variety of lock washers and modified thread designs. They prevent vibration, shock, and thermal cycle loosening, and seal threads from leakage and seizure.

Pipe Thread Sealants

Thread sealants prevent leakage of gases and liquids from all types of pipe connections. The potential for leakage exists in all fluid systems because joints tend to be "dynamic" due to vibration, changing pressure, and temperature.

Non-Curing Pipe Compounds

One of the oldest methods of sealing the spiral leak paths of threaded joint, pipe compounds are pastes made from oils and fillers. Advantages include the ability to lubricate joints and fill threads, although they provide no locking, which is suitable for plastic fittings. Disadvantages include the tendency to squeeze out under pressure, poor solvent resistance, and the inability to work on parallel threads.

Solvent-based Pipe Compounds

Another older method of sealing threaded joints, these compounds contain volatile solvents, which are released into the atmosphere during the drying process.

Advantages include the ability to lubricate and jam threads. Disadvantages include the tendency to shrink during cure as solvents evaporate, which may require the fittings to be re-torqued to minimize voids.

PTFE (Teflon®) Tape

PTFE is a chemically resistant lubricant. PTFE tape gives a good initial seal because the lubrication allows fittings to be wrenched very tight. PTFE tape resists chemical attack and is the only organic sealant allowed for gaseous oxygen.

Advantages include lubrication during assembly, resistance to chemicals, and suitability for plastic fittings. Disadvantages are that lubrication permits fittings to be overtightened and also permits them to loosen after assembly, especially fittings subject to dynamic loads, and that tapes are often banned in pneumatic and hydraulic systems due to shredding, which can clog key orifices and lead to maintenance problems.

Anaerobic Thread Sealants

Liquid sealant lubricates threads facilitating assembly. The paste consistency allows valves, elbows, and fittings to be re-adjusted for alignment. The sealant anaerobically cures after assembly to form a resilient temperature and chemically resistant thermoset plastic. The cured sealant unitizes the fitting, which eliminates relative thread movement and vibration loosening.

Anaerobic thread sealants are designed to increase the reliability of most types of metal fittings, including tapered thread, straight thread, and JIC and ORB fittings. Standard grades are suitable for common fluids and gases and are qualified to agency approvals, including UL, NSF, and others. Products designed for high-precision pneumatic and hydraulic fittings do not contain fillers to eliminate the risk of clogging fine filtration openings. Additional specialized versions are available for other requirements, including refrigerant and steam systems.

Advantages are that they lubricate during assembly, cure without shrinkage, seal regardless of assembly torque, seal to burst rating of pipe, provide controlled disassembly torque, do not cure outside joint for easy clean-up, are the lowest cost per sealed fitting, are easily dispensed on production lines, and are available in pre-applied grades. Disadvantages are that they are not suitable for oxygen sealing and strong oxidizing agents, are not suitable for sealing at temperatures above 392° F (200° C), and are not recommended for plastic fittings.

Gasketing

Gaskets prevent the leakage of liquids or gases by forming impervious barriers between mating flanges. Three types of gaskets are predominantly used to seal flanges.

Compression Gaskets

Compression gaskets form a seal because they conform to and fill surface irregularities when they are squeezed between flange surfaces. Convention compression gaskets are

Anaerobic Adhesives & Primers

Anaerobic Adhesives

Single component resins that cure at room temperature in the absence of oxygen and in contact with an "active" metal. An "active" metal is one that oxidizes, such as steel, brass, or iron. Newer anaerobic formulations now cure reliably on all metals, active and inactive.

Primers

Pre-applied to surfaces to speed the anaerobic cure time for older formulations on inactive surfaces. They are also used to speed the cure of all anaerobic products when metal parts are cold or gaps are large.

pre-cut from paper, rubber, cork, metal, and other materials. They are also available in sheet and cord stock forms that can be custom fit to a flange for assembly or repair.

Figure 14-2: This gasket has been die cut to match the outline of the gear reducer

Cured-In-Place gaskets (CIP) are an innovative type of compression gasket. A robotically controlled applicator precisely traces a bead of sealant onto one of the flanges. The pre-applied bead is then cured in seconds by exposure to a UV light. The result is an elastomeric material that adheres to the flange surface. Sealing is achieved through compression of the cured gasket during assembly of the flange. The gasket is permanently attached to one flange, which is ideal for sealing joints that must be opened frequently for service. CIP is effective for flexible and JIT (just-in-time) manufacturing.

Formed-In-Place Gaskets (FIP)

Formed-In-Place gaskets (FIP) are applied as a bead of sealant to one of the unassembled flange surfaces. Upon assembly, the bead conforms to the imperfections of the mated flanges, then cures and forms a durable seal. FIP gaskets are versatile and can be used on almost every type of flange. The key FIP advantage is that flanges fully mate, which eliminates the most common root cause of gasket failure: compression set (relaxation of a hard gasket after initial compression).

Figure 14-3: Liquid gasket material is being applied to a flange

The two most common types of FIP sealants are:

1. Anaerobic FIP sealants: Optimized for close-fitting rigid metal flanges. Provide a unique combination of high shear strength and environmental resistance.
2. RTV Silicone FIP sealants: High elongation elastomers optimized for flexible flanges such as stamped covers. RTV silicones cure through a reaction with atmospheric moisture to cure. They offer a unique combination of high gap-fill temperature resistance and a strong, flexible adhesive.

Retaining Cylindrical Metal Assemblies

An example of a "retained assembly" is a bearing mounted in an electric motor housing. Retaining compounds are anaerobic adhesives formulated to provide very high shear strength. They are suitable for use on metal components subject to a wide range of loading and environmental conditions. There are two distinct types of bonded cylindrical assemblies.

Bonded Slip Fit

Bonded slip-fit parts are assembled with a clearance, and the cured adhesive transmits the entire load. Key advantages are lower-cost machining tolerances, no created stress because there is no interference fit, and the ability to precisely align components before retaining compound is cured.

Bonded Shrink or Press Fit

With bonded shrink or press fit, the load is transmitted both by the cured adhesive and by friction between the parts due to the interference fit. There is no more than 25 percent metal surface-to-surface contact in the tightest interference fit. The liquid retaining compound fills the 75+ percent of inner space, then anaerobically cures to a high shear strength adhesive plastic. Key advantages are that bonded joints are stronger, the operating load is evenly distributed, fretting corrosion is prevented, and the joint is sealed. In either case, the adhesive is applied as a liquid, completely fills the joints, and cures to a tough thermoset plastic that unitizes the cylindrical components.

Key Terms

Anaerobic Adhesives: Cure in the absence of air and in the presence of metal.

RTV Silicone: Cure by reaction with atmospheric moisture.

Threadlocking: Prevents bolts from seizing or vibrating loose.

Pipe Sealants: Prevents leaks in threaded pipe joints.

Gasketing: Seals flanges and prevents leaks.

Retaining Compounds: Increase the strength of cylindrical joints.

Additional Information

Addresses and phone numbers for associations are provided in the Association Appendix.

Selection of Engineering Materials and Adhesives, Lawrence W. Fisher, P.E. © 2005

The Adhesive and Sealant Council, Inc., Bethesda, MD

Chapter 15

PT Accessories

Chapter Objectives

When you have finished Chapter 15, "PT Accessories," you should be able to do the following:

1. Describe the features and benefits of shafts seals.
2. Define the purpose and identify the types of drive tensioners.
3. Describe the features and benefits of retaining rings.
4. Explain the operation and variations of shaft-locking devices.
5. Understand the benefits and variations of different shaft collars.
6. Discuss the purpose and need for keystock.
7. Learn how O-rings seal out fluids or gases.
8. Describe the uses of various locknuts.
9. Learn how single-point lubricators help reduce downtime.

Introduction

This chapter covers components that are important accessories for drive systems. Accessories covered include shaft seals, tensioners and idlers, retaining rings, shaft-locking devices, shaft collars, keystock, O-rings, locknuts, and single-point lubricators.

Shaft Seals

Equipment with rotating, reciprocating, or oscillating shafts calls for radial-lip shaft seals that retain lubricants and exclude foreign matter (see Fig. 15-1). The seal surrounds the shaft, employing an interference fit that develops pressure between the shaft and the seal's contact surface. When the application requires the retention of grease or the exclusion of dirt, the elasticity of the rubber lip will create adequate sealing force. When oil is the lubricant used, the seal pressure must be assisted by the use of a spring. This is referred to as a spring-loaded seal.

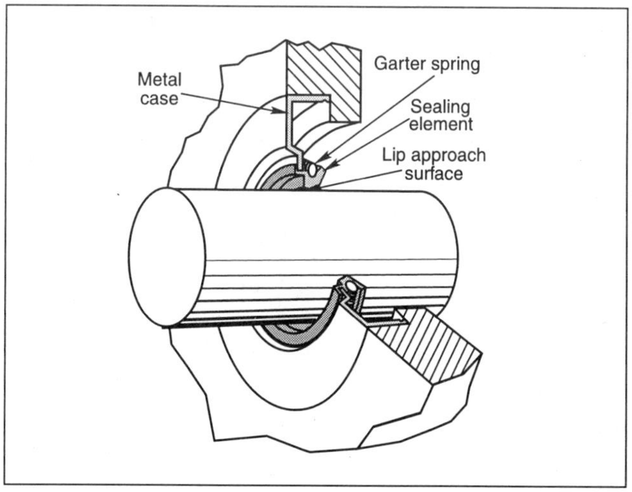

Figure 15-1: Cut-away view of typical spring-loaded oil seal.

Shaft seals are subject to a wide variety of lubricants, temperature ranges, fluid pressures, equipment misalignment, dynamic shaft runout, and shaft speed variations (see box: Shaft Eccentricity). The seals are available in a variety of configurations employing different materials to accommodate specific application requirements.

Performance Parameters

The lip of a seal rides on a thin oil film. Friction and shear in the oil film raise sealing-lip temperature. As lip temperature increases, the lubricating film becomes thinner and may break down. Excessive wear and temperature also cause premature failure of the seal. Excessively low temperature causes the sealing material to become hard and brittle, which can also result in excessive wear or leakage. Thus, both ambient and operating temperature ranges are important factors in selecting shaft seals.

Lip seals work best with steel or stainless-steel shafts. Nickel-plated surfaces are acceptable. Brass, bronze, aluminum alloys, zinc, and magnesium, however, should not be used except under unusual circumstances.

For optimum performance, the interference fit between the sealing lip and shaft surface must be precisely controlled. The portion of the shaft contacted by the seal should be hardened to 30^0 Rc to minimize shaft scoring. Excessive shaft hardness, however, tends to wear the seal.

The corner of a shaft edge can damage the sealing lip during installation; therefore, a burr-free chamfer or radius finish is recommended to prevent damage during assembly. Furthermore, shaft surface finish should be smooth enough to provide continuous contact between the sealing lip and shaft surface. Also, the shaft should retain enough lubricant to reduce friction between the sealing lip and shaft. For this reason, plunge grinding the shaft to a finish of 10 to 20 micro-in. with no machine lead is recommended.

Shaft Eccentricity

Shaft-to-bore misalignment (STBM) is the amount that a shaft center is offset from the bore center, a value that should be as small as possible. A high STBM increases friction and wears seals abnormally on one side. Dynamic run-out is the amount by which a shaft does not rotate about its own center. It may be caused by an out-of-round shaft, a bent shaft, or shaft imbalance.

For housings machined to accept the seal OD, bore depth (or depth to a retaining ring) should exceed seal width by at least 1/64 in. To prevent leakage, a bore finish of 125 micro-in. or less is recommended whenever lubricant pressure is present at the seal OD. If a fine housing-bore finish is not possible, coatings can be applied to the seal OD to effectively fill minor bore imperfections. The leading or entering edge of the bore should be chamfered to prevent seal damage during installation.

Seal Configurations

Springless seals for retaining highly viscous lubricants such as grease are employed at shaft speeds less than 2,000 fpm. For dust or fluid exclusion, the seal lip should face the contamination. A heavy-duty multi-lip seal provides good lubricant retention under severe dust applications. But, shaft speed is usually limited to 500 ft-min, and the dust lip should be lubricated.

Spring-loaded seals retain low-viscosity lubricants at speeds up to 3,600 fpm. Single-lip, spring-loaded seals are the most common type. Typical applications include automotive engines and transmissions, gear reducers, and various other industrial machines requiring the retaining of fluid in a housing.

Medium-duty dirt exclusion calls for a dual-lip, spring-loaded seal or two seals mounted back-to-back with lubrication between the lips.

A common seal case design includes a stiffening plate on the lubricant side of the seal. This plate does not assist in sealing, but it can prevent damage to the seal during installation. The secondary case provides extra strength and protection during assembly or operation.

Besides the standard metal case, some seals are available with a rubber OD. Housing dimension tolerances are slightly more relaxed than those for steel-case seals. The rubber OD seal may also be used when the housing bore has become slightly damaged and no longer meets the standards required for steel-case seals. The housing may be split, and/or material may be porous or of a soft material. The surface finish may be damaged, rough, or machined slightly out of round, which could allow fluid to leak from between the housing bore and seal OD. The rubber OD of a seal can act as a gasket to prevent leakage from around the seal OD.

Seal Materials

The following are the five common families of materials for lip seal elements:

- Nitriles
- Polyacrylates
- Silicones
- Fluoroelastomers
- PTFE

Seal material must be compatible with fluid used and be able to withstand the temperatures involved with each application. The capabilities of synthetic materials are such that leather and felt, two previously common materials, are very rarely used.

Nitriles are the most widely used shaft-seal materials. They have good oil resistance, wear resistance, and low-temperature properties, as well as low cost. Their upper temperature limit is about 225° F. Carboxylated nitriles are more expensive than standard nitriles, but provide better wear resistance.

Polyacrylic polymers are capable of operating at 300° F and are also compatible with most EP additives, whereas nitriles, in general, are not.

Silicones are more expensive than nitriles or polyacrylates, but perform well in the presence of water, inorganic acids and bases, diesters, and non-petroleum-based brake fluids. Silicones also operate well in temperatures from -100° to +325° F. Silicones are unacceptable in EP type oils.

Fluoroelastomer (E. I. DuPont, Viton™) lip seals are used in applications with temperature requirements of -40° to +400° F. It resists attack from most chemicals and lubricants that destroy nitrile, polyacrylates, and silicones.

PTFE is resistant to virtually all fluids within a temperature range of -100° to +450° F. PTFE is not, however, an elastomer—its consistency is more like wax than rubber—and does not follow shaft motion as well as most other seal materials. Moreover, it cannot be bonded readily to an elastomer or a metal case. Thus, PTFE seals are expensive, easily damaged, and call for careful installation procedures.

Other materials are used, but much less often than the foregoing. Urethanes, for example, are tough materials that are employed when the highest wear resistance at low temperature is required. Butyl is used to seal non-petroleum-based brake fluids and some solvents. But, both materials cost more than the common ones and offer no advantage in most general applications. Therefore, they are specified only when their somewhat specialized characteristics justify their higher cost.

Selecting a material calls for consideration of chemical resistance, temperature range, wear resistance, and cost. Final material choice is often a compromise, however, that will satisfy several of the most important application parameters. The most important performance factor is the lip seal's ability to maintain the proper radial force between the sealing lip and the shaft surface. Anything that increases the force shortens seal life; anything that reduces the force tends to allow leakage.

Drive Tensioners

Belts and chains should operate with a specified tension for optimum life. Too much tension accelerates wear and usually results in premature chain, belt, or bearing failure. On the other hand, too much slack causes chain and belt whipping, vibration, belt slippage, and uneven power surges. Again, premature component failure is the likely result.

To illustrate these power surges: if you grab a string, hold it tight at both ends, and pull, you may not be able to break the string. If, however, you let the string go slack by moving your hands close together, then suddenly jerk them apart while still grasping the string, the string snaps and breaks. This demonstrates how whipping action can damage a chain or belt by momentary power surges.

Belt manufacturers recommend that tension be set by adjustment of drive motor position or some similar action. They also generally discourage the use of tensioners in drives because the induced bending reduces belt life. For those applications where tensioners must be used, several types provide satisfactory results with a minimal loss in service life and performance.

In a drive with no tensioners, a belt or chain travels in a straight line as it leaves one pulley or sprocket and travels to another. A tensioner works by engaging the chain or belt and deflecting it out of a straight path; the chain or belt has to travel around the tensioner. The higher the tensioner force acting to deflect the chain or belt, the greater the tension.

Types of Tensioners

There are three basic types of drive tensioners: fixed, adjustable, and self-adjusting. Generally, tensioner frames are constructed of cast iron, ductile iron, or steel, depending on the manufacturer and type of tensioner. A hardened and ground steel shaft attached to the frame serves as a mount for a standard sprocket, sheave, or pulley with a bronze bearing press-fit into its bore. Most shafts have an internal grease channel and grease fitting on the end for lubricating the bronze bearing. Drives requiring higher peripheral speed use a rolling-element bearing (usually a needle bearing) instead of the bronze bearing.

Fixed Tensioners

Sometimes referred to as idlers or positioners, fixed tensioners (see Fig. 15-2) mount permanently to the drive framework and cannot be readjusted unless the tensioner is repositioned or remounted. When used as a positioner, a fixed tensioner simply routes a chain or belt around some obstruction that lies between the driving and driven sprockets or pulleys. The device establishes an initial drive tension, but this tension decreases as the drive wears, so some other means of tension adjustment should be provided. An adjustable motor base often fills this need.

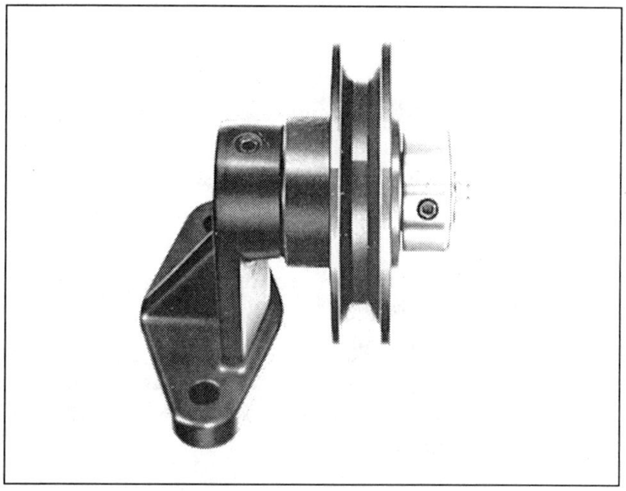

Figure 15-2: Typical fixed tensioner (positioner) shown with idler sheave installed.

Adjustable Tensioners

Nearly identical to fixed types, adjustable tensioners use an elongated mounting hole in the frame (see Fig. 15-3). When drive tension decreases due to wear, loosening the tensioner's mounting bolts, repositioning it, and retightening the mounting restores proper tension.

Another way of adjusting tension in the device shown (Fig. 15-3) is to loosen a fastener securing the two frame pieces, reposition the frame half with the pulley or sprocket, and retighten the fastener.

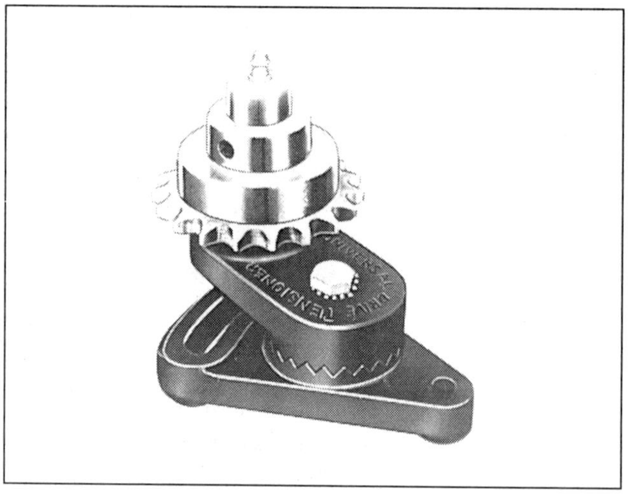

Figure 15-3: Adjustable tensioner with two means of repositioning idler sprocket.

Self-adjusting Tensioners

Once initial tension is set, self-adjusting tensioners are spring-loaded to adjust to variations in drive tension due to normal wear and load fluctuations. Manufacturers employ different techniques for accomplishing this, so tensioner designs should be evaluated for their suitability to a particular application.

In one design (see Fig. 15-4), tension is adjusted by rotating a threaded shaft to position a pulley or sprocket. The technique is especially well-suited for heavy-duty drives and provides a convenient means of loosening a chain or belt for replacement. With it, an enclosed drive can be re-tensioned from outside its enclosure.

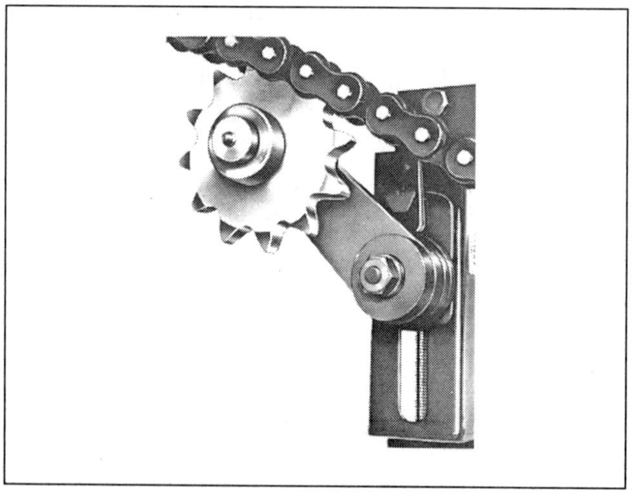

Figure 15-4: Spring-loaded, self-adjusting tensioner with chain or belt take-up adjustment.

Application and Installation

Select a tensioner for a particular application after determining the size and type (chain or belt) of drive needed from known parameters of speed, torque, loading cycles, service factor, etc. Next, determine the type of tensioner needed—fixed, adjustable, or self-adjusting—and how the tensioner will be mounted (vertically, horizontally, shaft-mounted, etc.). Manufacturers' catalogs assist in these procedures and provide information on specific applications and designs.

Chain and belt manufacturers publish very specific instructions on the idler position, the size of the idler, and the tension force that is created. These specifications should be followed for maximum component life and transmission of power. Tensioner manufacturers suggest that a tensioner be selected to exert a suitable range of force so that these specifications can be met.

As a general rule, a self-adjusting tensioner should be installed so that it will be deflected one-half of its total travel. For example, if the moveable arm will deflect 90 deg., it should be installed at a point of 45 deg. deflection. That way, if power surges occur, there will be enough travel in the tensioner to avoid excess force on the chain or belt. Also, as the drive wears, there will be enough travel in the tensioner to maintain a suitable tension.

Other guidelines for applying tensioners include:

- Always install the tensioner on the slack side of the drive.
- For bidirectional drives, mount tensioners on tight and slack strands of a drive or, preferably, use a twin floating tensioner (Fig. 15-5).

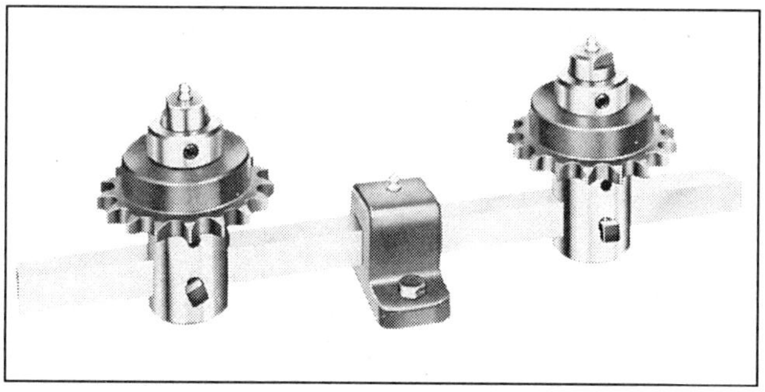

Figure 15-5: Twin-floating tensioner for bidirectional drives.

- Mount tensioners one-third of the center distance from the larger sheave or pulley when used on the inner surface of a belt.
- Reduce drive capacity when applying tensioners on the inside of nonsynchronous belts (those without teeth) because outward deflection from the tensioner reduces the arc of contact between the belt and sheaves.
- Avoid using tensioners on the back side of v-belts because the resultant back-bending shortens the life of the belt due to excessive chord stress.
- Position chain tensioners outside the perimeter of a strand of chain.
- Locate a tensioner one-third of the center distance from the smaller sprocket, pulley, or sheave when used on the outer surface of a chain or belt.
- Make sure at least three teeth of a positioner's idler sprocket engage the chain.

Retaining Rings

In most installations, the inner ring of a bearing is pressed onto a shaft, and the outer ring is installed with a tight push-fit in the housing bore. Despite this, installing bearings on shafts and in housings requires a means for preventing axial displacement of the bearing. The retaining members should be large enough to contact the straight face of the bearing, and when thrust is a factor, the retaining member must accommodate the thrust of the bearing.

Conventional bearing installations—using machined shoulders and various types of threaded fasteners, lock washers, and other devices—are bulky and expensive and add considerably to the size, weight, and complexity of a bearing assembly. Both the shoulders and the threaded areas require machining operations. Here, disassembly for adjustment or field service is labor-intensive and time-consuming.

Retaining rings (sometimes called snap rings) require less space than machined shoulders and nuts and can be used with smaller shafts and housings. The rings avoid the costs for producing shoulders and threads. Grooves for the rings often can be cut simultaneously with the shaft cut off or machined inexpensively in housings. Assembling the rings with pliers, applicators, and dispensers, or mechanized tools saves assembly time. The rings provide easy access to both sides of a bearing, can be removed quickly for field service, and are reusable.

Some ring types have radial sections wide enough to directly abut bearing races manufactured with corner radii or, as they are sometimes called, corner breakouts.

For bearings with corner radii exceeding a retaining-ring capacity, support rings are available as intermediate members. Support rings provide a shoulder that contacts the straight face of a bearing race and spreads the load over the face of the retaining rings. Support rings are made of high-carbon, heat-treated steel and are ground for close parallelism of the abutting surfaces. Refer to manufacturers' catalogs for detailed information.

Conventional Designs

External retaining rings are used to prevent bearing motion when it is mounted on a shaft. They are available in plated or non-plated spring steel and stainless steel.

Internal rings are used to prevent movement of a bearing when it is used in a bore or housing. They are available in plated or non-plated spring steel and stainless steel.

Whether internal or external, heavy-duty rings exhibit the highest thrust capacity and impact resistance. They have a high protruding shoulder for maximum contact surface with the face of the bearing and can accommodate a shoulder with a large corner radius without support rings.

Inverted rings provide the next-highest retaining shoulder. While substantially thinner than heavy-duty rings, they provide adequate shoulder height and thrust-load capacity for many bearing applications. However, basic external rings have higher thrust-load capacities. Support rings are recommended for bearings with large corner radii or high thrust-load requirements.

Miniature high-strength rings are similar to heavy-duty rings but do not have protruding areas with holes, which aids installation and removal. They are assembled with tapered pins and provide a tamper-proof shoulder for holding bearings on small shafts and studs.

Permanent-shoulder rings are compressed into a v-shaped groove to grip a shaft tightly and provide a 360 deg. shoulder. By replacing machined shoulders in expensive materials such as brass or copper, the rings make possible substantial savings on both materials and labor.

Spiral Retaining Rings

Spiral-wound retaining rings consist of two or more turns of rectangular-cross-section material having rounded edges. The material is coiled on its edge to provide a gapless ring with 360 deg. of retention (see Fig. 15-6).

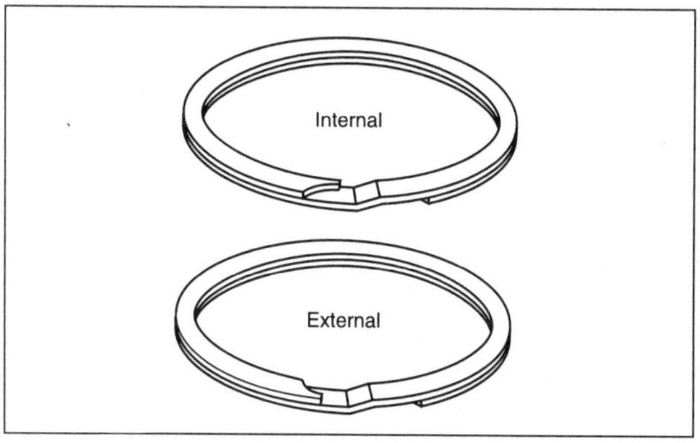

Figure 15-6: Internal and external double-wound spiral retaining rings.

There are two basic categories of spiral rings: medium-duty and heavy-duty. Both are available for external and internal applications. External rings have notches on the ring ends at the ID for removal; internal rings have notches on the OD.

Medium-duty rings offer economy and low space and weight requirements in applications with moderate thrust loads. Heavy-duty rings have thicker cross-sections and fit into deeper grooves. The combination of heavier ring cross section and deeper grooves gives heavy-duty rings about 30 percent more thrust-load capacity than medium-duty rings.

Spiral retaining rings can be installed or removed with hand tools, or, where production volume warrants, with automatic equipment. A uniform radial wall height is important when radial clearance is minimal. Ring thickness can be adjusted by specifying material thickness or number of turns.

Spiral retaining rings are used for purposes other than positioning a part or keeping several components assembled. Other functions include:

- Vibration dampers in gears.
- Oil dams in splines and mechanical couplings.
- Back-up rings for O-rings and packings.
- Oil or grease slinger rings.
- Labyrinth seals in pillow-block bearings.
- Location stops.
- Bumper impact absorbers.

Assembling Spiral Rings

Spiral retaining rings, being flexible, may be installed manually by spreading the leaves, inserting one ring's end into the shaft groove, and working the rest of the ring around the circumference until the last turn snaps into the groove. This method, however, is only practical when a few units are to be assembled. A tapered sleeve or mandrel and a plunger are generally used for high-volume assembly. This arrangement works well with manual or automated assembly. Installation fixtures should not overstress a retaining ring during assembly, however. An overstressed ring will not grip the groove bottom unless adequate compensation in the ring-free diameter has been made. A loss in thrust capacity could result.

Shaft-Locking Devices

Tapered Bushing Types

Tapered shaft bushings are among the most widely used means to mount or attach a part to a shaft. All bushing types use the principle of the wedge to clamp the bushing to the shaft. They are easy to install/remove and are interchangeable between manufacturers. Various types of bushings are offered for a variety of power transmission products and are widely accepted in the industrial market. They can be found in sprockets, sheaves, gears, timing pulley/sprockets, couplings, and conveyor pulleys.

The three types of bushings are QD, taper-lock, and split-taper. Within each bushing type is a series with a bore range that will mate only with components that utilize that same type and series of bushing. Each type of bushing has its advantages and limitations, and it is up to the user to decide which fits their application or needs.

QD Bushings

QD bushings are also known as a quick-disconnect bushing. This style of bushing uses a 3/4 in. per foot taper. The bushing has two sets of holes, one threaded and one unthreaded. When the mating part has the like arrangement, the bushings bolts may be tightened from either side. Standard mounting is when the bolts are mounted through the unthreaded holes of the bushing into the threaded holes of the mating part. When the bolts are mounted through the unthreaded holes of the part into the threaded holes of the bushing, this is known as a reverse mount. When dismounting the bushing is necessary, the bolts are taken out and screwed into the threaded holes to disassemble the parts.

Taper-Lock Bushings

Taper-lock bushings allow for a flush fit to the mating part, which requires less space on the shaft than the flanged bushing styles. This bushing uses set screws to fasten the bushing to the mating part. The taper is 1 11/16 in. per foot, which gives less clamping force than the other bushings. Bushings can be mounted from one side only. Most taper-lock sprockets are available with mounting from the opposite side of standard. This feature must be called out when specifying the mounting arrangement. This bushing is not recommended for shock loading or reversing applications.

Split-Taper Bushings

Split-taper bushings utilize a double split barrel, which increases concentricity and improves clamping. This bushing has a flange and can be mounted only one way. External key allows for higher torque capacity and aligns bolt holes between mating parts. Long barrel provides better support on the shaft. This bushing is only offered by a few sources.

The three styles of bushings are offered in standard and metric bores. Spline bores may be offered by certain manufacturers only. With all bushing styles, it is important to adhere to the manufacturer's recommendations regarding shaft tolerances and tightening torque of the fasteners. This will give the user the best results for their applications.

Fig. 15-7 shows the relative size of the QD, taper-lock, and split-taper bushings by comparing a like-size sprocket that has the same size chain and number of teeth for the comparative bushing styles.

Figure 15-7: Comparing bushings to sprockets

Keyless Locking Devices

Keyless shaft-to-hub locking devices allow components to be easily removed, repositioned, or mounted more securely and conveniently on a shaft than with keyways, setscrews, or locknuts. Keyless locking devices also allow repositioning a component on a shaft to adjust alignment, synchronize position of that component with another, or for timing functions. The arrangement avoids the backlash that is sometimes encountered with keyed connections. Also, overall cost is often less for low-volume applications because keyways, threads, and close tolerances are not required. Moreover, most keyless locking devices allow generous tolerances on shafts and hubs.

Principle of Operation

Manufacturers offer a variety of keyless shaft-to-hub locking designs. The majority employ an inner sleeve with a tapered OD and an outer sleeve with a tapered ID (see Fig. 15-8 and 15-9). The inner sleeve fits around the shaft, and the outer sleeve fits inside the hub bore of the pulley, sheave, gear, sprocket, or similar component. Tightening loading bolts forces the inner sleeve into the outer, which causes the inner sleeve to squeeze onto the shaft and the outer sleeve to press outward on the bore ID.

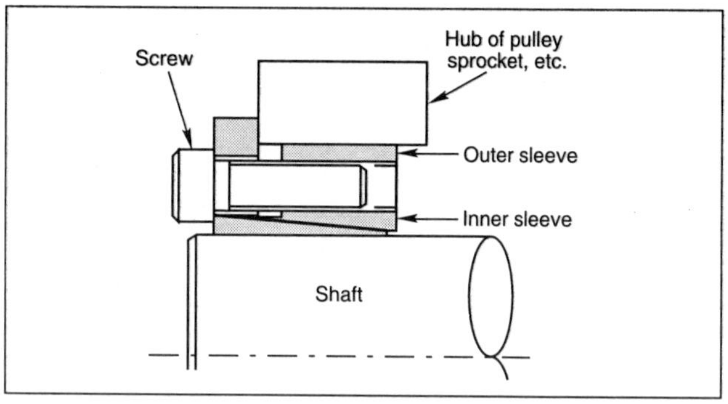

Figure 15-8: Sectional view of typical keyless shaft-to-hub internal locking device.

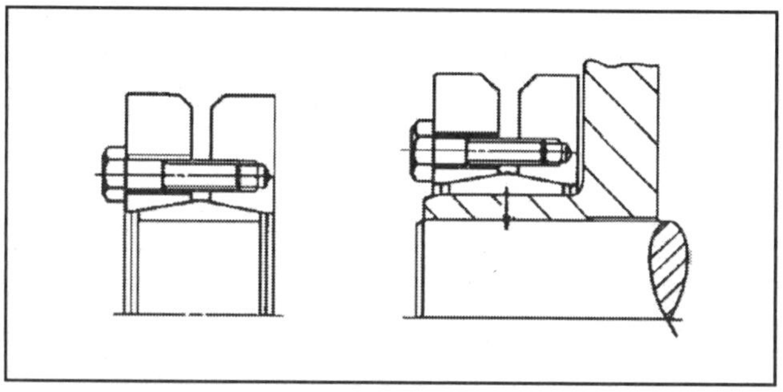

Figure 15-9: Sectional view of typical keyless shaft-to-hub external locking device. (Applied for transmission of very high torques, particularly when external clamping is advantageous and excellent concentricity is required)

Regardless of their design, keyless shaft-locking devices allow a shaft of a given diameter to transmit more torque than if it had a keyway. The reason is that the holding stress is uniformly distributed around the OD of the shaft and ID of the bore, instead of being concentrated at keyways.

Because all of the shaft-hub locking devices discussed here exert pressure between the hub and shaft, manufacturers' literature should be consulted for the minimum hub thickness required. A hub that is too thin could crack under the high pressure exerted by the locking device. If hollow shafting is to be used, minimum recommended shaft wall thickness should be checked.

Some types of keyless locking devices can be used with shafts and components that contain keyways or flats. Others require keyways and flats to be filled and rounded to match the shaft OD. Technical literature or the manufacturer should be consulted before attempting such a retrofit.

Tapered Versions

Three common variations employing the principle of tapered sleeves are illustrated in Fig. 15-10. The left-most version uses a pair of loading rings tapered on their ID and OD on facing sides. Holes drilled through one ring allow loading bolts to engage the other ring, which has threaded holes. Tightening loading bolts moves the rings closer together axially; loosening them moves the rings apart. Ring sections with tapers matching the loading rings are positioned between the tapered surfaces of the loading rings around the OD of the shaft and ID of the hub bore. Tightening the loading bolts, then, causes the tapered surfaces to force the outer ring sections against the hub bore and the inner ring sections against the shaft circumference.

Another variation, center, uses a single nut to force an elongated inner ring (collet) into an elongated outer ring (sleeve). One end of the collet is threaded; as a loading nut is tightened on these threads, it pushes axially on the sleeve and pulls the collet deeper into the sleeve. The tapered surfaces of the collet and sleeve act as a wedge to expand the sleeve against the hub bore and squeeze the collet onto the shaft.

Still another variation, right, employs narrow, tapered surfaces that form a helix—a configuration resembling matching tapered threads on an inner sleeve and outer sleeve. Forcing the outer sleeve over the inner sleeve causes a wedging action of the tapered surfaces at each thread.

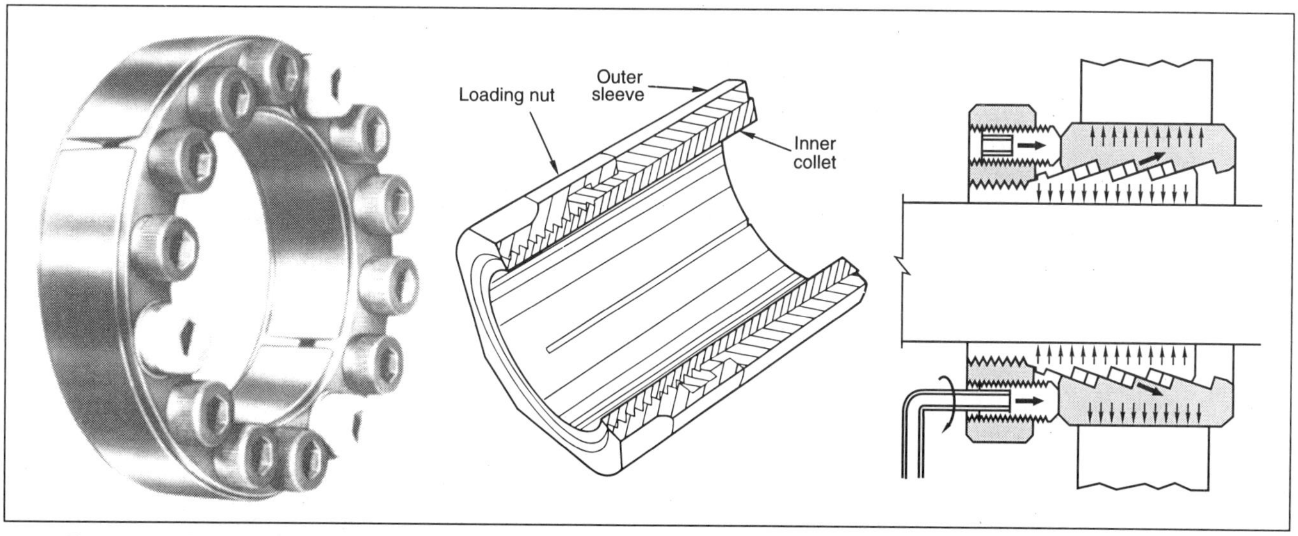

Fig. 15-10: Three common keyless shaft-hub locking devices that generate locking action through mating tapered surfaces.

Loading is accomplished by a nut threaded onto one end of the inner sleeve. Loading screws positioned parallel to the shaft are threaded into the loading nut. As the loading screws are tightened, they push on the outer sleeve, creating stress at mating tapered surfaces.

Other Versions

In a sense, forcing tapered surfaces together fills in the space between a shaft and hub bore. The greater the force driving wedges together, the greater the stress for locking the hub to the shaft. Other types of keyless locking devices employ a different principle. They rely on radial expansion of a collar that fits into the space between the hub bore and shaft.

One version has a sleeve that resembles a bellows or accordion. Loading bolts positioned parallel to the shaft fit into through-holes around the sleeve. As bolts are tightened at one end of the sleeve, they advance into threaded holes at the other end to compress the sleeve lengthwise in an accordion-like fashion. The action compresses and shortens the sleeve, expanding it radially, so the OD increases and the ID decreases. A result is an interference fit between the hub bore and shaft OD.

Shaft Collars

Shaft collars are used as a stop or location device to retain a part on a shaft. Collars are available in standard and metric as well as unthreaded and threaded versions. Unthreaded versions are the most common. There are three styles: solid, single-split, and double-split.

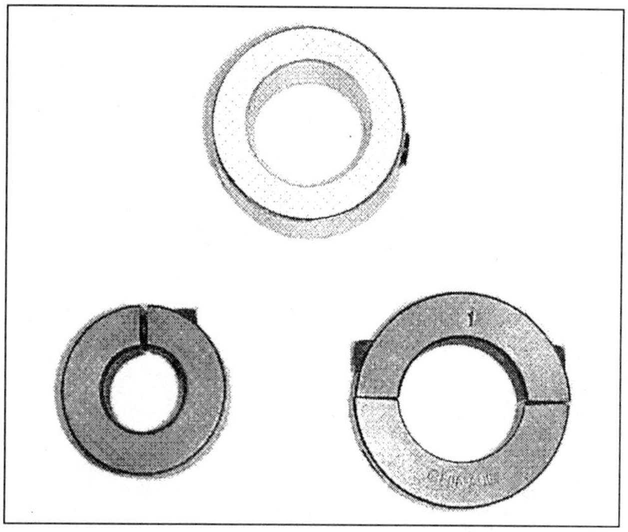
Figure 15-11: Shaft collars

Solid

The solid-style collar has a set screw to hold the collar to the shaft. This is also the most economical and common style.

Single-Split

The single-split collar allows greater clamping force and does not mar the shaft. Both the solid and single-split styles may require that other components on the shaft be removed to allow for installation of these collars.

Double-Split

The double split allows for positioning on a shaft without disturbing other components.

Collars are available with coatings, such as zinc or black oxide, as well as in different materials such as stainless steel or aluminum.

Keystock

Keystock or keys come in a variety of shapes and act as a torque transmitter between the shaft and the mating component. Keys fit into the keyways of the shaft and the component being attached to the shaft. Keys are held in place with a set screw when used with finished bore components or clamped to the shaft via a split-taper bushing. They should fit snug in the keyways to help prevent wallowing or excessive clearance between the key and keyways. Set screws must be tightened to the proper torque specification to help prevent loosening and loss of torque transmission.

Keyways are usually standard on power transmission products over 1/2 in. bore. Square keys are the most common for small bores. Rectangular keys are generally used when the bore size of the component is reaching up to its maximum bore and a shallow keyway is needed, or on larger bores, as indicated in Table 15-1. Step keys are used when the width of the keyways of the shaft and component to be attached do not match. Woodruff or Half Moon keys are used where a rounded keyway has been machined into the shaft to accept the half moon shape of the key.

Table 15-1

Bore Range	Keyseat	Bore Range	Keyseat
1/2 - 9/16	1/8 x 1/16	2 5/16 - 2 3/4	5/8 x 5/16
5/8 - 7/8	3/16 x 3/32	2 13/16 - 31/4	3/4 x 3/8
15/16 - 1 1/4	1/4 x 1/8	3 3/8 - 3 3/4	7/8 x 7/16
1 5/16 - 1 3/8	5/16 x 5/32	3 7/8 - 4 1/2	1 x 1/2
1 7/16 - 1 3/4	3/8 x 3/16	4 5/8 - 5	1 1/4 x 5/8
1 13/16 - 2 1/4	1/2 x 1/4		

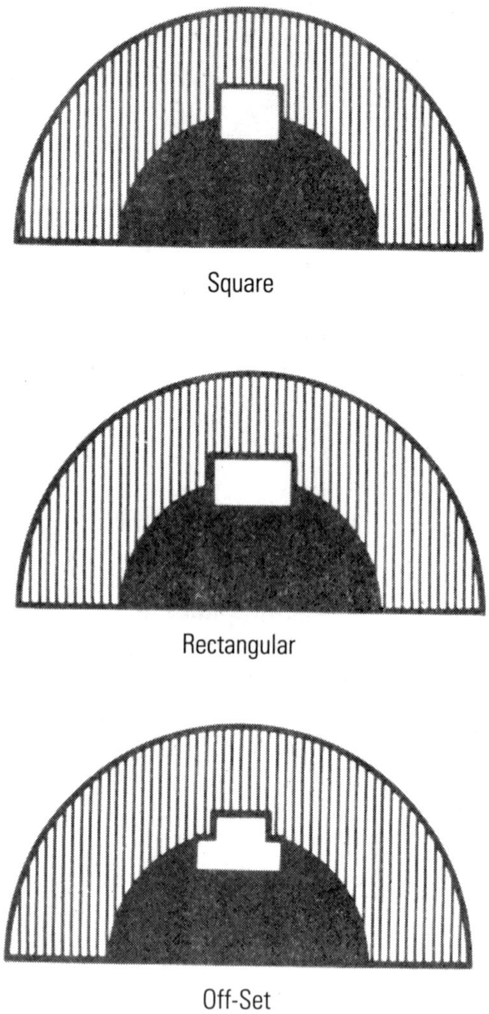

Figure 15-12: Keystock

O-Rings

An O-ring is a simple sealing device with a circular cross section, from which the "O" in its name is derived. O-rings operate in both static (stationary) and dynamic (moving) applications and tolerate differences in stack-up dimensions between mating parts. O-rings are most commonly molded in one piece from an elastomeric (rubber like) material.

O-rings are available in a variety of materials and durameters (hardness of the material) for many different applications. The most common material is Nitrile rubber, which has excellent oil and fuel resistance as well as good wear resistance. O-rings seal by blocking potential leak paths of fluid or gas between two surfaces.

The O-ring is most commonly placed in a machined groove in one of the surfaces to be sealed. As the surfaces are brought together, they squeeze the O-ring and deform the material to take the shape of the surfaces. The areas of contact between the O-ring and the surfaces create a barrier to block the fluid or gas and create a seal. O-rings are also used in non-sealing applications, such as drive belts, tension bands, and spacers.

Locknuts

Keyway Locknuts

By adjusting and controlling clearance and preload of bearings and other power transmission components, in addition to locking them to a position, retaining nuts together with lockwashers and keyed shafts or spindles improves the reliability of power transmission assemblies. These locknuts provide a limited number of locking positions as determined by the number of spanner slots in the locknut and the number of tangs on the lockwasher. A basic, standard, threaded series and a basic, metric, threaded series are promulgated in an ANSI/ABMA/ISO Standard. Their identifications correlate with the bore-size designation of metric dimension bearings. For instance, a bearing number ending in 06 will use a locknut number ending in 06.

Keyless Locknuts

To machine keyways has become unduly expensive. This has caused manufacturers to turn to an array of retaining nuts that lock by mechanical and/or prevailing torque means to avoid the need for keyways. These self-locking nuts allow infinite adjustment and preloading of power transmission assemblies, assuring alignment, balance, and specific clearances. The first developed and most popular are self-locking nuts with prevailing torque collars. Even more effective are retaining nuts in which the lock is developed by mechanically compressing and locking the retaining nut threads to the external thread of the shaft or spindle. This is done radially, axially, tangentially, or circularly, the choice being determined by the configuration and accessibility of the assembly.

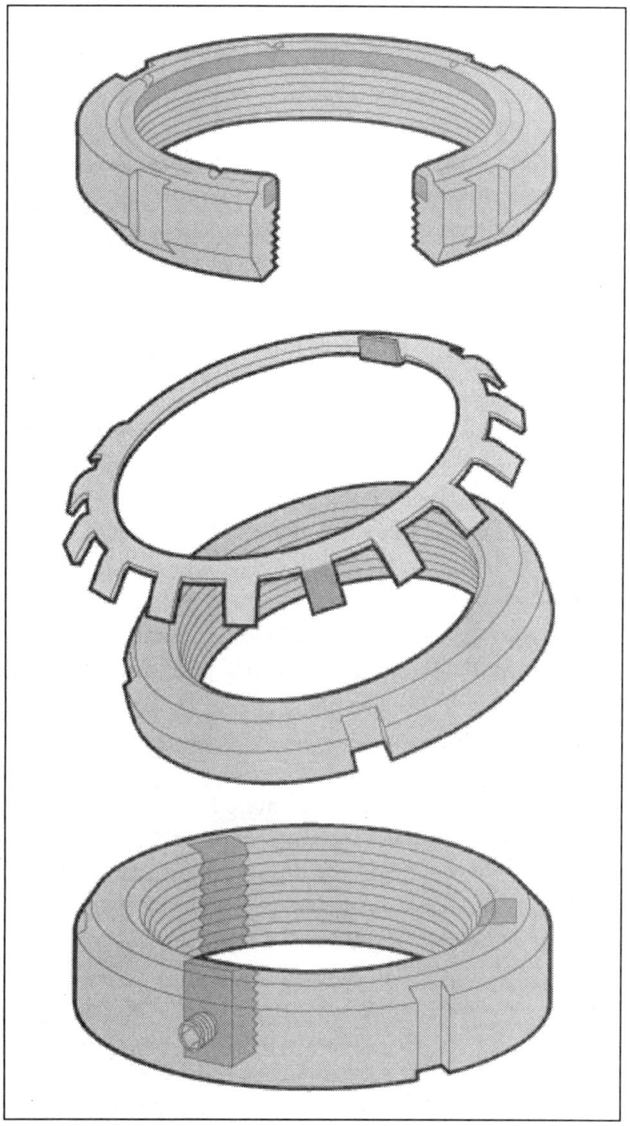

Figure 15-13: Locknuts

Single-Point Lubricators

Proper lubrication is essential to reducing equipment downtime. Tremendous cost savings can be obtained by reducing downtime created from under- or over-lubrication of machinery components. Single-point lubricators come in three common versions: spring-operated, gas-operated, and motor-powered.

Spring-Operated

Spring-operated versions generally offer the user different strengths of springs to meter the grease based on NLGI grease consistency and operating temperature. Reservoir size will be based on the shaft size and whether the machine is used intermittently or on a continuous basis. Many of these lubricators are refillable and have a clear-view reservoir for easy viewing of the grease level.

Gas-Operated

Gas-operated versions utilize a screw-in activator plug to create a chemical reaction to push on the piston and dispense the lubricant. Lubricant is continuously injected into the lubrication point. Different plugs allow for different metering rates, generally from one to twelve months. Temperature variations can cause the metering rates to fluctuate. Care should be taken to understand the temperature variations and the effects they will have on the application. These lubricators are usually offered in various reservoir sizes, depending on the manufacturer. Most are considered one-time-use lubricators.

Motor-Powered

Motor-powered units generally have a battery as the power source to dispense the lubricant at a preset interval. These units may offer a variety of settings to dispense lubricant at different rates. Most units are unaffected by temperature and environmental conditions. Some units have replaceable lubricant cartridges to allow the motorized portion of the lubricator to be used more than once. Reservoir sizes vary based on manufacturers' offerings. Some units have the ability to be integrated with the machinery's electronics to allow for lubricant dispensing only when the machine is in operation.

Additional devices such as automatic oilers are offered in a variety of sizes and styles. Many are used for drip feeding of bearings or come with brushes for lubrication of chains. Wicking units dispense oil based on the viscosity of the oil used and the dispensing rate of the wick. Most units are able to be refilled for continuous use on the machinery that they are intended to protect.

There are also a multitude of accessories offered by manufacturers for all the various lubricators previously described. It is important to decide which unit fits your budget or application.

Additional Information

Addresses and phone numbers for associations are provided in the Association Appendix.

"The Practical Handbook of Machinery Lubrication," by Tex Leugner, available from STLE.

Chapter 16

Lubrication

Chapter Objectives

When you have finished Chapter 16, "Lubrication," you should be able to do the following:

1. Describe viscosity.
2. Identify lubrication regimes.
3. Understand the fundamental concepts of lubrication.
4. Explain the use and application of lubricants.
5. Identify the proper replacement or substitution of lubricants.
6. Understand contamination and how to avoid it.
7. Understand safe storage and handling procedures.

Introduction

The science and technology of interacting surfaces in relative motion is called tribology. It is primarily concerned with the friction, wear, and lubrication of moving parts in instruments, machinery, manufacturing, and technology. Proper tribology decreases the consumption of scarce non-renewable natural resources and produces higher energy-efficiency gains and prolonged component life cycles.

Studies show that significant total potential annual savings can be attained when suitable tribology applications and specifications are indulged.

From practical experience, we know that adding a lubricant to a solid-solid contact will significantly reduce friction. The reduced friction leads to less wear, heat generation, and energy loss—all of which reduce operation costs and downtime. How lubricants provide these benefits will be explored in this chapter.

Lubrication

The primary function of a lubricant is to provide protection for moving parts—thereby reducing friction and wear of the machine. Cooling and debris removal are the other important benefits provided by a fluid lubricant.

Lubrication is used in almost every mechanical device, such as the automobile engine, including the piston (see Fig. 16-1) and gears (see Fig. 16-2).

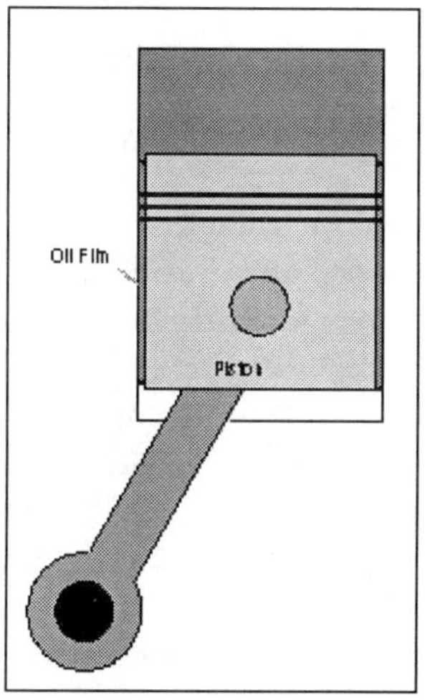

Figure 16-1: Automobile engine piston

Fluids and Viscosity

Simply put, a fluid is a material that is either a liquid or a gas, and fluids include air, water, and oil. Most lubrication is the result of a fluid film that is in between two solid surfaces that move relative to each other. The fluid film in the lubricated area can have a thickness ranging from a few nanometers (billionths of a meter) to hundreds of microns (millionths of a meter) thick. As a point of reference, a human hair has a diameter between 50 and 150 microns.

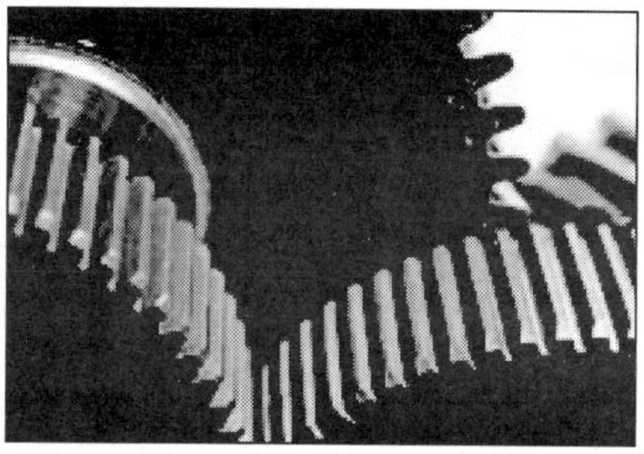
Figure 16-2: Automobile engine gears

The most important property of a lubricant is its viscosity. Loosely defined, the viscosity is the fluid's ability to resist motion. High viscosity means that a fluid is thicker and does not flow as easily. For example, molasses has a much higher viscosity than water, which has a much higher viscosity than air. The viscosity of oil is usually between that of water and molasses. A higher viscosity fluid will typically make a thicker film between the moving surfaces and support greater loads.

Of course, viscosity is not a constant property. Like most fluid properties, it depends on pressure and, more importantly, temperature. The oil in your car's engine has a high viscosity on a cold morning before the engine is started and a low viscosity after the engine heats up.

High viscosity does not guarantee a good lubricant, though. Chemistry of the fluid and conditions at the interface also determine the proper lubricant. These effects go beyond the scope of this chapter. For now, we will consider only oils.

To determine minimum base oil viscosity, the mean bearing diameter d_m in [mm], the bearing speed (in rpm), and the bearing temperature under standard operating conditions are used. The required minimum base oil viscosity for the example shown in Figure 16-3 is 40 °C (104 °F) 38 mm^2/s.

The actual base oil viscosity v should be [1] x 1 ... 4.

Legend

Bore. . . . . 340 mm
Outer diameter. . . . . . 420 mm
Mean bearing diameter . 380 mm
Speed. 500 min–1
Temperature . . . . 70 °C (158 °F)

NOTE: This diagram applies to mineral oils. Information for synthetic oils is available.

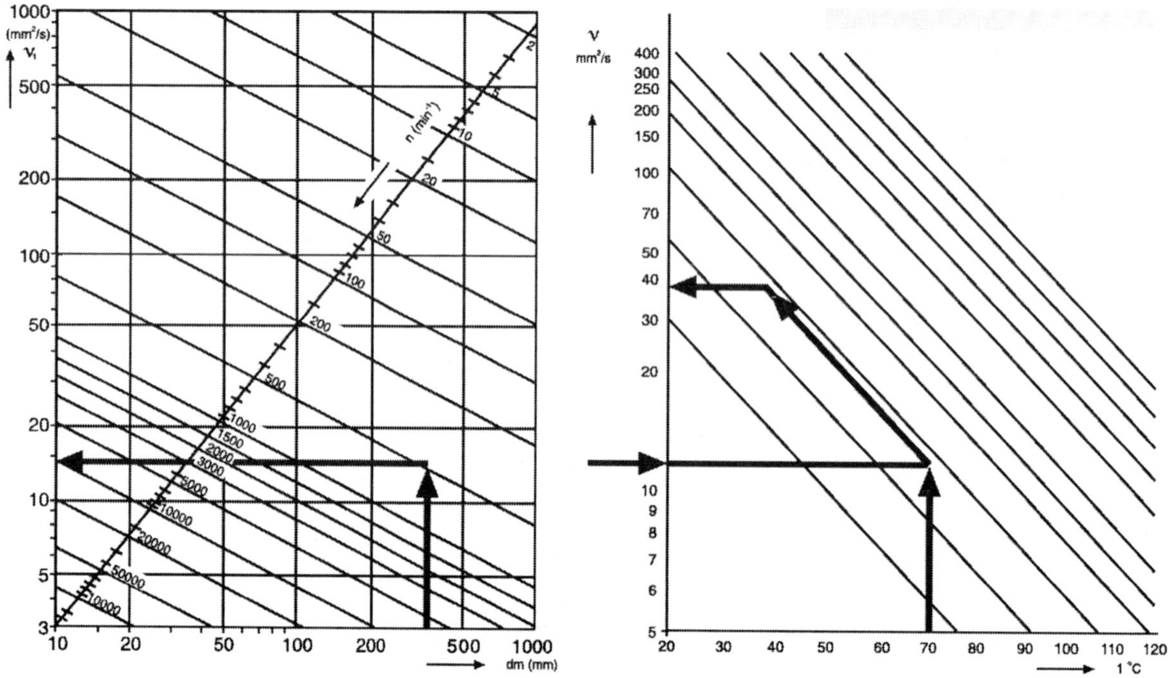

Figure 16-3: Example Determing Minimum Base Oil Viscosity

The following is generally used as a parameter indicating the expected lubricating condition:

- = / ^{1} = viscosity ratio
- = viscosity under standard operating conditions
- ^{1} = required minimum viscosity, depending on mean bearing diameter and speed

The Table 16-1 provides an overview of the anticipated lubricating conditions indicating whether antiwear additives, EP additives, or solid lubricants are required.

Even though synthetic oils can go to higher temperatures than mineral oil, they also have higher temperature limits, and at those limits, they have a relatively short operating life. The Table 16-2 shows those limits.

Temperatures and Oil Service Life

Lower oil temperatures lead to a longer oil life. At higher temperatures, oxidative degradation occurs much faster than at lower temperatures. The Figure 16-4 shows the expected oil service life of different base gear oils. The oil service life for the different chemistry base oils is published in several papers and studies. For the determination of an average oil service life, a large number of used oil analyses were checked. The oil samples were tested with respect to additive and base oil degradation, viscosity change, water, and solid particle content. The slope of the curves represents the so-called 10K rule. A temperature increase of 10K doubles the speed of chemical reaction.

TABLE 16-1 — Anticipated Lubricating Conditions	
€	Lubrication Condition
4 €	Full fluid-film lubrication
> 4 €	In the regime of full fluid-film lubrication + cleanliness + moderate loads = no fatigue
< 4 €	Mixed friction. Lubricants containing antiwear additives have to be used.
1 €	The basic rating life of the rolling bearing is achieved.
< 0.4 €	Mixed friction with increased solid contact; the lubricant has to contain EP additives or solid lubricants.

TABLE 16-2 — Oils Temperature Limits	
Oil Chemistry	Maximum Operating Temperature
Mineral Oil	80–100 °C (176– 212 °F)
Synthetic Hydrocarbons (SHC) Polyalphaolefines (PAO)	140 °C (284 °F)
Polyglycol Oils (PAG)	160–180 °C (320–356 °F)
Ester	180·°C (356 °F)
Silicone	200 °C (392 °F)
Perfluorinated Polyethers (PFPE)	260 °C (500 °F)

Gearbox oil sump temperature is the most important factor affecting the oil change interval. Sump temperature is influenced mainly by the ambient temperature and friction between the gears. Because of this, the expected oil change

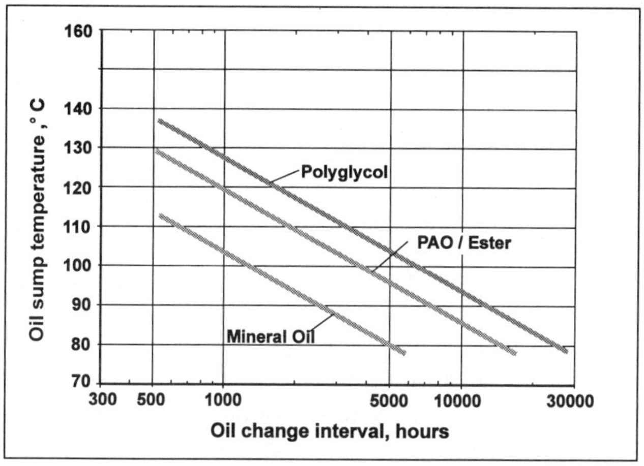

Figure 16-4: Expected Oil Service Life

interval for synthetic oils is better than mineral oil due to their better aging and thermal resistance as well as their reduced coefficient of friction.

It can be stated that the oil service life for polyalphaolefin oils is approximately three times longer than mineral oil, and that of polyglycol oils is five times longer than that of mineral oil's. Reduced wear also leads to an extension of oil service life because of the low solid particle content. Wear metals in the oil can act as a catalyst in the oxidation reaction.

Lubrication Regimes

The thickness of the fluid film determines the lubrication regime, or the type of lubrication. The basic regimes of fluid-film lubrication are:

Hydrodynamic lubrication

Hydrodynamic lubrication—Two surfaces are separated by a fluid film (like a car tire hydroplaning on water).

Elastohydrodynamic lubrication

Elastohydrodynamic lubrication—Two surfaces are separated by a very thin fluid film.

Mixed lubrication

Mixed lubrication—Two surfaces are partly separated and partly in contact.

Boundary lubrication

Boundary lubrication—Two surfaces are mostly in contact with each other even though a fluid is present.

In addition to fluid-film lubrication, there is solid film lubrication in which a thin solid film separates two surfaces. The fluid viscosity, the load that is carried by the two surfaces, and the speed that the two surfaces move relative to each other combine to determine the thickness of the fluid film. This, in turn, determines the lubrication regime. How these factors all affect the friction losses and how they correspond to the different regimes are shown on the Stribeck curve (see Fig. 16-6). Engineers required to evaluate lubricants, design bearings, and understand lubrication regimes use the Stribeck curve.

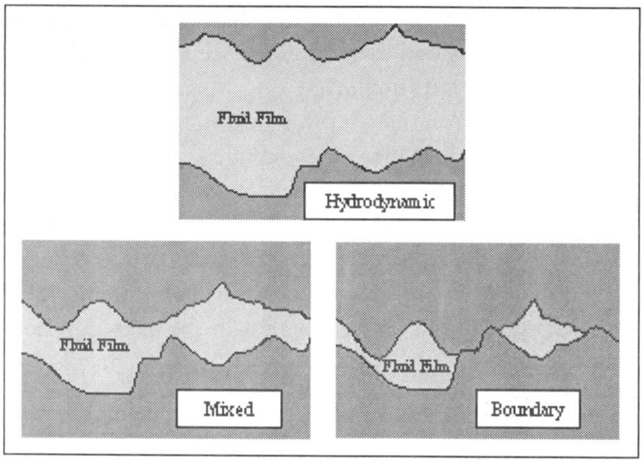

Figure 16-5: Lubrication regimes

Stribeck Curve

The Stribeck curve is a plot of the friction as it relates to viscosity, speed, and load. The vertical axis shows the friction coefficient. The horizontal axis shows a parameter that combines the other variables: µN/P. In this formula, µ is the fluid viscosity, N is the relative speed of the surfaces, and P is the load on the interface per unit bearing width. Basically, as you move to the right on the horizontal axis, the effects of increased speed, increased viscosity, or reduced load are seen. The zero point on the horizontal axis corresponds to static friction.

The combination of low speed, low viscosity, and high load will produce boundary lubrication. Boundary lubrication is characterized by little fluid in the interface and large surface contact. We can see on the Stribeck curve that this results in very high friction.

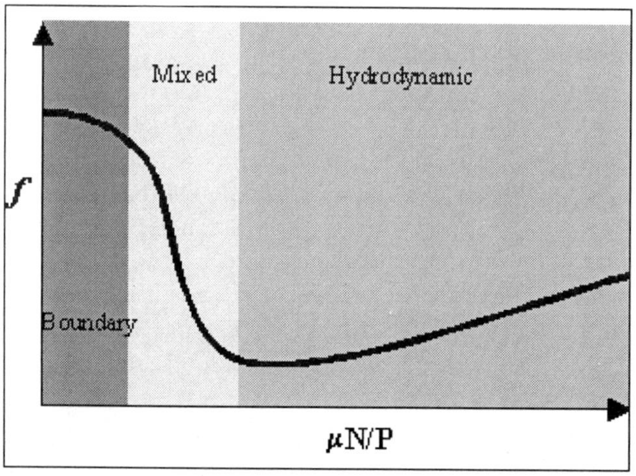

Figure 16-6: Stribeck curve

Machinery will see boundary lubrication at startup and shutdown (low speeds and thin film) before transition to hydrodynamic lubrication at normal operating conditions (high speeds and thick film). Inspection of the Stribeck curve will show that a machine will see the most friction and wear during startup and shutdown.

Note: The Stribeck curve is plotted in log-log format (see Fig. 16-7), so each tick represents a 10x increase over the previous interval.

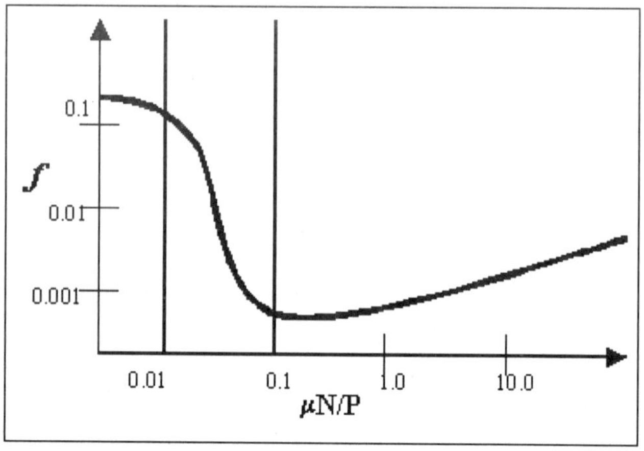

Figure 16-7: Stribeck curve plotted

Example Problems

Imagine a contact region between a moving and stationary surface with a load of 44.4 N/m (3 lb-ft) that is lubricated with oil at 20° C (µ = 0.104 Pa*s). Assume that the Stribeck curve applies to this contact.

a. What speed is necessary for the contact to see hydrodynamic lubrication (µN/P = 0.1)?

Solution:
µN/P = 0.1
N = 0.1*P/µ
N = 0.1*(44.4)/0.104
N = 42.7 m-s = 140 ft-s

b. Now, assume that the lubrication is well within the hydrodynamic regime so that µN/P = 0.2. What lubrication regime would be present if the load were suddenly increased to 222 N/m (15 lb-ft)?

Solution: This solution will have two parts. First, we must calculate the speed for μN/P = 0.2, and then we can calculate the μN/P value for that speed and a 222 N/m load.

$\mu N/P = 0.2$
$N = 0.2*P/\mu$
$N = 0.2*(44.4)/0.104$
$N = 85.4$ m-s = 280 ft-s
$\mu N/P = 0.104*85.4/222$
$\mu N/P = 0.04$

We can see from the Stribeck curve that increasing the load from 3 lb-ft to 15 lb-ft will cause the lubrication regime to go to mixed lubrication. This change will increase the friction and wear in the contact, wasting energy and money. The increased friction will also cause excessive heat generation, decreasing the viscosity and making the situation even worse.

Hydrodynamic Lubrication

In the previous discussion of the Stribeck curve, it stated that the presence of a full fluid film and no surface contact indicates hydrodynamic lubrication. Hydrodynamic lubrication gets its name because the fluid film is produced by relative motion of the solid surfaces and the fluid pressure increase that results.

Hydrodynamic Lubrication Fluid Film

To understand hydrodynamic lubrication, first look at Fig. 16-8. A surface will have tiny asperities or peaks that will come in contact with each other if two plates are placed together. If one of the plates were to slide over the other, then friction would increase, the asperities would break, and the surfaces would wear. In hydrodynamic lubrication, a fluid film separates the surfaces, prevents wear, and reduces friction.

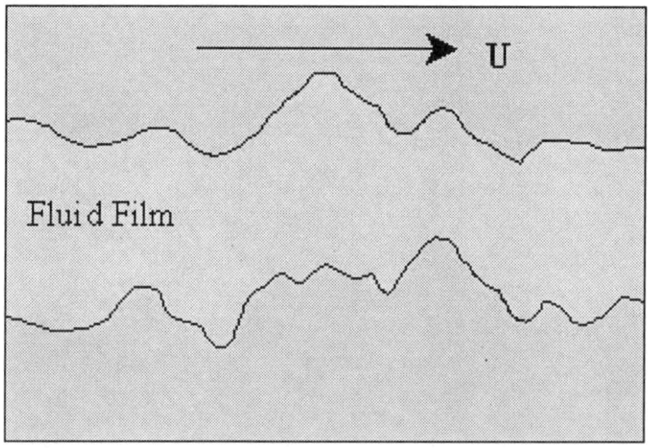

Figure 16-8: Hydrodynamic Lubrication Fluid Film

The hydrodynamic film is formed when the geometry, surface motion, and fluid viscosity combine to increase the fluid pressure enough to support the load. The increased pressure forces the surfaces apart and prevents surface contact. Therefore, in hydrodynamic lubrication, one surface floats above the other surface. The increase in fluid pressure that forces the surfaces apart is called hydrodynamic lift.

Hydrodynamic Lift

Consider two parallel plates with relative motion: if one surface is angled where the entrance area is slightly larger than the exit area, then a wedge-shaped gap is created. This is a converging gap and is the geometry necessary to produce hydrodynamic lift. Note that the difference between the inlet and outlet is extremely small (a few microns at most), so the surfaces will look parallel to the naked eye. Any figures in this chapter or any other source will be greatly exaggerated to illustrate the concept. Surfaces that are this closely matched create a conformal contact.

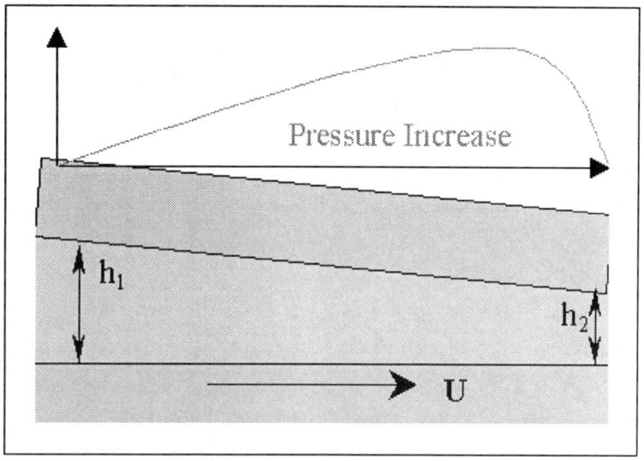

Figure 16-9: Hydrodynamic Lift

Whenever a surface moves over a fluid or a fluid flows over a surface, then the fluid immediately next to the surface will move at the same speed as the surface. So, if two surfaces move relative to each other and a fluid is present, then it will be dragged into the interface. A fluid that enters a converging gap in this manner will see a pressure increase as the gap converges, which creates hydrodynamic lift and forces the surfaces apart like a wedge.

Hydrostatic lift is present when a higher-pressure fluid is forced between two surfaces. In this case, the surface separation is caused by the static fluid pressure and can occur without surface motion.

The mathematical equation that describes the fluid pressure as it relates to surface motion, film thickness, and viscosity, the Reynolds equation, was developed by Osborne Reynolds more than 115 years ago. In its full form, the Reynolds equation is very complicated and difficult to solve; however, the equation can be simplified to solve many problems in lubrication. The Reynolds equation itself is beyond the scope of this chapter.

Hydrodynamic Bearings

A bearing is a machine component that supports or bears a load on a moving interface. Ball bearings, thrust bearings, and journal bearings are common examples of fluid-film bearings. Fluid-film bearings are divided among hydrodynamic, hydrostatic, and elastohydrodynamic bearings. Hydrodynamic bearings get load support by hydrodynamic lift. The most recognizable hydrodynamic bearings are slider bearings and journal bearings.

A simple description of a slider bearing is that of a block moving over a stationary surface on a thin fluid film. In a slider bearing, the moving surface will "slide" over the stationary surface—hence the name. This configuration is used to provide load support for a number of machines.

Now imagine that the converging gap is rolled up—the result would be a journal bearing (see Fig. 16-10). A journal bearing consists of a shaft (the journal) and a ring (the bearing). A journal bearing is used to support the load on a rotating shaft. The load causes the journal and bearing to be slightly offset so that a converging gap is created. As lubricating oil is fed into the bearing and is dragged by the shaft into the converging gap, the fluid pressure increases, and a hydrodynamic lift is created. After the fluid flows through the narrowest part of the gap, the fluid pressure decreases, and vapor pockets may form in the film (an adverse condition known as cavitation). The fluid added in the inlet replaces the oil that leaks out the ends of the journal bearing.

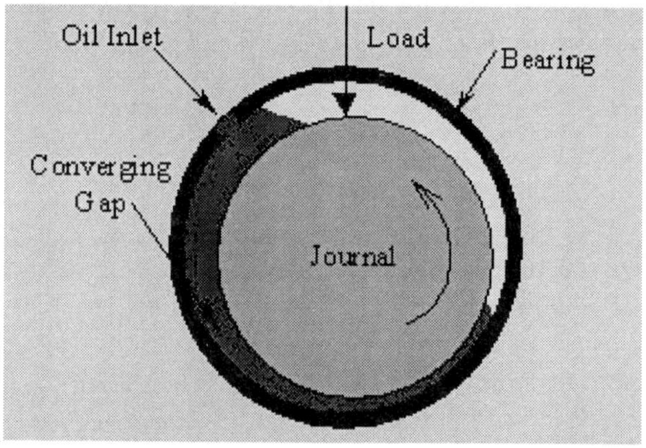

Figure 16-10: Journal bearing

Elastohydrodynamic Lubrication

"Thick" fluid film, low friction, and no wear are the defining characteristics of hydrodynamic lubrication, which generally occurs at conformal contacts. A lubricated nonconformal contact will experience elastohydrodynamic lubrication (EHD).

The classic description of a nonconformal contact is the ball-on-flat, as seen in Fig 16-11. The ball-on-flat is known as a Hertzian contact, which is a point contact with extremely high pressure. As an example, a 19 mm (3/4") diameter steel ball on a flat steel surface has a maximum contact pressure of 950 MPa (138,000 psi) for a 30 N (6.7 lb.) load. That is more than 9,300 times greater than atmo-

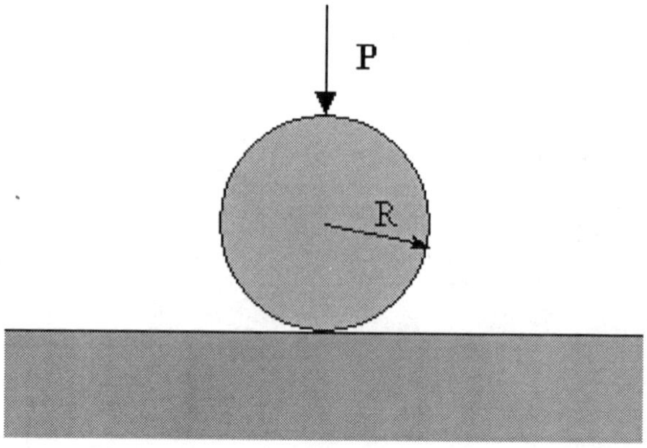

Figure 16-11: Nonconformal Contact

spheric pressure, which is a mere 14.7 psi. The nonconformal contact can produce pressures that are large enough to temporarily deform the solid steel surface.

The enormous pressure produced in a nonconformal contact causes some interesting behavior in oil. While the pressure is high enough to deform the solids, it will also affect the fluid viscosity. Under moderate conditions, the effect of pressure is hardly noticeable, but the EHD pressures are high enough to have a significant effect on the fluid viscosity. In fact, the oil in an EHD contact can become semi solid. This allows a very thin oil film to form and support the load. The science that studies the properties of fluids at these extremely high pressures is known as rheology.

Rolling Element Bearings

Rolling element bearings include many types of ball bearings and roller bearings and provide load support through elastohydrodynamic lubrication.

As the name suggests, roller element bearings have rolling elements that carry the load. The elements can take many shapes like balls or cylinders, but they will always have nonconformal contacts and elastohydrodynamic lubrication. The nonconformal contact is usually not a ball-on-flat because the rolling element may ride on a curved surface (the race). But, the curvature of the race is considerably less than the ball so that the contact is still nonconformal.

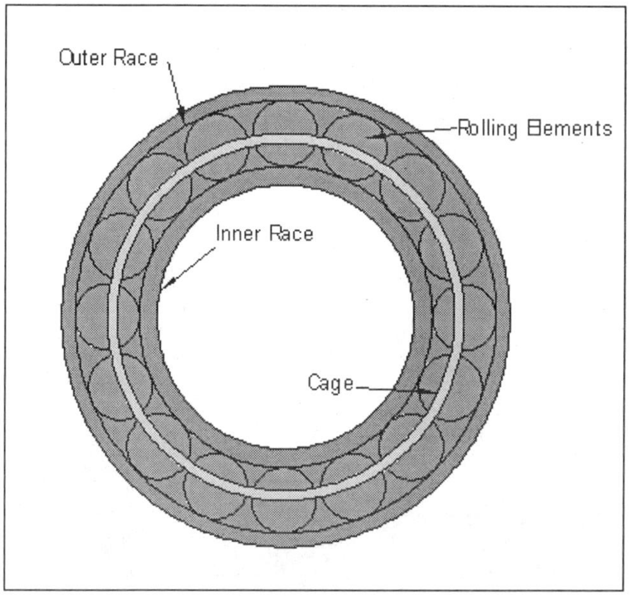

Figure 16-12: Rolling element bearings

The rolling elements and race of a rolling element bearing are made from hardened steel that is able to withstand the extreme pressures of the nonconformal contact. The bearing materials and increase in fluid viscosity allow rolling element bearings to smoothly and reliably support loads in a wide range of applications where a rotating shaft is present, including automobiles, pumps, compressors, and turbines.

Boundary Lubrication

Boundary lubrication occurs when the lubricating film is approximately the same thickness as the surface roughness such that the high points (asperities) on the solid surfaces contact. This is generally an undesirable operating regime for a hydrostatic or hydrodynamic bearing because it leads to increased friction, energy loss, wear, and material damage. But, most machines will see boundary lubrication during their operating lives, especially during startup, shutdown, and low-speed operation. Special lubricants and additives have been developed to decrease the negative effects of boundary lubrication.

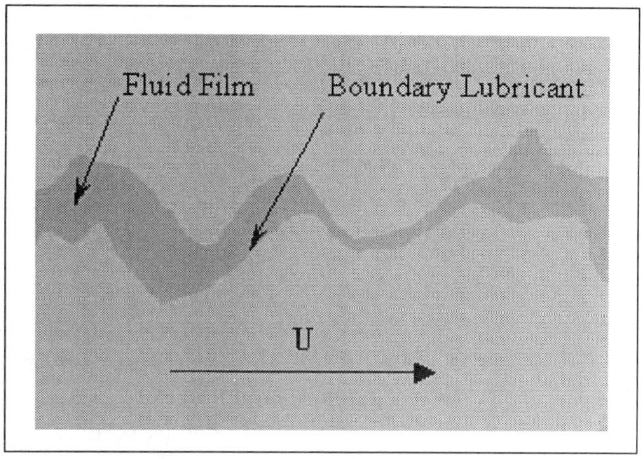

Figure 16-13: Boundary lubrication

Boundary lubricants generally have long, straight, polar molecules, which will readily attach themselves to the metal surfaces. The lubricant molecules will form a thick protective layer (see Fig. 16-14).

The thin layers keep the metal surfaces from contacting, but the boundary lubricant layers will contact each other, causing wear. The sacrificial wear of the lubricant layer will reduce metal wear and prolong the life of the machine.

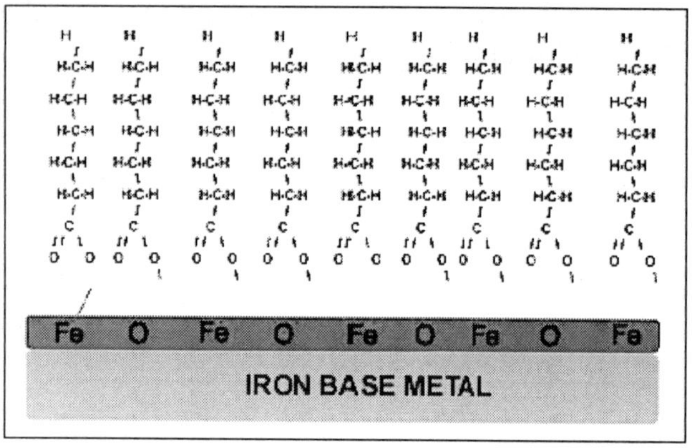

Figure 16-14: Diagram of molecules

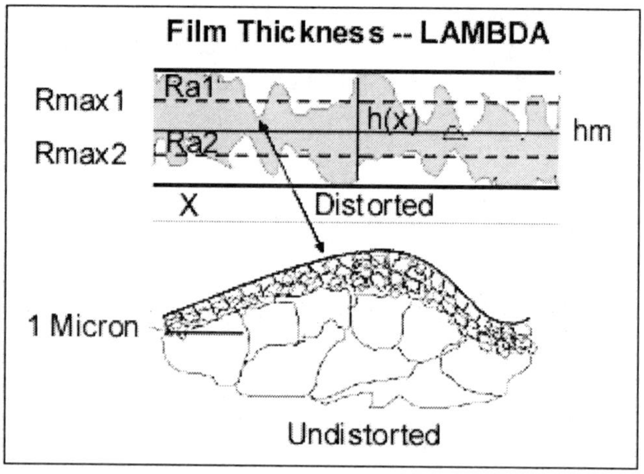

Figure 16-15: Micro-elastohydrodynamic lubrication.

Mixed Lubrication

Mixed lubrication occurs between boundary and hydrodynamic lubrication, as the name suggests. The fluid film thickness is slightly greater than the surface roughness, so that there is very little asperity (high point) contact, but the surfaces are still close enough to affect each other. In a mixed lubrication system, the surface asperities can form miniature nonconformal contacts. Nonconformal contacts lead to EHD; however, when dealing with asperities, not ball bearings, the effect is localized. This phenomenon is termed micro-elastohydrodynamic lubrication (see Fig. 16-15).

Grease

Grease consists of a solid soap such as calcium or lithium soap, or in some cases, a fine clay that forms a matrix in which a liquid lubricant is dispersed. The matrix does not aid lubrication but is a reservoir that releases lubricant to the contact area. The liquid lubricant can contain boundary and EP additives as well as solid lubricants such as graphite and molybdenum disulfide. Composition of grease is shown in Fig. 16-16.

Grease-Gelling Agents – Soap and Soap Complex

Thickeners, while not contributing much toward lubrication, impart unique properties to the grease, affecting its applicability in certain applications or environments. Of these, the lithium and so-called lithium-complex-thickened greases are by far the most common.

- Lithium – Most common, easy to manufacture, easy to store, good pumpability, resists dust and coal, flowability permits dirt to flow out.
- Calcium – Requires less regreasing, good water resistance, calcium soap aids lubrication.
- Aluminum – Highest resistance to water, chemicals, acids, (edible).
- Barium – High water resistance, somewhat toxic.
- Sodium – Fibrous, water-soluble.

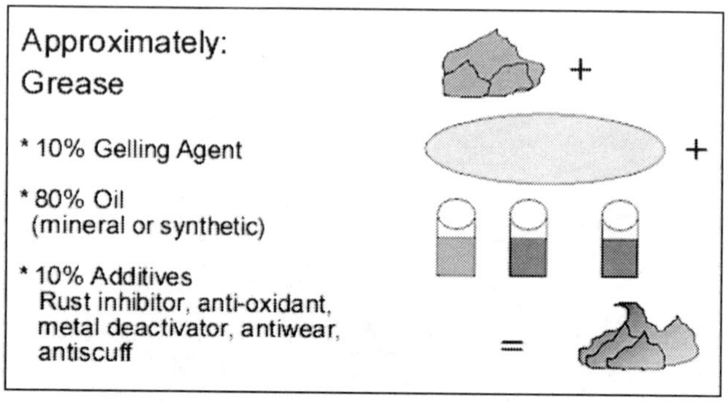

Figure 16-16: Composition of grease

Non-soap Greases

Another class of thickeners is the non-soap thickeners. These are usually used in high-temperature applications, causing the other types of thickeners to soften excessively. This can cause the grease to not stay in place or even cause it to lose its thickness permanently.

- Clays and Silica – Insoluble powders, silica, or platelets of clay. Chemically modified structures and surfaces are made usable as gelling agents for grease. These greases further increase the maximum usable temperature.
- Polyurea – Polyurea greases are called high-performance greases due to their broad range of performance attributes.

Solid Lubricants

Selection of Solid Lubricants

A wide variety of solid materials with inherent lubricating capabilities is available for use in solid-film lubricants. The most commonly used are molybdenum disulphide, graphite, and polytetrafluoroethylene (see table 16-3).

While these are the most common, we also see materials such as tungsten disulphide, boron nitride, lead oxide, antimony oxide, niobium selenide, lead, tin, silver, indium, fluorinated ethylene propylene (FEP), perfluoroalkoxy copolymer (PFA), etc.

No one formulation can satisfy all the requirements on a cost-effective basis. Properties that should be considered are the co-efficient of friction, load-carrying capacity, corrosion resistance (susceptibility to galvanic corrosion), and electrical conductivity. Further, one must consider the environment in which the solid-film lubricant must perform.

Environmental factors include temperature, pressure, humidity, oxygen content, radiation, etc. Each of the previous solid-lubricant materials has strengths and weaknesses in each, and corresponding compromises must be made. Molybdenum disulphide generally has the highest load-carrying capability and a corresponding low co-efficient of friction. However, in an oxidative atmosphere in excess of 400° C (750° F), it begins to decompose.

Graphite has high-temperature capability in an oxidative environment, but it tends to promote galvanic corrosion and will not function in high-vacuum environments. Fluorinated polymers such as polytetrafluoroethylene (PTFE) generally exhibit a low co-efficient of friction and are aesthetically quite suitable for formulations with colored pigments. However, the fluorinated polymers cannot sustain high loading, nor do they have good radiation stability.

TABLE 16-3

PROPERTIES OF COMMON SOLID LUBRICANTS

LUBRICANT TYPE	COLOUR	LOAD-CARRYING CAPABILITY (psi)	THERMAL STABILITY	TYPICAL PARTICLE SIZE	VACUUM SUITABILITY	MOISTURE SENSITIVITY
Molybdenum Disulphide	Grey-Black	>100,000	Good <399°C (<750°F)	2-6 micrometers	Yes	Detrimental
Graphite	Grey-Black	<50,000	Excellent <631°C (<1200°F)	2.5-10 micrometers	No	Necessary
Polytetra-fluoroethylene (PTFE)	White	<6,000	Fair <269°C (<500°F)	Sub-micronic	Yes	No effect

Applications of Solid Lubricants

Solid lubricants are used primarily as extreme pressure (EP), or anti-wear type, additives and are applied in one of three ways. The first and most popular application for solid lubricants, particularly molybdenum disulphide and graphite, is as an EP additive in grease formulations. The plate-like structure of these solid lubricant particles reduces friction by allowing the surfaces in motion to easily slide over each other.

This application finds uses, such as in pins and bushings. Graphite is particularly useful where moisture is present, and in fact, the presence of moisture is necessary to ensure graphite's full benefit as a friction reducer.

Molybdenum and PTFE is frequently used as an anti-wear additive blended with engine oils or other lubricants. There is some controversy over the effectiveness of these solid lubricants suspended in oil, but the practical evidence seems to suggest some benefit in the reduction of friction.

The third and perhaps most practical and successful method of applying solid lubricants is the bonding or impregnation of these friction modifiers onto the surface of various machine components, such as cylinder liners and piston skirts in large stationery natural gas engines. The bonded film of solid lubricant forms a sacrificial wear film on the surface of the component subjected to friction. The bonding or impregnation process can be continually repeated for most components. Critical in these applications is the method of bonding the solid films onto the surfaces to be treated.

Binder Selection

A wide range of resins are available for use as binders or bonding agents in solid-film lubricant formulations. The organic bonding agents are divided into two types:

- Air-dried – includes acrylics, alkyds, vinyls, and acetates.
- Thermosets – includes phenolics, epoxy phenolics, silicones, epoxies, polyamide-imides, and urethanes.

Inorganic bonding agents include silicates, phosphates, and ceramics.

As with the lubricating pigment selection, the desired physical properties and environment must be considered when selecting the proper product. In addition to the previously mentioned physical properties, solvent resistance, corrosion resistance, and cure temperature should be considered.

For example, the high temperature needed to cure some ceramics may have a deleterious effect on the metallurgical properties of the substrate. In general, the organic air-dry resins are less expensive to apply, amenable to field application, and due to the low cure temperature, have no deleterious effects on metallurgical properties of the substrate. However, the air-dry resin systems generally have much lower durability, wear life, and solvent resistance. The oven-cured thermoset resins usually have outstanding durability, solvent resistance, and wear life, but like all other organics, are subject to radiation damage, vacuum outgassing, limited high-temperature applicability, and liquid oxygen incompatibility.

Methods of Application

The methods of applying resin-bonded solid-film lubricants are nearly identical to those paint-like materials. However, resin-bonded, solid-film lubricants generally need a very tightly controlled film thickness. Unlike conventional paints, the normal recommended thickness is in the range of 0.0002 in. and 0.0005 in. for optimum wear and lubrication performance. But, other performance factors can require compromise. For example, a thickness of 0.0005 in. to 0.0020 in. may be needed for optimum corrosion resistance.

The major factors that govern the selection of application method of solid-film lubricants include the number of parts, size of parts, types of parts, film thickness, tolerance, and functional areas to be coated, including masking and overspray allowances. Once these parameters have been defined, it can be determined whether the coating should be applied by conventional spray equipment, electrostatic spray, dipping, roll coating, brush coating, etc. Generally, spray application with multiple coats provides the best lubrication.

Filtration

The number one cause of lubrication failure is contamination. This contamination can arise from outside sources such as water, dust, dirt, etc. Or, it can arise from internal sources such as wear debris. With respect to solid or particulate contamination, proper filtration is the first line of defense—sometimes the only line of defense.

Sadly, filtration is one of the least understood and poorest implemented tools in preventative maintenance. Even OEMs tend to cut costs when installing filtration on their equipment with the resulting loss of performance.

Most contamination problems are caused by particles that cannot be seen by the human eye. The particles that we most want to eliminate are those that are approximately the same size as the dynamic clearances found in the specific component being lubricated. To put this in perspective, particle size is measured in micrometers (microns). A micron is a millionth of a meter, or 39 millionths of an inch. The lower limit on visibility is about 40 microns, which is a very large contaminant. In many situations, we are trying to control particles as small as 8 microns (the size of a red blood cell), 2 microns (the size of bacteria), or even 1 micron and smaller in some specialized cases.

Knowing your equipment and the tolerances (clearances) of its components are critical to good filter application and selection. Filtration can be expensive, and excessive filtration can be a waste of time and money. The key is to work with your filter supplier to get the right filter for the right system and to operate it properly. The financial benefit will be obvious. Also, when charging new lubricant to a machine, be certain that the new lubricant does not have particulate contamination or that particulates are added during the filling operation.

Degradation and Analysis of Oils in Service

Oil degradation occurs by three methods:

1. Contamination (the number one cause of oil failure).

2. Depletion of additives in the oil.

3. Direct degradation of the base oil (usually due to heat, oxidation, and similar chemical reactions).

When oil degradation occurs to some point, it is unfit for use, and the machine will not operate properly and will begin to wear out prematurely. Most contamination can be handled by filtration as previously described, but at some point, base oil oxidation products that enter the oil cannot be removed by practical filtration. Additives become depleted and can no longer do their job. At this point, the only practical action is to drain the oil and thoroughly flush and clean the system. This is time consuming and costly. Most OEM-recommended drain intervals are based on relatively arbitrary time of operation and are inherently conservative. In order to maximize time and minimize shutdowns for oil changes, companies have instituted oil analysis programs on their equipment, which help them monitor the health of the lubricant and the machine being lubricated.

A well run oil analysis program will do the following:

- Detect abnormal wear in the equipment.
- Detect contaminants in the oil, regardless of the source.
- Determine the condition of the oil itself.
- Aid in determining the proper operating conditions for the equipment.

The benefits are the following:

- Detection of small problems before they become unmanageable.
- Reduction of overall operating costs.
- Reduction of unscheduled downtime by catching failures before they occur.
- Optimization of drain intervals.
- Determination of proper filtration techniques and equipment for the specific service conditions.

Storage of Lubricants

The proper storage of lubricants is also an often-neglected part of a good maintenance program. Storage is very much a cleanliness, safety, and environmental issue. The three Cs of storage are: avoiding Contamination of product and people, preventing Confusion of what the product is and its usage, and providing Containment to prevent environmental pollution.

In general, a single location for storage of all lubricants is preferred to help with inventory control. Facilities should accommodate the quantities stored, whether in cans, kegs, drums, tanks, etc. Most are in cans and drums, which should be in a covered storage area on racks, shelves, or pallets, as appropriate. Drums should be stored on their sides, especially if stored outside or in unheated areas where water can condense in the drum, causing contamination. Proper labeling of lubricants is critical to eliminate the possibility of a mix up. One significant source of contamination is transfer containers. For example, a dirty pail is used to transfer a small quantity of lubricant from the properly stored clean oil drum to the machine. Additionally, proper personal protective equipment should be worn by the operator when dispensing lubricants. Further, storage areas should be properly diked to contain spills, protecting the environment.

Aftermarket Additives

It is too common of a thought that, by adding some magic material to the lubricant of an industrial machine, misoperation, misalignment, mismanagement, poor maintenance, or human laziness can be cured. There is probably no more contentious subject in the lubricant market than that of aftermarket additives. Some of these products are, at best, of dubious quality or efficacy, while enough others add value in some applications to negate blanket statements about the use of such additives.

Generally, there are no specifications covering these additives. Machine manufacturers or lubricant manufacturers do not approve of the use of aftermarket lubricant products primarily because there is no control of their quality or use. Many OEMs and lubricant formulators will not honor warrantees if aftermarket lubricants are used. Modern lubricants, especially engine oils, have as many as 10 or 15 additives, very carefully blended into their products with emphasis on compatibility as well as efficacy in operation. Further, these additives comprise as much as 20 percent of the engine oil. Addition of an aftermarket additive could cause incompatibility of the additive package, causing catastrophic failure of the lubricant formulation. The example is analogous to adding lemon juice to milk, shaking it up, and suddenly producing buttermilk.

As previously stated regarding purchasing lubricants for expensive machinery, one should only purchase those lubricants that meet the OEM specification for the lubricant. Adding aftermarket additives defies that specification. Rather, an appropriate oil analysis program will help to get the most out of the lubricant and machine. In some cases, lubricant manufacturers will sell specific additive packages for a specific lubricant. However, this is only done when oil analysis indicates that this is the proper corrective action.

Selection Criteria

The single most important consideration is to use a lubricant that meets or exceeds the lubricant specified by the equipment manufacturer. Failure to do so can negate any warranties and could cause catastrophic failure of the machine.

As a general rule, lubricants are purchased via three methodologies, which are often blended somewhat to meet specific needs. The first step is to determine the needs of your company. Such needs include the volume of lubricants used, the type of equipment to be lubricated, how critical this equipment is to the operation, and in some cases, the geographical location of the operation. Defining these needs can be accomplished in several ways.

One option is to provide your lubricant supplier a complete and detailed list of the equipment that requires lubrication and then allow the supplier to decide which lubricants it wishes to provide.

A second option is to provide a list of the current lubricants being purchased and allow prospective suppliers to provide equivalents to those products being used. There is a risk that suppliers may not have exact equivalents that meet the specifications of the equipment manufacturer.

A third option is to provide a list of the lubricant specifications required for all the major/critical pieces of equipment. This may take some effort to compile. The lubricant supplier is then required to supply lubricants that meet these specifications and to certify that their products do so. In reality, most companies use a combination of these options, depending on the criticality of the equipment.

Key Terms

Lubrication
The reduction of friction and wear between two surfaces in relative motion by application of a lubricant.

Lubricant
An oil or grease placed between two surfaces in relative motion that reduces friction and wear between them.

Regimes
Lubrication types that define the relationship of the contacting surfaces in an operating condition. Hydrodynamic, Elastohydrodynamic, and Boundary.

Additional Information

Addresses and phone numbers for associations are provided in the Association Appendix.

Society of Tribologists & Lubrication Engineers (STLE), Park Ridge, IL.

"Basic Handbook of Lubrication" by the Alberta Section, 2nd edition, available from STLE.

"The Practical Handbook of Machinery Lubrication" by Tex Leugner, available from STLE.

Chapter 17

Vibration Analysis

Chapter Objectives

When you have finished Chapter 17, "Vibration Analysis," you should be able to do the following:

1. Understand the difference between acceleration, velocity, and displacement in relation to vibration analysis.
2. List the four key steps in a vibration analysis process.
3. Describe the relationship between time waveform and frequency spectrum.
4. List the three components used to measure vibration.
5. Identify five common faults and know their characteristics.
6. Understand the frequency spectrum analysis.
7. Comprehend the applications of vibration analysis.

Introduction

Industrialized companies, such as manufacturers and material processors, must be efficient in order to achieve high productivity from their machinery. In order to achieve and maintain competitive productivity levels, maintenance must be done on this machinery with minimum interruption to production. This is a difficult task, but a number of technologies are available to assist the maintenance professional. Detecting when a machine requires maintenance (fault detection), determining what corrective action needs to be taken, and the tools to perform precision maintenance, like balancing rotors, will be explained in this chapter.

All machinery vibrates during operation; "What is causing the vibration?" and "What is too much?" are questions that need to be addressed. A maintenance professional needs to consider these vibration technology questions to quantify, trend, and analyze data.

Analyzing vibration is imperative for the maintenance professional to effectively plan, schedule, and define repair activities to resolve problems with plant machinery.

In this chapter, we will discuss the following:

- Vibration fundamentals and measurement.
- Descriptions of fault types (diagnostics), which can be detected using vibration technology.
- Using vibration technology to perform precision balancing.
- Understanding a structure's vibration.

After reading this chapter, you should have a fundamental understanding of what vibration in machinery is and how measuring vibration can help you operate a maintenance program that supports production.

Theoretic Basis

Principals of Vibration Analysis

Vibration is defined as the repetitive, periodic, or oscillatory response of a mechanical system. Before addressing vibration-based, machine-fault diagnostics, some basic principals of vibration analysis will be provided in this section.

To illustrate the basic concepts, a spring-weight system, as shown in Fig. 17-1, will be used. The model consists of two elements: inertia (the weight) and elastic force (the spring). As the model is used for explanation, damping, which would typically occur, will not be considered here.

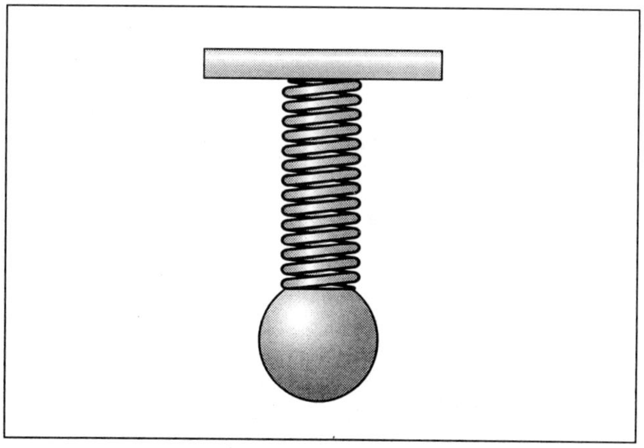

Figure 17-1: Spring-Weight System

Assume that the mass is constrained to move only vertically; therefore, the motion of this model could be described by a single coordinate. Three parameters could be used to describe the motion of the system: displacement, velocity, and acceleration. Displacement is the distance of the movement from the neutral position; it is commonly measured in millimeters (mm) and one thousandth of an inch (mil). Velocity is the rate of change of displacement; it is measured in millimeters per second (mm/s) and inches per second (inches/s). Acceleration is the rate of change of velocity; it is measured in units of gravity (g). Displacement is shown in Fig. 17-2, using the motion of Fig. 17-1. There are three major quantities to define the sine wave, as shown in Fig. 17-2. They are amplitude (A), frequency(ω), and phase (ϕ).

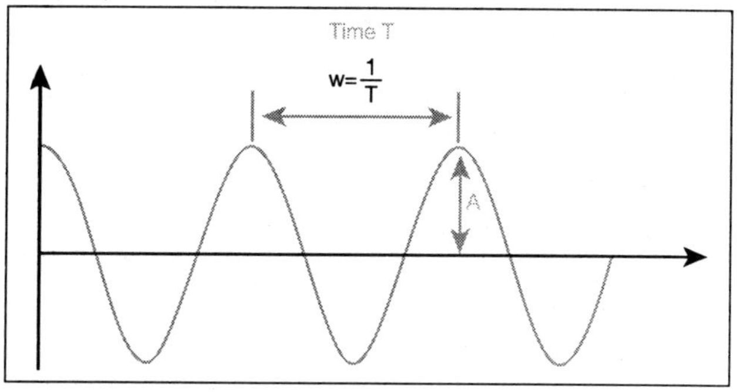

Figure 17-2: Displacement in waveform from Figure 17-1

The time required to complete one cycle or period of vibration is used to calculate the frequency of vibration. To calculate the frequency, the period is divided into 1. If the time is measured in seconds, the calculated frequency is Hz, or cycles per second. If the time is measured in minutes, the frequency is cycles per minute (CPM). The relationship between the two is as follows: 1 Hz = 60 CPM.

For example, if a motor is running at a frequency of 1,800 rpm, dividing 1 by 1,800 equals 5.6×10-4 minutes (the time it takes for one revolution) or converting to Hz by multiplying 5.6×10-4 by 60 tells us one revolution takes 3.3×10-2 sec.

Phase is defined as the relative position of a vibrating object with reference to another object, which is a fixed object or a vibrating object, at a given time. Phase, which is measured in degrees, is very useful for identifying some machine faults. To illustrate this concept, two spring-weight systems will be used. As shown in Fig. 17-3, assume that the two systems are identical, and initially, the mass of the first system is held at the highest point that the system can travel, and the mass of the second system is held at the lowest point that the system can travel. The two masses are released simultaneously.

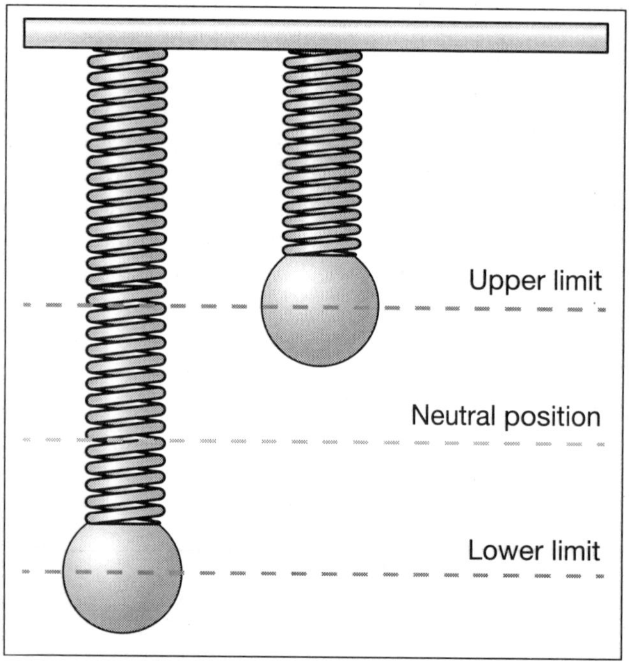

Figure 17-3: Spring-weight with two systems

When using displacement to present the motion of the system in Fig. 17-3, the waveform is shown in Fig. 17-4. Now, we can see that the phase difference between the mass of the first system and the mass of second system is 180 deg.; this is called "180 degrees out of phase."

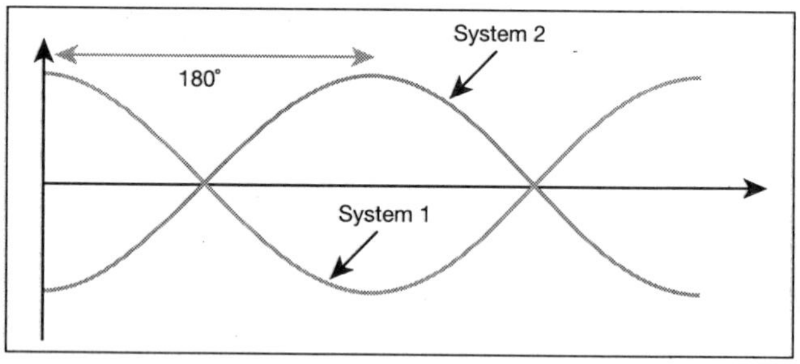

Figure 17-4: Displacement in waveform from Fig. 17-3

Vibration Analysis Procedure

The vibration analysis procedure can be organized into four major steps:

1. Converting the mechanical response of a machine to an electrical signal.
2. Digitizing and processing the collected signal.
3. Identifying the key features related to the machine faults.
4. Deciding if an abnormal condition exists and taking the appropriate action.

Measurement System

Measurement system is very important for vibration analysis. The measurement system includes three major components: transducers, the analog to digital (A/D) module, and the processing units. Transducers convert the mechanical vibration movement to the electrical signal. The A/D module digitizes the transducer's output signal. The processing unit processes the collected digital signal for vibration analysis.

Currently, vibration transducers could be used for measuring displacement, velocity, and acceleration. In vibration analysis applications, the displacement transducers measure only the relative displacement rather than absolute displacement, while the most commonly used velocity and acceleration transducers measure the absolute movement. The details on the three types of transducers will be introduced in this section.

Accelerometers are used to measure acceleration. As we learned in the theoretical section of this chapter, acceleration is the rate of change of velocity; it is measured in units of gravity (g). Accelerometers produce an electrical signal proportional to acceleration. The typical accelerometer, (see Fig. 17-5), is a device with a small weight mounted on a small quartz crystal or ceramic stack. When force is applied to the quartz or ceramic by the weight as it vibrates, a charge is generated in it. This charge is amplified by an internal amplifier in the accelerometer, producing an output voltage proportional to the amount and frequency of the vibration.

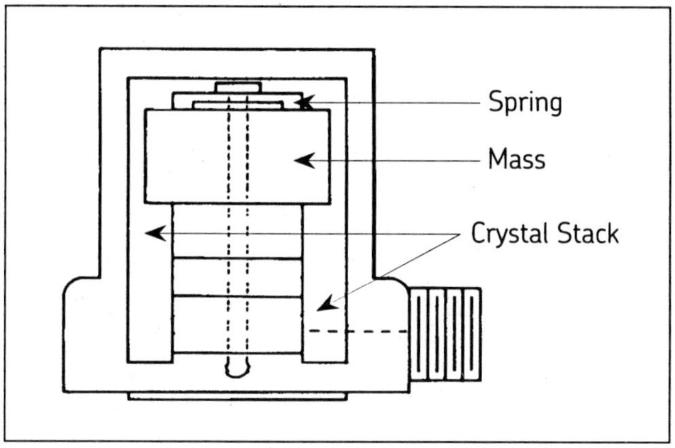

Figure 17-5: Typical accelerometer

Accelerometers must be supplied with a constant current source for operation of the internal amplifier. Modern devices such as data collectors and signal analyzers provide what is called Integrated Circuit Piezoelectric (ICP) power for accelerometers. In turn, the output of most industrial accelerometers received by the data collector or signal analyzer is specified in millivolts per unit of gravitational force, or mv/g. A wide variety of accelerometers are available; selection depends on the frequency range desired, sensitivity, temperature, and room available for mounting.

Generally speaking, there are two types of transducers used for measuring velocity: velometers and velocity pickups. Velometers are commonly used for velocity measurements; they are an accelerometer with additional electronics in the housing that integrate the signal from acceleration to velocity. A velocity pickup is demonstrated in Fig. 17-6. Internally, there is a spring-suspended, permanent magnet that is held motionless relative to a point in free space when the transducer is operating within the specified frequency range. As the wire coil cuts through the stationary magnetic field, a voltage proportional to the relative velocity of the magnet and coil is generated.
The inherent problem with a velocity transducer is that the signal tends to be very noisy. Also, being primarily mechanical devices, they tend to become unreliable over time.

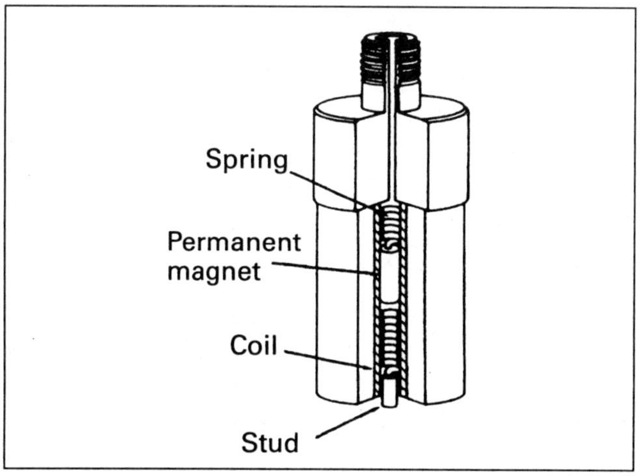

Figure 17-6: Typical velocity pickup

Proximity probes are used to detect relative displacement of equipment. A proximity probe, like the one shown in Fig. 17-7, is made by inserting a pickup coil in the end of a small housing. A high-frequency, alternating current is supplied to the coil, creating a magnetic field. As the shaft or roll moves, it causes a change in the magnetic field, producing a signal proportional to the motion. Proximity probes are used on fluid film bearings in turbomachinery, where it is important to measure motion of a shaft in a relatively feasible fluid film bearing inside a heavy housing. Because a proximity probe does not contact the shaft directly, it is sensitive to the surface finish of the shaft surface as well as residual magnetism. Proximity probes require gap calibration for each installation. This limits them to permanent installations.

The analog signal generated by a transducer must be digitized so it can be processed by computer software. This task is accomplished in the A/D module. After the signal digitization, the signal processing units could perform various digital signal processing algorithms for vibration analysis. In real-life application of vibration analysis, frequency domain analysis is the most widely used technique. This section will mainly

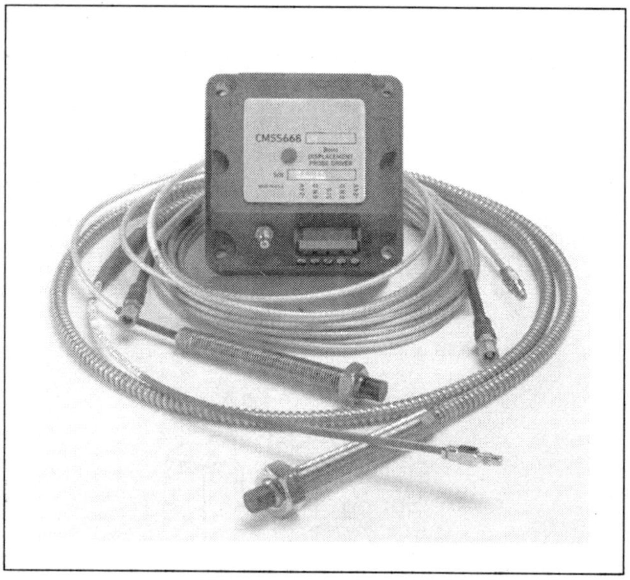

Figure 17-7: A typical proximity probe

discuss the general idea of the frequency domain analysis. To perform frequency domain analysis, time domain signal needs to be transformed into frequency domain. Fast Fourier Transform (FFT) is used to perform the calculation.

To demonstrate how the frequency contents inside a time waveform could be represented in the frequency domain, a time waveform consisting of four sine waves with different frequency will be used, as shown in Fig. 17-8. In Fig. 17-8, one could see that, from the time wave of the complex waveform, it is very difficult to tell how many frequency components there are, while frequency contents are clearly identified in the frequency domain.

Applying the FFT to the time waveform of a single-reduction gearbox running at 1,800 rpm with unbalance, we have a spectrum that has two peaks: unbalance and gearmesh (see Fig. 17-9). The vibration analyst looks for peaks in the frequency spectrum. The presence or absence of peaks helps the analyst determine the condition of the machine.

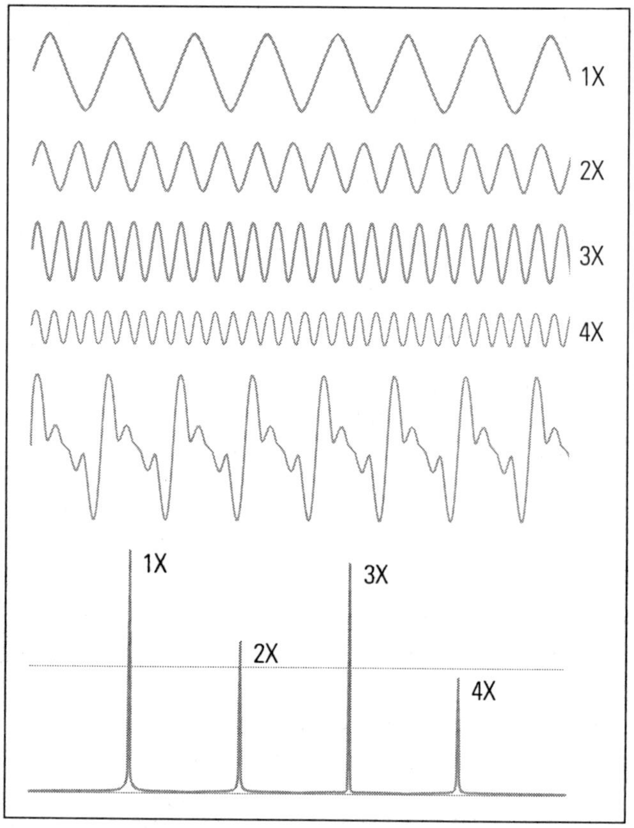

Figure 17-8: Frequency contents inside a time waveform represented in the frequency domain

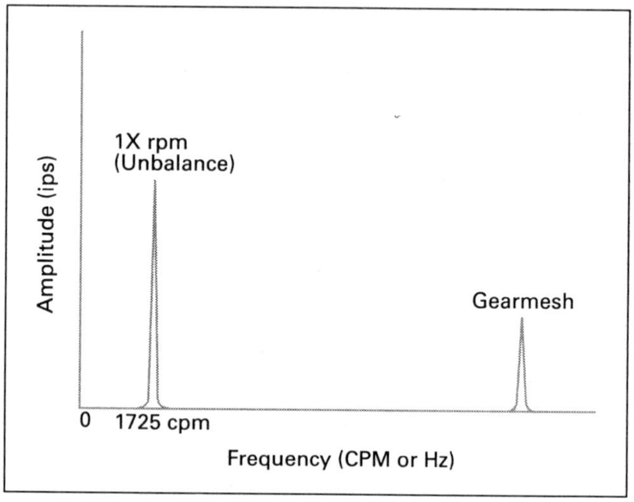

Figure 17-9: Gearbox spectrum showing unbalance and gearmesh frequencies

Rotating Machine-Fault Diagnostics

Rotating machine-fault diagnostics is a key component in rotating machine-condition-based maintenance. Fault diagnostics identify the faults inside the machine. There are two major types of fault diagnostics: a model-based approach and signal-based techniques. By applying the model-based approach, the dynamics of the monitored machine is analyzed, and the physical model is first created. By comparing the output of the model and the actual machine, machine faults have been detected and diagnosed. For signal-based techniques, signals collected form different types of sensors, and then various signal processing methods are used to extract the key features. Based on analyzing the key features, machine faults could be identified. Model-based approaches need an accurate model of the system and are computationally expensive. Signal-based techniques are the most widely used techniques; therefore, only the signal-based techniques will be discussed here.

The signals commonly collected in the machine-fault diagnostics include: oil-debris information, temperature, acoustic emission, and vibration. Vibration is the most commonly used as it contains abundant information and is relatively easy to collect, especially for an online collection system.

The following section does not include all the faults; one comprehensive reference to other faults may be found in references such as the "Vibration Fault Guide" published by Full Spectrum Diagnostics.

Unbalance

Unbalance occurs when the rotation about the center of mass does not match the geometric center of rotation determined by the bearings. Depending on the type of unbalance (static, couple, or dynamic), the axis of mass and geometry may or may not intersect. However, in all of these cases, the forces created by the unbalance result in a spectrum (see Fig. 17-10), with a high 1X peak. Confirmation of unbalance is done by taking another measurement 90 deg. apart on the same bearing. There should be a 90-deg. phase shift with a margin of ± 30 deg. To determine if the unbalance is static, couple, or dynamic, a comparison of phase angles between bearings allows the analyst to complete the diagnosis.

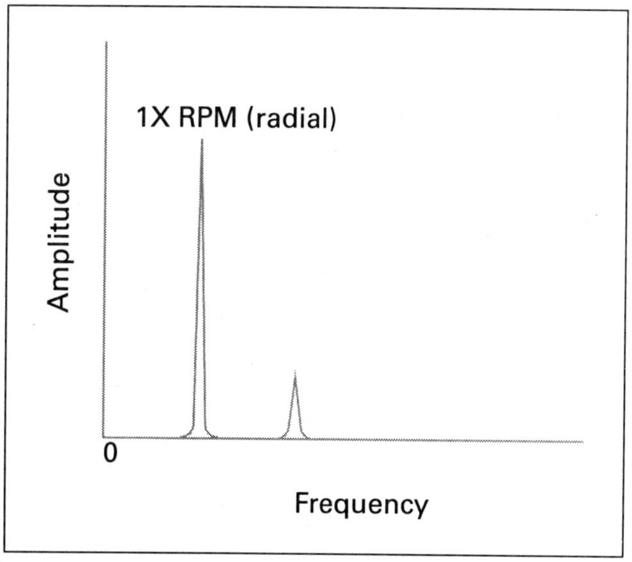

Figure 17-10: Unbalance spectrum

Misalignment

Misalignment occurs when the centerlines of two rotating shafts have a parallel, angular, or an angular offset at the point where power is transmitted (see Fig. 17-11). As the shafts try to find a common centerline, the abnormal forces produce excessive vibration levels and eventually cause premature failures of bearings, shafts, couplings, seals, etc.

If we connect two machines together using a flexible coupling and do not have the centers of the shafts aligned within tolerance standards, our vibration spectrum looks like the one shown in Fig. 17-12. In addition to a peak at the running speed of the machines (which could be measured in displacement, velocity, or acceleration depending on the speed), there are additional peaks. These peaks are at multiples of the machine's running speed.

If the two coupled machines are running at 1,800 rpm (CPM), the first peak is at 1,800 CPM, the second at 2X 1,800 CPM or 3,600 CPM, the third at 3X 1,800 or 7,200 CPM. These peaks are called harmonics because they are whole-number multiples of the running speed. The presence of harmonics is an important feature that can be used in the identification of many machinery faults.

In the majority of situations, misalignment is indicated by an excessive 1X rpm amplitude, an excessive 2X rpm, and harmonics of the running speed (see Fig. 17-13). The am-

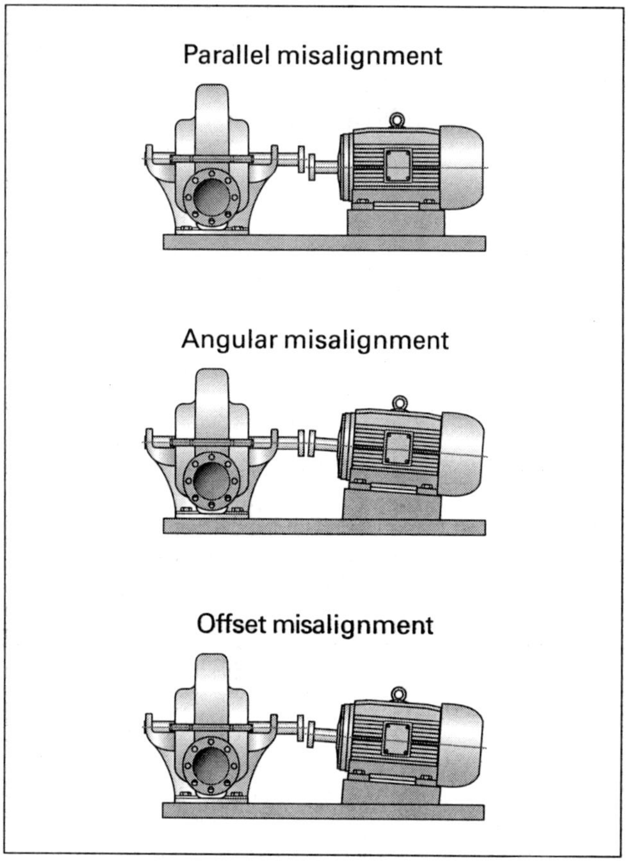

Figure 17-11: Example of shaft misalignments

plitudes and number of harmonics depend on the amount of misalignment as well as the type of coupling used. Comparison of the phase angle between the horizontal, vertical, and axial readings at 1X rpm is made to determine the type (horizontal, vertical, angular, or a combination) of misalignment.

Looseness

Looseness is usually caused by improper fit between mechanical components. In general, looseness could be detected by high 2X rpm or multiple harmonics of the shaft-rotating frequency. There are three major types of looseness. They are "Type A," "Type B," and "Type C." The following details the definition and the fault signatures of each type.

A basic looseness condition often called "Type A" (see Fig. 17-14) has a dominant peak at 1x rpm and can be misdiagnosed as unbalanced. The greatest response is typically localized at a single-machine component measured verti-

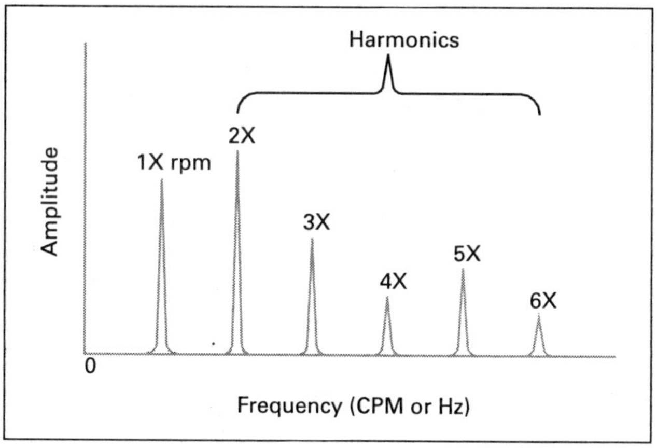

Figure 17-12: Misalingnment spectrum showing harmonics of running speed

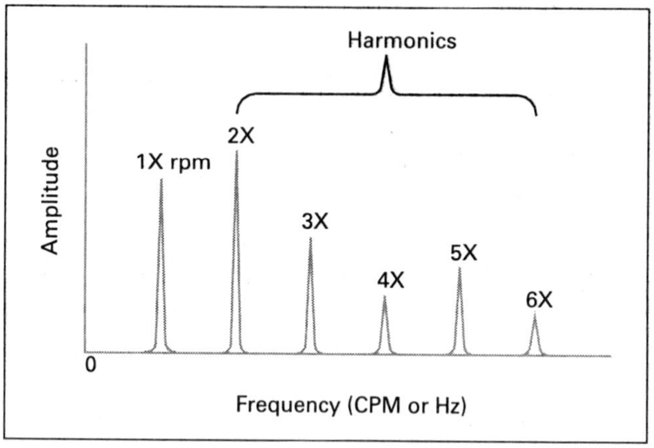

Figure 17-13: Misalignment spectrum

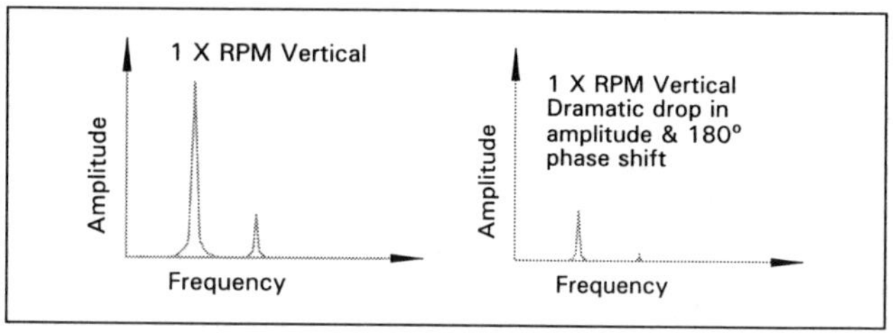

Figure 17-14: Type A Looseness spectrum

cally. This type of looseness is typically found at connections between machine components or bases.

Looseness is verified by phase analysis across the machine joints such as bolted connections, welds, and grout lines. The phase does not shift 90 deg in a horizontal to vertical comparison that typically signifies unbalance. But, the phase will shift across the loose joint by 180 deg. ± 30 deg., and a dramatic change in magnitude will occur.

Another type of looseness called "Type B" (see Fig. 17-15) is not as easy to diagnose. It can produce a spectrum that looks similar to misalignment with a high 2x rpm peak and frequently fractional sub-harmonics. If severe enough, it may also create alternating high and low harmonic peaks. Phase is erratic, and measurements across the suspected failed interface are used to diagnose this type of looseness. This looseness is usually caused by cracks in housings or pedestals, loose pedestal bolts, or faulty mechanical isolators.

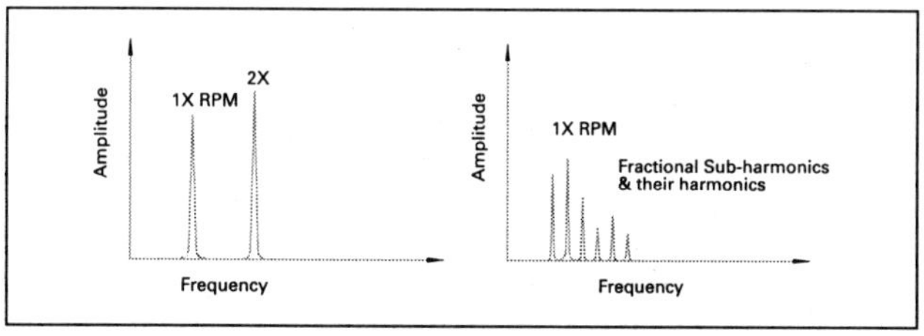

Figure 17-15: Type B mechanical looseness

"Type C" mechanical looseness (see Fig. 17-16) is caused by a poor fit between a coupling, hub, impeller, or antifriction bearing and its shaft or housing. It is recognized by numerous 1X rpm harmonics and a raised noise floor in the spectrum. A time waveform can be used to verify this type of looseness; look for truncation of peaks, sharp repetitive impact peaks, and random patterns between peaks. Acceleration amplitudes less than 2 g usually indicate misalignment, while those above 2 g indicate looseness.

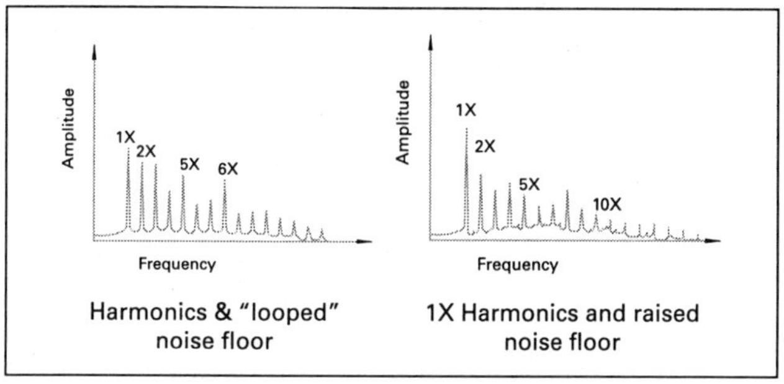

Harmonics & "looped" noise floor

1X Harmonics and raised noise floor

Figure 17-16: Type C mechanical looseness

Phase readings will be unstable, and the 90-deg. phase shift from horizontal to vertical commonly found in unbalance will not occur.

Rolling-Element Bearing Faults

Defects that occur in rolling-element bearings are usually caused by damage and occasionally by normal wear. When a rolling element passes over the damage point, an impact force could be generated. The impact force will cause a ringing of the bearing and support structure at the natural structural frequency. The resulting defect peaks that occur in a vibration spectrum are directly related to the bearing geometry (raceway-pitch diameter, rolling-element diameter, number of rolling elements, and thrust angle).

Four coefficients determined by the geometry can be calculated and used to determine the location of the defect: inner raceway, outer raceway, rolling element, and cage. These coefficients are available from all of the major bearing manufacturers.

- BPFO – Ball Pass Frequency Outer raceway
- BPFI – Ball Pass Frequency Inner raceway
- BSF – Ball or roller Spin Frequency
- FTF – Fundamental Train Frequency

These coefficients are multiplied by the 1X rpm to determine the defect frequencies in the spectrum. These non-synchronous peaks (a non-synchronous peak will not be a whole number multiple of running speed) are one of the important features that allow an analyst to identify rolling-element bearing defects. For demonstrational purposes, fault signatures of each component of the rolling-element bearing are shown in the following figures.

A single defect on a rolling element (see Fig. 17-17) will contact both the inner and outer raceways every revolution, typically generating a dominant 2X BSF frequency.

The raceways of the inner and outer rings in an antifriction bearing are designed with curvatures that match the rolling element shape. When a rolling element in a bearing passes over a defect in one of the raceways, a peak is created in the spectrum (see Fig. 17-18) for a given bearing, and the frequency of the outer ring peak will always be lower than the inner ring peak.

The Fundamental Train Frequency (FTF) is always lower than the rpm of the shaft or housing (outer ring rotation). For inner ring rotation, the FTF defect frequencies fall in the range of 0.33X–0.48X rpm; for outer ring, 0.52X–0.67X rpm. This type of defect, shown in Fig. 17-19, may

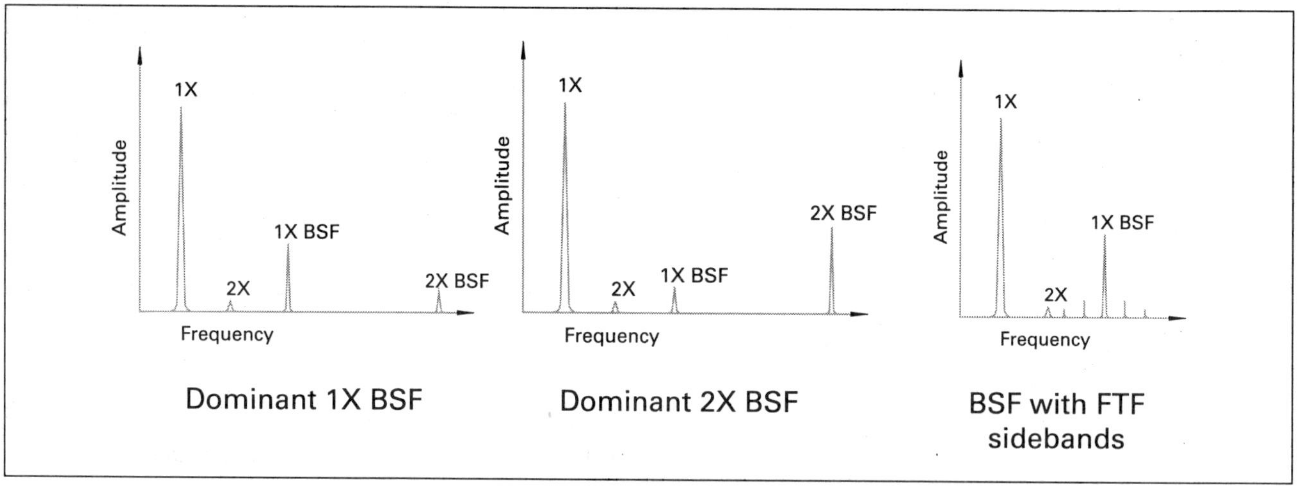

Figure 17-17: BSF fault - velocity spectrum

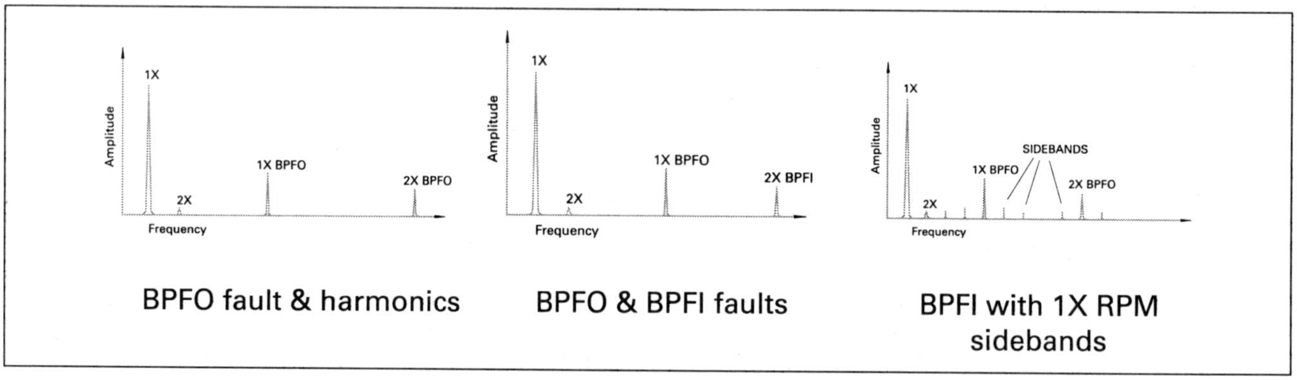

Figure 17-18: Raceway faults - velocity spectrum

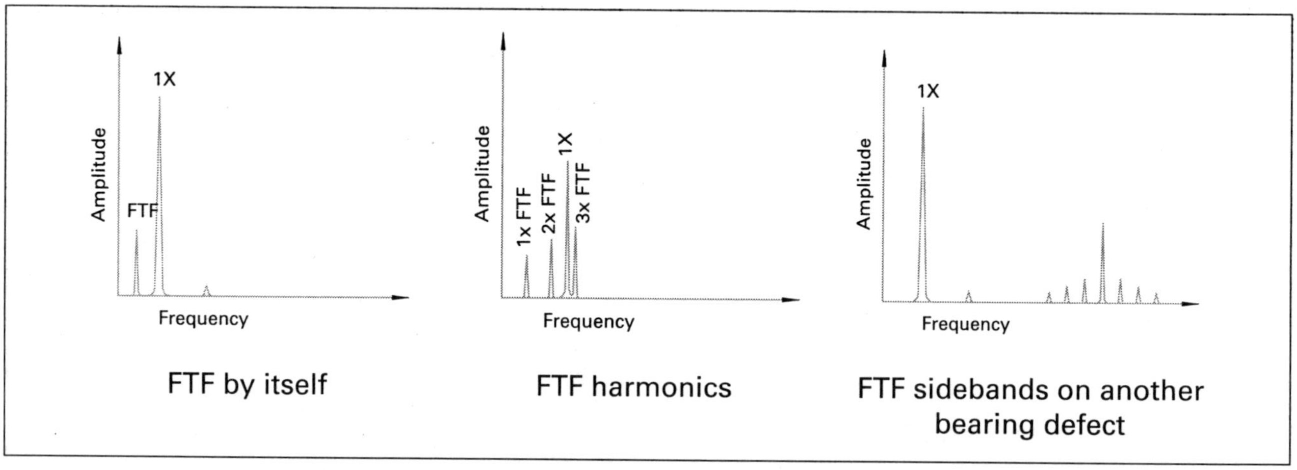

Figure 17-19: FTF defect - velocity spectrums

be present as an individual peak, harmonics, or sidebands around another spectral peak.

Because damage can, and frequently does, occur on more than one component of a bearing, combinations of spectral peaks caused by the different components can be found in a spectrum from a single bearing. However, rolling-element bearing deterioration follows a fairly predictable pattern. Generally speaking, most bearing failures progress through four stages; each of which has distinguishing characteristics.

The earliest indication of bearing distress often begins in the high frequency ultrasonic or acoustic emission frequency range(s). The damage begins either on or just below the surface(s) and may not be visible to the unaided eye. As the damage progresses, the high frequency energy increases in response to impact, inadequate lubrication film, or developing subsurface faults.

An ultrasonic demodulated spectrum is often used to detect this early stage of deterioration. Demodulation is a signal processing technique used to remove the low frequency signal from an incoming signal and process the remaining signal into a form that can be displayed in a low frequency spectrum. This removes all the information caused by what are considered low frequency events like unbalance, misalignment, looseness, motor defects, bent shaft, etc. The remaining very high frequency but low amplitude signal caused by bearing defects is analyzed by constructing an envelope around them and taking an FFT of the envelope. Various signal processing algorithms are used to accomplish this: PeakVue (CSI), GSE Spectrum (ENTEK/IRD), and Acceleration Enveloping (SKF). These systems use a variety of filters and signal processing algorithms to separate the low amplitude/high frequency content from the high amplitude/low frequency content allowing the vibration analyst to detect defects far in advance of failure. So, when a defect is first found using one of these algorithms, typically about 2 percent of the L10 bearing design life remains.

Gear Faults

Major gear faults include gear-tooth wear, gear eccentricity, backlash problems, and cracked, chipped, or broken gear teeth. The most significant contents in the gear vibration spectrum are the gearmesh frequency, its harmonics, and sidebands due to the modulation phenomena. Sidebands are uniformly spaced lower amplitude peaks on either side of a peak in a spectrum. For demonstrational purposes, Fig. 17-20 shows the sidebands located around the gearmesh frequency. A sideband is created when one frequency is modulated by another. In Fig. 17-20, the contact of one gear with another generates a peak in the spectrum at a frequency of 1X the shaft rpm x the number of gear teeth. If a gear running at 1,000 rpm has 47 teeth, the gearmesh frequency is 1,000 x 47 = 47,000 CPM. Because of small variations in the manufacture of the gear, there is a slight change in the amplitude of the gearmesh frequency for each revolution. This creates sideband peaks in the spectrum spaced 1,000 CPM away from the gearmesh frequency.

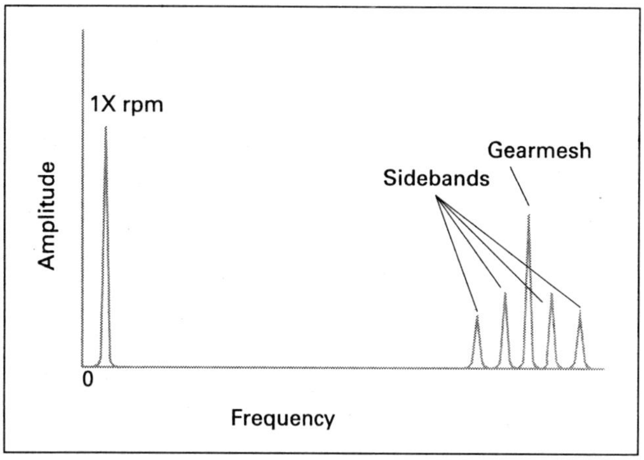

Figure 17-20: Gearmesh frequency with sidebands

Note that, when gears mesh, gearmesh frequencies and sidebands around them will be generated even for the healthy gears. Thus, the gearmesh frequencies are not a gear fault characteristic frequency as is the bearing fault frequencies. However, when the gears are properly assembled and in healthy conditions, the amplitude of the gearmesh frequencies and its sidebands should be low. Usually, the increment in the number and amplitude of the sidebands indicate a gear fault condition.

Guidelines

Machinery-fault diagnosis began as an offshoot of balancing as the instrumentation was developed. The analyst or maintenance professional is often called to answer the question, "How much vibration is too much?" Various charts and tables were developed from empirical data as guides to help evaluate a machine's overall health and energy consumption.

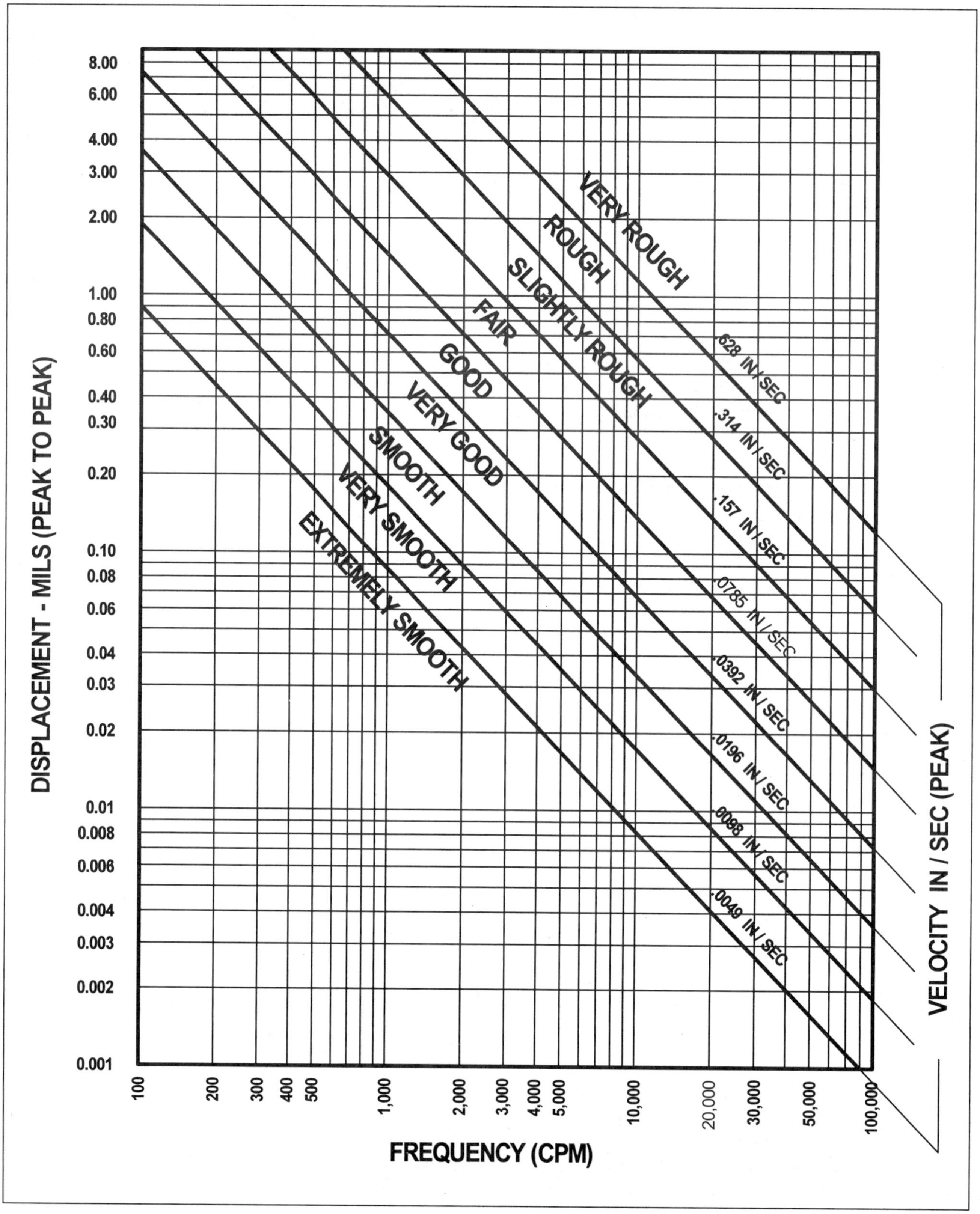

Figure 17-21: Vibration severity

One of the more familiar graphs is the vibration severity chart; see Fig. 17-21. This chart defines categories ranging from extremely smooth to very rough but does not tell us what is normal for a particular type of machine. An acceptable vibration level for a cooling tower fan would be unacceptable for a precision grinder.

Several features of the chart in Fig. 17-21 should be noted: In terms of displacement (the left-hand side), tolerable vibration levels decrease with increasing frequency. But, the allowable vibration, when measured in velocity using the diagonal lines, remains constant as the speed increases. Because the chart was designed for filtered readings, it should not be applied to overall vibration levels but to the vibration amplitude at running speed. This and most other similar charts were constructed for typical machines with a casing to rotor weight ratio in the order of 5:1. Failure to observe these limitations can lead to incorrect conclusions.

A more recent chart derived from ISO standards (see Fig. 17-22) better addresses the type and mounting of machines using velocity measurements. Because frequency is inherent in the velocity measurement, unfiltered readings may be used.

Vibration Severity	Velocity Range Limits & Machine Class Designations ISO Standard 2372-1974			
Inches / Second (Peak)	Small Machine Class I	Medium Machine Class II	Large Machines: Rigid Supports Class III	Large Machines: Flexible Supports Class IV
0.020				
0.030	Good	Good		
0.040			Good	Good
0.060	Satisfactory			
0.100		Satisfactory		
0.160	Unsatisfactory		Satisfactory	
0.250		Unsatisfactory		Satisfactory
0.400			Unsatisfactory	
0.620				Unsatisfactory
1.000				
1.560	Unacceptable (Danger)	Unacceptable (Danger)		
2.500			Unacceptable (Danger)	Unacceptable (Danger)
3.950				

Machine Classes: ISO 2372-1974

I.	0.250 ips:	Small machines, Electric motors up to 20 hp.
II.	0.395 ips:	Medium machines, Electric motors 20 to 100 hp.
III.	0.623 ips:	Large / Rigid support machines.
IV.	1.000 ips:	Large / Flexible support machines / Turbomachinery.
V.	1.560 ips:	Reciprocating / Rigid support machines.
VI.	2.500 ips:	Reciprocating / Flexible support machines.

Figure 17-22: Current ISO standards for unfiltered overall vibration levels

We have described five common of the more than 16 faults that may be encountered by a vibration analyst. Other faults an analyst may encounter include, but are not limited to, the following:

- Rotor eccentricity
- Rotor rub
- Bent shaft
- Drive belt misalignment
- Journal bearing faults
- Aerodynamic and hydraulic problems
- Fluting
- Beat frequencies
- Barring and corrugation

There are many excellent sources of information on the characteristics of these and other faults that may be used to aid diagnosis. Some of the sources of information are listed at the end of this chapter.

Balancing

In this section, we will discuss the following:

- Why is balancing important?
- Description of three types of unbalance.
- What needs to be checked prior to balancing?
- Basic steps for balancing.
- Balance specifications.

As discussed, one of the most common types of faults detected when using vibration analysis is unbalance of the rotating components.

Achieving and maintaining good balance on a piece of rotating equipment is important for a number of reasons:

- The forces introduced when a piece of machinery is unbalanced will contribute to early wear of the machinery, specifically shaft seals and bearings.
- Unbalance forces may cause vibration that will affect other pieces of machinery.
- Unbalance forces can be transmitted into normal working spaces as vibration or noise, which may result in a poor work environment.
- Unbalance forces will contribute to increased energy consumption.

Unbalance can occur for a variety of reasons:

- Introduced during the manufacture of the rotating component.
- Introduced from damage, contamination, corrosion, or erosion during operation.
- Introduced during maintenance.

Improving the balance of a rotating component is not difficult with the proper equipment and training. The following will discuss the different types of unbalance. Also, methods of correction and guidance of standards will be offered to correct the unbalances.

The simplest type of unbalance is known as static unbalance. Static unbalance occurs when the center of mass is not the same as the center of rotation. Static unbalance is familiar to everyone in a situation such as putting new tires on a vehicle and adding wheel weights for balance correction and smooth operation.

Figure 17-23: Wheel and correction weights

The force and vibration generated by the difference in the center of rotation and center of mass is also a function of speed of rotation. The force generated will increase exponentially as the speed of rotation increases.

The equation to calculate the force is:

$$F = \frac{4mr\pi^2}{T^2}$$

Where m is the mass, r is the radius, and T is the period for one revolution.

Consider a rotor where an unbalance condition exists with the center of mass (equivalent to 100g) displaced from the center of rotation by 0.1 meters. Figure 17-24 shows the exponential increase in force as the speed of rotation increases.

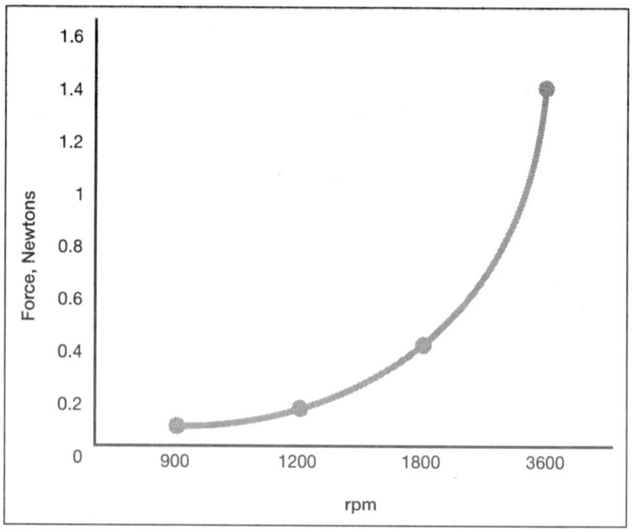

Figure 17-24: Exponential increase in force due to imbalance with rotating speed

Unbalance, as a force, also contains a position that is measured from a reference on the shaft, typically in degrees opposite the direction of rotation. This position is known as the "phase of vibration."

Figure 17-25 depicts a static unbalance condition to the rotating object, which has a center of mass 10 cm away from the center of rotation and 45 deg. from the zero-deg. reference. The force it generates will increase exponentially with the speed of rotation, and the magnitude of vibration will also increase with speed.

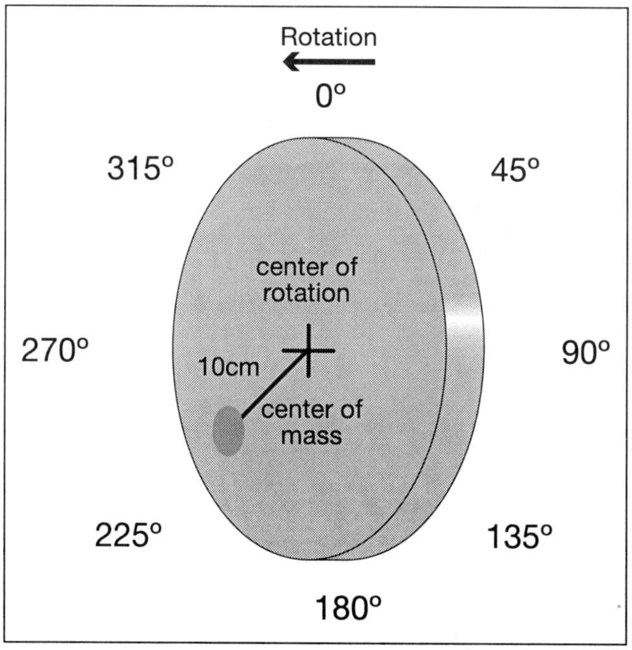

Figure 17-25: Static unbalance

Static unbalance is the simplest case, which, when present in more than one rotating plane, will produce the two other types of unbalance conditions: couple unbalance and dynamic unbalance.

Couple unbalance exists when two ends of the rotating mass have static unbalance conditions that are out of phase with each other, causing one end of the mass to be pulled in the opposite direction from the other end at the same time.

Figure 17-26 depicts pure couple unbalance condition where the two ends of the rotating mass will produce forces, which are 180 deg out of phase with each other, causing the ends of the mass to be pulled in opposite directions. When the vibration is measured on either end of the rotating mass, it will also be out of phase.

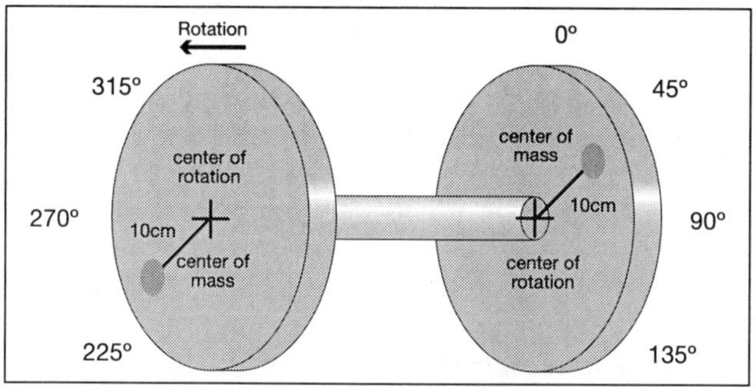

Figure 17-26: Couple unbalance

Note: In pure couple unbalance, the entire rotating mass is in static balance, where the center of mass and the center of rotation are in the same plane; however, significant forces are produced, resulting in vibration when the object is rotating.

The final case of unbalance is dynamic unbalance, where both static and couple unbalance conditions exist in the rotating mass. Dynamic unbalance occurs when the static unbalance at either end of the rotating mass is different either in mass, center of mass away from the center of rotation, or in relative phase angle to each other. Dynamic unbalance is the type that must be corrected most often.

Figure 17-27 depicts dynamic unbalance. The mass on the left end is larger at a greater distance from the center of rotation and at a different relative position from our zero-deg. reference angle.

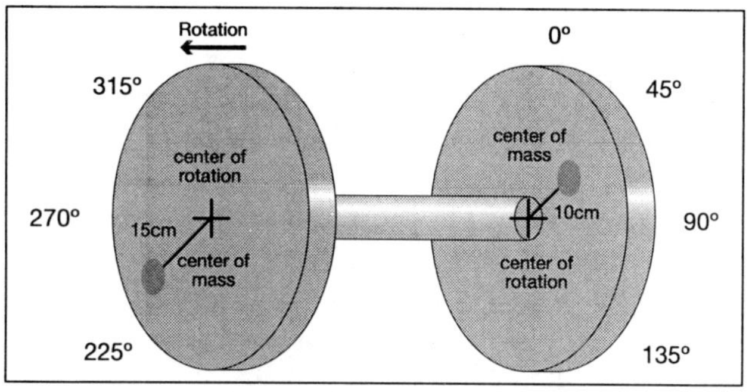

Figure 17-27: Dynamic unbalance

In a dynamic unbalance condition, the entire rotating mass will have a force acting away from its rotating center line while at the same time have unequal forces acting in different directions at either end of the rotating mass.

Correcting the balance of a rotating piece of machinery is not difficult and can be done either in the field or in a shop balancing unit. Regardless of whether balancing takes place in the shop or the field, improving the balance of rotating mass is done by measuring the amount and location of the unbalance on either end of the rotating mass. Once the unbalance condition is quantified, it is corrected by adding or removing mass at calculated positions and quantities.

Shop balancing takes place during manufacture and assembly or after rebuilding a piece of equipment in a repair facility, such as a motor rebuild shop. Shop balancing is done using a dedicated balancing machine, which has a calibrated response to unbalance. An example of a shop balancing machine is depicted in Fig. 17-28.

Figure 17-28: Shop balancing machine

In a shop balancing machine, the mass is rotated and the amount and position of unbalance is calculated. Based on the distance between available correction locations, the balancing machine will calculate how much mass needs to be added or removed from each location. As corrections are carried out, additional trim runs may take place until the balance of the rotating mass is acceptable.

Field balancing is normally done to correct an unbalance condition after maintenance or due to damage to the rotating mass. Field balancing is carried out using portable vibration analysis equipment that has the ability to measure phase.

Balancing in the field requires additional considerations beyond those in the shop.

The primary consideration is safety. Lock-out/Tag-out procedures should be followed as the person doing the balancing and those who may be assisting will be working on machinery where guards have been removed as well as making mass corrections to the rotating components.

Some basic checks need to take place prior to attempting the balance correction:

- Inspect the rotating components, including belts and couplings, for damage, corrosion, or erosion.
- Check all hold-down fasteners for proper torque or tightness.
- Check the condition of foundations and flexible mountings, if present.
- Clean all rotating components.

The basic checks are necessary to eliminate other potential faults that may exist within the machine that may make the balancing attempt unsuccessful or only mask the root cause of the unbalance.

Figure 17-29: Portable vibration analysis equipment configured for balancing

After the machine checks are completed, the balancing work can proceed. Typically, the steps are as follows:

- Set up phase reference equipment, and mark the zero-deg. reference.
- Locate your vibration sensor(s). Two vibration sensors are often used for measurement of dynamic unbalance conditions.
- Start the machine, and record the vibration amplitude and phase at 1x rpm.
- Shut down the machine, and add a trial mass. The trial mass is required to calculate the machine's vibration response to unbalance. Make sure that the trial mass is well fastened and will not come off when the machine is started.
- Start the machine, and record the new vibration amplitude and phase. As a rule of thumb, a 30 percent change in magnitude and phase will indicate a good trial run for calibration.

- More than one trial run will be required to correct a dynamic unbalance condition. Mass will need to be added on both correction planes at different times and recorded in different trial runs.
- Trial mass can either be removed or left in place for balancing, but this must be considered in the calculation.
- Shut down the machine, and allow the field balancing equipment to calculate the initial correction mass and location.
- Add or remove the mass at the locations specified.
- Start the machine, and measure the vibration and phase. A trim run may be required to achieve the desired vibration level.

The tolerance and specification for balance is normally related to ISO 1940-1:2003. This is available for purchase online at (www.iso.org/catalogue). The specification relates the recommended amount of residual unbalance per rotor weight at rotating speed.

The six categories on the chart can be generally applied against typical industrial machinery. The following are some examples:

1. G0.4– Precision grinding machinery.
2. G1.0– Machine spindles for roughing operations.
3. G2.5– Steam and gas turbines, medium and large electric armatures.
4. G6.3– Fans, pumps, general machinery parts.
5. G16– Drive shafts, cardan shafts.
6. G40– Car wheels.

ISO balance grades are meant to quantify the level of residual unbalance in a rotating mass and provide a standard that can be used by design and field engineers.

There are two main factors that are considered when assessing the balance grade:

- The entire rotating mass:
  Commonly, mass is measured in the metric system as Kilograms (Kg) or grams (g). The previous ISO chart has converted mass units into pounds (lbs) or ounces (oz).

- The speed of rotation:
  The speed of rotation is important, as we saw in our previous example, where the same residual unbalance caused an exponential increase in the amount of force generated as the speed increased. Figure 17-30 shows speed on the bottom axis.

The ISO chart is applied as follows for supply of new equipment or repair of existing:

- Rotating mass of a fan is 200 lb.
- Rotating speed = 1800 rpm
- Recommended balance grade is G6.3.

In this example, G6.3 is the required grade. The intersection of the G6.3 line with 1,800 rpm on the X axis occurs with 0.0009 in-lb/lb on the Y axis. A further calculation needs to take place to determine the allowable residual unbalance:

0.0009 lb-in/lb x 200lb = 0.18 lb-in

The balance report for the shaft assembly should specify the following:

- The speed at which the balancing was carried out.
- The target grade, in this case G6.3.
- The amount of residual unbalance, in this case 0.18 lb-in.
- In repair situations, the amount of weight added or removed should also be included.

Vibration Analysis of Structures

The application of the technology to perform an operating deflection shape (ODS) analysis or modal analysis requires advanced training and experience. In the next section, we will only provide an overview and discuss the following:

- Operating deflection shape analysis.
- Modal analysis.
- Technology requirements for both types of analyses.

Frequently, maintenance professionals will encounter a machine that gets classified as a "bad actor." Typically, bad actors are machines that have frequent and repeated failures of the same type. In other cases, machines will experience high vibrations that prevent them from operating at the design-rated throughput without affecting product quality. In either case, these machines may require a failure analysis

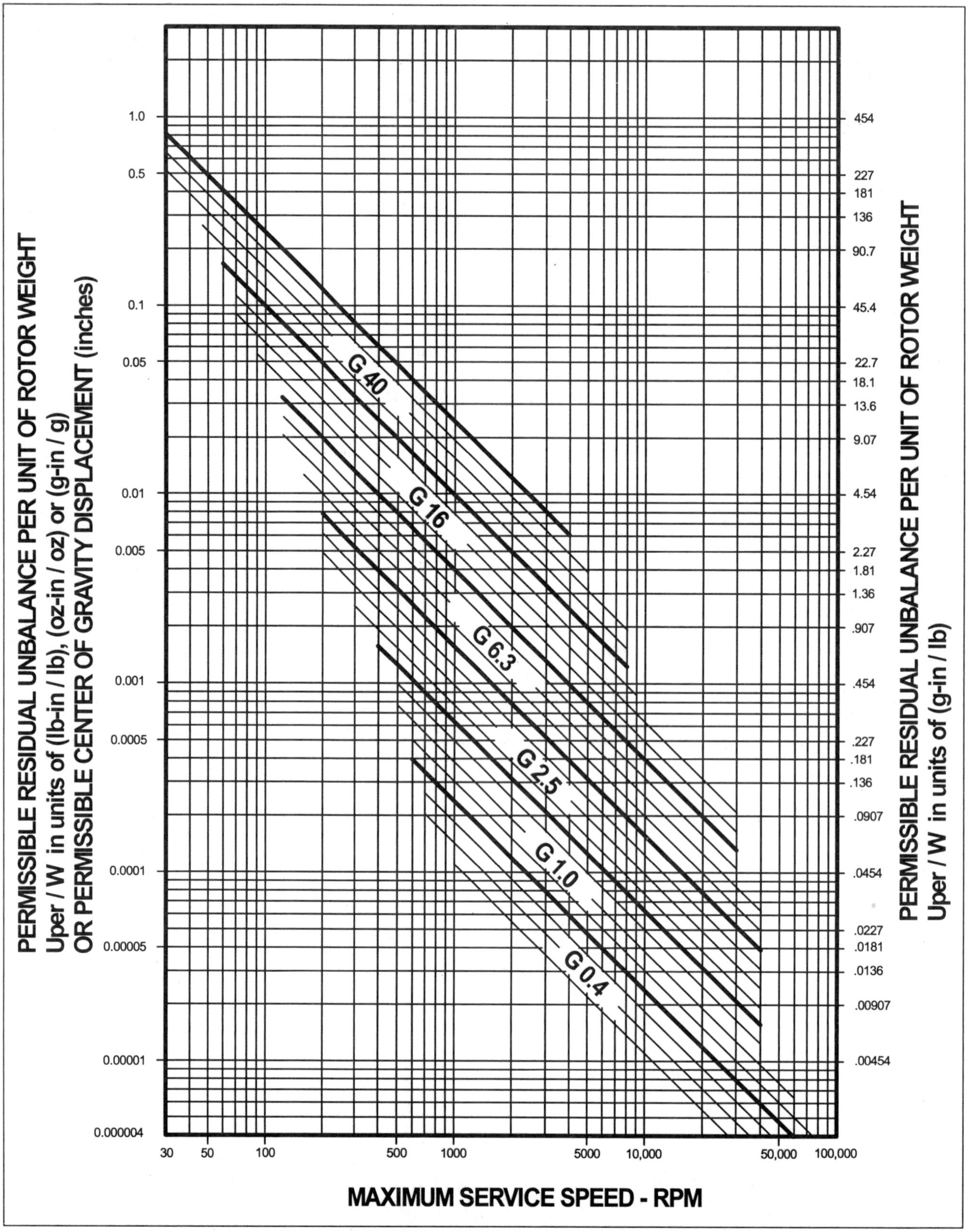

Figure 17-30: ISO Balance quality grades

and, subsequently, an investigation into the operating dynamics of the machine structure.

Vibration analysis uses the FFT algorithm to visualize and measure the vibration. Operating deflection shape and modal analysis utilize a Frequency Response Function (FRF) algorithm to measure and visualize the movement of a structure in response to force.

The dynamic response of a structure will vary between locations around the structure depending on the mass, stiffness, and damping characteristics of the structure. The response will also vary with the location and type of input force.

For example, consider how sitting on the edge of a mattress will cause the mattress to move at the other edges by some amount. Jumping on the edge will cause more movement. Jumping on the middle of the mattress will cause a different response at the edges. Operating deflection shape analysis and modal analysis quantify this behavior using the FRF, while modern software allows us to visualize the relative motion.

While both technologies use similar equipment, there are a couple of key differences.

Operating Deflection Shape (ODS) Analysis

- Uses the FRF to compare the relative amount of movement at various points around a structure, but the response of the structure to a force cannot be quantified.
- The energy for the structural response comes from the operation of the machine.
- The speed of the machine and the energy consumed by the machine must remain constant over the period of collection in order for the analysis to be accurate.
- Requires collection of a minimum of two vibration signals simultaneously.
- When using two transducers, one stays at a fixed location; the other is moved to the locations of interest. If multiple channels are used for collection, then one will need to be selected as the reference.

Operating deflection shape analysis is typically used to understand how a structure is moving when in operation. Often, it helps maintenance professionals to understand why rotating components are failing prematurely.

An example could be a coupling failing due to misalignment. Precision alignment and thermal growth measures indicate that alignment is well within tolerance for the coupling. Operating deflection shape analysis is applied and indicates a weak supporting structure causing misalignment to take place when the machine is operating.

Modal Analysis

- Uses the FRF to compare the response of the structure to a measured input force.
- Produces a quantified measure of a structure's dynamic response.
- The force is typically supplied by an instrumented force impact hammer or a calibrated shaker.
- The machine or structure must be de-energized when taking the readings.
- The response to the measured force is collected by an accelerometer, which is moved around the locations of interest or by multiple accelerometers simultaneously.

Modal analysis is typically used to understand how a structure moves in response to a given input force. Modal analysis is often applied when validating a finite element analysis, FEA. Finite element analysis and modeling is a computer-based method of modeling the dynamic behavior of a design before production. Often a working prototype is made before full scale production takes place. Modal analysis is then applied to the prototype to validate the FEA model or to provide input for changes in the FEA model. In other cases, modal analysis is used when proposing structural modifications to already installed machinery where decisions must be made on the addition of mass, stiffness, or damping to eliminate high-amplitude vibrations that occur due to the structure's natural frequency.

An example could be a large section of piping that is installed into a building moves significantly, resulting in work areas experiencing vibration. No apparent reason for the pipe vibrating can be found. Modal analysis study is done on the piping, and the piping is found to have a natural frequency coincident with a rooftop fan running at a speed of 1200 rpm. The fan speed cannot be changed, so the pipe must be stiffened or damped. The modal analysis study will be able to determine the location and quantity of stiffness or damping required to shift the natural frequency of the pipe.

Vibration analysis of structures uses advanced techniques requiring specialized equipment and training to perform, as discussed; the two common techniques are operating deflection shape and modal analysis. To recap, operating deflection shape analysis shows the relative motion between points on a structure during operation, while modal analysis shows the structural response to measured input force. Both techniques are powerful tools that are applied to understand the dynamics of a structure and help the maintenance professional determine solutions for bad-actor machines.

Summary

Predictive maintenance techniques determine when maintenance is needed based on the condition of the machine. It greatly extends a machine's lifetime, increases productivity, and reduces costly shutdowns and expensive repairs. Vibration analysis is an essential tool in predictive maintenance.

In this chapter, basic principals were covered to illustrate the theories on which vibration analysis was built. The procedure used to analyze vibration and the measurement systems used to collect and process the vibration signals were explained. Three major industrial applications of using vibration analysis techniques were introduced. They are rotating machine fault diagnostics, balancing, and structure analysis. For rotating machine fault diagnostics, five common faults were introduced: unbalance, misalignment, looseness, rolling element bearing faults, and gear faults. The typical fault signatures and the techniques used to diagnose them were explained. For balancing, three types of unbalance and the way to balance them were discussed in detail. Balancing of rotating equipment is important from design through operation and maintenance. Maintaining good balance of rotating machinery extends the life of the machine as well as reduces the amount of energy required for operation. The most common type of unbalance found is dynamic unbalance, and it can be easily corrected in the field or by using shop balancing machinery. Most specifications for balance reference the ISO balance quality grades, and they can easily be applied as a design requirement or as a correction requirement. For structure analysis, an overview of operating deflection shape analysis and modal analysis was provided. Advanced training and experience are needed to perform an operating deflection shape analysis or modal analysis.

All running machines in the real world produce some degree of vibration. By analyzing the vibration of a machine, machine faults could be diagnosed, and guides on how to correct the faults could be provided.

Vibration analysis plays a critical role in our industrial world. This chapter has explained the vibration analysis in a practical way. The tools and techniques provided in this chapter could be used by vibration analysts for their basic vibration analyses.

Key Terms

Acceleration
The rate of change in velocity of an object with time. See "g."

Acoustic emission
As materials are subjected to events that result in a sudden change or movement, they emit energy in the form of high frequency vibrations. These are usually due to a defect-related condition. The technique of listening to these "acoustic emissions" is used to locate defects as they occur, providing early warning of an impending failure. Typically, acoustic emission systems operate in the 1 KHz to 2 MHz range. In the case of bearing fault detection, these ranges may be considerably lower depending on the manufacturer of the equipment. See Ultrasonic or acoustic frequency emission.

Amplitude
The maximum value of a vibration caused by a machine. It may be expressed in units of displacement (mils – peak to peak), velocity (ips – peak), or acceleration (g rms).

Axis of rotation
The geometric center of a shaft/rotor. An axis of rotation about which all parts of a rotor are symmetric.

Center of mass
An axis about which the mass of a rotor is equally distributed.

Correction plane
Also called "balancing plane"; a plane perpendicular to the shaft axis of a rotor in which correction for unbalance is made.

CPM
A unit of frequency – cycles per minute. See frequency.

Damping

A reduction of vibration by dissipating energy. Types of damping include viscous (oil), coulomb (friction), and solid (elastomer).

Displacement

A quantity that specifies the change in position of a body or particle with respect to a reference frame. Vibration displacement is usually measured in units of mils (one mil equals 0.001 in.).

Fast Fourier Transform (FFT)

A mathematical process to change a complex signal from one form (time waveform) into another (frequency) that can be displayed as a series of sine waves.

Finite Element Analysis (FEA)

A machine or structure is represented by a geometrically similar model consisting of multiple, linked, simplified representations of discrete regions called finite elements. Equations of equilibrium, in conjunction with physical characteristics of the elements, are applied to each element, and a system of simultaneous equations is constructed. The equations are solved for unknown values using the techniques of linear algebra or nonlinear numerical schemes to determine how a machine or structure will respond to the forces applied to it.

Forcing frequency

The frequency at which a force is applied to a machine or structure. The force may be a result of a machine fault or generated as a result of normal operation (i.e., hydraulic or impact).

Frequency

The number of cycles occurring during a given time period. It may be expressed in Hz (cycles per second), CPM (cycles per minute), or rpm (revolutions per minute). Frequency is the reciprocal of a period or cycle.

Frequency domain

Graphical display of a complex signal as various vertical lines in a graph with amplitude on the vertical scale and frequency on the horizontal scale. This representation is also called a spectrum of a signal.

Frequency response function

An algorithm to measure and visualize the relative movement between points on a structure in response to a force.

g

A measure of acceleration amplitude related to the acceleration due to the earth's gravity (386 in/sec2).

Harmonic

A sinusoidal vibration whose frequency is an integer multiple of its fundamental frequency.

Hz

A unit of frequency – cycles per second. See frequency.

IPS

A measure of frequency – inches per second.

L10 bearing design life

The number of hours a population of bearings will run before 10 percent of them have failed.

Mil

A unit of vibration displacement equal to 0.001 in. See displacement.

Modal analysis

Modal analysis is typically used to understand how a structure moves in response to a given input force. Modal analysis is often applied when validating a finite element analysis, FEA. Finite element analysis and modeling is a computer-based method of modeling the dynamic behavior of a design before production. Often a working prototype is made before full scale production takes place. Modal analysis is then applied to the prototype to validate the FEA model or to provide input for changes in the FEA model. In other cases, modal analysis is used when proposing structural modifications to already installed machinery where decisions must be made on the addition of mass, stiffness, or damping to eliminate high-amplitude vibrations that occur due to the structure's natural frequency.

Mode shape

The vibrating shape of a structure from an applied force at a specific rotational speed or frequency. Each mode is defined by the modal frequency (natural frequency), modal damping (amplification), and mode shape (three-dimensional deflection of the structure).

Natural frequency
The free vibration frequency(ies) and mode shape(s) of a structure or machine that exists independently of others in the same machine. Natural frequencies are determined by the mass stiffness, geometry, and mounting.

Operating Deflection Shape (ODS)
The shape of a machine or structure as a result of the forces applied to it during operation. It is usually displayed as an animated computer model. These animations are very useful for diagnosing machinery faults. It does not allow determination of resonance or natural frequencies.

oz – in/lb
A description of rotor unbalance per unit of rotor weight used in ISO balance tolerances. Other units may also be used: lb-in/lb, oz-in/oz, gram-in/gram or gram-in/lb.

Period
The time required for one complete rotation of a shaft or single cycle oscillation of an event. A period is the reciprocal of frequency.

Phase angle
The timing difference between two signals. It may be measured by the shift between two signals or by the time shift between an event and the detection of a reference signal (key phasor). It is usually expressed in zero–360 deg.

Plane
A surface perpendicular to the axis of rotation.

Resonance
The amplification of a natural frequency by a matching forcing frequency. A highly directional amplitude increase and 90-deg.-phase shift occur at resonance.

Running speed
The rotating speed of a shaft in a machine. Measured in rpm (revolutions per minute), which is identical to CPM.

Sideband
Two or more spectrum peaks located at equally spaced intervals on both sides of a carrier frequency peak. They are low frequency in comparison to the carrier frequency and are created when a low frequency event modulates a high frequency event such as an eccentric low speed gear modulating the high frequency gear mesh.

Spectrum
An array of vibration amplitudes displayed in accordance with frequency. It is also called the frequency domain, a signature, or FFT.

Sub-harmonic
A sinusoidal vibration with a frequency that is a fractional multiple of its fundamental frequency (i.e., 1/2 rpm, 1/3 rpm, 1/4 rpm, etc.).

Time trace or time waveform
A plot of vibration amplitude vs. time showing the amount of vibration and the amount of time required to complete one or more cycles. It is useful for diagnosing certain equipment faults, especially those involving impacts.

Ultrasonic or acoustic emission frequency range
Ultrasonic vibration frequencies are generally considered those in the range from approximately 5,000 to 30,000 Hz; this range may vary slightly between manufacturers. Acoustic emission frequencies are in the range of 1 kHz to 1 MHz.

Ultrasonic demodulated spectrum
A vibration spectrum created by using filters and a mathematical algorithm to remove the low frequency components from a vibration signal and identify the very high frequency but low amplitude repetitive signals. The demodulation techniques are very useful to identify lubrication and rolling element bearing problems.

Unbalance
A condition when vibration is caused by unequal weight distribution in a rotor.

Unbalance – Couple
Occurs when the principal mass axis intersects the shaft axis at the center of gravity. More practically, a condition when the vibration in a rotor is caused by two heavy spots located 180 deg. apart on opposite ends of the rotor.

Unbalance – Dynamic
A combination of static and couple unbalance when the shaft axis and the principal mass axis do not coincide or touch. It is the most common unbalance condition.

Unbalance – Residual

Unbalance remaining after a rotor has been balanced. It is expressed in units of weight at a given radius per unit of rotor weight in balance tolerance charts. See oz-in/lb.

Unbalance – Static or Mass

A condition when the center of gravity axis is displaced parallel to the shaft axis. It can be corrected with a weight exactly opposite the heavy spot.

Velocity

The rate of change in position of an object with time. See IPS.

Appendix

Career Opportunities in the PT/MC Industry
Common Conversion Factors
Common Formulas
Associations That Interface with the PT Industry
Credits
Index

Career Opportunities in the PT/MC Industry

Power transmission and motion control (PT/MC) manufacturers and distributors serve the production and replacement needs of equipment manufacturers and others who use their products. Nearly every product manufactured worldwide uses PT/MC products.

The purpose of this section is to provide insights into the many challenging and rewarding career opportunities in the PT/MC industry. Individual success depends upon a multitude of skills, education, talent, and personal characteristics.

Industrial distributors are the local source wholesalers of PT/MC products. Many industrial plant and facilities professionals turn to distributors for product information and solutions. Manufacturers and distributors sell bearings, belt drives, chain drives, clutches and brakes, conveyors and components, couplings and U-Joints, gears, hydraulic and pneumatic products, linear motion devices, adjustable-speed drives, controls, sensors, accessories, and many other value-added services.

Manufacturers and distributors are differentiated by size and focus. A manufacturer competes on an international basis, while a distributor competes as a local business or as a branch of a regional or national company. Manufacturers and distributors can have a broad line, carrying a full range of products, or they can be specialized and manufacture or stock particular product types.

Job Roles

Both manufacturers and distributors have human resource needs in (1) finance and accounting, (2) marketing and sales, and (3) operations. It is probable that multiple tasks will be required of PT/MC professionals and that broad knowledge and skills in all areas will be required.

Typical career paths include accountant, application specialist, branch manager, counter salesperson, credit manager, customer service representative, engineer, field sales representative, information systems manager, inside salesperson, marketing, operations manager, outside salesperson (account manager), product manager, product specialist, purchaser, receiving and shipping clerk, sales manager, and warehouse manager. The following are position descriptions for the most widely available jobs.

Application Specialist

This position provides product knowledge and application assistance to the sales department and customers on assigned products. Their responsibility is to promote the sales of their product lines, maximizing sales and gross profit dollars. They work with the sales staff and the customer to determine the customer's exact need and to apply the appropriate products and technology solutions.

Branch Manager

This position is responsible for the branch's profitability. They oversee sales, pricing, inventory, and customer relations; prepare budgets; forecast sales; perform sales functions; handle customer complaints and service problems; supervise staff; and assume responsibility to meet company goals.

Counter Salesperson

This position sells and services customers at the sales counter, providing product information and pricing. They arrange displays and inform customers of promotions and quantity pricing. They assist in filling orders from the warehouse, keep catalogs up to date, handle returns, and keep their supervisor informed of customer needs and requests.

Credit Manager

This position develops and administers credit policies. They advise management of credit extensions and disseminate credit information to salespeople. They exercise control of customers who have credit problems and slow pay accounts.

Customer Service Representative

This position is responsible for the proper entering, filing, and pricing of orders. They properly handle customer complaints and claims against suppliers and trucking companies. They edit and check orders for proper pricing. The CSR obtains samples, sales literature, catalogs, and price sheets for all sales personnel.

Engineer

Engineers provide product information and applications assistance to the sales team and customers on assigned projects. They determine the customer's exact needs and apply appropriate engineering calculations, components, and equipment to solve problems.

Field Sales Representative

This position is responsible for the direct sales of the company's product in a defined territory, generating profitable sales on new and existing accounts. These persons call on customers, provide product information, and quote customer prices. They are responsible for identifying business opportunities to maintain sales levels and to develop new accounts. Field sales reps coordinate sales efforts with inside salespeople and provide a flow of data on competitors and suppliers to product managers.

Information Systems Manager

This position oversees accounting, billing, payroll, inventory control, and other business procedures through the use of data processing equipment and electronic commerce devices. They stay informed of current technology and select software programs for company use. They oversee the activities of their company information centers and prepare reports for management as directed.

Inside Salesperson

Inside salespeople sell products and services to customers over the phone. They handle orders and provide product information, pricing, and delivery information. They assist outside salespeople and coordinate activities with shipping, credit, and billing. They handle customer complaints, prepare quotes, and perform other duties as requested by management.

Marketing

This position assumes responsibility for achieving sales goals by developing profitable marketing plans. They develop marketing strategies, advertising, and incentive promotions. They coordinate activities directly and indirectly with sales personnel. They participate in the development of budgets, advertising, major marketing activities, market research, and sales training. They may make field sales calls with outside salespeople and trainees.

Operations Manager

Operations managers' concerns include six decision categories: the product/service mix, transformational processes, capacity, inventory, work force, and quality. They are responsible for planning, creating, managing, and evaluating the performance of transformational events that result in added value. A key to business survival is effective control of all operations.

Outside Salesperson

This position represents both the supplier and distributor to customers and prospects in specified marketing areas and accounts. They call on customers and provide solutions. They make calls with vendor representatives in selling new products and introducing new lines. They prospect for new accounts; increase their knowledge and skills; handle customer complaints; submit reports; use modern business technology; obtain credit information; make special deliveries; and forward information on competitors' activities, new products, promotions, as well as sales leads to management.

Product Manager

Product managers are responsible for managing a location's designated product line to maximize sales and profit margin goals. This includes product promotions, inventory control, price quotations, and any other information required. Product managers negotiate local competitive costs and returns of product; they collaborate with suppliers and communicate sales advantages to other company personnel. They also monitor local marketplace trends and provide support for the sales force by working together with customers.

Purchasing Manager

Purchasing managers direct the purchasing of all products as well as the purchase of office supplies, equipment, and service contracts. They develop systems and methods that are essential to the purchasing function. They may administer inventory controls and levels, approve invoices for payment, assist in customer service, interview vendor representatives, and recommend product purchases. Purchasers expedite special orders and assist in taking the annual inventory.

Sales Manager

This position is responsible for the planning, direction, and control of the personal selling activities of a business unit, including recruiting, selecting, training, equipping, assigning, routing, supervising, and motivating. Sales management involves three interrelated processes: (1) forming a strategic sales plan, (2) implementing a sales program, and (3) evaluating and controlling the sales force performance.

About PTDA

Power Transmission Distributors Association (PTDA) is an international trade organization serving the power transmission/motion control (PT/MC) industry—a 28 billion dollar industry in the United States alone. Wherever the production facility and whatever the product, PT/MC products are essential to the manufacturing, power transmission, and motion control process.

Industries served: agriculture, chemicals, electric/electronic, fabricated metals, food and beverage, instruments, lumber, mining, pulp and paper, primary metals, petroleum and coal, printing and publishing, rubber and plastic, grain, textiles, and transportation.

Customers served: PTDA members provide products and value-added services to plant and maintenance professionals throughout all industries. They also serve the original equipment manufacturer of motion control systems.

About PTDA Members

More than 170 distributors representing more than 3,400 locations throughout North America and 11 other countries, and nearly 200 leading manufacturers of PT/MC products are members.

PTDA distributor members are highly qualified to help end users maximize their uptime, efficiency, and profitability. In addition to providing on-site, one-on-one technical support, they have the specialized product knowledge base to advise on nearly all product applications. The PTDA distributor serves as a one-stop shopping resource, providing a broad range of products dedicated to keeping industrial machinery "in motion."

PTDA manufacturer members are committed to selling power transmission/motion control products and services through distribution. They work closely with their distributor partners to better serve the needs of the industrial customer.

Common Conversion Factors

PREFIXES USED IN THE METRIC SYSTEM

Prefix	Symbol	Equivalent
giga	G	1,000,000,000, (10^9)or 1 billion
mega	M	1,000,000, (10^6)or 1 million
kilo	k	1,000, (10^3) or 1 thousand
centi	c	0.01, or one-hundreth
millionth	m	0.001, (10^{-3}) or one-thousandth
micro	m	0.000,001, (10-6) or one-millionth

Approximate Densities Of Common Materials

Material	Density, lb/ft^3
Gases	
Air	0.075
Oxygen	0.08378
Hydrogen	0.005
Nitrogen	0.073
Liquids	
Water	62.0
Sea water	64.0
Ethyl alchohol	49.2
Kerosine	50.0
Gasoline	45.0
Metals	
Aluminum	169
Bronze	549
Copper	555
Gold	1,200
Cast iron	443
Lead	709
Magnesium	109
Carbon steel	487
Stainless steel	480
Engineering plastics	
ABS, general purpose	64
Acrylics	74
Nylon	71
Phenolic	87
Polycarbonates	75
Polyesters, unreinforced	86
Polyethylene, medium density	58
PVC	89
Others	
Concrete	144
Anthracite coal	100
Bituminous coal	83

Volume	To	Multiply by
cubic centimeters	cubic inches	0.061
	cubic feet	3.53×10^{-5}
	gallons, U.S.	2.64×10^{-4}
cubic inches	cubic centimeters	16.4
quart	liters	0.946
	pints	2
	fluid ounces	32
liters	gallon, U.S.	0.264
	quarts	1.06
	fluid ounces	33.8
	cubic inches	61.0
	cubic centimeters	1,000
cubic feet	cubic inches	1,730
	cubic centimeters	4,390
	liters	28.3
	fluid ounces	958
barrels, U.S., liquid	gallons, U.S.	31.5
board feet	cubic feet	1/12
gallons, British	gallons, U.S.	1.20
gallons, U.S.	Canadian, U.K.	0.833
	cubic feet	0.134
	cubic inches	231
	ounces	128
	liters	3.79
ounces, U.S. fluid	cubic centimeters	29.6
	cubic inches	1.80
bushels, U.S.	bushels, U.K.	0.969
	cubic feet	1.24
	gallons, U. S.	9.31
	liters	35.2

Volumetric Flow Rates	To	Multiply by
gallons, U.S./min	liters/sec	0.00843
	cubic feet/min	0.134
cubic feet/sec	gallons, U.S./min	449
	gallons, U.K./min	374
	liters/sec	1,700
liters/sec	cubic feet/min	2.12
	gallons, U.S./min	15.9
	gallons, U.K./min	13.2

Temperature	To	Multiply by
degreesFahrenheit	degrees Celsius	C = 5/9 (F-32)
degrees Rankine	degreesCelsius	R = F + 459.69
		F = (9/5)C + 32
degrees Fahrenheit	degreesKelvin	K = C + 273.16

Energy	To	Multiply by
BTU	Joules	1.055×10^{3}
Calories, gram	Joules	4.18

Pressure	To	Multiply by
pascals	pounds/square inch	0.000145
	pounds/square foot	0.0209
	newtons/square meter	1
pounds/square inch	atmospheres	0.0680
	pounds/square foot	144
	pascals	6985
	foot of water at 60 F	2.30
	grams/square centimeter	70.3
	millimeters of mercury, 32 F	51.7
pounds /square foot	kilograms/square meter	4.88
atmospheres	lb/square inch	14.7
	pounds/square foot	2117
	pascals	101,325
inches of water, 60 F	pounds/square inch	0.036
	pounds/square foot	5.20
	pascals	249
foot of water, 60 F	pounds/square inch	0.433
	pounds/square foot	62.4
	pascals	2986
inhces of mercury, 68 F	feet of water, 68 F	1.131

Force and Torque	To	Multiply by
newtons	pounds	0.225
newton-meters	pound-feet	0.738
ounce-inches	pound-feet	0.0052
	newton-meters	0.0071
	pound-inch	0.063
pound-inches	pound-feet	0.083
	newton-meters	0.130
	ounce-inches	16
pound-feet	newton-meters	1.36
	pound-inches	12
	ounce-inches	192
	poundal-feet	32.2
	dyne-centimeters	1.36×10^{7}

Area	To	Multiply by
square centimeters	square inches	0.155
square inches	square centimeters	6.45
square feet	square inches	144
	square centimeters	929
square meters	square feet	10.8
	square inches	1550
square kilometers	square miles	0.386
acres	square feet	43,560
	square meters	4,047
circular inches	square inches	0.785

Length	To	Multiply by
inches	centimeters	2.54
centimeters	inches	0.394
feet	inches	12
	centimeters	30.5
meters	inches	39.4
kilometers	miles	0.621
	feet	3280
miles	kilometers	1.61
	feet	5280
microns	inches	39.4×10^{-6}
Angstrom units	inches	3.94×10^{-9}

Viscosity	To	Multiply by
pound-sec/square feet	centipoises	47,880
reyns	centipoises	6.90×10^{5}
centistokes	square feet/sec	1.076×10^{-5}

Motor-Constant Conversions

To convert from	To	Multiply by
ounce-inch-sec	ounce-in/rpm	0.105
	newton-meter-sec	7.062×10^{-3}
ounce-inch/rpm	ounce-inch-sec	9.55
	newton-meter-sec	0.0674
newton-meter-sec	ounce-inch/rpm	14.83
	ounce-inch-sec	0.0142
	ounce-inch-sec	9.55
	newton-meter-sec	0.0674
newton-meter-sec	ounce-inch/rpm	14.83
	ounce-inch-sec	0.0142

Weight, Mass, Inertia	To	Multiply by
pounds, mass	kilograms	0.454
	ounces	16
kilograms	pounds, mass	2.21
	ounces	35.3
tons, short	metric tons	0.907
	kilograms	907
	pounds	2,000
metric tons	tons, short	1.10
	kilograms	1,000
	pounds	2,205
pounds, weight	slugs, mass ($lb\text{-}sec^2/ft$	0.031
pound-square foot	kilogram-square meter	0.042
	kilogram-meter-sec^2	0.00430
dynes	pounds	2.25×10^{-6}

Power	To	Multiply by
horsepower	Btu/hour	2,545
	foot-pounds/sec	550
	ounce-inches/sec	1.056×10^{-5}
	kilowatts	0.746
watt	Joules/sec	1
kilowatts	horsepower	1.34

Specific Gravity Of Fluids

Liquid	Specific gravity	Temp, C
Acetone	0.79	20
Alcohol	0.79	20
Benzene	0.90	0
Carbon disulfide	1.29	0
Ether	0.74	0
Gasoline	0.67	—
Glycerin	1.26	0
Kerosene	0.82	—
Mercury	13.6	—
Milk	1.0	—
Oils: castor	0.97	15
cotton seed	0.93	16
olive	0.92	15
Sea water	1.04	15
Turpentine	0.87	—
Water	1.0	4

Mechanical Properties Of Common Materials

Material	Equivalent	Ultimate			Yield point, tension (psi)	Modulus of elasticity, tension or compression (psi)	Modulus of elasticity, shear (psi)	Weight (psi) (lb/in.^{3})
		Tension	Compression[a]	Shear				
Steel, forged-rolled								
0.10-0.20 C	SAE 1015	60,000	39,000	48,000	39,000	30,000,000	12,000,000	0.28
0.20-0.30 C	SAE 1025	67,000	43,000	53,000	43,000	30,000,000	12,000,000	0.28
0.30-0.40 C	SAE 1035	70,000	46,000	56,000	46,000	30,000,000	12,000,000	0.28
0.60-0.80 C		125,000	65,000	75,000	65,000	30,000,000	12,000,000	0.28
Nickel	SAE 2330	115,000	---	92,000	---	30,000,000	12,000,000	0.28
Cast iron								
Gray	ASTM 20	20,000	80,000	27,000	---	15,000,000	6,000,000	0.26
Gray	ASTM 35	35,000	125,000	44,000	---	---	---	0.26
Gray	ASTM 60	60,000	145,000	70,000	---	20,000,000	8,000,000	0.26
Malleable	SAE 32510	50,000	120,000	48,000	---	23,000,000	9,200,000	0.26
Wrought iron	---	48,000	25,000	38,000	25,000	27,000,000	---	0.28
Steel cast								
Low C	---	60,000	---	---	---	---	---	0.28
Medium C	---	70,000	---	---	---	---	---	0.28
High C	---	80,000	45,000	---	45,000	---	---	0.28
Aluminum alloy								
Structural, No. 350	---	16,000	5,000	11,000	5,000	10,000,000	3,750,000	0.10
Structural, No. 17ST	---	58,000	35,000	35,000	35,000	10,000,000	3,750,000	0.10
Brass								
Cast	---	40,000	---	---	---	---	---	0.30
Annealed	---	54,000	18,000	---	18,000	---	---	0.30
Colddrawn	---	85,000	---	---	---	15,500,000	6,200,000	0.30
Bronze								
Cast	---	22,000	---	---	---	---	---	0.31
Colddrawn	---	85,000	---	---	---	15,000,000	6,000,000	0.31
Brick, clay								
Grade SW	ASTM	---	3,000 min.[b]	---	---	---	---	0.72
Grade MW	ASTM	---	2,500 min.					
Grade NW	ASTM	---	1,500 min.					
Concrete, 1:2:4, 28 days	---	---	2,000	---	---	3,000,000	---	0.087
Stone	---	---	8,000	---	---	---	---	0.092
White oak								
Parralel to grain	---	---	7,440	2,000	4,760[c]	1,780,000	---	0.028
Across grain	---	800	---	---	1,320[c]			
White pine								
Parallel to grain	---	---	4,840	860	3,680[c]	1,280,000	---	0.015
Across grain	---	300	---	---	550[c]			

a. The ultimate strength in compression for ductile materials is usually taken as the yield point. The bearing value for pins and rivets may be much higher, and for structural steel is taken as 90,000 psi.
b. Average for five bricks.
c. Proportional limit in compression.
Source: S.I. Heisler, *The Wiley Engineer's Desk Reference*

Common Formulas

Duty-cycle calculation

The RMS (root mean square) value of a load is one of the quantities often used to size power-transmission components. The RMS value of the load can be in any unit—HP, amps, ... It is a time-weighted calculation:

$$L_{RMS} = \sqrt{\frac{L_1^2 t_1 + L_2^2 t_2 + \ldots + L_n^2 t_n}{t_1 + t_2 + \ldots + t_n}}$$

Where: L_{RMS} = RMS value of the load, which can be in any unit
L_1 = Load during time period 1
L_2 = Load during time period 2, etc.
t_1 = Duration of time period 1
t_2 = Duration of time period 2

Convert linear to rotary motion

N = V / (0.262D)

Where: N = Speed of shaft rotation (rpm)
V = Velocity of material (ft/min)
D = Diameter of pulley or sprocket (in.)

Horsepower

1 HP = 33,000 lbs-ft/min

1 Watt = 1 Newton x 1 meter/1 sec.

HP = T x rpm/63,025

Where: HP = Horsepower
T = Torque
rpm = Revolutions per minute
63,025 = Constant

Efficiency

E = Power Out/Power In

Pumps

P = QHS/(3,960 m)

Where: P = Power (HP)
Q = Flow rate (gal/min)
H = Head (ft)
S = Specific gravity of fluid
m = Pump efficiency

Fans and blowers

P = Qp/(229m)

Where: P = Power (HP)
Q = Flow rate (ft3/min)
p = Pressure (lb/in.2)
m = Efficiency

Force and Torque

T=FD

Where: T = Torque
F = Force
D = Distance

Circumference of a Circle

C = πD

Where: C = Circumference
π = 3.1416
D = Diameter

Area of a Circle

A = πr2

Where: A = Area
π = 3.1416
r = Radius

Power

$$P = FD/T$$

Where: P = Power
F = Force, lbs
D = Distance
T = Time

Linear Motion

Maximum Rotational Speed

$$MRS = DN$$

Where: MRS = Maximum rotational speed
D = Shaft diameter
N = Rotational speed

Life Calculation

$$L = (Ca/(Fa \times fw))^3$$

Where: L= Life, total revolutions or total hours or running distance
Ca = Basic dynamic load capacity
Fa = Applied axial load
fw = Load factor based on operating conditions

Drive Torque

Drive Torque = Operating Torque = Drag Torque

Where: Operating Torque = $((Fa \times 1)/(2\pi E))$

Where: Fa = Applied load
1 = The lead
E = Ball screw efficiency

Where: Drag Torque = Friction force, supplied by manufacturer

Designing of drives

Variable Speed Belts Speed Ratio Change

$$P_d = \frac{B_b}{2\mathrm{Tan}\frac{\alpha}{2}} - H_b$$

Where: P_d = Pitch diameter variation
B_b = Belt top width
H_b = Belt thickness
α = Sheave groove angle

Belt drives

To calculate belt length in an open belt drive:

$$L_o = 2c + 1.57(D+d) + (D-d)^2/4c$$

To calculate belt length in a crossed-belt drive:

$$L_x = 2c + 1.57(D+d) + (D+d)^2/4c$$

Where: c = center distance, in.
d = small pulley diameter, in.
D = large pulley diameter, in.

Chain drives

To calculate chain length:

$$L = 2C + (N + n)/2 + (N - n)^2/39.44C$$

Where: L = chain length, pitches
C = Shaft centers, pitches
N = Number of teeth in large sprocket
n = Number of teech in small sprocket

Formulas related to DC electric-motor performance

In a DC motor, the magnitude of reverse voltage generated in the armature is defined by the back EMF constant, K_b:

$$K_b = 0.74K_t$$

Where K_t = the torque constant, the torque produced per unit armature current in oz-in./amp or N-m/amp

The constant that defines braking characteristics of a motor with shorted leads, K_d:

$$K_d = (K_b K_t)/R_t$$

Where R_t is terminal resistance, ohms

The time required for the armature or winding current to reach 63.2 percent of its steady-state condition is the time constant, electrical, T_e.

$$T_e = La/R_t$$

Where L_a = winding inductance (henrys)

The time required for an unloaded motor to reach 63.2 percent of its final velocity afterapplying armature or winding voltage is the time constant, mechanical, Tm.

$$T_m = (J_m R_t)/K_t K_e$$

Where: J_m = motor inertia, oz-in.-sec^2
K_e = voltage constant, volts/rad/sec

Formulas for Hydraulics and Pneumatics

Sizing Cylinders

Force

$$F = PA$$

Where: F = Force
P = Pressure, psi
A = Piston area

Speed

$$V = QA$$

Where: V = Velocity
Q = Flow, gallons per minute
A = Piston area

Transmission Sizing

Corner Horsepower

$$Hc = FtV/3{,}600n$$

Where: Hc = Corner horsepower, kW
Ft = Maximum vehicle tractive force, N
V = Maximum vehicle speed, Km/h
n = Final drive efficiency, percent

Transmission Corner Horsepower

$$Ht = TtN/9{,}500$$

Where: T = Theoretical torque at maximum system pressure, N-m
t = Torque efficiency, percent
N = Maximum transmission speed, rpm

Pipe and Tubing Line Sizing

Average Pipeline Velocity

$$Q = 2.4d2V$$

Where: Q = Flow rate, gallons per minute
d = Tube internal diameter, inches
V = Velocity, feet per minute

Pipe Wall Thickness

$$Tt = pdoM/2s$$

Where: Tt = Wall thickness, inches
p = Pressure, lbs per square inch
do = Outside diameter, inches
M = Safety factor
2 = Constant
s = Material tensile strength, psi

Filter Performance

Beta Ratio

$$x = Nu/Nd$$

Where: x = Beta filtration rating for particles larger than x microns
Nu = Number of particles larger than x microns upstream of the filter
Nd = Number of particles larger than x microns downstream of the filter

Associations that Interface with the PT Industry

The following organizations create standards or otherwise provide industry forums in North America for mechanical, fluid, and electronic power transmission components and systems or related materials and equipment.

The Adhesive and Sealant Council, Inc. (ASC)

7101 Wisconsin Ave, Ste. 990
Bethesda, MD 20814
(301) 986-9700
Fax: (301) 986-9795
www.ascouncil.org

American Bearing Manufacturers Association (ABMA)

(Formerly the Anti-Friction Bearing Manufacturers Association)
2025 M St., NW, Ste. 800
Washington, DC 20036
(202) 367-1155
Fax: (202) 367-2155
www.abma-dc.org

American Gear Manufacturers Association (AGMA)

1001 N Fairfax Street, Suite 500
Alexandria, VA 22314-1587
(703) 684-0211
Fax: (703) 684-0242
www.agma.org

American National Standards Institute (ANSI)

1899 L Street, NW, 11th Floor
Washington, DC 20036
(202) 293-8020
Fax: (202) 293-9287
www.ansi.org

ASTM International

(Formerly American Society for Testing and Materials)
100 Barr Harbor Rd.
West Conshohocken, PA 19428
(610) 832-9500
Fax: (610) 832-9555
www.astm.org

Association for High Technology Distribution (AHTD)

N19 W24400 Riverwood Dr.
Waukesha, WI 53188
(262) 696-3645
Fax: (262) 696-3646
www.ahtdmembers.org

The Association for Hose and Accessories Distribution (NAHAD)

(Formerly National Association of Hose and Accessory Distributors)
105 Eastern Ave., Ste. 104
Annapolis, MD 21403
(410) 263-1014
Fax: (410) 263-1659
www.nahad.org

Association for Rubber Products Manufacturers (ARPM)

7321 Shadeland Station Way, Suite 285
Indianapolis, IN 46256
Phone: (317) 863-4072
Fax: (317) 913-2445
www.arpminc.org

Bearing Specialists Association (BSA)

800 Roosevelt Rd., Bldg. C, Ste. 312
Glen Ellyn, IL 60137
(630) 858-3838
Fax: (630) 790-3095
www.bsahome.org

Canadian Standards Association International (CSA)

178 Rexdale Blvd.
Toronto, ON, Canada M9W 1R3
(416) 747-4000
Fax: (416) 747-4149
www.csa-international.org

Conveyor Equipment Manufacturers Association (CEMA)

5672 Strand Ct., Suite 2
Naples, FL 34110
(239) 514-3441
Fax: (239) 514-3470
www.cemanet.org

Electrical Apparatus Service Association (EASA)

1331 Baur Blvd.
St. Louis, MO 63132
(314) 993-2220
Fax: (314) 993-1269
www.easa.com

The FPDA Motion and Control Network

(Formerly Fluid Power Distributors Association)
105 Eastern Ave., Ste. 104
Annapolis, MD 21403
(410) 940-6347
Fax: (410) 263-1659
www.fpda.org

Industrial Supply Association (ISA)

100 N. 20th St., 4th Floor
Philadelphia, PA 19103
(866) 460-2360
Fax: (877) 460-2365
www.isapartners.org

Institute of Electrical and Electronics Engineers (IEEE)

3 Park Ave., 17th Floor
New York, NY 10016
(212) 419-7900
Fax: (212) 752-4929
www.ieee.org

Material Handling Equipment Distributors Association (MHEDA)

201 U.S. Highway 45
Vernon Hills, IL 60061
(847) 680-3500
Fax: (847) 362-6989
www.mheda.org

Material Handling Industry of America (MHIA)

8720 Red Oak Blvd., Ste. 201
Charlotte, NC 28217
(704) 676-1190
Fax: (704) 676-1199
www.mhia.org

Mechanical Power Transmission Association (MPTA)

5672 Strand Ct., Suite 2
Naples, FL 34110
(239) 514-3441
Fax: (239) 514-3470
www.mpta.org

National Association of Electrical Distributors (NAED)

1181 Corporate Lake Dr.
St. Louis, MO 63132
(314) 991-9000
Fax: (314) 991-3060
www.naed.org

National Electrical Manufacturers Association (NEMA)

1300 N. 17th St., Ste. 1752
Rosslyn, VA 22209
(703) 841-3200
Fax: (703) 841-5900
www.nema.org

National Fluid Power Association (NFPA)

3333 N. Mayfair Rd., Ste. 211
Milwaukee, WI 53222
(414) 778-3344
Fax: (414) 778-3361
www.nfpa.com

National Institute of Standards and Technology (NIST)

100 Bureau Dr., Stop 1070
Gaithersburg, MD 20899
(301) 975-6478
Fax: (301) 975-1630
www.nist.gov

National Lubricating Grease Institute (NLGI)
4635 Wyandotte St., Ste. 202
Kansas City, MO 64112
(816) 931-9480
Fax: (816) 753-5026
www.nlgi.org

NIBA-The Belting Association
6737 W Washington St., Suite 1300
Milwaukee, WI 53214
(414) 389-8606
Fax: (414) 276-7704
www.niba.org

Power-Motion Technology Representatives Association (PTRA)
16A Journey, Ste. 200
Aliso Viejo, CA 92656
(949) 859-2885
Fax: (949) 855-2973
www.ptra.org

SAE International
400 Commonwealth Dr.
Warrendale, PA 15096
(724) 776-4841
Fax: (724) 776-0790
www.sae.org

Society of Manufacturing Engineers (SME)
One SME Drive
Dearborn, MI 48128
(313) 425-3000
Fax: (313) 425-3412
www.sme.org

Society of Tribologists and Lubrication Engineers (STLE)
840 Busse Hwy.
Park Ridge, IL 60068
(847) 825-5536
Fax: (847) 825-1456
www.stle.org

The Vibration Institute
6262 Kingery Hwy., Ste 212
Willowbrook, IL 60527
(630) 654-2254
Fax: (630) 654-2271
www.vibinst.org

Credits

PTDA would like to thank the following for their contribution:

The Adhesive and Sealant Council, Inc. (www.ascouncil.org)
Altra Industrial (www.altramotion.com)
Association for Rubber Products Manufacturers (www.arpminc.org)
The Balancing Systems Group, Inc.
Baldor/Dodge (www.baldor.com)
Bonfiglioli USA (www.bonfiglioliusa.com)
CEMA - Conveyor Equipment Manufacturers Association (www.cemanet.org)
Emerson - Control Techniques (www.controltechniques.com)
Michael Gulas
ITW Polymers Adhesives, N.A. (Devcon) (www.itwadhesives.com)
Jamaica Bearings Co., Inc. (www.jamaicabearings.com)
Klüber Lubrication NA (www.kluber.com)
Lane Conveyors & Drives, Inc. (www.lane.us.com)
Loctite Industrial, Henkel Corporation (www.loctite.com)
Motion Industries Inc. (www.motionindustries.com)
Nidec Motor Corporation (www.usmotors.com)
Nord Drive Systems, Inc.(www.nord.com)
Oiles America Corporation (www.oiles.com)
Osborn (www.osborn.com)
PBC Linear (www.pbclinear.com)
Rexnord Industries, LLC (www.rexnord.com)
Rollon Corp. (www.rollon.com)
Rotator Products Limited (www.rotatorproducts.com)
SKF USA (www.skf.com)
Society of Tribologists and Lubrication Engineers (www.stle.org)
The Timken Company (www.timken.com)
Tribology ABC (www.tribology-abc.com)
U.S. Tsubaki Power Transmission (www.ustsubaki.com)

Index

A

ABMA (American Bearing Manufacturers Association) ... 27, 39, 40, 42, 313, 367
Absolute encoders ... 283
AC adjustable-speed drives ... 243, 256
AC Brushless Servo (a.k.a. servo) ... 222
AC motors ... 225, 227, 228, 232, 233, 234, 237, 241, 256, 263, 269
Acceleration ... 350
Accumulation conveyors ... 101, 104
Accuracy Grade ... 199
Acoustic emission ... 350
Actuation methods, clutch and brake ... 87, 94
Actuator Motors ... 214
Adhesive and Sealant Council, Inc. (ASC) ... 300, 367, 371
Adjustable tensioners ... 306
Adjustable-Frequency Drives ... 241, 249, 250, 252, 261, 262
Adjustable-speed drives ... 117, 234, 243, 245-266, 269, 285
Adjustable-speed-drive functions ... 252, 253-255
Aftermarket Additives ... 328
AGMA (American Gear Manufacturers Association) ... 151, 156, 165, 170, 171, 367
AIAG (Automotive Industry Action Group) ... 120
American Bearing Manufacturers Association (ABMA) ... 27, 39, 40, 42, 313, 367
American Gear Manufacturers Association (AGMA) ... 151, 156, 165, 170, 171, 367
American National Standards Institute (ANSI) ... 367
Amplitude ... 350
Anaerobic Adhesives & Primers ... 298
Anaerobic Sealants ... 295, 297-300
Anaerobic Thread Sealants ... 298
Analog regulators, DC drive ... 255, 257, 262, 264
Analog Versus Digital Regulators ... 257
Angled-shaft gears ... 15-9163
Angular-Contact Bearings ... 26, 160, 208
Applications of Solid Lubricants ... 326
Applications Suited to Plain Bearings ... 23
Area of a Circle ... 363
Assembling Spiral Rings ... 308
Association for High Technology Distribution (AHTD) ... 367
Association for Hose and Accessories Distribution (NAHAD) ... 367
Association for Rubber Products Manufacturers (ARPM) ... 58, 59, 367
ASTM International ... 367
Automated Electrified Monorails ... 107
Automated Storage and Retrieval Systems (AS/RS) ... 118, 119
Automatic restart ... 250
Automatic tuning ... 250
Automotive Industry Action Group (AIAG) ... 120
Autotransformer Starters ... 272
Axial load, bearing ... 31, 203, 207, 209, 364
Axial-Piston Motors ... 186
Axial-Piston Pumps ... 177
Axis of rotation ... 350

B

Backlash, gear ... 158, 163, 208, 222, 341
Balancing ... 127, 144, 333, 343-347, 350, 371
Ball Bearing Slides ... 214
Ball bearings ... 24-28, 34, 37, 41, 42, 200-208, 214, 215, 322-324
Ball recirculation system ... 206
Ball screws ... 205-210, 220, 223
Ball-screw jacks ... 219, 220
Banded Belts ... 50
Bar and Pin Chains ... 76
Barloop Chains ... 76
Base-mounted reducers ... 167
Bearing Lubrication ... 22, 33, 38
Bearing Specialists Association (BSA) ... 367
Bearing standards, ABMA ... 39, 40
Bearing-and-Cross Joints ... 131
Bearings ... 19-42
Bearings, Rolling Element ... 24, 214, 323
Belt conveyors ... 97, 99, 102, 103, 109, 110, 116
Belt drives ... 102, 214, 247, 364
Belt Length ... 49, 52-54, 57, 247, 364
Belt tension ... 46, 47, 231
BEU (British Engineering Units) ... 227, 264
Bevel gears ... 160-162, 165, 170
Binder Selection ... 326
Bipolar Coils, Step Motors ... 240
Blender ... 245
Block Couplings (Elastomeric Element Type) ... 134
Block-and-Pin Joints ... 131
Board-Level Computers ... 279

Bonded Shrink or Press Fit 300
Bonded Slip Fit 299, 300
Bonded-Rubber Shear Couplings (Elastomeric Element Type) 132
Bonded-Shear Couplings (Elastomeric Element Type) 139
Boundary lubrication 319, 320, 323
Brake Valve 181
Brakes 91-96, 102, 107-109, 109, 231, 251
Braking 251
Breakage 165
Breakdown torque 233, 234
Breathers, gear 153, 168, 230, 238
British Engineering Units (BEU) 227, 264
British Standard Sprockets 81
Brushless DC Motors 225, 237
Brushless Servomotors 238, 264
Brush-Type DC Motors 236, 237, 255
Brush-Type Servomotors 238
Bubbler sensor 284, 285
Bucket Elevators 109, 114, 115
Bulk Belt Conveyors 99, 110
Bulk handling equipment 99, 109, 110
Bushed chain 75

C

Caged needle roller bearing 30
Cam Roller 200, 202
Cam-Type and Load-Following Bearings 31, 39
Canadian Electrical and Electronic Manufacturers Association (CEEMA) 233, 242, 293
Canadian Standards Association (CSA) 274, 289, 293, 367
Capacitive 288
Capacitor-start motors 235
Cardan U-Joints 131, 347
Carousel Conveyors 108
Cast Chains 75
Cast Combination 75
CEEMA (Canadian Electrical and Electronic Manufacturers Association) 233, 242, 293
Center of mass 350
Centrifugal Clutches 94
Centrifugal Compressors 179
C-face motors 88, 230
Chain attachments 61, 68
Chain Couplings 129, 130
Chain drives 63-84, 364
Chain drives, advantages and disadvantages 63
Chain drives, performance ratings 66
Chain lubrication 82, 83
Chain-driven roller conveyor 100
Chains and Cables 210
Chip conveyors 116
Circuit Breakers 252, 269, 271, 274, 275
Circumference of a Circle 363
Classes and Characteristics of Fuses 275
Classical V-Belts 48, 49
Closed-loop adjustable-speed drives 250
Clutch-brakes 87, 92, 95
Clutches 87-96
Combination Starters 274, 275
Common Types of Sensors Used in Automation Systems 288
Commonly Used PT Components 102
Composite Couplings 147
Compound and stabilized shunt-wound 237
Compound drive 247
Compounds, non-curing pipe 298
Compounds, solvant-based pipe 298
Compression Donut Couplings (Elastomeric Element Type) 134
Compression Gaskets 298
Compressors 178, 179
Conductors 188
Connecting link, chain drive 64, 130
Connectors 188
Constant-horsepower loads 228, 246, 249
Constant-torque loads 228, 249
Contactor 269-272
Contactor Life 270
Contoured Diaphragm Coupling 144
Control devices 95, 275-280
Control enclosures 289-292
Controlled speed range 251
Controls 222
Convention pre-cut compression, gaskets 298, 299
Conventional Designs 307
Conversion Factors Appendix 359-362
Convert linear to rotary motion 363
Conveyor Equipment Manufacturers Association (CEMA) 368
Conveyor Standards 120
Corded Tire Couplings (Elastomeric Element Type) 133
Correction plane 350
Corrosion Resistant Chain 66
Counterbalance Valve 181
Couplings 125-151
CPM 350
Critical torque levels 233

PTDA

Cross-belt sorter ... 117
Crossed Roller ... 204
CSA (Canadian Standards Association) ... 274, 289, 293, 367
CSI (current-source inverter) ... 257, 258
CSI Drives ... 258
Cured-in-place, gaskets ... 298-300
Current-source inverter (CSI) ... 257, 258
Current-source inverter drives ... 258
Customization ... 121
Cycloidal Disc Speed Reducers ... 167
Cylinders ... 183
Cylindrical roller bearings ... 28

D

Damping ... 351
DC adjustable-speed drives ... 255, 256
DC and Can Stack ... 221
DC motors ... 221, 235-238, 255
DC Stepper (a.k.a. stepper, step motor) ... 221
Degradation and Analysis of Oils in Service ... 327
Design A B, C, and D motors ... 233
Designing of drives ... 364
Detachable Link Chain ... 77
Detent torque, step motor ... 240
D-flange motors ... 230
Diaphragm Compressors ... 179
Diaphragm Coupling Design Principles ... 143
Diaphragm Coupling Operating Principles ... 143
Diaphragm couplings ... 136, 142, 143, 145
Diaphragm Couplings (Metallic Element Type) ... 142
Differential Plus Integral Plus Proportional Control ... 292
Digital regulators, dc drive ... 255, 257, 259, 262
Directional Control ... 181, 182
Disc brakes and clutches ... 87, 89
Disc Coupling Design Considerations ... 141
Disc Coupling Operating Principle ... 140
Disc couplings ... 127, 128, 136, 137, 140, 141, 147, 149, 150, 167
Displacement ... 333, 334, 342, 343, 348, 351
Donut couplings ... 134
Double Helical Gears ... 157
Double Pitch Roller Chain ... 67
Double Plus Sprockets ... 80
Double Single Sprockets ... 81
Double V-belts ... 48
Double-Acting Cylinders ... 184
Double-pitch roller chain ... 67
Double-reduction chain drives ... 84, 166
Double-Split ... 312
DP enclosures ... 229
DPBV enclosures ... 230
Drawn-cup needle bearing ... 30
Drive Chain ... 73
Drive Ratings ... 249, 254, 256, 261
Drive tensioners ... 305-307
Drive Torque ... 364
Drive torque, ball-screw ... 209, 364
Dual-sided synchronous belts ... 55
Duty Cycle/Life Cycle ... 197
Duty-cycle calculation ... 363

E

Eddy Current Couplings ... 146
Eddy-current clutches and brakes ... 90, 91
Eddy-current drives ... 245, 246, 253, 254
Efficiency ... 15, 178
Elastohydrodynamic Lubrication ... 319, 322-324
Elastomeric Block Couplings (Elastomeric Element Type) ... 140
Elastomeric couplings ... 126, 132, 133, 148
Elastomeric Donut Couplings-Unclamped (Elastomeric Element Type) ... 134
Elastomeric Flexible Element Couplings ... 132
Electric Actuation, clutch and brake ... 94
Electric linear motors ... 211
Electrical Apparatus Service Association (EASA) ... 368
Electrical drives ... 245
Electrified monorails ... 107
Electromagnetic Actuation ... 95
Electromagnetic clutches and brakes ... 87, 90, 95
Electromechanical Actuators ... 213
Enclosed belted mechanical drives ... 247, 248
Enclosed gear drive ... 165
Enclosed gears ... 165-169
Encoders ... 222, 282, 283
Energy ... 12, 13, 251, 261, 360
Energy Saving Opportunity with Variable Speed Drive ... 261
Engineered drive systems ... 265
Environment ... 171, 198
Environmental ... 263
Environmentally Acceptable (EA) ... 188
Example Problems ... 320
Extended-pitch chain ... 67
Extreme Service Environments ... 24

F

Face gears ... 162
Fans and blowers ... 363
Fast Fourier Transform (FFT) ... 336, 351
Ferrous materials ... 163
Filter Performance ... 191, 366
Filtration ... 190, 191, 326, 327, 366
Finished Bore ... 78
Finite Element Analysis (FEA) ... 349, 351
Fire-Resistant ... 188
Fixed Tensioners ... 305
Flange Couplings ... 125
Flanged cartridge blocks ... 36
Flat Belts ... 45, 59
Flat way, linear bearing ... 200
Flat-disc traction drive ... 248
Flat-Face O-ring ... 189
Flat-race bearings ... 26
Flexible couplings ... 125-128, 148
Floor Conveyors ... 100
Flow ... 178
Flow Capacity ... 182
Flow Coefficient (Cv) ... 183
Flow-capacity, valves ... 182
Flow-Control Valves ... 181, 182
Fluid Couplings ... 145-147
Fluid film lubrication ... 319
Fluid Motors ... 185, 186
Fluid power ... 175, 191
Fluid-Power Standards ... 191
Fluids ... 188
Fluids and Viscosity ... 317
Fluids, hydraulic ... 175, 188
Flux Vector Control Drives ... 259-261
Follower inputs ... 251
Food and Beverage Processing ... 104
Force ... 9, 150, 184, 360, 363, 365
Force and Torque ... 363
Force Sensors ... 286
Forcing frequency ... 351
Formed-In-Place Gaskets (FIP) ... 299
Formulas for Hydraulics and Pneumatics ... 365
Formulas related to DC electric-motor performance ... 365
Four-quadrant operation ... 251
Fractional Horsepower V-Belts ... 49
Frame Ratings ... 270
Frames ... 227
Frames and Enclosures ... 238
Frequency ... 232, 241, 336, 339, 342, 351, 352
Frequency Domain ... 335, 336, 351
Frequency response function ... 351
Friction ... 9, 13, 46, 59
Friction clutches and brakes ... 89, 90
Full-film lubrication ... 22
Full-Voltage Starters ... 269
Function block programming ... 277
Functions and Options ... 250
Fuse Classes ... 275
Fused Disconnects ... 274
Fuses, control and sensor ... 275

G

g ... 351
Gain error, signal conditioning ... 289
Gasket, cured-in-place ... 299
Gasket, formed-in-place ... 299
Gasketing ... 298, 299
Gas-Operated ... 314
Gear Classification Standards ... 165
Gear couplings ... 128-130, 138-141
Gear Couplings (Mechanically Flexible Type) ... 128
Gear Faults ... 341
Gear lubricants ... 169
Gear materials ... 163, 164
Gear Pumps ... 176
Gear Racks ... 159
Gear ratings ... 169-171
Gear Specifications ... 164
Gear Tooth Failures ... 164
Gearboxes ... 165
Gearmotor ... 108, 121, 167-170
Gear-on-Gear Motors ... 186
Gear-on-Gear Pumps ... 176
Gears ... 149, 150, 155-171, 176, 186, 341
Gears with Angled Shafts ... 159
Gear-within-Gear Motors ... 186
Gear-within-Gear Pumps ... 176
General Purpose Couplings ... 128
General Services Administration (GSA) ... 120
Grease-Gelling Agents – Soap and Soap Complex ... 324
Greases ... 324, 325, 369
Greases: Non-soap ... 325
Grid Couplings (Mechanically Flexible Type) ... 130
Grooved-race bearings ... 26, 27
GSA (General Services Administration) ... 120

H

Half-Step and Microstep Control 239
Half-step control, step motors 239
Hardware & software, vibration analysis 346
Harmonic 351
Hazardous-location motors 230
Heat treatment 163
Heavy and Super Series Chains 65
Helical Gears 157, 159
Helical Offset Tooth 57
Helical-Gear Speed Reducers 166
Herringbone Gears 157
Hertz (Hz) 227
High Tech Coated Chains 70
High-Angle Gear Couplings/Gear Spindles (Mechanically Flexible Type) 139
Higher Reduction Ratios 166
High-Performance Gear Couplings (Mechanically Flexible Type) 138
History 54
Hoists 108, 109
Hoists and Winches 108
Holding torque, step motors 240
Hollow Pin Chain 67
Horsepower 363
Horsepower ratings, motors 237, 254, 261, 270
Hose, fluid power 187, 188, 191
H-Type 55
Hunting 259
Hybrid Step Motors 239
Hydraulic Actuation 95
Hydraulic actuation, clutch and brake 95, 96
Hydraulic and Pneumatic Filters 190
Hydraulic pumps 175, 180
Hydraulics, fluid power 175, 191
Hydrodynamic Bearings 322
Hydrodynamic Lift 321
Hydrodynamic lubrication 319, 321
Hydrodynamic Lubrication Fluid Film 321
Hydrostatic drives 186, 187
Hydrostatic lubrication 323
Hypoid gears 162
Hysteresis clutches and brakes 90, 91
Hz 351

I

I/O capability, PLC 222, 277
Idler Sprockets 81
IEC (International Electrotechnical Commission) 293
IEC 61131-3 279
IEC Control Enclosures 291
IEC standards 227, 242, 270
IEEE (Institute of Electrical and Electronics Engineers) 293, 368
Inching or threading 252
Inclined and Vertical Screw Conveyors 113
Incremental Encoders 282
Indicators 280
Indicators, control 280
Induction Motors 241
Induction motors, single-phase 234
Inductive 288
Industrial Computers 279
Industrial Supply Association (ISA) 368
In-floor tow line 100
Input / Output Capability 277
Input and Output Ratings 227
Institute of Electrical and Electronics Engineers (IEEE) 368
Insulated-gate, bi-polar transistor (IGBT) 249, 264
Internal Gears 157
International Electrotechnical Commission (IEC) 293
International Organization for Standardization (ISO) 151, 294
International Standards 293
International Standards (BS, BSI, CENELEC, DIN, EN, IEC, ISO, KEMA, NF, SEMKO, SEV, SI, VDE) 242
Inverted-tooth chain 156
Inverter 234, 257
Involute tooth form (gears) 156
IPS 351
IPS (inches per second) 351
ISO-9000 294
Isolation, signal conditioning 288

J

Jaw Clutches 91
Jaw Couplings 135
Jaw Couplings (Elastomeric Element Type) 135
JIC (Joint Industry Council) 120

JIC 37 Degrees 189
Jogging 252
Joint Industry Council (JIC) 120

K

Key phasor 352
Keyless Locking Devices 309
Keystock 312
Keyway Locknuts 313

L

L10 bearing design life 351
Labyrinth seal 235, 236
Ladder logic programming 277
LAN (local area network) 252
Lead 20, 208
Lead/Acme Screw 210
Leaf Chain 69
Length Thru Bore 78
Limit Switches 284
Line Sizing 189
Linear actuators 213-215
Linear bearings 34, 223
Linear bushings 201
Linear Encoders 222
Linear Force Motors 212
Linear guides 201, 202
Linear Induction Motors 212
Linear Motion 364
Linear Motor 211
Linear Slides 214
Linear Stage 215
Linear Step Motors 213
Linear variable differential transformer (LVDT) 286
Linearity, signal conditioning 288, 289
Linear-motion systems 195, 221, 223
Links, pin and roller 64, 65
Liquid Level Sensors 284
Liquid sealants 298
Liquid-Filled Couplings 146
Liquid-level sensors 284, 285
Load 197
Load and Life Ratings 40
Local area network (LAN) 252
Locked-rotor torque 233
Locknuts 313
Looseness 338, 339
Lubricant selection 328
Lubricant storage 327, 328
Lubricants 324, 325, 329
Lubrication 21, 171, 329
Lubrication of Linear Bearings 35
Lubrication Plugs and Filled Grooves 23
Lubrication Regimes 319
LVDT (linear variable differential transformer) 286

M

Machinery fault diagnosis 341
Machine-screw jacks 219
Magnetic 288
Magnetic Chip Conveyor 116
Magnetic Type Couplings 145
Magnetically Latched 217
Magnetic-particle clutches and brakes 90
Magnetostrictive Actuators 218
Material Handling Equipment Distributors Association (MHEDA) 368
Material Handling Industry of America (MHIA) 368
Maximum Bore Diameter 78
Maximum Rating 183
Measurement System 334
Mechanical Actuation 94
Mechanical actuation, clutch and brake 87, 94
Mechanical arrangement 263
Mechanical drives 245
Mechanical Power Transmission Association (MPTA) 368
Mechanical-lockup clutches 87
Mechanically Flexible Couplings 128, 138
Memory 277
Message Displays 280
Metal Bellows Couplings (Miscellaneous Type) 137
Metallic Beam Couplings (Miscellaneous Type) 137
Metallic Element Couplings 128, 136, 138
Metallic Plain Bearing 20
Metallic-beam couplings 137
Metallic-disc couplings 136, 140, 141
Metals Processing 104
Methods of Application 326
Microstep control, step motor 239
Mil 351
Miniature Linear Actuators 214
Misalignment 337
Miscellaneous General Purpose Coupling Types 137

Miscellaneous Special Purpose Coupling Types 145
Mixed lubrication 319, 324
Mixed-film lubrication 324
Mixer 245
Modal Analysis 349, 351
Mode shape 351
Modifications 230
Monorail Conveyors 107
Motor enclosures 246
Motor frame size 231, 263
Motor Heating 228
Motor Insulation 228
Motor overload protection 274
Motor Protection 231
Motor starters 269
Motorized reducer 167, 168
Motorized Trolleys 109
Motor-Powered 314
Motors 215, 221
Motor-starter performance 274
Mounted bearings 36
Mounted Plain Bearings 37
Mounted Rolling-Element Bearings 37
Mounting 167
Mounting position 171
Mountings 231
MPTA (Mechanical Power Transmission Association) . 368
Multi-motor drives 252
Multiple Axis 215
Multiple Convoluted Diaphragm Coupling 144
Multiple Flat Diaphragm Couplings 145
Multiple Strand Sprockets 79
Multiple Tooth Clutches 92
Multiple-tooth clutch 92
Multispeed motors 234

N

Narrow V-belts 48, 49
National Association of Electrical Distributors (NAED) 368
National Electrical Code (NEC) 293
National Electrical Manufacturers Association (NEMA) 368
National Fluid Power Association (NFPA) 368
National Industrial Belting Association (NIBA) 58, 59, 369
National Institute of Standards and Technology (NIST) 368
National Lubricating Grease Institute (NLGI) 369
Natural frequency 352
Nature of load 170
NEC (National Electrical Code) 293
Needle Roller Bearings 30
NEMA (National Electrical Manufacturers Association) 293, 368
NEMA, CSA, and UL Control Enclosures 289
Network communications 252
Networking 279
NIBA-The Belting Association 58, 369
Nomenclature 54
Nominal Flow Rating 183
Nominal Rating 190
Noncompensated Flow Controls 181
Non-Curing Pipe Compounds 298
Nonferrous materials 164
Nonmetallic materials 164
Nonmetallic Plain Bearing 20
Nonsoap grease 325
North American Standards 293

O

Offset Coupling/Schmidt Coupling (Miscellaneous Type) 145
Offset error, signal conditioning 289
Offset link, chain drive 64
Oil, degradation 327
Open- and Enclosed-Track Overhead Conveyors 107
Open Frame 217
Open gears 155
Open or Closed-Loop 250
Open-belted mechanical drives 247
Open-loop 250
Open-loop adjustable-speed drives 250
Operating Deflection Shape (ODS) 352
Operating Deflection Shape (ODS) Analysis 349
Operator displays 280
Operator Interface Units 280
Optical linear encoder 221-223
O-rings 312, 313
Oscillating shaft applications 23
Other Versions 311
Output performance 262
Overhead conveyors 106, 107
Overhead Trolley Conveyors 106
Overload Relays 270
Overloads 229, 231, 270, 271

P

P&F (power and free) conveyors ... 106
Packaging Machinery ... 119
Palletizers ... 117
Parallel-Shaft Gears ... 155-159
Part-Winding Starters ... 272
Performance Parameters ... 303
Period ... 352
Permanent magnet ... 236
Permanent Magnet Step Motors ... 239
Permanent split-capacitor motors ... 235
Permanent-magnet motors ... 236
Personal Computers ... 279
Petroleum-Based ... 188
Phase angle ... 352
Photoelectric – Diffuse ... 288
Photoelectric – Retroreflective ... 288
Photoelectric – Thru Beam ... 288
Photoelectric Sensors ... 287
PID Control ... 292
Pillow blocks ... 36
Pin and Bushing Couplings (Elastomeric Element Type) ... 135
Pipe and Tubing Line Sizing ... 366
Pipe Thread Sealants ... 298
Pipe Threads ... 189
Pipe threads, sealing ... 189, 298-299
Pipe, fluid power ... 175
Piston Pumps ... 176
Pitch, chain drive ... 64, 67-81
Plain Bearings ... 200, 215
Plain Bore ... 78
Plain-bearing lubrication ... 22
Plane ... 352
Planetary gear systems ... 157
Plastic Chains ... 71
Plastic Engineering chains ... 77
Plastic flow ... 165
PLCs (programmable logic controller) ... 252, 277
Plugging switch ... 271
Pneumatic Actuation ... 95
Pneumatic Conveyors ... 115
Pneumatics, compressors ... 178-180
Polyphase Motors ... 233
Poppet Mechanisms ... 180
Ports ... 181
Position ... 182
Position Control ... 215
Power ... 364
Power and Free Conveyors ... 106
Power Semiconductors ... 249
Powered Mechanical Storage Systems ... 118
Power-Motion Technology Representatives Association (PTRA) ... 369
Pre-applied fastener coatings ... 297
Precision (a.k.a.: accuracy, repeatability) ... 196
Preload ... 199, 207, 208, 223
Preset speed ... 252
Pressure angle, gear ... 158
Pressure Compensated Pumps ... 177
Pressure Sensors ... 285
Pressure-Compensated Flow Controls ... 181
Pressure-Control Valves ... 180
Primary-Resistor Starters ... 271
Primers ... 298
Principals of Vibration Analysis ... 333
Principle of Operation ... 310
Process-related failures ... 165
Profile Rail (a.k.a.: square rail, linear guides) ... 201
Programmable controllers ... 276
Programmable logic controller (PLC) ... 252
Programming Languages ... 278
Proportional and Servo Valves ... 182
Proportional Control ... 292
Proportional Plus Integral Control ... 292
Proportional valves ... 182
Protection ... 252
Proximity Sensors ... 284
PT Components ... 107
PTFE ... 298, 304, 325
PTFE (Teflon®) Tape ... 298
Pull-up torque ... 233
Pulse-width modulated (PWM) drives ... 258
Pulse-width modulated (PWM) inverters ... 257
Pump rating factors ... 178
Pumps ... 363
Pushbuttons ... 275
PWM Drives ... 258

Q

QD Bushings ... 309
QD Sprockets ... 80

R

Rack and Pinion ... 211
Rack-and-Pinion Actuators ... 185
Rack-and-pinion gear arrangement ... 159, 211
Radial ball bearings ... 24, 25
Radial Ball Bearings (Precision Ground) ... 24
Radial lip seal, gear ... 168, 169
Radial load, bearing ... 24-26
Radial-Piston Motors ... 186
Radial-Piston Pumps ... 177
Ramp and Roller Clutches ... 93
Reciprocating Compressors ... 179
Recirculating Element ... 200
Reduced-Inrush Starters ... 272
Reduced-Voltage Starters ... 271
Reducing Valves ... 180
Reduction ratio ... 166
Reflective path, photoelectric-sensor ... 287
Regimes ... 329
Relays ... 276
Relief Valves ... 180
Resolvers ... 283
Resonance ... 240, 352
Resonance, step motor ... 240
Resonance, vibration analysis ... 333
Response terminology ... 260
Retained assembly ... 26, 299
Retaining Cylindrical Metal Assemblies ... 299
Retaining rings ... 307, 308
Reversing Starters ... 271
Revolutions per minute (rpm) ... 11, 227, 352
Ribbed Couplings ... 126
Rigid couplings ... 125
Ringing ... 259
Rod Size ... 184
Rod-end Bearings ... 38
Rodless Cylinders ... 184, 215
Rodless Mechanical Actuators ... 216
Roller Bearings ... 28
Roller Chain ... 64
Roller Conveyor Chain ... 74
Roller Conveyors ... 100
Roller Screw ... 209
Roller Slides ... 214
Roller-ramp clutch ... 93
Roller-screw jacks ... 220
Rolling Element Bearings ... 24, 214, 323
Rolling-Element Bearing Faults ... 339
Rope Belts ... 46
Rotary Actuators ... 185
Rotary Encoders ... 222
Rotor dynamics ... 337
Round roller clutch ... 93
Round Shaft (a.k.a.: round rail, linear bearings, linear ball bushings) ... 201
RPM (revolutions per minute) ... 11, 12
R-Type ... 57
Rubber block couplings ... 132
Rubber-tire couplings ... 133
Running speed ... 352

S

SAE Flange ... 190
SAE International ... 369
SAE Straight-Thread ... 190
Screw Conveyor Flighting ... 112
Screw Conveyors ... 111
Screw Jacks ... 219
Seal Configurations ... 304
Seal Materials ... 304
Sealants & Adhesives ... 297, 367
Seals ... 35
Seals and Breathers ... 168
Seals, bearing ... 40
Seals, gear ... 186
Selection Criteria (Linear Guidance) ... 199
Selection of Solid Lubricants ... 325
Selection Procedures ... 213
Self-actuating Clutches ... 96
Self-adjusting Tensioners ... 306
Self-Lube Roller Chain ... 65
Self-lubricated Bearing Plastics ... 23
Self-Lubricated Plain Bearings ... 22
Self-lubricating bearings ... 201
Sensors ... 281-288
Sequence Valves ... 181
Series-wound ... 236
Service Factor ... 229
Servo Actuators ... 184
Servo Drive Ratings ... 264
Servo Valves ... 182
Servodrives ... 264
Servomotors ... 238
Servovalves ... 218
Shaded-pole motors ... 234
Shaft and Housing Fits ... 40

Shaft collars ... 311
Shaft Eccentricity ... 303
Shaft locking devices ... 309, 310
Shaft misalignment ... 126
Shaft orientation ... 170
Shaft seals ... 303
Shaft-mounted reducers ... 121, 167
Shoe brakes ... 90
Short-Stroke Actuators ... 216
Short-stroke linear actuators ... 216
Shot-Filled Fluid Couplings ... 147
Shunt-wound ... 236
Side Bow Chain ... 68
Sideband ... 352
Signal Conditioning ... 288
Silent Chain ... 70
Silicone-Filled Fluid Couplings ... 147
Single point lubricators ... 313
Single-Acting Cylinders ... 183
Single-Phase Induction Motors ... 234
Size Ratings (Flow Capacity) ... 183
Sizing Cylinders ... 184, 365
Skew tooth gears ... 162
Sleeve Couplings ... 125
Slew range, step motors ... 240
Sliding-Block Couplings (Elastomeric Element Type) ... 136
Slip ... 233
Slip, motor ... 233
Small and Large Roller Double Pitch Sprockets ... 79
Small Package Sorting ... 117
Society of Manufacturing Engineers (SME) ... 369
Society of Tribologists and Lubrication Engineers (STLE) ... 369
Solid Lubricant Liners ... 23
Solid lubricants ... 325, 326
Solid-State Starters ... 272
Solid-State Switches ... 276
Solvent-based Pipe Compounds ... 298
Sortation Conveyors ... 117
Special Flight Types ... 112
Special-Purpose Couplings ... 138
Specialty Roller Chains ... 72
Spectrum ... 352
Speed (a.k.a.: velocity, acceleration, motion profile) ... 196
Speed ratio ... 170
Speed reducers ... 165-167
Speed-Torque Characteristics ... 240
Speed-Torque Curves ... 229
Spherical Roller Bearings ... 31
Spiral bevel gears ... 161
Spiral Retaining Rings ... 308
Spiral-jaw clutch ... 92
Split roller-bearing assemblies ... 38
Split-phase motors ... 235
Split-Taper Bushings ... 309, 312
Spool Mechanisms ... 180
Sprag Clutches ... 93
Spring couplings ... 137
Spring-loaded oil seal ... 303
Spring-Operated ... 314
Spring-Return Cylinders ... 183
Sprocket Caliper Diameter ... 77
Sprocket Hub Style ... 79
Sprocket Outside Diameter ... 78
Sprockets ... 77-81
Spur Gears ... 155, 158
Square-jaw clutch ... 92
Stabilizing adjustments ... 259
Stable responses ... 259
Standard Chain Attachments ... 68
Starter Performance Overview ... 274
Start-stop range, step motors ... 240
Static Load Capacity ... 214
Static Load Limitations ... 199
Steel Split Sprockets ... 80
Step motors ... 239
Storage of Lubricants ... 327, 328
Straight bevel gears ... 160
Strain gauges ... 286
Stribeck Curve ... 320
Stroke ... 199
Structural and rotor analysis ... 350
S-Type ... 56
Sub-harmonic ... 352
Surface fatigue ... 164
Synchronous Belts ... 54
Synchronous speed ... 232
Synchronous-belt pulleys ... 54
Synthetic Oils ... 188
System Response and Stability ... 259, 260

T

Tabletop Chain ... 71
Tachometer Generators ... 281
Take-up units, bearing ... 36, 37
Tangential Spring Couplings ... 137
Tapered Bushing Types ... 309
Tapered Roller Bearings ... 32

Tapered Versions 310
Taper-Lock Bushings 309
TAPER-LOCK® Sprockets 80
TEAO enclosures 230
TEFC enclosures 230
TEFV enclosures 230
Telescopic Slides 204
Temperatures and Oil Service Life 318
TENV enclosures 230
The FPDA Motion and Control Network 368
The Vibration Institute 369
Thin-section Bearings 28
Threadlocking 297
Thrust Ball Bearings 26
Thyristors 249
Tight and Slack Strands 66
Time trace or time waveform 352
Timed acceleration 253
Timing Belt 210
Tip-tray sorter 117
Torque 9, 10
Torque control 253
Traction Drives 162, 163, 248, 249
Transistors 249
Transmission Sizing 187, 365
Trapezoidal-tooth belts 54, 55, 196
Travel (a.k.a.: stroke, overtravel, envelope) 196
Trolley conveyors 106
Trolley, Beam Lift Sections, and Spurs 109
Tubular 217
Twin-floating tensioner 307
Types and Features 211
Types of Bucket Elevators 114
Types of Loads 228
Types of Tensioners 305
Typical Applications 245

U

U-joints 123-152, 357
U-Joints (Mechanically Flexible Type) 131
UL (Underwriters Laboratories, Inc.) 293
Ultrasonic demodulated spectrum 341, 352
Ultrasonic/acoustic emission frequency range 341, 352
Unbalance 337, 343, 344, 352
Unbalance – Couple 352
Unbalance – Dynamic 352
Unbalance – Residual 353
Unbalance – Static or Mass 353
Underwriters Laboratories, Inc. (UL) 293
Unground ball bearings 28
Unit Handling equipment 99
Unit Systems 227
Universal joints 131
Unloading Valves 181
Urethane Dual Flex Diaphragm Coupling 133
Urethane Tire Couplings 132

V

Valves, fluid power 180
Vane Actuators 185
Vane Compressors 178
Vane Motors 186
Vane Pumps 176
Variable Displacement Pumps 177
Variable Speed Belts 51
Variable-Reluctance Step Motors 239
Variable-speed drive 51, 261
Variable-torque loads 254
V-Belt Length 52
V-belt pulleys 14
V-Belt Types 47-50
V-Belts 46-52
Vee Guide Wheels 202, 203
Velocity 196, 353
Vertical carousel 118
Vertical Conveyors 104
Vertical reciprocating conveyors 104
Vibration analysis 331-354
Vibration Analysis Procedure 334
Viscosity 317
Vision Systems 288
Voice-Coil Motors 212, 213
V-ribbed Belts 50, 51

W

Wear 164
Weather-Protected Machines 229
Wedge V-Belts 48, 49
Welded Chain 75
Wheatstone bridge circuit 286
Work 12
Workstations 281
Worm gears 159, 160, 164, 166, 167, 169
Worm-gear Speed Reducers 165-167, 170

Wound-field motors.................................... 236, 252, 255
Wound-rotor drives..................................... 245, 246, 254
WP enclosures.. 229
Wrap-spring Clutch.. 92
Wye-delta Starters 272, 273

X

X-Y Tables... 215

Z

Zerol bevel gears.. 160-162
Zero-pressure accumulation, conveyors 104